M000249030

AutoCAD® 2008
for Architecture

AutoCAD® 2008
for Architecture

ALAN JEFFERIS
MICHAEL JONES
TEREASA JEFFERIS

autodesk Press

THOMSON

DELMAR LEARNING™

Australia · Canada · Mexico · Singapore · Spain · United Kingdom · United States

autodesk® Press

AutoCAD® 2008 for Architecture
by Alan Jefferis, Michael Jones, and Tereasa Jefferis

Autodesk Press Staff

Vice President, Technology and Trades ABU:
David Garza

Director of Learning Solutions:
Sandy Clark

Managing Editor:
Larry Main

Senior Acquisitions Editor:
James Gish

Senior Product Manager:
John Fisher

Marketing Director:
Deborah S. Yarnell

Channel Manager:
Kevin Rivenburg

Marketing Coordinator:
Mark Pierro

Production Director:
Patty Stephen

Production Manager:
Stacy Masucci

Production Technology Analyst:
Thomas Stover

Senior Content Project Manager:
Elizabeth C. Hough

Art Director:
Benj Gleeksman

Editorial Assistant:
Sarah Timm

Cover Images:
Getty Images

COPYRIGHT © 2008 Thomson Delmar Learning, a division of Thomson Learning Inc. All rights reserved. The Thomson Learning Inc. logo is a registerd trademark used herein under license. Autodesk, AutoCAD and the AutoCAD logo are registered trademarks of Autodesk. Delmar Learning uses "Autodesk Press" with permission from Autodesk for certain purposes.

Printed in Canada
1 2 3 4 5 XX 11 10 09 08 07

For more information contact
Thomson Delmar learning
Executive Woods
5 Maxwell Drive, PO Box 8007
Clifton Park, NY 12065-8007
Or find us on the World Wide Web at
www.delmarlearning.com

ALL RIGHTS RESERVED. No part of this work covered by the copyright hereon may be reproduced in any form or by any means—graphic, electronic, or mechanical, including photocopying, recording, taping, Web distribution, or information storage and retrieval systems—without the written permission of the publisher.

For permission to use material from the text or product, contact us by
Tel.   (800) 730-2214
Fax   (800) 730-2215
www.thomsonrights.com

Library of Congress
Cataloging-in-Publication Data:

Jefferis, Alan.
  Autocad 2008 for architecture /
Alan Jefferis, Michael Jones, Tereasa
Jefferis.
    p. cm.
  Includes index.
  ISBN-13: 978-1-4283-1161-9
  ISBN-10: 1-4283-1161-0
1. Architectural drawing--Data
processing.
2. Architecture--Computer-aided
design. I. Jones, Michael, 1964–
II. Jefferis, Tereasa. III. Title.
  NA2728.J43867 2008
  720.28'40285536--dc22

                          2007023868

ISBN-13: 978-1-4283-1161-9
ISBN-10: 1-4283-1161-0

NOTICE TO THE READER

Publisher does not warrant or guarantee any of the products described herein or perform any independent analysis in connection with any of the product information contained herein. Publisher does not assume, and expressly disclaims, any obligation to obtain and include information other than that provided to it by the manufacturer.

The reader is expressly warned to consider and adopt all safety precautions that might be indicated by the activities herein and to avoid all potential hazards. By following the instructions contain herein, the reader willingly assumes all risks in connection with such instructions.

The publisher makes no representation or warranties of any kind, including but not limited to, the warranties of fitness for particular purpose or merchantability, nor are any such representations implied with respect to the material set forth herein, and the publisher takes no responsibility with respect to such material. The publisher shall not be liable for any special, consequential, or exemplary damages resulting, in whole or part, from the readers' use of, or reliance upon, this material.

CONTENTS

CHAPTER 13   PLACING PATTERNS IN DRAWING OBJECTS 459

PREFACE

AutoCAD 2008 for Architecture is a practical, comprehensive text that is easy to use and understand. This text provides a user-friendly environment to master skills that will lead to a productive career. The content is designed to teach the skills needed to master the 2D drawing commands used on construction-related drawings. The content can be used as presented to gain an understanding of the use of AutoCAD. This text can also be used with older releases of AutoCAD, but it is best suited for use with AutoCAD 2007, 2006, 2005, and 2004. All concepts are presented using architectural examples so that each command can be correlated to a specific skill that will be required of architects, engineers, and CAD technicians. Commands and drawing skills are presented in a sequence that is easy to understand and follow with each chapter building on the skills presented in the previous chapter. Other specific features of this text include the following:

- How a skill will relate to office practice

- Drawings by professional architects, engineers, and residential designers are used as illustrations

- Chapter exercises which review chapter contents in addition to reinforcing previous material

- Drawing chapters that contain drawing problems that will reinforce the chapter concepts

Individual chapters will provide knowledge to master the following drawing skills:

- Drawing lines and geometric shapes

- Using varied linetypes and widths

- Editing basic geometric shapes

- Combining skills to produce drawings such as floor and foundation plans, elevations, details, and sections

- Editing an entire drawing to produce other similar drawings

- Understanding dimensioning methods and techniques

- Placing text within a drawing

- Creation of varied text and dimensioning styles

- Creating symbols libraries to increase drawing speed and efficiency

INTRODUCTION TO DRAWING COMMANDS

AutoCAD 2008 is a very powerful program that contains hundreds of commands to meet a variety of drawing needs. In this text, only the 2D command options that are used within the construction field will be explored. Several methods are available to execute each drawing command. Data entry can be made by using toolbars, shortcut menus, dynamic input, keyboard, or the drop-down menu. The selection method will depend on individual preference. Each command will be presented initially through the method that provides the fastest access. For many commands this will be use of the dynamic input or keyboard, with available options used to supplement the basic procedure. With the keyboard presented as the primary option, the entire command sequence can be presented, displaying the many options that are available within a specific command.

Commands, options, or values that must be typed at the dynamic display or command prompt are listed with **BOLDFACE** capital letters. Commands will be displayed as they will appear at the dynamic display unless noted otherwise. Once the command is entered, pressing the ENTER or return key is required and will be represented by the symbol ENTER throughout the text. Command sequences will be presented in the following manner:

> **Command:** *Click the* **Line** *icon on the* **Draw** *toolbar (Or type* **L** ENTER.*)*
> Specify first point: **4,8** ENTER
> Specify next point or **8,8** ENTER
> Specify next point or ENTER

PREREQUISITES

Although this text can serve as a reference for experienced professionals, no knowledge of architecture or construction is required to learn the command structure of AutoCAD. Some of the advanced projects do require knowledge of basic residential architectural standards.

E.RESOURCE

E.resource is an educational resource that creates a truly electronic classroom. It is a CD-ROM containing tools and instructional resources that enrich the classroom and make preparation time shorter. The elements of e.resource link directly to the text and combine to provide a unified instructional system. With e.resource you can spend your time teaching, not preparing to teach. You can find the Online Companion at: *http://www.autodeskpress.com/onlinecompanion.html*. When you reach the page, click on the title *AutoCAD for Architecture*.

Features contained in e.resource include:

- Instructors Guide: This contains an outline for a one semester course, chapter summaries, and answers to chapter exercises and quizzes.

- Chapter Notes: Chapter Rationale, Objectives, and Teaching Hints prepare you to present the course.

- PowerPoint Presentation: Slides that provide the basis for a lecture outline that helps you to present concepts and material. Key points and concepts can be graphically highlighted for student retention.

- Video & Animation: AVI files, listed by topic, which allow you to view a quick video illustrating and explaining key concepts.

- Exam View Computerized Test Bank: Questions of varying difficulty are provided in true/false and multiple-choice formats, so that you can assess student comprehension.

ABOUT THE AUTHORS

Alan Jefferis has been a drafting instructor at Clackamas Community College for 29 years. His professional experience includes 8 years of drafting experience with structural engineers, and 28 years as a professional building designer. He is currently the owner of Residential Designs, and is author of Residential Design, Drafting and Detailing, and co-author of *Architectural Drafting and Design, Print Reading for Architectural and Construction Technology, Commercial Drafting and Detailing, AutoCAD for Architecture Release 2007, 2006*, 2005, 2004, 2002, and *AutoCAD 2002 and the Fundamentals of Mechanical Drafting*. All are available through Delmar Publishers.

Tereasa Jefferis is a graduate of Clackamas Community College and the Mechanical Systems Institute. She has been an architectural drafter for 11 years and has experience in architectural drafting, commercial plumbing drawings, and with HVAC system design. She designed several of the homes that are shown throughout this text. She currently works as an independent contractor with multiple structural engineering firms.

Mike Jones is the CAD Systems Manager, Department Chairperson, and a CAD instructor at Clackamas Community College, which is an Autodesk Premier Training Center in Oregon City, Oregon. Mike has introduced new users and experienced professionals to the commands and options of AutoCAD for over 25 years. His training center experience is enhanced by 27 years of professional drafting and CAD system management experience in private industry.

ACKNOWLEDGMENTS

The quality of this text has been greatly enhanced by the donations of many professional offices and individuals within private and municipal agencies. My special thanks to:

Kenneth D. Smith

Kenneth D. Smith Architects, A.I.A.

El Cajon, CA

LeRoy Cook and Sonya Fakelman

Clackamas Community College

Oregon City, Oregon

Walt Cheever
South Central Technical College
N. Mankato, MN

Shelly Fry
CPCC Corporate Center
Charlotte, NC

Aaron Rumple
St. Louis Community College
St. Louis, MO

Dean Urevig
Dunwoody Institute
Minneapolis, MN

Sylvia Sullivan
College of the Canyons

Joseph Yamello
Greater Lowell Regional Vocational
Technical School

Robert Kamenski
Lenape A.V.T.S.

Walter Eric Lawrence
Gwinnett Technical Institute

All of these people have made substantial contributions to the success of this text. We're also especially thankful for the work and efforts of Tricia Coia, John Fisher, Sandy Clark, Stacy Masucci, and Sarah Timm of Delmar Publishers.

No one has been more helpful and supportive than Janice Ann. Thanks for the help and support. But most importantly, thanks to "him who is able to keep you from falling, and to present you before his glorious presence without fault and with great joy."

TO THE STUDENT

AutoCAD 2008 for Architecture is designed for students in architectural, engineering, or construction programs who are learning AutoCAD. The contents are presented in an order that has been tested within the classroom. Information is based on common drafting practices from the offices of architects, engineers, and designers. To make the best use of the information within the text, students should follow these few helpful hints.

READ THE TEXT

I know it sounds simple, but if you'll read a chapter prior to sitting at a computer, you'll find your drawing time will be much more productive. Read through the chapter and make notes of things that you're not quite sure of. With the reading complete, now try to mentally work through the commands that you've been reading about.

STUDY THE EXAMPLES

One of the many features of this text are the many illustrations and examples of how to complete each command sequence. Carefully study and compare the command sequence with the illustration of how to complete the command.

PRACTICE AT A WORKSTATION

Once you've read the chapter and studied the illustrations, complete the command sequences at a computer workstation. Practice each command several times and don't be afraid to explore all of the options within a command. Always compare commands using icons and keyboard options. Just because icons are good for one command, doesn't make them the best for all commands. Because each chapter is based on the previous chapters, don't hesitate to reread past material. Above all else, don't be afraid to experiment. It's very common for new computer users to be filled with a fear of breaking something. If you save your drawings often, there is very little that can go wrong by experimenting. Relax and have fun.

Exploring AutoCAD Tools and Displays

INTRODUCTION

This chapter will introduce methods for controlling the work environment, starting AutoCAD, and learning the tools and displays of AutoCAD. These include:

- Drawing area
- Workspaces
- Title bar
- Menu bar
- Status bar
- Command line
- Scroll bars
- User coordinate icons
- Cursors

Command selection methods are introduced:

- Toolbars
- Menus
- Keyboard entry
- Dynamic input
- Dialog boxes
- Tool palettes
- Shortcut menus
- Function keys
- Control keys

Commands to be explored in this chapter include:

- **Line**

- **Arc**

- **Erase**

- **Help**

- **Save**

You're about to start exploring the major commands used to create 2D drawings using the most popular computer-aided drafting program on the market. It will only take a few minutes of exploring to realize that AutoCAD offers lots of options, and the task of learning the program might seem a bit overwhelming. Relax! It's really not that bad. CAD drafting, AutoCAD in particular, tends to become addictive. AutoCAD may seem frustrating for the first few weeks, but I think you'll love it. An understanding of the tools you'll be using will help to eliminate the natural frustration that comes with learning. This chapter introduces key components of the hardware and the software that you'll be using to create and control your drawings.

CONTROLLING THE WORK ENVIRONMENT

Although becoming a proficient AutoCAD user will provide many exciting opportunities, spending long hours at a workstation also has its pitfalls. You might start to notice problems with your eyes, back, and wrists. Fortunately, most physical problems arising from computer use can be minimized or eliminated by proper workstation configuration, good work habits, and good posture.

EQUIPMENT

You might have little control over your workstation at school, but if you're setting up your own workstation, it should meet the following requirements:

- Your chair should be adjusted to a height that allows your feet to be supported on an inclined footrest or the floor. The chair should provide a tilt adjustment so that your back is firmly supported as you work. It should also have an armrest to provide support for your wrists. The height of the chair should allow your wrists to remain level as you type.

- The monitor should be placed so that the top of the monitor is level with your line of eyesight and between 18 and 30 inches from your eyes. Lighting should be placed so that it will not produce a glare on the display screen.

- Use a desk with a keyboard drawer that allows your wrists to remain level.

- Use an ergonomic keyboard, a foam wrist support, or forearm supports.

WORK HABITS

You can have a state-of-the-art workstation, but if you don't maintain good work habits, your body will wear out from fatigue. To keep your productivity high and maintain your sanity, you'll need to control your work habits and posture.

- Avoid looking at a monitor for long periods of time. To ease eyestrain, several times an hour look away from the monitor and force your eyes to focus on an object several feet away from your workstation.

- Avoid resting your hand on the mouse when you're not actually using it. When you are using the mouse, avoid resting your hand on the desk. Get in the habit of keeping your wrist level, or adjust the armrest of your chair to provide support.

- Take time to stretch your legs, neck, shoulders, back, and wrist several times each hour. In addition to reducing tension throughout your body, a stretch provides an excellent opportunity to stop focusing your eyes on the monitor.

- Most importantly, take time to walk away from your workstation at least once an hour. It's easy to get caught up with a deadline and convince yourself that you just don't have time for a break, but short breaks will help keep you productive and eliminate drawing errors.

STARTING AUTOCAD

The easiest method of starting AutoCAD is to double-click the **AutoCAD** button from the desktop. You can also start AutoCAD by double-clicking the **AutoCAD** program button in the **Programs** menu, or by using the **Run** menu. Each of these methods can be accessed by selecting the **Start** button. Double-clicking the **AutoCAD** button will display the drawing screen shown in Figure 1.1a.

Tip: Depending on how your machine is configured, the 3D drawing screen, shown in Figure 1.1a, or the 2D annotation screen, shown in Figure 1.1c, may be displayed. This chapter will assume that your machine has been configured to display the screen shown in Figure 1.1b. Methods of changing the display will be introduced later in this chapter. Several toolbars may also be displayed throughout the drawing area with titles that start with ET. For now, these toolbars can be removed from the screen by selecting the Close button in the upper right corner of the toolbar. The options contained in these dialog boxes and toolbars will be explored later in this chapter.

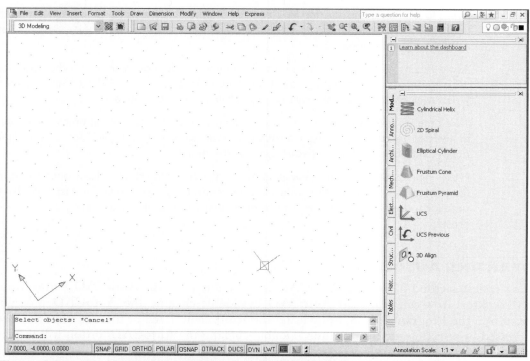

Figure 1.1a *The 3D AutoCAD display showing the **Dashboard** and **All Tool Palettes** and the **Standard**, and **Layers** toolbars on top of the drawing display. The display can vary greatly depending on how AutoCAD has been configured.*

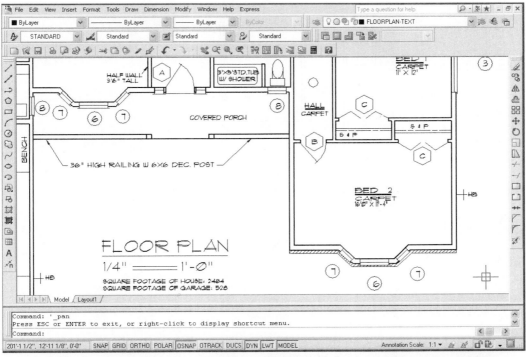

Figure 1.1b   *The AutoCAD Classic display shows the **Standard**, **Styles**, **Properties**, and **Layers** toolbars.*

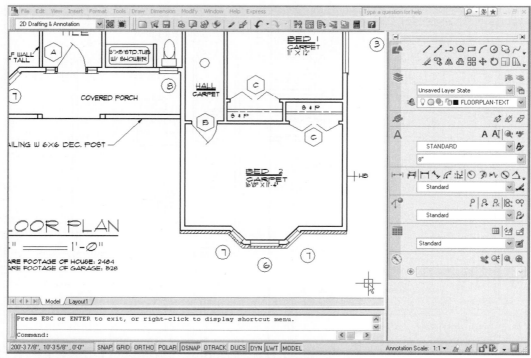

Figure 1.1c *The 2D Annotation workspace should be used when text will be added to a drawing. Appropriate menus, toolbars, tool palettes and dashboard panels are automatically displayed as the 2D drafting and Annotation workspace is displayed.*

The drawing area is used to create and display drawings. Each portion of the display will be the subject of the balance of this chapter. The exact settings displayed as AutoCAD is opened will vary depending on how the program has been configured. The drawing area contains several important tools for controlling AutoCAD. To use AutoCAD effectively, you need to understand each of the methods of interacting with the software. This includes knowledge of the drawing area displays, toolbars, types of AutoCAD menus, common components of a dialog box, tool palettes, function keys, and control keys. Don't let all of the different terms confuse you. Each will become quite easy to use as you explore the program.

Tip: Push a few keys and explore some options. As long as you keep liquid away from the keyboard, there is very little you can do to hurt the hardware, software, or yourself.

EXPLORING THE DRAWING AREA TOOLS AND DISPLAYS

AutoCAD offers the user many tools and displays to enter information into the program, and to determine the status of the current drawing file. These tools include the workspaces, drawing area, title bar, menu bar, command line, status bar, scroll bars, user coordinate icon, and cursors.

WORKSPACES

Figures 1.1a, 1b and 1c presented the methods used by AutoCAD to view a project. These displays, or workspaces, are sets of menus, toolbars, and palettes that are grouped and organized to provide a custom, task-oriented drawing environment. When a workspace is used, only the menus, toolbars, and palettes that are relevant to a task are displayed. In addition, several workspaces will automatically display the dashboard, a palette with task-specific control panels. 2D drawings should be completed using the AutoCAD Classic workspace shown in Figure 1.1b or the 2D Annotation workspace shown in Figure 1.1c. Other than in chapter 1, commands in chapters 2 through 12 will be introduced using the classic workspace. The 2D Annotation workspace will be used in examples in chapter 13 and all following chapters. The 3D workspace will not be covered in this text. Methods of changing the workspace will be introduced later in this chapter.

DRAWING AREA

Central to each of the AutoCAD tools is the drawing area, where the material that you enter is displayed. The size of the drawing area varies depending on which workspace is displayed, how many toolbars or dialog boxes are displayed, and the size of your monitor. With the AutoCAD Classic workspace current, the drawing area is the black area surrounded by gray areas. Methods of altering the colors of each portion of the display will be introduced in future chapters.

MULTIPLE WINDOWS

One of the features of Microsoft Windows–based programs is that several programs can be opened at the same time. One of the features of AutoCAD is the ability to open several drawing projects in one AutoCAD session. Figure 1.2 shows an example of a drawing session with two drawings displayed. AutoCAD refers to the side-by-side arrangement as vertical tiles. As with other windows, the arrangement and size of the drawing window can be altered. Selecting **Window** from the menu bar will provide options for altering the arrangement of each window. The arrangement of windows can be altered at any point during the drawing session. Later chapters explore the process for opening and working with multiple files.

TITLE BAR

A title bar is located across the top of every window. It displays the program icon and name of the current program as well as the name of the current file. The title bar serves as a handle to drag the window to a different location on the screen. Notice that in Figure 1.2 two drawings are opened and displayed in two windows within the AutoCAD program.

The AutoCAD title bar is at the top of the drawing screen, and each drawing window has its own title bar. If multiple drawings are open, the title bar of the active window is highlighted with a blue bar, while inactive windows are displayed in another color. The name of each drawing session is listed in the title bar. Notice in Figure 1.2 that the active drawing on the left is named "Elevations" and the inactive drawing on the right is named "Floor." At the far right end of the title bar are the Minimize, Maximize, Restore, and Close file buttons. These buttons function similarly to other Windows-based programs. You'll notice that sometimes there are two rows of buttons. Although they function in a similar manner, the buttons at the end of the title bar control the display of the current program window. The buttons on the lower row control the current drawing file.

Tip: It's important to remember that if you select the Minimize button and remove the file from the screen, the file still exists—it's just been placed out of your view. It hasn't been saved or destroyed, just moved for convenience so you can do something else.

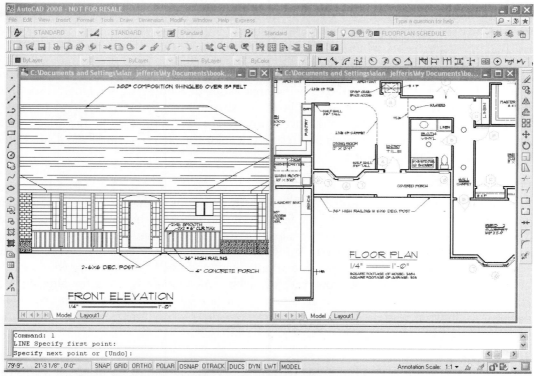

Figure 1.2 *AutoCAD allows multiple drawings to be opened. The name of each drawing is listed in the title bar of each window. The active drawing (ELEVATIONS) is displayed in a dark blue title bar, and the inactive drawing (FLOOR) has a light blue title bar. The cursor is only displayed in the active window.*

MENU BAR

The menu bar is located below the title bar and is used to display menus specific to the open program. The menu bar for AutoCAD can be seen in Figure 1.3. Each word of the menu bar represents a separate menu with options related to the key word. Moving the cursor to one of these words and clicking will display the corresponding menu. Figure 1.4 shows the **Draw** menu for AutoCAD. Some commands such as **Arc** have a solid black triangle on the right side. The triangle indicates that moving the cursor to this command will produce another menu of options related to that command. Moving the cursor to the **Arc** listing will produce the menu shown in Figure 1.5. Select the **Format** menu with a left click of the mouse. Notice that the **Plot Style** listing appears different from other listings. This option is inactive and is not currently available for selection. Inactive options only become active as other options are activated.

Title Bar

Figure 1.3   *The menu bar for AutoCAD allows access to many of the commands and controls of AutoCAD.*

Figure 1.4   *The **Draw** menu of AutoCAD shows a list of names that represent individual commands for creating objects with AutoCAD.*

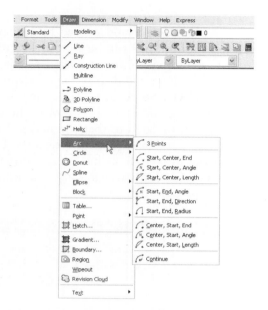

Figure 1.5 *Listings in a menu followed by a triangle will produce a submenu. The **Arc** submenu is displayed by moving the cursor over **Arc**.*

Four important tools are located at the right end of the Menu Bar including:

- Help edit box – Allows a specific subject to be entered into the edit box. The info Center can then be used to obtain help related to the specified subject.

- InfoCenter – Contains key features of the help menu including the Users Guide, Command Reference, Customization Guide, and the New Features Workshop.

- Communication Center – The **Communication Center** is an interactive feature that delivers information via the Internet. When connected, information is sent from Autodesk to allow you to stay informed about general product information, to provide product support information, general announcements, and articles and tips relative to the software being used.

- Favorites – Allows InfoCenter Search and Communication Center settings to be set and saved in the appropriate dialog box.

These tools will be explored at the end of this chapter as methods for obtaining Help are explored further.

COMMAND LINE

The command line displays the prompt **Command:** and is one of two ways AutoCAD allows you to enter your request using the keyboard. When a command is entered, the request and the required response will be displayed on the command line. The **Dynamic Display** places a small command line by the cursor. Methods of controlling both display methods will be introduced later in this chapter. As you explore AutoCAD, it's important to remember that the command line displays prompts for completing commands. Even if you select commands by using a toolbar or menu, AutoCAD displays prompts and a command history in the command window. The command line also shows the result of selecting a command by picking a button. If the **Line** command is started by selecting the **Line** icon from the **Draw** toolbar that is currently located on the left side of the screen,

    Specify first point:

is displayed at the command line. AutoCAD is now waiting for your response.

 Tip: New AutoCAD users often fail to read the Command prompt and miss AutoCAD's attempt to communicate with them. An alternative to inputting information through the Command prompt is to use Dynamic Input. Take time to read each method of input if AutoCAD does not react as you expect it to.

Expanding the Command Line

The Command prompt is one of the key methods for interacting with AutoCAD. By default, the command window displays two lines of text. Many AutoCAD users prefer to have more lines of text displayed so that an entire command sequence can be seen at one time. Additional lines of command text can be viewed by picking and holding the top border of the command box, and then dragging the top of the box to a new location. A scroll bar is added to the window to allow viewing of previous commands. Up to 100 lines of text can be displayed, but a display of over 35 lines will totally fill the drawing area. The scroll bar on the right edge of the display can be used to show additional lines of text.

Accessing Command Line Data

AutoCAD provides command line users help with spelling and name recognition when entering command or system variable names. If you are unsure of the name of a specific command, begin typing what you think is the name of the command. Once three letters have been entered, pressing the TAB key will list possible options.

A second aid available at the command line is the Recent Input Function that is available at the command line. If you're drawing a series of related lines, rather than entering the coordinates in multiple times, AutoCAD allows you to right-click to display a shortcut menu, and select the option of **Recent Input**. The most recent values for length and angle are stored here and can be reselected to avoid retyping the information.

A third alternative to expanding or using the command display is to totally remove it from the drawing. Although it provides a valuable method for interacting with AutoCAD, it may slow down your drawing time, because it often requires that you look away from the drawing area as you input commands. Methods for removing the command line will be introduced later in this chapter once methods of using the dynamic display are explored.

STATUS BAR

The status bar is located below the command line. Key elements of the status bar can be seen in Figure 1.6a and include the coordinate display and toggle buttons that can be used as aids for placing objects in a drawing. The coordinate display is used to track the location of the cursor through the drawing area. This feature will be explored in chapter 3. The drawing aids in the status bar affect the placement or display of your drawings. Each button will be explained in detail in chapter 2. Options are **ON** when the button appears to be pressed in. Buttons include the following:

SNAP—Toggles an invisible grid ON/OFF for placing lines at precise intervals. These invisible markers can be used to establish modular patterns to aid in drawing layout.

GRID—Toggles a visible grid ON/OFF for placing modular objects. The grids serve as guides for drawing layout.

ORTHO—Toggles the ability to draw lines that are vertical and horizontal (ON) or at any angle (OFF). With Ortho **ON**, rectangular objects can be easily drawn. With the **OFF** setting, irregular shapes can be created.

POLAR—Toggles the polar tracking feature that will describe the position of the cursor in distance and angle relative to another point. By default, the tracking angle is set to a 5° increment.

OSNAP—Toggles the ability to select drawing objects such as endpoints, midpoints, intersections, or ten other options. When **ON**, the cursor automatically snaps to one of the specified points of existing objects as the cursor is moved toward those objects.

OTRACK—Toggles the tracking feature that allows specific angles to be tracked along a projected path based on an object endpoint as if the line were extended. If you've drawn a line at 26°, this feature will allow a new line to be projected from the existing line at the specified angle.

DUCS—Toggles between allows/disallows for the Dynamic UCS. This feature is used as an aid to create objects on a planar face of a 3D solid. 3D drawings will not be discussed in this text.

DYN— Toggles a command line referred to as **Dynamic Input,** which uses a floating command display near the cursor.

LWT—Toggles the display of lineweight. When **ON**, assigned relative lineweights are shown on the display. When **OFF**, all lines are shown using the same lineweight.

MODEL/PAPER—Toggles between model and paper space. For example, when in model space, click **MODEL** to change to paper space. As you explore the basics of the AutoCAD drawing area, think of model space as the area used to create a drawing and paper space as the area where the drawing is prepared for plotting.

Figure 1.6a *The status bar contains the coordinate display, buttons for controlling drawing aids and the controls for annotative scaling. The exact display will depend on the settings of the **Tray Settings** dialog box, which is displayed by right-clicking while the cursor is over an open area of the status bar.*

Selecting the **Status Bar Menu** button (the down arrow) allows the display of the buttons on the status bar to be altered. A right-click on any blank portion of the status bar will also access the display. Options preceded by a check mark are displayed on the status bar. To remove an option from the status bar, select the checked option. As the option is selected, the display is removed, and the option is removed from the status bar. Right-click on the status bar, and then select the desired option to restore an option.

Tip: Since Dynamic UCS will not need to be altered for 2D drawings this would be an excellent time to remove it from the status bar. The Dynamic UCS button will be removed from the status bar for the balance of this text.

The status bar also contains icons for three annotative controls, and five drawing and program controls. Each can be seen in Figure 1.6b. The Annotative controls will be explored in chapter 15.

- The **Annotative scale** display shows the current scale factor for paper and model space and allows a new scale to be selected.

- The **Annotative visibility icon controls the display of annotative objects for the current scale only.**

- The **Annotative Autoscale icon automatically adds scales to the annotative objects when the annotative scale changes.**

- The **Lock** icon is displayed to indicate if specific toolbars and windows are locked. This icon controls locks for the size and position of toolbars and windows. Information regarding the use of locks will be introduced later in this chapter as toolbars are explored.

- The **Trusted AutoCAD DWG** icon indicates that you are working on a drawing with an accepted digital signature. Advanced features allow a drawing to be assigned a digital signature to restrict viewing options. This feature lies beyond the needs of new CAD operators.

Where is this located

- The **Clean Screen** icon removes all of the toolbars above the drawing area to maximize the drawing area. Click the icon to restore the original screen display. For your initial drawings you'll typically want the toolbars displayed.

Other icons that may be displayed on the status bar depending on drawing content include:

Manage Xrefs—Displayed on the status bar if the current drawing has an attached xref (referenced drawing). Click the icon to access the **Xref Manager**. Right-click the icon to select **Xref Manager** or **Reload Xrefs**.

CAD Standards—Displayed on the status bar if the current drawing has an associated standards file. This icon shows a balloon message and an alert when a standards violation occurs. Click the icon to audit your drawing. Right-click the icon to configure the CAD Standards settings or to audit your drawing.

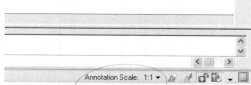

Figure 1.6b *The status bar contains icons for annotative scaling controls and additional drawing and program controls*

MODEL AND PAPER SPACE TABS

At the bottom of the drawing area are the **Model** and **Layout** tabs. Drawings are created in model space at full scale. If a wall is to be 6" wide, the lines that represent the walls are drawn 6" apart. If you're drawing a building that is 300' long, the drawing area can be adjusted so that the entire structure can be seen. Figure 1.7 shows a floor plan displayed in the drawing area. Paper space is used to display drawings using various layouts or to plot drawings that have been reduced to a scale such as 1/4"=1'–0". With 6" wide walls, the lines will be 1/8" apart. Paper space allows a 300' long building to fit on a 24" × 36" sheet of paper. Figure 1.8 shows the plan from Figure 1.7 in paper space. This is how the drawing will appear when plotted. Making a quick print will be explored in chapter 3, and plotting to scale will be explored in chapter 25.

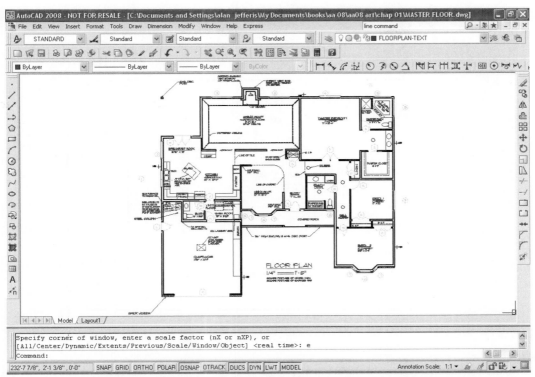

Figure 1.7   *Drawings are created in model space at full scale. Enlarging the drawing area will show the entire floor plan.*

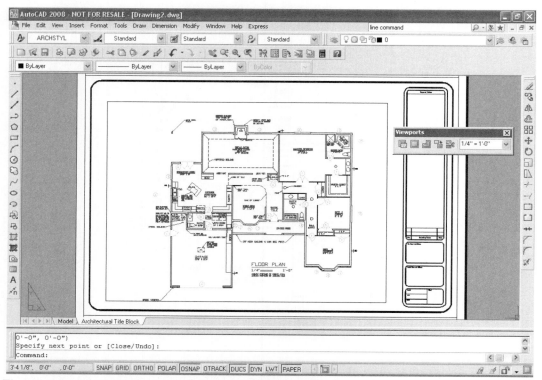

Figure 1.8 *The floor plan from Figure 1.7 is shown in paper space. Paper space scales a drawing to fit within the limits of a sheet of paper for plotting.*

 Tip: Keep the Model tab as the current tab as you explore AutoCAD in this chapter. Uses for model and paper space layouts will be further discussed through the balance of this text, as they are needed. As you're exploring, remember, model space is for drawing, paper space is for plotting at a specific scale.

USER COORDINATE SYSTEM ICON

The **User Coordinate System** icon describes the current coordinate system and orientation of the drawing area. AutoCAD drawings are imposed over an invisible coordinate system based on X, Y, and Z coordinates. AutoCAD uses both a fixed world coordinate system (WCS) and a movable user coordinate system (UCS). Each method will be explained in later chapters. In its standard setting, as you draw in AutoCAD, you're working in model space. When a drawing is to be plotted, it is plotted using paper space. Figure 1.9 shows the model and paper space icons.

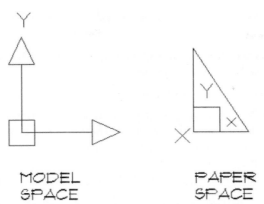

Figure 1.9   *The User Coordinate System (UCS) icons for model and paper space.*

 Tip: As you explore AutoCAD, the important thing to remember is that the icon is a reminder of what mode you are currently using to create your drawing. Unless your instructor tells you otherwise, work in model space as you create a drawing, and work in paper space to plot a drawing to scale.

SCROLL BARS

The scroll bars consist of a vertical bar located on the right side of the drawing area and a horizontal bar located below the drawing area. Scroll bars are used to alter the material viewed in the drawing area. Scroll bars are one of many ways to alter the display in the drawing area or to move through a long list of options. Future chapters will introduce faster methods of altering the drawing display.

CURSORS

The term *cursor* refers to the pointing device that is placed on the screen to indicate the current mouse position. The cursor for the AutoCAD drawing area is crosshairs. As the mouse is moved, the location of the crosshairs is moved. The cursor provides a method of entering the location for drawing objects. When a command such as **Line** is started, the cursor is the primary method of selecting the beginning and ending points for a line.

Just as with other Windows programs, the cursor changes shapes depending on where it is located. When the cursor moves outside the drawing area, the crosshairs turn into an arrow. It can also appear as an hourglass, pointing finger, flashing vertical line, double arrow, or four-sided arrow.

Altering the Crosshairs Size

Depending on the complexity of the drawing being completed, some operators like to adjust the size of the cursor. Crosshairs that extend across the screen will aid in projecting objects from one view to another. The size of the cursor can be altered using the following method:

1. Select **Options** from the **Tools** menu (See Figure 1.3). This will display the **Options** dialog box.

2. Select the **Display** tab. Use the slider bar in the **Crosshairs size** edit box or enter a value such as 100 to alter the cursor size.

3. When adjusted, select the **OK** button to return to the drawing area.

The default cursor size is 5. The number represents the percentage of the screen the cursor will occupy. Select and hold the slider bar with the left mouse button and move it to the right to increase the size of the cursor. Entering a value of 100 will cause the lines forming the crosshairs to cross the entire screen.

DYNAMIC INPUT

Dynamic Input uses a floating command line placed beside the cursor to display command prompts. This allows the command line to be closed, increasing the size of the drawing area. To remove the command line from the drawing area:

1. Select the **Tools** button.

2. Select **Command Line.**

3. Select the **Yes** button.

To restore the command process, select **Command Line** from **Tools** menu .Selecting **DYN** from the status bar activates **Dynamic Input**. This is a toggle control that will display and remove the Dynamic Input display. Once activated, no changes occur until a command is started. Press **L ENTER** to start the Line command. The Dynamic Input display will now appear similar to Figure 1.10. Information displayed with Dynamic Input will vary based on the command being executed. Controls for the dynamic display will be introduced as the various commands that affect the display are explored.

 Tip: Dynamic Input can be used in conjunction with the command line, without the command line, or be toggled off and not used at all. As a beginner, use both display methods and see where you are most comfortable looking for command prompts.

206-940-6589

206-783-8822
www.wilsontaxexperts.com
ACCOUNTING
TAX &
Wilson

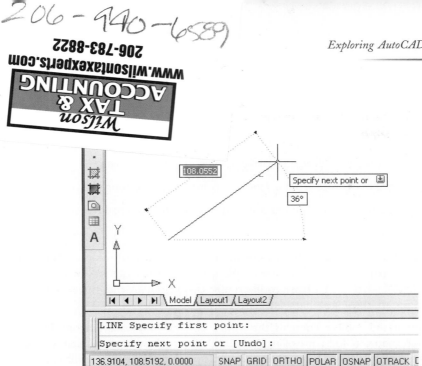

Figure 1.10 *The **Dynamic Input** display for the **LINE** command. Information displayed can be altered using the **Dynamic Input** tab that is displayed by selecting **Drafting Settings** from the **Tools** menu found in the **Standard toolbar**.*

COMMAND SELECTION METHODS

AutoCAD contains hundreds of commands related to drawing or editing a drawing. Five options are available for entering commands, including selecting buttons from toolbars, keyboard entry, menus, dialog boxes, and shortcut menus. Consideration will also be given to using function and control keys to alter drawing settings or alter commands.

Tip: Throughout the balance of the text, commands will be presented in the order that tends to be the fastest method of entry. For most commands, the fastest entry method is by selecting a button from a toolbar or by keyboard entry. Remember that the key word is fastest, not best. As you're learning each command, take the time to explore each option. Each method has its advantages.

As you begin to work with AutoCAD menus, you will see common terms used repeatedly. You will need to understand these so you can progress through AutoCAD. Common terms include the following:

    Click—To use a pointing device to make a selection from a menu or in a dialog box.

    Command—An instruction to be carried out by the computer, such as **Line** or **Arc**.

    Default—A value that will remain constant until a new value is entered. Default values are shown on the command line in angle brackets: < 3 >.

    Enter—ENTER is used to denote pressing the appropriate key (key names vary, depending on the keyboard manufacturer) after typing a request at the command line.

9057 Greenwood Ave. N., Seattle, WA 98*

Escape—ESC is used to exit a command.

Option—A portion of a command that requires a selection.

Pick—To use a pointing device to make a selection in the drawing.

Select—To make a choice from a menu of commands or options.

Specify—To enter a value, such as a coordinate on the command line.

TOOLBARS

A toolbar contains buttons that represent commands. In its default display, the **Standard, Styles, Layers**, and **Properties** toolbars are displayed above the drawing area just below the menu bar, and the **Modify** and **Draw** toolbars are displayed on the sides of the drawing screen. Key features of these toolbars are as follows:

- The **Standard** toolbar contains the names of frequently used buttons. In addition to common methods for controlling AutoCAD, the toolbar contains standard Microsoft buttons such as **Open, Save, Print, Cut, Copy**, and **Paste**. The toolbar also contains the **InfoCenter, Communication Center**, and the **Favorites** icons.

- The **Styles** toolbar shows the current Text Style, Dimension Style, Table Style, and Multileader Style settings and provides controls for each.

- The **Properties** toolbar can be used to set the properties of objects such as color, linetype, lineweight, and the Plot Style control.

- The **Draw** toolbar contains buttons to represent each of the major drawing commands, including **Line, Rectangle, Arc, Circle, Ellipse, Polygon**, and **Point**.

- The **Modify** toolbar contains buttons to represent commands that can be used to edit drawing objects, such as **Erase, Copy, Mirror, Offset, Array, Move**, and **Rotate**.

- The **Layers** toolbar contains controls for altering the display of drawing layers. Each control will be introduced in chapter 5.

Displaying Toolbars

Because of their ease of use and availability, the toolbars will become the primary method of selecting commands for most AutoCAD users. AutoCAD offers a wide variety of toolbars that can be displayed to meet various drawing needs. Use the following steps to display additional toolbars:

1. Move the cursor to a blank area beside one of the existing toolbars, and right-click. This will produce a menu showing the available types of menus in your program.

2. Move the cursor over the word **ACAD.** This will display the available toolbars that can be displayed in AutoCAD.

3. Select **Dimension** in the list to produce the **Dimension** toolbar shown in Figure 1.11.

Each of the toolbars listed will be discussed as specific commands are introduced throughout the text.

Tip: Depending on settings made when AutoCAD was installed, other menus may be available including Express, Custom and AutoCAD Impression. Express toolbars provide additional controls for many of the standard toolbars. Selecting Custom allows you to create your own toolbars with your favorite commands. Express toolbars will be introduced in each chapter where they apply. The Custom and AutoCAD Impression toolbars lay beyond the scope of this chapter. As you gain experience, you may want to use the HELP menu to create your own toolbars using Custom.

Figure 1.11   *The **Dimension** toolbar is displayed by right clicking a blank space beside an existing toolbar, and then selecting **Dimension** from the **ACAD** listing.*

Executing a Command with a Toolbar

As the cursor is moved from the drawing screen to a toolbar, it changes from crosshairs to the arrow cursor. The cursor can now be used to select the desired button. You'll also notice as the cursor passes over, or rests on a button on the toolbar, that the icon takes on a 3D appearance. Move the cursor to the button in the left corner beside the drawing window. As the cursor is placed above the button, the title of the button will be displayed below the cursor, similar to the display in Figure 1.12. The title is referred to as a tooltip. Tooltips provide a written display of the results that will occur if the button is selected. In this case, the cursor is over the **Line** button. Select the **Line** button with a single-click to start the **Line** command. This will display the following prompt on the **Dynamic Input** display:

Specify first point:

The program is now waiting for you to select a starting point for a line, as well as showing you the current location of the cursor. Move the cursor and select an endpoint for a line by clicking the select button on the mouse. A display similar to Figure 1.13 will be shown. As the endpoint is selected, the Command prompt is changed to request the next point of the line. AutoCAD is waiting for you to select the line endpoint. Move the cursor to a new location and pick a point. The prompt will continue to request additional points after each new point is entered until you choose to stop drawing lines. To terminate the **Line** command, press ENTER or click the right mouse button and select the **Enter** or **Cancel** option. When in the middle of a command, a right click will produce a shortcut menu similar to Figure 1.14. Selecting the **Enter** or **Cancel** option will end the **Line** command. The other options of this menu will be explored in chapter 3 as the **Line** command is explored further.

Figure 1.12   *Resting the cursor over a button provides a written description of the button's function called a tooltip.*

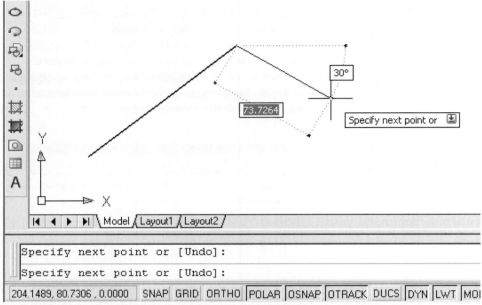

Figure 1.13   *As the cursor is moved, the dynamic display will provide information about the line being displayed. The information in the display will vary depending on the command being executed.*

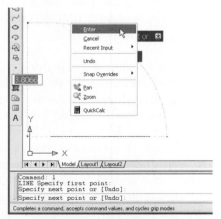

Figure 1.14   *Right-clicking while in the middle of the **Line** command will produce a command-mode menu. Selecting **Enter** or **Cancel** will terminate the **Line** command at the last point that was specified. Other options in this menu will be explored as you progress through the text.*

To start another line in a different location, press ENTER, which automatically restarts the last command sequence. Repeating this process will allow an unlimited number of line segments to be drawn without having to select the **Line** button for each new line. Take time to enter AutoCAD and draw several different continuous line segments. Once you've drawn several lines, end the command, and then reenter the **Line** command by pressing the ENTER key. Now draw several additional individual line segments. This process can be used with each of the commands represented on the **Draw** toolbar.

Although the commands will be discussed in detail in later chapters, take time to explore the **Arc, Circle** and **Rectangle** commands on the **Draw** toolbar. As you're exploring, you might also want to investigate the **Erase** command. Selecting the **Erase** button on the **Modify** toolbar will allow you to select objects and remove them from the drawing screen. Use the following sequence to erase a line:

- Select the **Erase** icon from the **Modify** toolbar.

- Select the object to be removed by moving the cursor to the object and left clicking.

- Continue to select objects to be erased as desired.

- Right-click to end the selection process and remove the selected objects from the drawing.

As you're exploring various toolbars, notice that several toolbar buttons have a small triangle in the lower right corner. Clicking and holding down the cursor on these buttons will produce a flyout menu, which is a submenu of the selected menu. Figure 1.15 shows the flyout menu for the **Zoom Window** button of the Standard toolbar. Options on this toolbar will be explored in chapter 8.

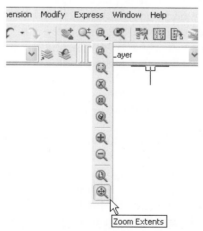

Figure 1.15 *Clicking and holding a button that contains a triangle will produce a submenu. The cursor can then be used to select one of the related listings.*

Altering Toolbars

Toolbars are used to enter commands because they are extremely flexible. In addition to being able to display multiple toolbars on the screen, you can move, dock, float, resize, and alter the contents of toolbars to allow for more convenient usage.

Moving Toolbars   As a toolbar is opened from the ACAD menu, it is placed randomly in the drawing area. A toolbar can be moved to any convenient area by placing the cursor in the title bar of the menu, and then clicking and holding the select button as you move the mouse. Release the select button when the toolbar is in the desired location.

Floating Toolbars   A toolbar is referred to as a floating toolbar once it is moved from the border of the drawing display. When a toolbar is floating, it can be resized, docked, or its contents can be altered. If the **Dimension** toolbar is activated, it will be randomly placed in the drawing area, making it a floating toolbar. Selecting and holding the title bar allows the toolbar to be moved to any desired location. The same is true of docked toolbars. Move the cursor to the top border of the **Draw** toolbar (the two parallel lines). Press and hold the select button and move the toolbar into the drawing area. The toolbar is now said to be floating.

Resizing Toolbars   If the toolbar interferes with the drawing display, the size of a toolbar can be altered. Placing the cursor anywhere on the border of the toolbar and dragging it in the direction you want the bar to be resized alters the shape of a toolbar.

Docking and Closing Toolbars   A toolbar is docked by placing the cursor over the title bar, and then selecting and holding down the select button. The toolbar can now be moved to a position above, below, or on either side of the drawing area. The border of the toolbar will change to a dashed line when it is placed near the drawing border to indicate that it will return to its full size when docked. An alternative to docking a toolbar is to totally remove it. A toolbar can be removed from the screen while it is floating by selecting the **Close** button in the upper right corner. To restore the window, right click the empty space by the docked toolbars, and select the desired toolbar from **ACAD** menu.

Refloating Docked Toolbars   To float a docked toolbar, select, hold and drag the grab bar on the left end of the toolbar to the desired location.

Altering Toolbar Content   It might seem unimportant now, but one of the nice features of AutoCAD is that it allows you to modify existing toolbars or to create a new one with your favorite buttons in it. Along with changing the size or location of toolbar buttons, you can also add or delete buttons from a toolbar. Because the purpose of this chapter is to explore major tools, , this specific skill will not be further explored. Use the help menu that will be introduced later in this chapter to learn to customize toolbars.

Hiding Toolbars   AutoCAD allows all of the toolbars surrounding the drawing area to be hidden using a feature named **Clean Screen**. Select **Clean Screen** from the **Tools** menu to toggle between the default setting and a display with all of the toolbars hidden. CTRL + 0 can also be used to toggle between displays. Active toolbars will be restored by reselecting **Clean Screen** or by reentering CTRL + 0.

Locking Toolbars   Once the toolbars and windows have been adjusted to your satisfaction, AutoCAD allows each to be locked so that they can't be accidentally moved or closed.

Currently the toolbars are unlocked allowing all of the normal features to be completed. Pressing the select button, or right clicking while the cursor is over the lock icon on the right end of the status bar, will display a shortcut menu with several options. Select **ALL,** and then select **LOCKED** from the flyout menu to lock all toolbars and windows in their current positions. Pressing the **CTLR** button provides a temporarily override. Repeating the process that was used to lock the toolbars, but selecting the **UNLOCKED** option restores the ability to adjust the toolbars.

 Note: Leave the lock in the UNLOCKED position to aid you in experimenting. Chapter 4 will introduce a method to restore all features to their original settings.

KEYBOARD ENTRY

The dynamic display and command line allow you to enter your request to AutoCAD using the keyboard. Throughout this text, commands will be discussed by providing examples of toolbar button and keyboard entry. To enter a command by keyboard, type the name of the command and press the ENTER key. To draw a line, type **Line** ENTER at the Command prompt. The text will automatically be entered at the **Dynamic Input** display and the Command prompt. For most commands, an alias can be used to start the command. To start the **Line** command, type **L** ENTER. Once ENTER is pressed, the command line display will be altered. The command sequence is as follows:

```
Command: L ENTER
LINE Specify first point:
```

Use the cursor to select a starting point for the line just as you did with any of the other method of entering the command.

 Note: AutoCAD is not case sensitive. You can use all caps, all lower case or a combination.

MENUS

As the cursor is moved into the menu bar, a key method of interacting with AutoCAD is presented. The menu bar displays the names of 11 (12, if the express menus are loaded on your machine) menus used to control AutoCAD. As with other Windows menus, selecting a specific name will produce a display of that menu. Selecting a menu displays a complete listing of related commands. Move the cursor to highlight **Draw**. Select **Draw** with a single click to display the menu shown in Figure 1.16.

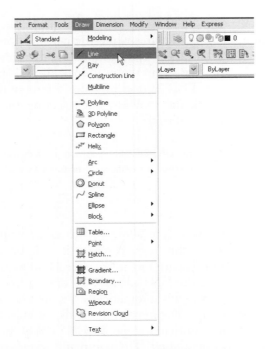

Figure 1.16 *Moving the cursor to the menu bar and selecting **Draw** will display the **Draw** menu.*

Once in the **Draw** menu, you'll notice that three different types of commands are listed:

- Typical commands, such as **Draw,** where the command is selected and executed.
- Commands, such as **Arc,** with a triangle on the right edge produce a submenu of related commands or options.
- Commands, such as **Hatch,** followed by ... (ellipsis) produce a dialog box.

Select **Line** from the menu. This will remove the menu from the screen and start the **Line** command. The command works the same as when you entered the command using the **Line** button on the **Draw** toolbar or by typing **L ENTER**.

Exit the **Line** command and return to the **Draw** menu. Move the cursor to rest over **Arc** in the **Draw** menu to display the submenu shown in Figure 1.5. Select the **3 Points** option. This will remove the menu, return the crosshairs, and allow the selected command option to be executed.

The command sequence to draw a three-point arc is:

```
Select 3 Points from Arc in the Draw menu.
Specify start point of arc or  (Select an arc start point with
   the cursor.)
Specify second point of arc or (Select a mid point for the arc.)
Specify end point of arc: (Select the end point of the arc.)
```

As the third point of the arc is selected, the arc will be displayed in the drawing area and the Command prompt awaits your next entry.

Open the **Draw** menu. A third type of display is found in this menu. Notice that several options are followed by an ellipsis (...). Selecting a command such as **Hatch...** with a single click will produce the dialog box shown in Figure 1.17. Using dialog boxes will be introduced later in this chapter and explained throughout the balance of the text as specific commands are introduced. To exit the **Boundary Hatch** dialog box, move the arrow to the **Cancel** button, and press the select button to return to the drawing screen.

Figure 1.17   *Selecting a menu option that is followed by an ellipsis (...) will produce a dialog box. Dialog boxes allow the functions performed by certain commands to be adjusted.*

DIALOG BOXES

Dialog boxes provide a convenient way to adjust certain commands, options, or portions of the drawing program. When a dialog box is selected, the crosshairs of the drawing area changes to an arrow. The arrow can be used to select an option by placing the arrow in the box or button by the desired option and pressing the select button. More specific information regarding the actual use of each dialog box will be discussed as each command is introduced. Common components of a dialog box are seen in Figure 1.18. Common features include **OK** and **Cancel** buttons, radio buttons, scroll bars, check boxes, edit boxes, image tiles, and alerts.

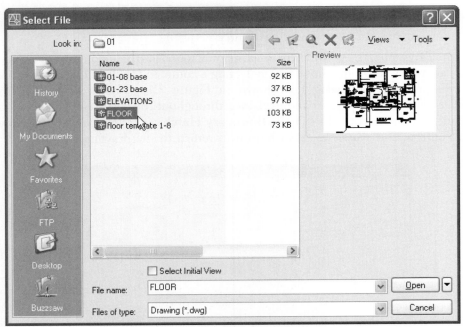

Figure 1.18 *Common components of a dialog box include buttons, radio buttons, scroll bars, check boxes, edit boxes, image tiles, and alerts.*

Ok/Cancel

To return to the drawing and remove the dialog box, select either the **OK** or the **Cancel** button.

- The **OK** option updates the dialog settings and removes the box from the drawing area.

- The **Cancel** option removes the box from the screen, but it will also delete any changes you might have made. Pressing ESC has the same effect as selecting **Cancel**.

Buttons

Several buttons are often included in a dialog box in addition to **OK** and **Cancel**. Examples are seen in Figure 1.19. The **Plot** dialog box can be displayed by clicking the **Plot** button on the **Standard** toolbar. Clicking one of these buttons in a dialog box will cause the selected option to be performed.

- Buttons with a wide border are default options.

- A button with a name followed by ... will display another dialog box that must be addressed before proceeding.

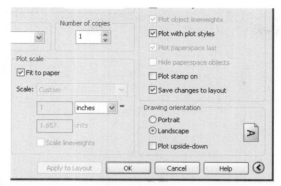

Figure 1.19 *Radio buttons such as Portrait and Landscape are mutually exclusive choices that can be found in the lower corner of the Plot dialog box.*

Radio Buttons

Radio buttons are buttons that list mutually exclusive options. Examples can be seen in Figure 1.19. **Portrait** and **Landscape** are radio buttons found in the **Plot** dialog box. Each will present a different paper orientation for plotting, but only one of the options can be used per plot. Notice in Figure 1.19 that several of the **Plot options** are not displayed with black text. Gray options are inactive options and can't be selected until another option has been completed. Select one of the active or black options by moving the cursor to the desired button and pressing the select button.

Check Boxes

Check boxes serve as toggle switches. Open the **Drafting Settings** dialog box by selecting **Drafting Settings** from the **Tools** menu. Once it is open, select the **Object Snap** tab to produce the display shown in Figure 1.20. In the **Drafting Settings** dialog box, each of the Object Snap modes is represented by a check box. Each mode is independent of the other listings. With the **Endpoint** mode active, the cursor will automatically jump to the nearest endpoint of existing lines as the cursor is moved. Each of these options, as well as many others, will be presented throughout the text. For now, remember that if a box has an "x" or check in it, that option is activated. If the box is empty, the option will be inactive.

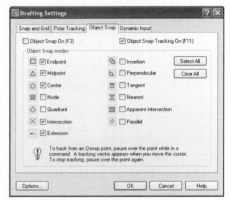

Figure 1.20 *Check boxes are used to control menu options. Object Snap check boxes are displayed in the **Object Snap** tab of the **Drafting Settings** dialog box.*

Text Boxes

A text box is an entry box that allows information to be entered by keyboard. Figure 1.18 shows an example of an edit box. The **Select File** dialog box is displayed by selecting **Open** from the **File** menu. Information can be selected from a list or typed into the box, in this case the **File name** edit box. If errors are made in typing, the cursor can be moved using the left and right arrow keys. Activate the text in the box for editing by pressing ENTER.

Edit Keys

Several keys can be used to aid input in a dialog box. These include text keys, arrow keys, and control keys. These functions include the following:

> TEXT—Symbols from the keyboard are inserted at the cursor. As a letter is inserted, the cursor and existing text move to the right. When the cursor reaches the right side limit, the cursor remains and text is moved to the left.

> LEFT ARROW—Moves the cursor to the left without changing the existing text.

> RIGHT ARROW—Moves the cursor to the right without changing the existing text.

> DEL—Deletes the character at the cursor and moves all remaining text to the left. The cursor will remain stationary.

> BACKSPACE—Deletes the character to the left of the cursor while the cursor and all remaining text is moved to the left.

Image Boxes

An image box is a portion of a dialog box that displays an image of the object or pattern that is available. Figure 1.18 shows the image box used to show the drawing that will be opened.

SHORTCUT MENUS

Similar to most Microsoft Windows programs, AutoCAD contains menus that provide rapid access to options for tasks that are underway. Click the right mouse button to display these shortcut menus. The contents of the menu will change depending on when the menu is selected. If you're in the middle of a drawing command and right-click, the menu will be different than if you right-click while trying to edit a drawing. Five basic shortcut menus are available within AutoCAD. Each can be customized. The default settings of these menus are as follows:

- The default menu is displayed when a right-click is performed with the cursor in the drawing area prior to starting a command sequence.

- The command-mode menu appears when a right-click is performed in the middle of a command sequence. Options specific to the command will be displayed.

- The edit-mode menu is displayed with a right-click if objects have been selected, but no command is in progress.

- Right-click with the cursor over a toolbar to produce a list of all toolbars.

- Right-click while over the command line, layout tabs, or status bar to produce different context-specific menus.

- Right-click while over a tool palette to display different versions of the context menu, which allows palettes to be created, customized, hidden, renamed or deleted.

- Other menus will be displayed based on the cursor location when a right-click is performed.

TOOL PALETTES

 AutoCAD allows frequently used symbols (blocks), patterns, commands and custom tools supplied by third party developers to be saved and displayed on tool palettes. Tool palettes are similar to a toolbar in that they can be docked and resized without being closed. Palettes are displayed by:

- Selecting the **Tool Palettes Window** icon from the **Standard** toolbar

- Selecting **Tool Palettes** from **Palettes** of the **Tools** menu.

Figure 1.21 shows the **Architectural** tab of the palette. Other common tabs that will useful for architectural drawing include the **Command**, **Hatches**, and **Annotation** tabs. Selecting the tab that represents the desired symbol library alters the display of symbols. Once on the desired tab, the display can be altered by using the scroll bar or by using the **Pan** command (the hand symbol on the Standard toolbar). With the hand displayed, press and hold the select button, and move the mouse up or down to scroll through the symbols.

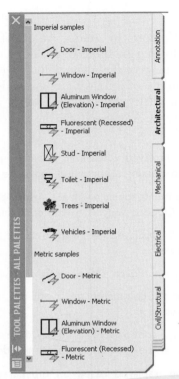

Figure 1.21 *The **Architectural** tab of the **All Palettes** contains predefined symbols that can be inserted quickly into a drawing. Right-click on the covered tabs at the bottom of the menu to view different palettes.*

Methods of creating, viewing, using and managing palettes will be introduced in later chapters as blocks, **DesignCenter**, **Properties** and **dbConnect** are introduced.

Palettes are also referred to as modeless dialog boxes. Unlike regular dialog boxes, modeless dialog boxes can be displayed even if other tools or commands are active. Similar to a docked toolbar, palettes can be hidden when not in use to maximize the drawing area. To hide a palette, click the minus button in the top left corner. This will hide the palette, but leave the title bar displayed. The palette can be redisplayed by moving the mouse over the **Auto-hide** button or any part of the title bar.

Rather than hiding the tool palettes, AutoCAD allows palettes to become transparent. In the default setting, palettes are not transparent. Right-clicking the title bar of a palette or selecting the **Properties** icon at the bottom of the palette will display the **Palette** shortcut menu. Selecting **Transparency** will display the **Transparency** dialog box shown. Sliding the scroll bar to the right increases the transparency of the palette. Slide the bar to the desired position and select the **OK** button. As the palette is redisplayed, the drawing beneath the palette can now be viewed.

Dashboard

A dashboard is a special palette that displays buttons and controls associated with either the 2D Drafting & Annotation workspace or the 3D Modeling workspace. With the exception of Figures 1a and 1c, each of the displays for this chapter has been of the 2D classic workspace. Figure 1.1b shows the 2D Drafting & Annotation workspace with the dashboard docked on the right edge of the drawing area. Notice that the **Draw** and the **Modify** toolbars are removed from the drawing area, but their icons are displayed in the upper area of the dashboard. Similar to a framers tool belt, dashboards group common tools together for speed and convenience. Dashboards display common tools specific to the workspace in one area and eliminate the need to display multiple toolbars. Figure 1.22 shows the 2D dashboard. Specific areas will be explored as each command associated with the dashboard is explored further.

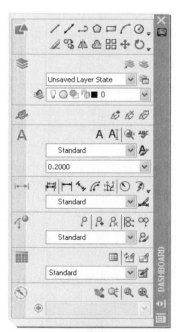

Figure 1.22 *The 2D Drafting & Annotation dashboard groups controls for drawing, editing, layers, scaling, text, dimensions, multi-leaders, tables, and navigation.*

FUNCTION KEYS

Function keys are located across the top of the keyboard. Ten of these keys are used by AutoCAD to provide access to commands. Each is used as a toggle from ON to OFF and will be explained further in chapter 5. Other keys can be programmed to perform specific commands. AutoCAD uses the following function keys:

F1—Help. Selecting this key displays the Help window tab at the bottom of the screen. Picking the tab will display the Help menu. Using the Help options of AutoCAD will be discussed later in this chapter. Select the **Cancel** button or the **Close** button to remove the display from the screen.

F2—Flip Screen. Pressing F2 in the middle of a drawing will display a listing of the commands that were used to generate the drawing over the top of the drawing area. Press F2 a second time and the text window will be removed from the drawing area.

F3—Object Snap. Snap allows the cursor to be moved to specified points on drawing objects such as an endpoint or midpoint of a line. Later chapters will explore how to set the object snap settings. Pressing F3 will toggle Object Snap **ON/OFF**.

F4—Tablet mode. This function key serves as a toggle if a digitizing tablet is used. A digitizing tablet can be used as an alternative to a mouse for selecting objects from menus and controlling the crosshairs.

F5—Isoplane crosshairs mode. This is used when creating isometric drawings. This button toggles between the top, left, and right planes of an isometric square. These planes will be discussed as isometric drawings are explored.

F6—Toggles the UCSDETECT display **ON** or **OFF**.

F7—Grid mode. In later chapters you will learn how to create a visible grid to aid in drawing layout. The grid can be displayed or removed by pressing the F7 key. Grids can also be toggled **ON/OFF** by selecting the GRID button on the status bar.

F8—ORTHO. ORTHO can be used to control the movement of the cursor. When it is ON, only vertical and horizontal lines can be drawn. When OFF, lines at any angle can be drawn. These options will be explored in chapter 5.

F9—Snap mode. This key activates the ability to move the crosshairs within the drawing area at specific intervals. Snap will be explained in chapter 5. Snap can also be toggled ON/OFF by selecting the SNAP button on the status bar.

F10—This key toggles Polar Tracking **ON/OFF**.

F11—This key toggles Object Snap Tracking **ON/OFF**.

F12—This key toggles Dynamic Input **ON/OFF**.

 Note: Remember that function keys can be used at any time throughout the life of a drawing including in the middle of a command.

CONTROL KEYS

Like many other software programs, AutoCAD uses control keys to perform common tasks. These jobs are achieved when the control key is pressed in conjunction with a specified letter key. Many of these tasks duplicate the jobs performed by the function keys. The control keys are located in the lower corners of the keyboard and may be labeled CTRL. The control key functions are:

CTRL + A—Selects all objects in drawing for editing

CTRL + B—Snap mode toggle **ON/OFF**

CTRL + C—**Copyclip** command copies selection to the clipboard

CTRL + SHIFT + C—Copies objects to the clipboard with a base point

CTRL + D—Toggles Dynamic UCS **ON/OFF**

CTRL + E—Toggles the crosshairs in isoplane position to either left/top/right

CTRL + F—Toggles running **OSNAP** mode **ON/OFF**

CTRL + G—Grid toggle **ON/OFF**

CTRL + H—Toggles the **PICKSTYLE** value **ON/OFF**

CTRL + I—Toggles **COORDS ON/OFF**

CTRL + J—Repeats the last command

CTRL + L—ORTHO mode toggle **ON/OFF**

CTRL + M—Repeats the last command

CTRL + N—Executes the **New** command

CTRL + O—Open command, used to open a new file

CTRL + P—Displays the **Plot** dialog box

CTRL + R—Toggles between viewports

CTRL + S—**Save** command

CTRL + SHIFT +S—Displays the **Saveas** dialog box

CTRL + T—**Tablet** mode toggle **ON/OFF**

CTRL + V—**Pasteclip** command, pastes Clipboard locations to a specified location

CTRL + SHIFT +V—Pastes information from the Clipboard as a Block

CTRL + X—**Cutclip** cuts selected objects to the clipboard

CTRL + Y—Cancels the preceding **Undo** action

CTRL + Z—**Undo** command

CTRL + [—Cancels the current command

CTRL + PAGE UP—Moves to the next Layout tab to the left of the current tab

CTRL + PAGE DOWN—Moves to the next Layout tab to the right of the current tab

CTRL + 0—**Clean Screen** toggles between a clean and current screen

CTRL + 1—Toggles the display of the **Properties** window

CTRL + 2—Toggles AutoCAD **DesignCenter**

CTRL + 3—Toggles the display of **Tool Palettes**

CTRL + 4—Toggles the display of **Sheet Set Manager**

CTRL + 5—Toggles the **Info Palette**

CTRL + 6—Toggles **dbConnect Manager**

CTRL + 7—Toggles **Markup Set Manager**

CTRL + 8—Toggles the **QuickCalc** calculator palette

CTRL + 9—Toggles the command window

GETTING HELP

One of the best features of AutoCAD is that it offers free advice with no ridicule. You're going to be exposed to hundreds of commands, options, and menus, and sometimes it's easy to feel lost or overwhelmed. If you can avoid panic in the first 30 seconds you'll do fine. Your chances for success are even better if you remember that there has been a **Help** button in every menu you've looked at. Selecting the **Help** or the **InfoCenter** button will provide instant access to the AutoCAD help display. The help display offers an explanation of nearly every button, command, option, variable and setting in AutoCAD. The help that is provided will depend on when and where you ask for help. In addition to the help button provided in each dialog box, help is available by selecting **Help** from the menu bar, selecting the F1 key, selecting the **Help** icon from the standard toolbar, typing **HELP** ENTER, or by accessing the InfoCenter. Each method will place the Help menu on the program bar.

USING THE HELP MENU

Help is available for all areas of AutoCAD by selecting the **Help** icon from the Standard toolbar. Picking the **Help** listing from the menu will produce the AutoCAD **2008 Help** menu shown in Figure 1.23. The Help menu can also be accessed in the middle of a command sequence. When the **Help** is accessed in the middle of a command, it automatically provides command-related help. Asking for help in the middle of the **Line** command will produce the Help display shown in Figure 1.24. The F1 key can also be used in the middle of a command to provide command-specific information. To get help in the middle of the **Line** command, make the following entry:

```
Command: L ENTER (Or select the Line button.)
Specify first point or (Select the desired starting point.)
Specify next point or (Select the Help icon.)
```

The **AutoCAD Help** display shown in Figure 1.24 is now displayed. The menu contains a brief description of the Line command. Click the **Close** button in the **Help** title bar to close the Help window and return to the drawing screen.

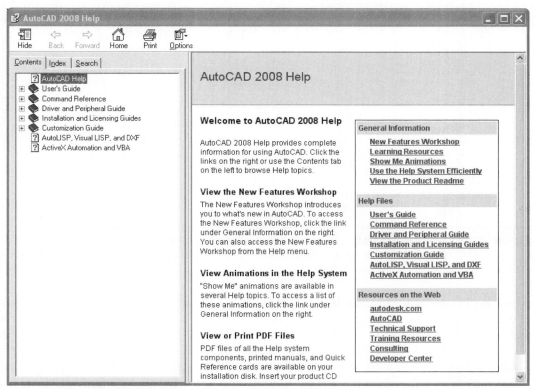

Figure 1.23   *Clicking the **Help** button on the **Standard** toolbar, pressing the F1 key, or typing **Help** ENTER at the command line will produce the Help window.*

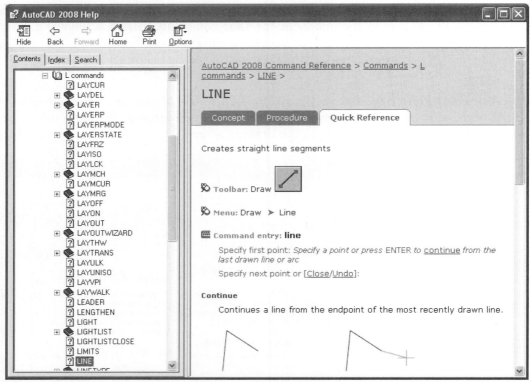

Figure 1.24 *Selecting the **Help** command in the middle of a command will produce command-specific help. The **Quick Reference** tab of the **Line help** menu produces a detailed description of the command.*

The left portion of the display is used for finding information in the Help window. The tabs at the top of the left screen provide the methods for using the Help window, and the right side of the window is where the help text will be displayed. In the top left corner of the Help window is the **Hide** button. Selecting this button will hide the left half of the Help window. Once the desired material has been displayed, the window can be reduced. Selecting **Show** will restore the left portion of the Help window, allowing additional use of the Help window. Selecting the Back button will restore the previous help screen. The Help window offers help in the areas of **Contents, Index, Search,** and **Ask Me.**

CONTENTS

The **Contents** folder of the Help window can be used to obtain information about the Help program or individual portions of the AutoCAD program. Notice in Figure 1.23 that several listings are displayed in the **Contents** list, preceded by a closed book. The **Contents** folder contains a **Users Guide** to provide instruction on using each aspect of the program. The **Command Reference** portion of the folder provides information on each command used by AutoCAD. Clicking a menu item will provide a brief description of the selected item on the right side of the window. Double-click a menu item to provide a submenu of the contents. Use the following sequence to receive information about the **LINE** command:

1. Display the **Help** menu using one of the methods introduced in this chapter.

2. Double-click the **Command Reference** listing and then double-click **Commands** to display a list of commands presented in alphabetical order. To find information on the **LINE** command, select the **L** Commands listing from the display.

3. Select **LINE** from the listing of **L** Commands. This will produce the display shown in Figure 1.24, and provide a brief description of the **LINE** command.

Depending on your needs, the information displayed on the right side of the window can be shown in three formats:

- **Concepts**—Selecting this option provides an explanation of what the expected outcome of the command will be.

- **Procedures**—Provides a brief description of how to access the command and a step-by-step method of using the command.

- **Quick Reference**—Provides an overview of the command and a list of options that affect the command and provides a listing of related commands.

INDEX

Selecting the **Index** tab will produce a menu that is like the index found at the end of a book. If you enter the first few letters of a command or option, the Help menu will display a list of commands and sources of information about that command. Entering **line** in the edit box will produce the display shown in Figure 1.25. This display box can be used to select information about the command. Selecting the **LINE** command listing and selecting the **Display** button will produce a listing of the desired information.

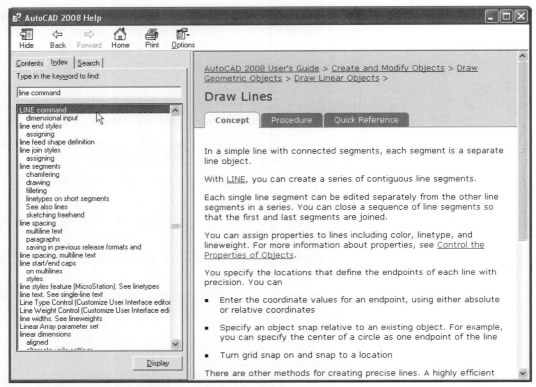

Figure 1.25 *Selecting the **Index** tab will produce a menu similar to an index at the end of a book. As commands are typed in the **Keyword** edit box, the display is altered to reflect the entry.*

SEARCH

The **Search** portion of the Help produces a search based on words entered in the search box. Entering the word **LINE** in the search box and press the **Search** button. For this example, **How do I draw lines** was entered. Figure 1.26 shows the display of related subjects. Notice that the number of matching topics is listed, as well as the name, location and rank of the options. Selecting **Show All Results** will list all of the related references.

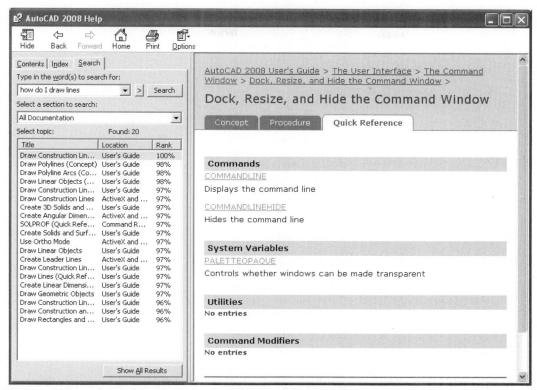

Figure 1.26   *Selecting the **Search** tab will allow specific questions to be entered into the help menu.*

INFOCENTER

The **InfoCenter** can be used to search multiple sources such as Help, the New Features Workshop, specified files, or a single file or location. The search process is similar to using the **Search** feature of the **Help** menu. Enter a key word or a question into the menu bar and then press ENTER or click the SEARCH button. This will start the search the contents of multiple Help resources as well as any additional documents that have been specified in the InfoCenter Settings dialog box. Search settings can also be altered through the **CAD Manager Control Utility**. The search results are displayed as links on the InfoCenter Search Results panel. Select the desired link to display the topic, article, or document. Figure 1.27 shows the menu options that resulted from entering LINE COMMAND in the search box.

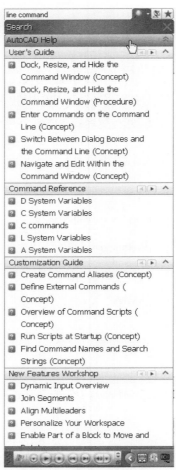

Figure 1.27 *Entering a key word or a question into the menu bar and then pressing ENTER or clicking the SEARCH button will start the search of the contents of multiple Help resources.*

COMMUNICATION CENTER

The AutoCAD **Communication Center** offers direct connection to Autodesk. Open the Communication Center by clicking the icon in the right side of the status bar. Selecting the Settings button allows you to select your country and how often you would like updates. The status bar icon displays a bubble message whenever new information is available. The Communication Center provides the following types of announcements:

General Product Information—Allows you to stay informed about company news and product announcements and provides feedback directly to Autodesk.

Product Support Information—Provides the latest news from the Product Support team at Autodesk including when Live Update maintenance patches are released.

Subscription Information—Provides subscription program news as well as links to e-Learning Lessons for Autodesk subscription members. **Articles and Tips**—Provides notification when new articles and tips are available on Autodesk Web sites.

CAD Manager Channel—Allows information (RSS feeds) to be published by your CAD manager.

Featured Technologies and Content—Offers the opportunity to learn more about 3rd party developer applications and content.

The items that display on the Communication Center panel can be customized using the *Specify InfoCenter Settings*. Figure 1.28 shows the default listings of the Communication Center.

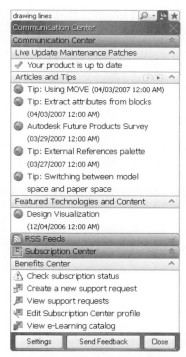

Figure 1.28   *The **Communication Center** offers direct connection to and from Autodesk. The type of information to be displayed can be altered to meet individual needs.*

EXPLORING THE SAVE COMMAND

Throughout this chapter you've explored several methods of controlling commands. The **LINE** command was used as toolbars, menus and keyboard entry for commands were explored. The lines you've created thus far will be destroyed if you end your drawing session without saving them. Chapter 4 will explore how and where files can be saved. This chapter will introduce the procedure for saving a drawing file.

If you have not already done so, enter the drawing area and draw several lines. Because this is a chapter for exploring, use the **Draw** toolbar to draw a circle and a rectangle. You have now created a work of art that must be saved for future generations to enjoy. Use the following procedure to save the drawing to the hard drive:

1. Select **Save** from the **File** menu. This will display the **Save Drawings As** dialog box shown in Figure 1.29.

2. Select the arrow in the **Save in** box. This display allows the storage destination for the file to be selected. Select the C: drive for storage on a hard drive or A: for storage on a diskette.

 Warning: Do not save to a diskette in the middle of a drawing session. Only save to a diskette at the end of a drawing session. Chapter 4 will explain why.

3. The name of Drawing1 is currently displayed. To alter the name, click in the edit box, move the cursor so that Drawing 1 is highlighted (in a blue box.), and type the desired name, in this case, **EXE1-1**.

4. Select the type of file to be used if the drawing is to be saved to be compatible with older releases of AutoCAD.

5. Click the **Save** button. The dialog box will be removed from the screen, the drawing will be saved, and the drawing area will be restored.

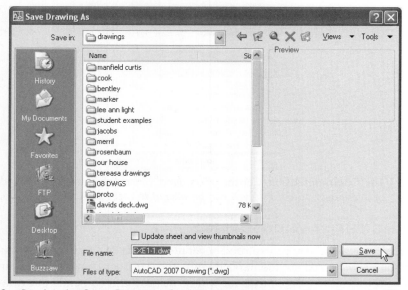

Figure 1.29 *Display the* **Save Drawing As** *dialog box by selecting* **Save** *from the* **File** *menu.*

ENDING A DRAWING SESSION

You can continue to add objects to this drawing for as long as you want. When you want to end the drawing session, save the drawing again by selecting the **Save** button to update the file. To exit the drawing session, click the **Close** button on the title bar or select **Exit** from the **File** menu.

Tip: Remember that there are two Close buttons in the upper right hand corner. The lower (black) Close button closes the current drawing but leaves AutoCAD open. The upper (red) Close button closes the current drawing and closes AutoCAD.

WHAT IT ALL MEANS

By now, you might be feeling overwhelmed by all the bells and whistles of AutoCAD. In this chapter you've been introduced to several methods of entering a command, and have explored a multitude of buttons, boxes, tool palettes, and controls for altering the commands and options. Each has been introduced with the **Line** command as an example. By the time you explore each of these controls your brain might be on overload.

- The good news is that you don't have to memorize every command, option, and control method. You only have to remember your favorites and the ones you use on a regular basis. Use the **Help** window, the InfoCenter, and this book to help keep track of options that you don't use on a regular basis. Tagging pages will really aid your memory.

- The other key point of this chapter is that there are usually several ways to do one thing. Generally, one method is as good as another, but don't limit yourself to the first method you find.

- A final tidbit to take from this chapter is, please ***don't be afraid to explore***. You're not going to hurt the computer, nor will it hurt you. Explore and enjoy!

CHAPTER 1 QUIZ

DIRECTIONS

Answer the following questions with short complete statements. Type your answers using a word processor.

1. Place your name, chapter number, and the date at the top of the sheet.

2. Type the question number and provide the answer in the form of a statement that includes part of the question. You do not need to write out the entire question.

Warning: Some of the questions have not been covered in the reading material and will require the use of the help menu and for you to do some exploring to answer the questions.

I. List three sources for obtaining help with AutoCAD.

_____

_____

2. Select the **My Computer** button from the startup display. Select the **Printer** button and list which printers or plotters are configured to the current workstation.

_____

_____

3. Explain the difference between a click and a double click.

_____

_____

4. List the four screen areas displayed after you have entered AutoCAD.

_____

_____

5. Is a click or a double-click required to start AutoCAD using the **Programs** menu?

_____

_____

6. Describe key functions of the left and right mouse buttons.

_____

_____

7. List three different methods that can be used for selecting commands and tell how they are accessed.

_____

_____

8. You've displayed a dialog box with 40 listings. Describe two methods that can be used to move through the list to the end.

_____

_____

9. List four sources for obtaining help within the AutoCAD **Help** menu.

_____

_____

10. Explain the procedure to float the **Draw** toolbar.

_____

_____

11. Explain the following terms:
Default
Pick
Command

_____

_____

12. List in order the first eight commands on the **Draw** menu.

_____

_____

13. You have just finished drawing a line and would like to draw an **arc**. Explain how to get the **arc** options using the pull-down menu.

_____

_____

14. What is the last **Arc** option listed in the menu?

_____

_____

15. List the options in the **File** menu.

_____

_____

16. You've displayed the **Insert** menu, but you wanted the **View** menu. How can you get **View** to be displayed?

_____

_____

17. List the buttons that are seen on the status line.

_____

_____

18. Is **Ellipse** on the **Draw** menu a command or an option? How can you be sure?

_____

_____

19. A button showing a paintbrush is on the **Standard** toolbar. What is the button, and what does it do?

_____

_____

20. List the toolbars shown by default as AutoCAD is loaded.

_____

_____

21. What key will execute each of the following requests?
Display grid ORTHO Snap mode

_____

_____

22. Someone has asked you what the last three commands were that you used on your drawing. How could you get a record if you don't remember?

_____

_____

23. List five items that can be found in a dialog box.

_____

_____

24. Search through each menu and list the location of these items:
Erase
Tablet
Toolbars

_____

_____

25. What letter key would be needed with the control key (CTRL) to complete the following options?
Grid toggle
Save
Pick style
Coordinate display

_____

_____

26. List and describe five types of shortcut menus of AutoCAD.

_____

_____

27. How many sides will be drawn if the default value is selected for a polygon?

_____

_____

28. The **Draw** toolbar has been removed from the screen. Explain the procedure to display the **Draw** toolbar.

_____

_____

29. Describe the process to display a hidden toolbar.

_____

_____

30. You've discovered, after terminating a command that you actually wanted to draw one more line. What is the quickest way to reenter the **Line** command?

_____

_____

CHAPTER 2

Creating Drawing Aids

INTRODUCTION

This chapter will introduce methods for

- Entering the Drawing Environment
- Starting a new drawing using Wizards, Templates, and Start From Scratch
- Opening an existing drawing
- Controlling drawing parameters from within a drawing

Commands to be introduced include:

- **Qnew**
- **Open**
- **Partialopen**
- **Partiaload**
- **Units**
- **Limits**
- **Zoom**
- **Grid**
- **Snap**
- **Pan**

In chapter 1, you opened AutoCAD, examined each workspace, and explored key features related to the drawing screen. Selecting the **Qnew** (Quick New) button from the **Standard** toolbar will also allow a new drawing to be started. It is the equivalent to taking a blank sheet of paper with no other tools and starting to draw. This system works well for simple sketches or schematic drawings but is not suitable for the construction industry. This chapter will introduce methods of starting a drawing session with preset drawing parameters in addition to exploring methods for controlling the drawing environment.

ENTERING THE DRAWING ENVIRONMENT

The **Create New Drawing** dialog box shown in Figure 2.1 offers four alternatives to help set up a drawing file. Options include **Open a Drawing, Start from Scratch, Use a Template**, and **Use a Wizard** to organize drawings. The alternative you use will depend on how you manage a drawing session and on personal preference. Each startup method has advantages for drawing setup.

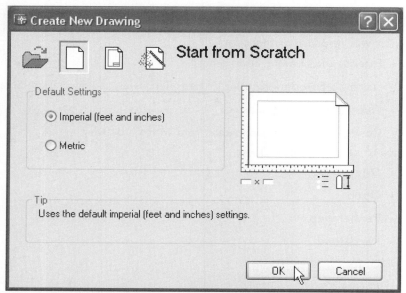

Figure 2.1 *The **Create New Drawing** dialog box can be displayed as AutoCAD is entered to provide help in setting a drawing parameter.*

1. **Open a Drawing**—Opens a drawing file created in a previous drawing session. This option should be used to open the file **EXE 1-1** that was created in chapter 1. Using this option allows editing or additions to be made to the existing drawing.

2. **Use a Wizard**—Starts a drawing session when you know the scale or the paper size that will be needed to plot the project. This will allow you to use the setup assistance from AutoCAD. This is an excellent option to use for the design stage of a drawing. Drawing wizards offer helpful prompts to guide in drawing setup but do not provide title blocks. Later chapters will help you provide a title block to drawings created without a template. **Use a Wizard** offers the **Quick Setup** and **Advanced Setup** options for determining how the new drawing environment will be established. Each option will be explored in the next section of this chapter.

3. **Use a Template**—Allows predetermined drawing parameters to be used to control the drawing environment. This is an excellent design tool to use when you have planned the drawing requirements and can choose the space required to display the

needed views. Templates can include the company logo or similar items that can be created and saved for use on each drawing. Once you've created and stored your own template, the drawing parameters will be preset to meet your standards each time you start a new drawing. Chapter 4 will help you create your own template.

4. **Start from Scratch**—Starts a drawing with minimal drawing controls. This option is the equivalent of drawing on a blank sheet of paper with no title block. It is best suited for users who like to plan as they go. Use **Start from Scratch** for sketching, line drawings that will not be drawn to scale, or for drawings that do not require a high level of accuracy. This option also works well when you want to establish drawing parameters as they are needed. Although this option provides some basic drawing controls, you will need to add many other controls and variables to increase your drawing efficiency.

OPEN AN EXISTING DRAWING

Much of your time in an office will be spent adding to or editing existing drawings. The **Open a Drawing** option of the **Startup** menu is similar to Figure 2.2 and is displayed when an AutoCAD session is entered. It shows the names and paths of the most recent drawings that have been created. Once a file is selected, the file size, and the date and time that the drawing was modified will be displayed. Selecting one of the filenames will display the image file for that drawing. Once you've found the desired file, double-click the title to open the file. Highlight the filename with a single click, and then select the **OK** button to open a file. If you're working at a computer, select the file created in chapter 1 titled **EXE1-1**. Select the **OK** button to close the dialog box and open the drawing that was created in chapter 1.

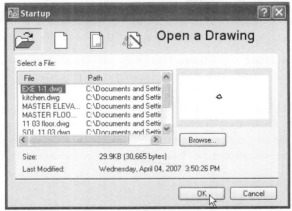

Figure 2.2   *Select the **Open a Drawing** button in the **Startup** dialog box to display a list of drawings that have been recently edited. The last five drawings that you have worked on will be displayed in the **Select a File** list. Click the **Browse** button to search for files that are not listed.*

The display of drawings to be opened can also be altered using the **Browse** option. This option is the same as when the **Open** button is selected from the **Standard** toolbar. Select the **Browse** button to find an existing drawing file that is not listed in the **Startup** dialog box. Selecting **Browse** will produce a **Select File** dialog box similar to Figure 2.3. The display will vary depending on the number of drives, the file contents and how the display has been configured. **Browse** can display the contents using thumbnails (image tiles) of folders and files contained in the specified location or the traditional detailed listing associated with other Windows products. The display is based on the setting in the **Views** menu located in the upper right corner of the **Select File** dialog box. With the **Details** option current, the filename, file size, and the date and time of the last drawing session are displayed. Display options also include selecting files based on usage (History), from My Documents, Favorites, FTP sites, Desktop and from the Internet (Buzzsaw). Selecting one of the filenames will display the image file for that drawing. Once you've found the desired image, open the file by selecting the filename with a single click, and then selecting the **Open** button.

The **Select File** dialog box can be used to list and select a drawing file from any of your storage locations. By default, files will be selected from the folder that was last accessed. Use the following steps to open an existing file that is not listed in the **Select File** listing:

1. Select the arrow beside the **Look in** box with a single click to open the display of available storage locations. Select the desired drive or folder that contains the drawing file to be opened.

2. Move to the desired drive and select the file to be opened.

   - Select the desired drive to be searched to display a list of the drawing folders contained in the selected drive.

   - Double-click a folder to display a list of files contained in that folder. Highlighting a file with a single click will display an image of the file in the **Preview box**.

3. Once you've found the desired file, select the **Open** button to bring the file into the drawing area. You can also open a drawing by double-clicking the file name.

 Note: Remember AutoCAD allows multiple drawings to be opened. An existing drawing can also be opened in the middle of another drawing session by selecting the **Open** button on the **Standard** toolbar, or by selecting **Open** from the **File** menu. Each method will produce the **Select File** dialog box shown in Figure 2.3.

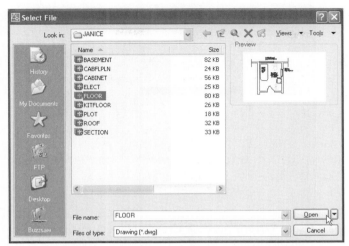

Figure 2.3 *Use the* **Look in** *list to define the location of the desired file.*

Open an Existing Drawing with Windows Explorer

An existing AutoCAD drawing can be opened using Windows Explorer. Find the desired file and double-click the file name from the listing in Windows Explorer. This will open the drawing file even if AutoCAD has not been opened. Double-clicking on the file name will open AutoCAD and display the drawing.

 Note: This method should not be used if you're working on a network that contains more than one AutoCAD program. Double-clicking a drawing will open the drawing using the last AutoCAD program that was used. You might find your drawing open in a session of AutoCAD Civil Design, Revit Architecture, Autodesk Inventor or AutoCAD Architecture.

DRAWING SETUP USING A WIZARD

A wizard is an excellent method for starting a new drawing session if you know the size of paper the project will be plotted on. AutoCAD provides the **Quick Setup** wizard and the **Advanced Setup** wizard. Figure 2.4 shows an example of the **Use a Wizard** menu. The method chosen will depend on how much help you would like. Choosing **Quick Setup** will assist you in setting the drawing units and the drawing area. The **Advanced Setup** will provide six startup prompts.

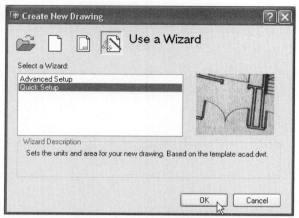

Figure 2.4 *Selecting **Use a Wizard** from the **Create New Drawing** dialog box provides two choices for drawing setup.*

Quick Setup

The **Quick Setup** option produces a drawing area in model space. This option works well for creating a drawing, but it requires some adjustment for plotting. Selecting **Quick Setup** allows the selection of the drawing units and the drawing area. Figure 2.5 shows the page for setting the **Units** using the **Quick Setup** wizard.

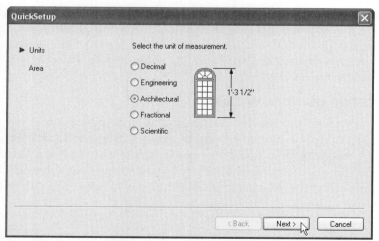

Figure 2.5 *The **Units** page of the **Quick Setup** wizard is used to control how drawing units are measured.*

Setting the Drawing Units   One of the unique features of AutoCAD is that as you work in model space you will be drawing at full scale. The drawing will not be scaled until you change to paper space to plot your drawing. If you want to draw a structure that is 300' long, you will draw it full size, and then choose a scale as you plot, using AutoCAD's paper space option. This will be covered in later chapters. The **Drawing Units** dialog box allows the measuring method to be set throughout the drawing session. In its default settings, your computer is set to measure lines in four-place decimal inch units. You can confirm this by looking at the coordinate display in the lower left corner of the status bar or at the coordinates displayed on the dynamic display. To describe the current location of the crosshairs, two four-place decimal numbers are used. Most architectural projects are better suited to the Architectural option. This will measure objects in feet and inches. Options for unit settings and their display include the following:

Decimal—15.0000—This option can be used to create drawings in decimal or metric units. Metric units might be needed on government construction projects. If decimal dimensions are used, they should be expressed as two-place decimals unless the project requires a higher degree of accuracy.

Engineering—1'–3.5000"—These units are best suited to civil drawings such as site plans, topography drawings, bridge and dam construction or other drawings dealing with large land areas.

Architectural—1'–3 1/2"—Architectural units express distance in feet and inches and are used for drawing most construction documents.

Fractional—15 1/2—This option can be used on drawings that express measurements in fractions such as cabinet drawings.

Scientific—1.5500E+01—This option is suitable for drawings used in chemical engineering. This unit of measurement is suitable when very large or very small measurements must be expressed on a drawing. Numbers are expressed using a base number and a multiplier. The E+01 represents a base number multiplied by 10 to the first power.

The units can be set by placing the cursor in the desired setting and pressing the select button. This will place a black dot in the button, indicating the selection is activated. Before proceeding, select the type of units that are appropriate for an architectural drawing such as a floor plan. Select the **Next** button to continue the drawing setup.

Setting the Drawing Area   Selecting the **Next** button will produce the display shown in Figure 2.6, showing a default value of 1' × 9". The drawing area should be based on what you expect to draw. For now, accept the default value and select the **Finish** button. This will display the AutoCAD drawing screen.

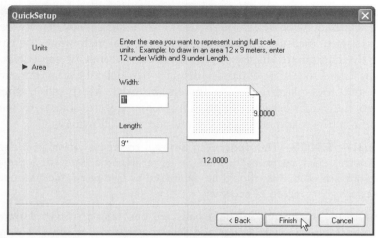

Figure 2.6 *Setting the area defines how much space is displayed on the screen. Set the area slightly larger than what you think you will need to complete the drawing. The size can be altered throughout the life of the drawing using the **Limits** command.*

Advanced Setup

Selecting this option will display the page shown in Figure 2.7. The **Advanced Setup** wizard will help you set five controls, including the units of measurement, method of angle of measurement, starting point of angle measurement, direction of angle measurement, and drawing area. The **Units** page is displayed as the **Advanced Setup** is opened. The default value is decimal. Choose the value that best meets the need of the drawing to be started using the methods described using the **Quick Setup** wizard.

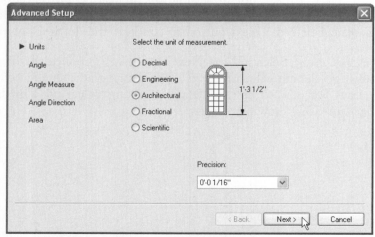

Figure 2.7 *Selecting the **Advanced Setup** wizard option will aid in setting five drawing aids. Although the display appears similar to the **Quick Setup** wizard, the advanced wizard allows the precision of the units to be set.*

Unit Precision   The **Precision** edit box is located at the bottom of the **Units** page. For Architectural units, the default precision is 0'–0 1/16". This will work well for drawing a small detail, but will be of little help for drawing a 300' long structure. The drawing precision can be changed throughout the drawing, but should be kept as large as possible to facilitate drawing speed.

To change the setting, select the arrow on the right side of the box. This displays the **Precision** list shown in Figure 2.8. Selecting the down arrow will allow the degree of accuracy to be selected. Settings range from whole units of 0'–0" through 1/256". For now, select the **0'–0"** setting by highlighting it with the cursor. This will return you to the **Units** page and allow you to continue the drawing setup. When you're satisfied with the unit settings, select the **Next** button to continue the drawing setup.

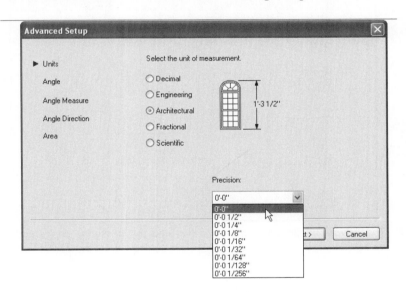

Figure 2.8   *Selecting the **Precision** arrow button displays the **Precision** list. Accepting the default value displays units with an accuracy of 1/16". Selecting 0'–0" with the mouse select button will change the precision value, close the menu and measure units in whole inches when the wizard is complete.*

Angles   Once the **Units** are selected, and the **Next** button is selected, the **Angle** page for controlling the units for measuring angles is shown. See Figure 2.9. Most architectural drawings can be completed easily using decimal units to measure angles. Options for measuring angles include the following:

> **The Decimal Degrees—90.00**—This option is best suited for plan views such as the drawing shown in Figure 2.10. It will display angles in degrees and decimal parts of a degree.

> **Deg/Min/Sec—90d00'00"**—This option is best suited for land measurements, such as site plans. This option will display angles in degrees, minutes, and seconds.

Grads—100.00g—Grads is the abbreviation for a gradient. One hundred grads equal one-quarter of a circle. This option of angle measurement is rarely used with architectural drawings.

Radians—1.57r—A radian is an angular unit measurement where 2p = 360° and 90° is p/2 radians. This option of angle measurement is rarely used with architectural drawings.

Surveyor—N0d00'00"E—This option presents surveyor's units, which are used to represent land, street, or sewer layouts, with angles displayed as bearings displayed as <N/S> <angle> <E/W>. The angle is based on north or south and will always be less than 90°.

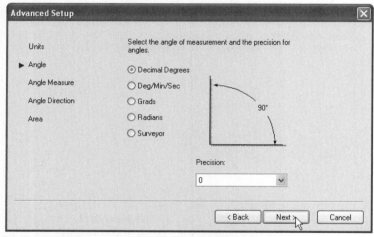

Figure 2.9 The **Advanced Setup** menu allows the units for measuring angles and the precision to be selected.

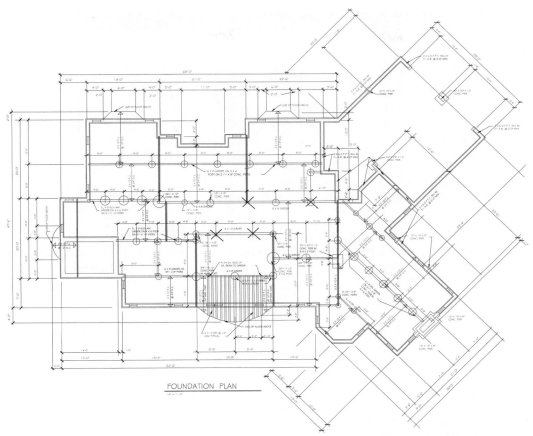

Figure 2.10   *Angles on a drawing such as this foundation plan can be measured conveniently using the **Decimal Degrees** angle option.*

The method of angle measurement and precision is set using the same methods used to set the drawing units. The method of angle measurement can be altered at any time throughout the life of the drawing with the methods that are presented at the end of this chapter. For now, accept the default option of **Decimal Degrees**. Selecting the down arrow by the **Precision Edit** box will provide options ranging from zero to eight decimal places with the default of 0. Normally, two-place accuracy is sufficient for architectural drawings while civil drawings might require a greater degree of accuracy. Accept the default by clicking the **Next** button to proceed to the next page of the wizard.

Angle Measurement   The **Advanced Setup** wizard allows the starting point of angle measurement to be selected. The default option is east. Options are shown in Figure 2.11. Figure 2.12 shows the visual representation of the angles. Standard AutoCAD angle layout places 0° at the three o'clock position and then moves counterclockwise.

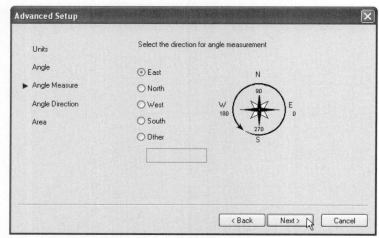

Figure 2.11 *The default starting point for measuring angles is east. You can select one of the other three quadrants, or any point in between, by choosing* **Other** *and providing an angle in the edit box.*

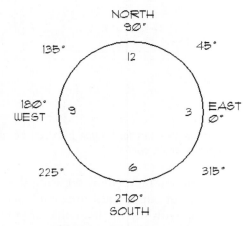

Figure 2.12 *A visual representation of the angle menu comparing compass and angular measurement.*

You can specify that angle measurement start in any location. To change the starting point of angles to north, select the **North** radio button. An angle other than one of the four compass points can be selected by choosing the **Other** button, and then entering the desired starting angle in the edit box. Entering a starting angle of 45° will place 0° halfway between north and east. When the desired starting point has been selected, click **Next** to allow the wizard to continue.

Angle Direction   Figure 2.13 shows the **Advanced Setup** wizard page for selecting the directions that angles are measured. The default setting for angle measurement is counterclockwise. Selecting the **Clockwise** button will change the direction of angle measurement. For now, accept the default. Remember that the defaults are there because they fit the needs of most users. The direction, like each of the other variables, can be altered at any time throughout the life of the drawing. Once the direction is selected, click the **Next** button to proceed.

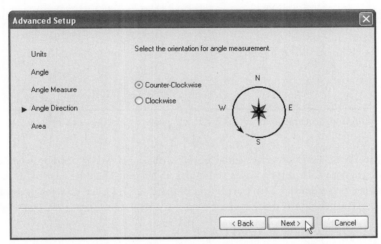

Figure 2.13   *The **Advanced Setup** wizard allows the direction for angle measurement to be entered. For most drawings, the default setting of counterclockwise works well.*

Area   The final setting of the **Advanced Setup** wizard allows the drawing area to be specified using the display shown in Figure 2.14. In the default setting, the drawing is 1' × 9". Enter the values for the width and length of the area of the intended drawing to be completed in the appropriate edit box. Enter the desired value, and click the **Finish** button. Clicking the **Finish** button closes the wizard, establishes each of the parameters that were entered, and returns you to the drawing area.

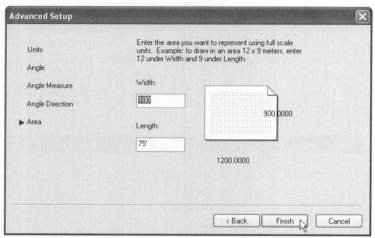

Figure 2.14 *Entering drawing units of 100',75' will allow a floor plan to be started.*

 Note: New CAD users are often fearful of making a wrong decision while using a wizard. With each setting it was stated that the setting could be adjusted throughout the life of the drawing. Don't make light of that fact. There are no wrong settings, only inconvenient settings. If the current setting is troublesome, change it, don't fight it.

DRAWING SETUP USING A TEMPLATE

Drawing templates with predetermined drawing values can be used to start a drawing. Each template contains values for drawing units, limits, and other values that can be altered throughout the life of a drawing. A template supplied by AutoCAD can be customized to meet specific requirements and saved for future use. Select the **Use a Template** button from the **Create New Drawing** dialog box to use a template to start a drawing session. This will produce the display shown in Figure 2.15.

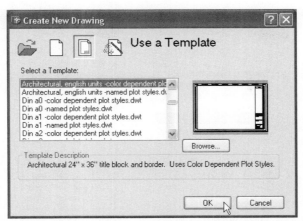

Figure 2.15 *Selecting the **Use a Template** option will display a listing of various templates contained in AutoCAD. The Architectural, English units template provides a good base drawing for most architectural projects.*

Selecting a Template

The template options are listed in groupings indicating the drafting standard used, the size of the title block in the preset layout, and the plot style. Listed drafting standards include Acad, Ansi, Architectural, Din (based on the German Institute of Standardization), Gb, Generic, ISO (International Organization for Standardization), JIS (Japanese Industry Standard) and metric templates. Architectural templates are based on feet and inch measurements. Acad templates are also based on feet and inch measurements but do not contain a title block. ANSI and generic templates provided with AutoCAD are based on decimal units of measurement. Din, ISO, and Jis templates are based on metric measurements.

Click the **Browse** button to display the **Select a template file** dialog box similar to Figure 2.16. This box contains a listing of the templates contained in the default AutoCAD Drawing Template File. By altering the location in the **Look in** edit box, you can use this dialog box to search for templates stored in other folders, other drives, other computers on the network, or on the Internet.

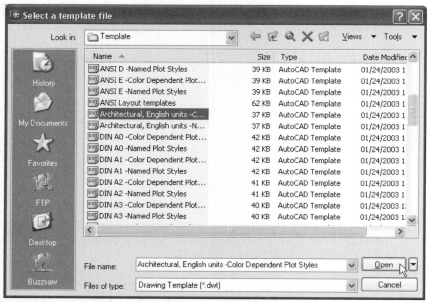

Figure 2.16 *Selecting* **Browse** *will display the* **Select a template** *file dialog box containing a listing of the templates contained in the default AutoCAD Drawing Template folder. This dialog box can be used to search for templates stored in other folders, other drives, other computers, on a network, or on the Internet.*

Scroll through the list of templates and highlight the Architectural, English units-Color Dependent Plot Styles.dwt. Clicking the **Open** button will produce a drawing display shown in Figure 2.17. The drawing area approximately 36" x 24" with units measured in architectural units with 1/16" precision. The limits can be adjusted as needed.

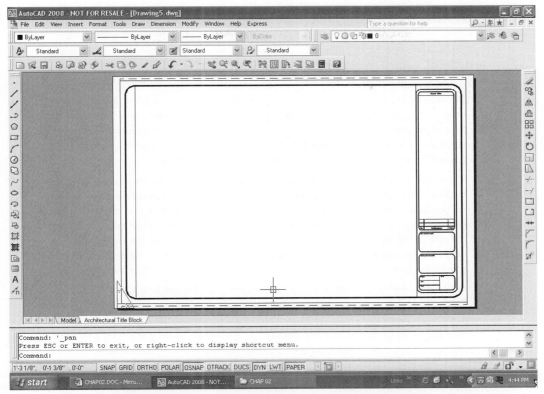

Figure 2.17 *The **Architectural, English units-Color Dependent Plot Styles.dwt** template is suitable for basic architectural or engineering projects. Chapter 4 will introduce options for creating and saving your own template.*

Rather than using the templates supplied with AutoCAD, you can develop your own template to meet specific needs of a project, your class, or your employer. In addition to each of the drawing controls that were introduced in this chapter, your template could include common text fonts, linetypes, layers, and other related features. Each of these components will be introduced throughout the text. As each is introduced, features that you find useful should be added to your template. Methods for creating and saving templates will be covered in chapter 4. For now, if you choose to use a template to enter the drawing area, accept the default and click **OK**.

Note: As the Architectural.dwt template is entered, you're in paper space. To create a drawing, select the **Model** tab at the lower left corner below the drawing area. This will remove the title block and produce a blank drawing screen. Now draw a few lines.

Select the **Architectural Title Block** tab and the lines that you drew will be displayed on the template. Click the **PAPER** button in the status bar (not the tab at the bottom of the drawing area) to display the drawing in the template in model space. If your drawing does not fit into the template, click the **Pan Realtime** button (the hand) on the **Standard** toolbar. Pan allows the drawing to be moved to the desired location. After you select Pan, a hand will be displayed in the drawing area. Press the select button and move the cursor in the direction you would like to move the drawing.

When you're satisfied with the location, press ENTER to exit the **Pan** command. If the drawing is too big to be seen in the template, click the **Zoom** button (the magnifying lens) on the **Standard** toolbar. After you select **Zoom**, a magnifying lens is displayed in the drawing area. Press and hold the select button and move the cursor down to reduce the size of the object. Press and hold the select button and move the cursor up if you would like to enlarge the drawings. When you're satisfied with the location, press ENTER to exit the **Zoom** command. The full power of these commands will be introduced later.

When the drawing is centered in the borders, press the **Model** button in the status bar. This will remove the thick black line that surrounds the drawing, and change the button name to **Paper**. Now press the **Model** tab at the bottom of the drawing area. This will remove the template and only display the original drawings. You've just loaded a template, created a drawing, prepared it in paper space for plotting, and returned to model space for further editing. Save the drawing as **EXER 2.1**.

START FROM SCRATCH

Selecting **Start from Scratch** from the **Create New Drawing** dialog box allows a drawing to be started with a minimum of predetermined values. Setup is allowed in Imperial or Metric units. Imperial units are displayed in four-place decimal units with drawing limits of 12" x 9". Rather than being prompted by a wizard or restricted by a template, **Start from Scratch** allows drawing controls to be set as needed using a dialog box or from the Command prompt. Using the **Imperial** setting (the default value) and selecting the **OK** button will produce the drawing display and allow work to begin with units measured in inches.

STARTING A NEW DRAWING FROM WITHIN AUTOCAD

In addition to the methods you've just explored, AutoCAD allows a drawing to be started from inside an AutoCAD drawing without harming the current drawing. Use one of the methods introduced earlier in this chapter to start a new drawing. Once the drawing is started, draw a triangle. Now open a new drawing using one of the following methods:

- Click the **QNew** button from the **Standard** toolbar

- Select **New** from the **File** menu.

- Type **NEW** ENTER at the Command prompt.

Each method will display the **Create New Drawing** dialog box, hide the existing drawing file, and begin the process of starting a new file. Start the new drawing using any of the startup methods. Your original drawing is removed from the drawing area, and a new drawing screen is displayed.

An existing drawing can also be opened from inside AutoCAD. Existing files can be opened using one of the following methods:

- Click the **OPEN** button on the **Standard** toolbar

- Select **Open** from the **File** menu.

  - The names of the drawing files that have been opened are displayed at the bottom of the **File** menu.

  - The number of files listed is controlled by altering the number displayed in the **Number of recently-used files to list** found on the **File Open** area of the **Open and Save** tab of the **Options** dialog box. Open the display by selecting **Options** from the **Tools** drop-down menu.

Use the **LINE** command to draw a rectangle in the new drawing. Open a third drawing and draw a circle. Click the **Minimize** button, and the original drawing will be displayed. Click the **Minimize** button for this drawing, and the display will resemble Figure 2.18. The drawing tabs of each of the open drawings are displayed. Several new drawings can be started, or you can open several existing files in one AutoCAD session. Although this might not seem important now, several of AutoCAD's features will make this very important. These features will be addressed throughout the balance of the text. For now, remember that two or more drawings can be worked on in the same drawing session without having to open AutoCAD twice.

 Note: New students often open AutoCAD several times to create multiple drawing sessions. Resist the urge to do this. Having several sessions of AutoCAD open at once will slow your machine, which may cause you undue frustration. You'll know if you have multiple copies of AutoCAD opened if you see more than one AutoCAD tab on the task bar. Open one session of AutoCAD, and then open multiple drawings within that session.

Figure 2.18   *Click the **Minimize** button, and the current display is minimized to display the drawing tabs for all current drawings. Select the **Maximize** button on the desired tab to restore a drawing.*

DISPLAYING MULTIPLE WINDOWS

As with other Windows programs, AutoCAD provides choices on how open windows and drawings is displayed. Individual program or drawing windows within AutoCAD can be arranged in horizontal, vertical, or cascading tiles. You can display multiple drawings by selecting **Window** from the menu bar, as shown in Figure 2.19. The lower portion of the menu will display a listing of the current open drawing layouts and place a checkmark by the current drawing. Selecting the name of the desired view will remove the current drawing and display the selected drawing. Choose a display for multiple windows by selecting one of the options from the top portion of the menu.

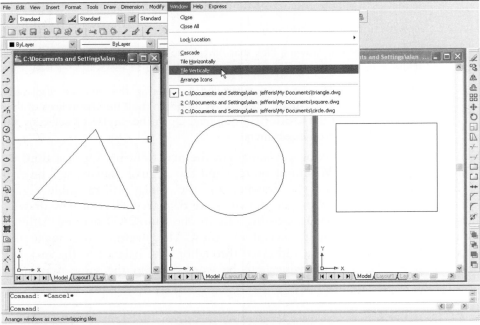

Figure 2.19 *Multiple layouts can be displayed using the options from the **Window** menu. This display shows the three open drawings using the **Vertical** layout. Notice the title block of the active drawing is a darker color than those of the inactive drawings. An inactive drawing can be made active by moving the cursor to the desired drawing and pressing the select button.*

Cascading Windows

Selecting **Cascade** from the **Window** menu displays the drawings similar to those shown in Figure 2.20a. Cascading drawings are stacked over each other in descending order. Drawings are displayed in the order that they are selected. To display one of the lower drawings, click any visible portion of the desired drawing to move it to the top of the stack. Clicking the title bar and holding the select button down allows the window to be dragged anywhere on the screen.

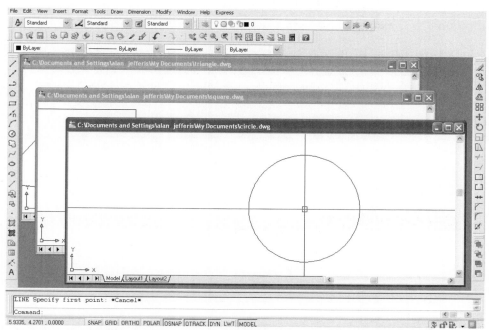

Figure 2.20a *Selecting **Cascade** from the **Window** menu displays drawing windows that are stacked over each other in descending order.*

Tile Displays

The **Tile** option allows drawings to be arranged in horizontal (Figure 2.20b) or vertical displays (Figure 2.19). Tiled windows permit each window to be viewed at the same time. Just as with other programs, if you're unhappy with the size of a window, alter the size by selecting and dragging the edge to the desired size.

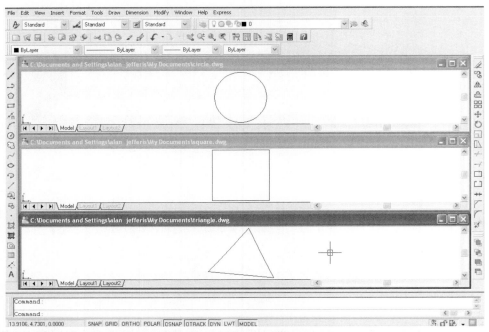

Figure 2.20b *The **Tile Horizontal** option displays drawing windows in horizontal rows.*

WORKING WITH PARTIAL FILES

AutoCAD offers additional methods of working with drawing files. Although each method is beyond your current needs as you explore AutoCAD, you should at least be aware of the **Partialopen** and **Partiaload** commands. Both commands can be used when you work with large files to allow only a specific portion of the drawing to be accessed. The projects that you work on throughout this text will typically be too small to benefit from **Partialopen**. On extremely detailed drawings, **Partialopen** will allow specific information to be viewed in the drawing area. If you're working on a complex floor plan, the plan can be divided into zones, and then each zone can be saved by what AutoCAD refers to as a view. **Partialopen** allows you to open a specific zone of the floor plan and display the framing information, while hiding the electrical or plumbing information. See chapter 4 for an explanation of how each command can be used to start a drawing session.

Note: You now have several methods that can be used to start a drawing session. Just as important as how you start a drawing session is where the file that you want to open is located. When you're working with existing files, always use files stored on your hard drive to open the drawing session. AutoCAD needs workspace as you work on a drawing session. If the drawing and the workspace exceed the storage space on the disk, you might encounter unwanted results. If the drawing is stored on a network, you'll rarely have a shortage of workspace, but your hard drive is generally faster than the network drive. If the most current file is saved on a diskette or in a network folder, copy the file to the hard drive prior to opening the file.

CONTROLLING DRAWING PARAMETERS FROM WITHIN A DRAWING

No matter what method is used to begin a drawing, the drawing parameters can be altered at any time. Parameters can be altered through dialog boxes, menus, or the keyboard.

SETTING UNITS

The **Drawing Units** dialog box shown in Figure 2.21 can be used to change the type and precision of drawing units. The **Drawing Units** dialog box can be displayed by selecting **Units** from the **Format** menu.

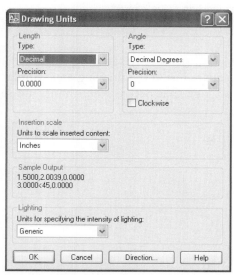

Figure 2.21   *Drawing units such as unit precision can be altered at any time throughout the life of the drawing using the **Drawing Units** dialog box. Display the dialog box by selecting **Units** from the **Format** menu.*

The drawing units are set by selecting the desired style from the **Type** drop-down list. The **Precision** edit box can be used to display and adjust the unit precision based on the accuracy needed for the drawing. The dialog box can also be used to select the method for measuring angles and for setting the angle precision. Angle precision accuracy is set using the same methods as for changing the unit precision. Selecting the **Direction** button will display the **Direction Control** dialog box shown in Figure 2.22. This dialog box allows the angle starting point and the angle direction to be adjusted by selecting the appropriate radio button or highlighting the desired accuracy. Once it is selected, the accuracy will be displayed in the edit box and the menu will be closed.

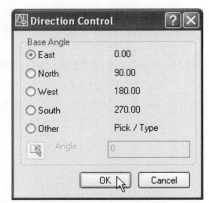

Figure 2.22 *Selecting the **Direction** button in the **Drawing Units** dialog box allows the orientation for angle direction to be altered.*

SETTING LIMITS

Changing the limits alters the size of the drawing displayed on the screen. Before you change the limits, draw a square inside the drawing area, similar to Figure 2.23a. This will help you to visualize the change that occurs when limits are set.

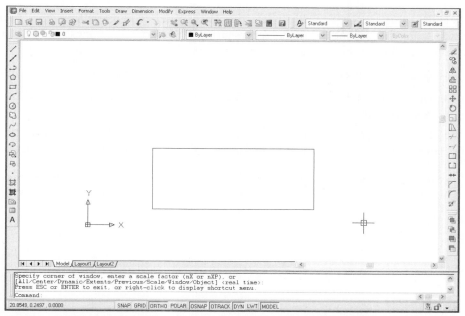

Figure 2.23a   *With the drawing limits set at the default units of 1'–0",0'–9", an object might appear quite large.*

To change the size of the screen that can be viewed, select **Drawing Limits** from the **Format** menu or type **LIMITS** ENTER at the command line. Each method of entering the command will produce the display:

```
Reset Model space limits:
Specify lower left corner or [ON/OFF] <0'-0",0'-0">:
```

Acceptable responses are ON, OFF, and ENTER.

- **ON** will keep the current values (0'–0",0'–0") and activates limit checking. Limit checking will not allow objects to be drawn that are outside the limits. Think of limit checking as a border on your drawing paper to keep your project on the paper.

- **OFF** will disable limit checking but retain the limit values for future use of limit check.

- **ENTER** retains the current location of the lower left drawing limit of (0'–0", 0'–0"). This will be the most typical response. Pressing ENTER produces the prompt:

```
Specify upper right corner (1'-0",0'-9"):
```

The default value is 1'–0",0'–9", with the first number representing the horizontal dimension and the second number representing the vertical dimension. This is your opportunity to change the size of the drawing area to fit your current project.

- Type **15',12'** ENTER, and the Command prompt will return with no apparent change.

- Type **Z** ENTER (Z represents ZOOM) at the keyboard, then **A** ENTER (the A represents all).

Now the effect of your new limits can be seen in Figure 2.23b. Remember that the box is still the same size but it appears smaller because the area to be viewed has increased.

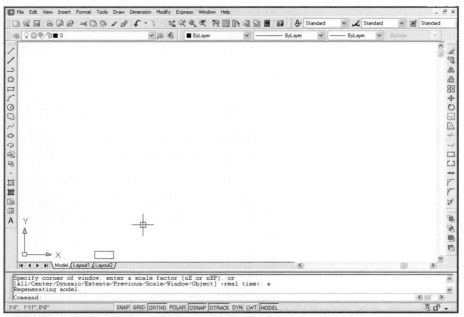

Figure 2.23b *By changing the location of the upper right corner you allow viewing of a much larger area. In this example, the limits are set to 15',12'. The rectangle is the same size as the rectangle shown in Figure 2.23a, but the viewing area has been enlarged.*

SETTING GRIDS

Another helpful tool for drawing layout is a grid. The **GRID** command will produce a visible grid of dots at any desired spacing. The grid size can be adjusted throughout the life of the drawing, depending on the size of the object being designed. The grid can be displayed or removed to aid in visual clarity, but it will not be produced when the drawing is plotted. Figure 2.24 shows an example of a drawing with the grid display.

The grid can be set by dialog box or from the command line. The current status is displayed on the status bar. Once the values have been set, the grid can be toggled **ON/OFF** by selecting the **GRID** button on the status bar. The grid can also be toggled **ON/OFF** in the middle of another command by using the **F7** key or by pressing **CTRL+G**.

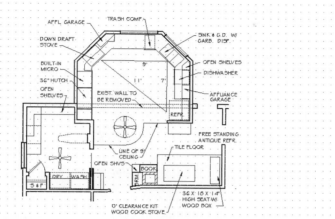

Figure 2.24   *A grid displayed with a value of 6" can be helpful in determining approximate sizes on a floor plan.*

Assigning Grid Values Using a Dialog Box

Grid values can be set using the **Snap and Grid** tab of the **Drafting Settings** dialog box that is accessed by selecting **Drafting Settings** from the **Tools** menu. The dialog box shown in Figure 2.25, can also be displayed by placing the cursor on the **GRID** or **SNAP** button on the status bar, right-clicking, and then choosing **Settings** from the shortcut menu. Once the **Drafting Settings** dialog box is displayed, grids are set using the following procedure:

1. Select the **Snap and Grid** tab.

2. Enter the desired value for the X (left-to-right) grid spacing.

3. Enter the desired value for the Y (top-to-bottom) grid spacing. If you want the Y value to match the X value, move the cursor to the Y spacing edit box and click the select button. Pressing ENTER will also match the X and Y values and close the dialog box. To enter a different Y value, enter the desired value.

4. Click the **OK** button to accept the new values and return to the drawing.

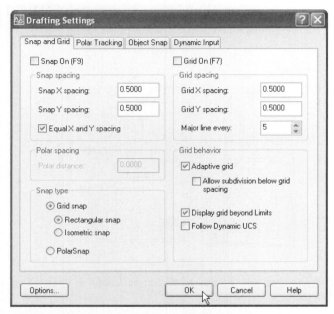

Figure 2.25 *Values for the grid can be altered on the **Snap and Grid** tab of the **Drafting Settings** dialog box by right-clicking the **GRID** button on the status bar and then selecting **Settings**. Grids can also be altered by selecting **Drafting Settings** from the **Tools** menu and then selecting the **Snap and Grid** tab.*

The dialog box also offers methods of altering the angle used to display the grid. Type the desired angle for the grid in the **Angle edit** box located in the **Snap** column. Even though it is listed as a **Snap** control, it will also control the **Grid** rotation. Figure 2.26 shows an example of a floor plan drawn with a grid rotated to 45°. Selecting the **Isometric Snap** setting will establish a grid suitable for isometric drawings. Isometric drawings will be discussed later in this text.

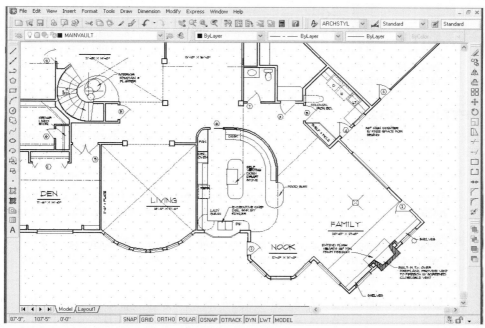

Figure 2.26   *A rotated grid will ease the layout of an irregular shape.*

SETTING SNAP

When working on a modular component, it is helpful to be able to control the accuracy of lines entered with the mouse. The **SNAP** command creates an *imaginary* rectangular grid that controls placement of the drawing cursor in precise intervals. The distance between cursor movements is defined as snap resolution. Snap intervals can be set at any interval and can be changed throughout the life of the drawing. A change in the snap grid value affects only the points entered after the change. Points entered prior to the change in spacing might no longer line up with the new snap location. The size of the grid setting and the size of the object to be drawn should be considered when setting the snap value. If the grid is set at 2', a snap value of 6" or 12" will be helpful in producing accuracy. If a detail of a beam to post connection is drawn, a snap of 1/2" might be used to accurately move through the drawing. If a 200' long structure is to be drawn, a 24", 60" or 120" snap will be more helpful than the 1/2" accuracy used for a detail.

Snap can be set using the **Snap and Grid** tab of the **Drafting Settings** dialog box. Snap is adjusted using the same methods that were used to adjust the grid. Entering an angle in the **Angle edit** box controls the display angle for **Snap and Grid**. Entering a value in the X or Y **Base edit** box allows the base point for the snap pattern to be established. Once the values have been set, snap can be toggled **ON/OFF** by clicking the **SNAP** button on the status bar. Snap can also be toggled **ON/OFF** in the middle of another command by using the **F9** key or by pressing **CTRL + B**.

Note: Coordinating the grid and snap will provide great accuracy and speed. If the grid value is set at zero, the grid spacing automatically adjusts to the snap resolution. To specify the grid spacing as a multiple of the snap resolution, type X **ENTER** after the grid value. For instance, if the snap value is 2", typing 6X **ENTER** at the Command prompt produces a grid of 12".

CHAPTER 2 QUIZ

Answer the following questions with short complete statements. Type your answers using a word processor.

1. Place your name, chapter number, and the date at the top of the sheet.

2. Type the question number and provide the answer in the form of a statement that includes part of the question. You do not need to write out the entire question.

 Warning: Some of the questions have not been covered in the reading material and will require the use of the help menu. You may also have to do some exploring to answer the questions.

1. What command will set an invisible grid to aid layout?

_____

_____

2. List five options of the Units.

_____

_____

3. What is the default unit for angle measurement?

_____

_____

4. What is the default for angle direction?

_____

_____

5. Your boss would like a drawing of a street and sewer layout based on the surveyor's notes. Describe the process for setting up the proper method of two-place angle measurement by keyboard.

_____

_____

6. List three methods to toggle between **Grid** ON/OFF.

_____

_____

7. List three methods to toggle between **Snap** ON/OFF.

_____

_____

8. What is the main difference between **Grid** and **Snap**?

_____

_____

9. You've just entered AutoCAD and don't want to establish any drawing parameters. How can you get to the drawing screen?

_____

_____

10. When would **Start from Scratch** be a useful method of entering a drawing?

_____

_____

11. What does **Area** control?

_____

_____

12. A site plan must be drawn with angles measured in a format such as N42° 30'–30" E. How can this be done if you forgot to adjust the angle measurement as you set up the drawing?

_____

_____

13. A friend wants to draw a house but has no idea how big it will be. What would you recommend setting for drawing limits as the drawing is started?

_____

_____

14. How can you use the **Advanced Setup** wizard to start a drawing?

_____

_____

15. How is the **Drafting Settings** dialog box accessed?

_____

_____

Drawing and Controlling Lines

INTRODUCTION

This chapter will introduce methods for:

- Entering the **Line** command, including toolbars, keyboard, and the **Draw** menu
- Controlling the placement of lines, including absolute coordinates, relative coordinates, and polar coordinates
- Controlling and editing lines with the tools of the status bar

Commands to be introduced include:

- **Line**
- **Close**
- **Snap**
- **Grid**
- **Ortho**
- **Polar**
- **Osnap**
- **Erase**
- **Oops**
- **Undo**
- **Redo**
- **Plot**

Now that you're able to set up drawing units and limits, basic drawing components can be explored. The main components of the drawings you'll be working with will consist of lines. Chapter 1 introduced the **Line** command using the **Draw** toolbar, **Dynamic Input**, the keyboard, and the **Draw** menu. This chapter introduces methods for making effective use of the **Line** command, as well as methods for locating and altering lines.

LINES

The basic components of every drawing are lines, which are used to represent topography, grades, pipelines, easements, walls, roads, property lines, and an endless list of other construction materials. Figure 3.1 shows the lines required to represent a portion of a framing plan for a multilevel steel structure. Although many of these lines are represented with different line patterns or thickness, each can be drawn using the **Line** command. Lines are used to represent different types of information within a drawing. Establishing and controlling varied linetypes will be introduced in later chapters.

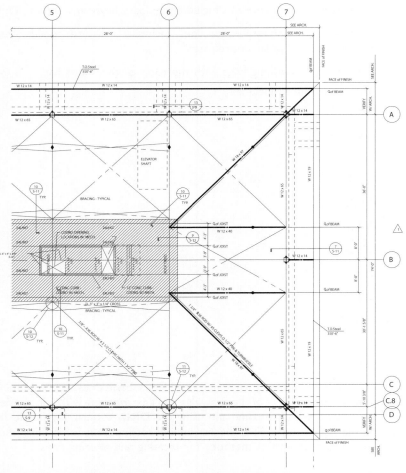

Figure 3.1 *Lines of varied widths and patterns are used on construction drawings to represent varied materials. Methods of changing linetype and line width will be discussed throughout this chapter. (Courtesy of Van Domelen/Looijenga/McGarrigle/Knauf Consulting Engineers.)*

ENTERING THE LINE COMMAND

The **Line** command is accessed by toolbar, keyboard, or menu. Each method was introduced in chapter 1. To start the command, select the **Line** button on the **Draw** toolbar. This will produce the prompt:

>   Specify first point:

The coordinate display for the current cursor location is displayed on the dynamic display, with a similar prompt in the command line. Move the cursor to where you would like to start a line and click the select button. The "first point" will become the beginning point of a line segment. As the first point is selected, the prompt displays

>   Specify next point or:

Move the cursor to the desired location and click the select button to specify the endpoint. As the cursor is moved, a rubber-band line is displayed from the first point to the crosshairs location. This can be seen in Figure 3.2. The rubber-band line will help you to visualize the placement of the line to be drawn and to experiment with its placement. The rubber-band line will remain as the crosshairs are moved until the next point is selected or the **Line** command is terminated.

Several other pieces of information can also be found on this screen as seen in Figure 3.2. Above a horizontal line, and on the left side of a vertical line, is the **Dynamic Distance Input** that displays the current distance between the selected end points. Below the cursor is the **Dynamic Angle Input** display that lists the current angle of the proposed line. A **Down** arrow is displayed at the end of the dynamic input display listing options for completing the command. Selecting the **Down** arrow key will display the options of **Undo** and **Close**. **Close** is not listed until two or more lines have been drawn. Each option will be explored shortly.

Once a line is drawn, there are four options for ending the **Line** command. Each method will draw the indicated line and terminate the command.

- Press ENTER to draw the indicated line and then exit the command.
- Press the SPACEBAR to exit the command without drawing a line segment.
- Right-click and press ENTER to produce the same results as pressing the SPACEBAR.
- Right-click and select **Cancel** to draw the indicated lines and terminate the command.

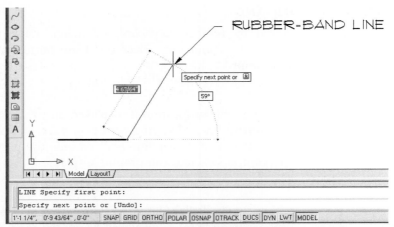

Figure 3.2 *When a "**next point**" is provided, a rubber-band line extends from the "**first point**" to the present crosshairs location. This line indicates the line that will be drawn if the current crosshairs location is used. Information is also provided in the dynamic display to indicate the length of the line and the current angle of the line.*

COMMAND LINE INPUT

A second method of entering commands is by using the keyboard. Command entry is displayed at the Dynamic Input or the command line or both. The balance of this text will assume the use of both, but will show the input as it would be in the Dynamic Input. The sequence for entering the **Line** command is as follows:

```
Type L ENTER to start the Line command
Specify first point: (Specify an end point)
Specify next point or: (Specify an end point)
```

The **Line** command functions exactly the same no matter how it is entered.

MENU ENTRY

A third method for entering a command is to select **Line** from a menu. When you select **Line** from the **Draw** menu, the menu will disappear and the dynamic display is added to the cursor. The command prompt will show

```
Specify first point:
```

Notice that four other options for creating lines are presented in the **Draw** menu—**Ray**, **Construction Line**, **Multiline**, and **Polyline** commands. Each line entry method will be presented in future chapters.

No matter which method is used to start the **Line** command, a request for the "first" and "next" points is displayed at the **Dynamic Input** display and at the command line if it has been left in the **ON** setting. The first and next points can be entered with the cursor, or by keyboard. Specific coordinates can be entered by keyboard rather than by selecting the points with the cursor. Methods of selecting line and point by keyboard will be discussed later in this chapter.

 Note: You've entered the **Line** command using three different methods. As you start to experiment with drawing lines, you should draw objects combining all entry methods. Each method works well to initially start the command. The easiest way to repeat a command is to press ENTER or the SPACEBAR. Once in the **Line** command, pressing ENTER a second time will pick the end point of the last line drawn as the start point for the new line.

LOCATING LINES USING COORDINATE ENTRY METHODS

As you complete drawings in this chapter, it is important to understand the user coordinate system (UCS) of AutoCAD. The UCS is based on a Cartesian coordinate system, dividing space into four quadrants. Figure 3.3 shows the division of space based on the Cartesian coordinates. Points in these quadrants are located in turn by using their points in a horizontal (X) and vertical (Y) direction along the plane. Points are always specified by the X coordinate followed by the Y coordinate. As you begin to draw, AutoCAD will ask you to specify a "first point." As you move to the "next point," the dynamic length and angle displays, as well as the coordinate display at the left end of the status line will reflect the crosshairs movement. To activate coordinate display in the status bar, press F6, or CTRL+D.

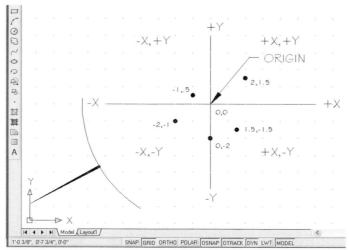

Figure 3.3   *The division of space using the Cartesian coordinate system. The intersection of the X and Y coordinates is assumed to be 0,0, which is the lower left corner of the screen. The grid has been moved to the center of the screen to represent negative X and Y locations.*

Because the fields of engineering and architecture deal with very specific locations, a means of controlling the point placement is desirable. Using absolute, relative, or polar coordinates for point entry methods will allow specific points to be selected.

 Note: If you have removed the command line, it can be redisplayed by selecting **Command Line** from the **Tools** menu.

ABSOLUTE COORDINATES

The absolute coordinate system locates all points from an origin assumed to be 0,0 using the Cartesian coordinate system. The 0,0 origin point is assumed to be the same 0,0 point that the limits, snap, and grid are based on, although it can be changed. The axes for the X and Y coordinates intersect at 0,0. Each point on the screen is located by a numeric coordinate based on the original axis. Objects are drawn by entering X and Y coordinates separated by a comma. To enter absolute coordinates, the **DYN** option must be **OFF**.

In the example in Figure 3.4 a line is drawn between 2,2 and 2,4. Based on the cursor location and the current coordinate display, the next "to point" is 4,4. To complete the drawing in Figure 3.5 use the following command sequence:

```
Command: L ENTER (Or click the Line button on the Draw toolbar.)
LINE Specify first point: 4,1.5 ENTER
Specify next point or [Undo]: 6,1.5 ENTER
Specify next point or [Undo]: 6, 3.5 ENTER
Specify next point or [Close/Undo]: 5,5 ENTER
Specify next point or [Close/Undo]: 4,3.5 ENTER
Specify next point or [Close/Undo]: C ENTER
Command:
```

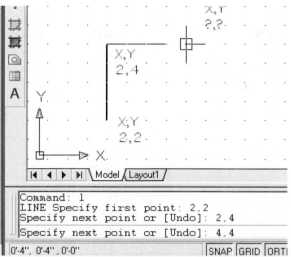

Figure 3.4   *With absolute coordinates, the line drawn starts at 2" over (+X) and 2" up (+Y) from the origin and extends to a location 2" over (+X) and 4" up (+Y). **DYN** must be toggled **OFF** to draw using absolute coordinates.*

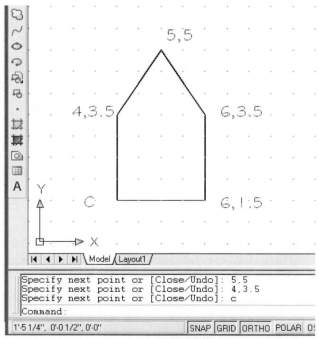

Figure 3.5   *A simple shape and the absolute coordinates required to describe it. Each set of coordinates describes a location relative to 0,0.*

Notice that **C ENTER** was typed as the last command, rather than numeric coordinates. **C** represents **Close**. If a sequence of lines will form a closed polygon, the last line can be drawn without providing endpoints or coordinates for the termination point. Typing **C ENTER** at the "next point:" will close the polygon precisely. Selecting the **Down** key at the end of the dynamic display will display a menu with the options of **Close** and **Undo.** Selecting the **Close** option will close the polygon. Chapter 7 will offer other drawing tools to select exact points of a drawing object as alternatives to **C ENTER.**

> **Note:** An alternative to using the **C** option to close the object is to use the Object Snap (OSNAP) option of AutoCAD. As the cursor is moved near the end of the original line, a box is placed around the end of the original line. (The color of the box can be altered in the **Options** dialog box by selecting the **AutoSnap Settings** area of the **Drafting** tab.) The box indicates that the nearest endpoint has been selected to become the "next point." Object Snap will be discussed in detail in chapter 7.

Because absolute coordinates are based on an origin of 0,0 they are usually not convenient for most architectural and engineering projects. Many CAD users find relative coordinates more suited to a construction project.

RELATIVE COORDINATES

Relative coordinates locate "to points" based on the last point instead of the 0,0 origin. Relative coordinates are entered using a similar method as absolute coordinates. Absolute coordinates are entered as X,Y. Relative coordinates entry method will depend on the **DYN** setting. With **DYN** set to **ON**, relative coordinates become the default entry method. With the **DYN** setting **OFF**, relative coordinates are entered as @ X,Y at the command line. With relative coordinates, no matter where you enter the coordinates, you will need to be mindful of positive and negative values.

> **Note:** Keep the **DYN** setting **ON** to save time and keep the setting relative to the last drawn point.

Figure 3.6 shows an object drawn using relative coordinates. Positive coordinates will move the crosshairs to the right or above the origin point. Negative coordinates will move the crosshairs down or to the left of the last entry point. The object in Figure 3.6 was drawn using the following command sequence:

```
L ENTER (Or click the Line button on the Draw toolbar)
Specify first point: 4,1.5 ENTER
Specify next point or 2,0 ENTER
Specify next point or 0,2 ENTER
Specify next point or –1,1.5 ENTER
Specify next point or –1,–1.5 ENTER
Specify next point or 0,–2 ENTER
Specify next point or ENTER
```

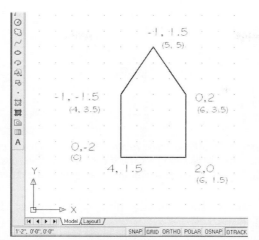

Figure 3.6   *Relative coordinates describe the location of the next point based on the last point. With **DYN** setting **On**, relative coordinates can be entered at the dynamic input or command prompt. With the **DYN** setting **OFF**, the coordinates must be preceded by the @ symbol. The value **@2,0** ENTER would draw a horizontal line 2" to the right of the start point. Relative coordinates are a common entry choice for many projects. For faster results, keep the **DYN** setting **ON**.*

With **DYN** set to **ON**, AutoCAD enters the first point based on absolute coordinates with the subsequent coordinate points entered by relative coordinates. The final coordinates can be omitted and the object closed using **Object Snap** or by typing **C** ENTER.

POLAR COORDINATES

Polar coordinates use the last entry point as the origin and combine a distance and an angle. This is useful on drawings such as sections for drawing a roof, or site plans with angled property lines. With **DYN** set to **ON**, relative polar coordinates are entered as

3<27.5 ENTER

This would produce a line 3" long at a 27.5° angle away from the horizontal, as shown in Figure 3.7. To draw the opposite side of this roof truss would require typing **3<332.5** ENTER. The location of line 2 at 332.5° is determined by subtracting 27.5° from 360°. Remember, polar coordinates are based on the current origin of the crosshairs. In Figure 3.8 the location of line 2 was determined by entering a negative number, in this case –27.5°. With **DYN** set to **OFF**, the @ symbol must be entered before the coordinates for an entry such as @3<275 ENTER. To complete the triangle started in Figure 3.8, use the Object Snap feature of the cursor or type **C** ENTER.

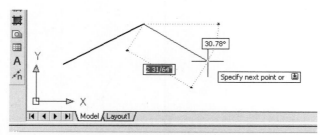

Figure 3.7 *Polar coordinates describe a point relative to the last cursor postion using a length and an angle. With the **DYN** settings **ON**, the < symbol precedes the angle specification in degrees. Polar coordinates are useful in drawing angled sides of a parcel of land.*

Figure 3.9 shows an object drawn using polar coordinates. The required command sequence would be as follows:

```
Command: L ENTER (Or click the Line button on the Draw toolbar.)
Specify first point: 4,1.5 ENTER
Specify next point or 2<0 ENTER
Specify next point or 2<90 ENTER
Specify next point or 1.8<124 ENTER
Specify next point or 1.8<236 ENTER
Specify next point or (With OSNAP ON, move the cursor to the
    origin point and press the pick button or type  C ENTER.)
```

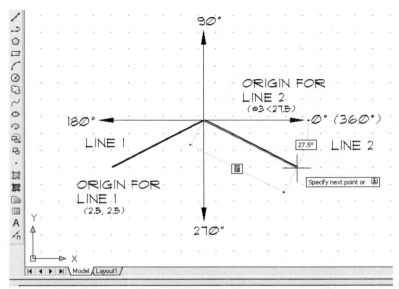

Figure 3.8 *Specifying angles using polar coordinates. The first line was created by using an angle of 27.5°. The second line was created by using an angle of −27.5°.*

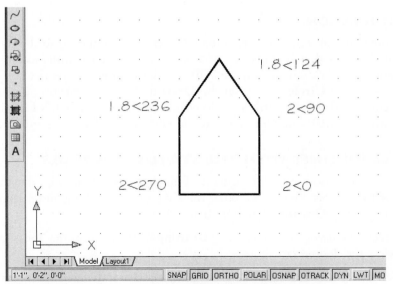

Figure 3.9 *The use of polar coordinates to establish a simple shape.*

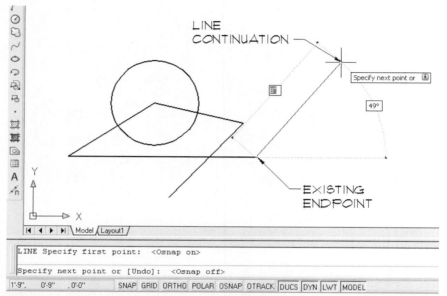

Figure 3.10 *A line can be continued from the end of the last line drawn by pressing ENTER for the "first point."*

LINE CONTINUATION

Once the **Line** command has been terminated, it can be resumed from the end of the last line segment that was drawn. This can be done even if other commands were used. The horizontal line segment drawn in Figure 3.10 was drawn prior to the circle. The circle was drawn by clicking the **Circle** button on the **Draw** toolbar and selecting a center or radius point. This command will be further explored in the chapter 6. If ENTER is pressed for the "first point," the new line will begin at the end of the last line segment.

CONTROLLING LINES WITH THE STATUS BAR TOOLS

The status bar was introduced in chapter 1. This discussion will help you make practical application of these tools. Snap, grid, polar tracking, object tracking and dynamic input can each be set using the **Drafting Settings** dialog box.

- Access the dialog box by selecting **Drafting Settings** from the **Tools** menu, and then select the desired tab.

- Settings for each can be adjusted by right-clicking on the desired button and then selecting the **Setting** option.

SNAP

Although the snap grid is invisible, it can be a valuable tool for placing lines and other drawing elements. Snap should be adjusted to a modular size of the object being drawn. When you work on a floor plan, a snap setting of 12" would be helpful for the initial layout of the walls. A setting of 4" allows lines to be easily placed as the plan is being fine-tuned. Snap is most helpful if you adjust the value throughout the life of the drawing to meet specific needs. Keep in mind that when snap is **ON**, you might have a difficult time drawing angled lines. Snap can be toggled **ON/OFF** using the **SNAP** button on the status bar at any point of a command sequence.

GRID

The visible grid should be used to plan and lay out a drawing. Setting the grid value as a multiple of the snap setting will greatly aid in placing drawing objects. The grid serves as a visual reference as well as an aid in placing lines. A grid of 24" and a snap of 6" will greatly aid in the layout of most residential plan views. The 24" spacing of the grid will provide a reference to keep track of object sizes, and the snap will provide a useful grid for placing lines. A grid of 5 or 10 feet may be more appropriate for larger commercial plans. Grid can be toggled **ON/OFF** using the **GRID** button of the status bar at any point of a command sequence.

ORTHO

Most of the structures that you will draw are composed of perpendicular lines. The ORTHO command will allow you to place the end of the rubber-band line in an exact horizontal or vertical position based on the current snap or grid pattern. If the snap and grid are rotated, ORTHO will respond by producing perpendicular lines based on the new grid. Toggle ORTHO **ON/OFF** by clicking the **ORTHO** button on the status bar. The residence shown in Figure 2.26 has a grid set to 45° for the right edge of the structure. As ORTHO was used on the right half of the structure, lines were automatically set to 45°, 135°, 225°, and 315°.

POLAR

Activating polar tracking will produce a display at specified intervals to aid in placing lines. When the cursor is moved to a pre-selected angle, the display will show the distance and the angle of the line in progress, as well as a projection of that line. The projection indicates where the line would be placed if it were lengthened. Figure 3.11 shows an example of a polar display for placing a line. The display can be a helpful alternative to entering coordinates at the keyboard. Settings can be adjusted on the **Polar Tracking** tab of the **Drafting Settings** dialog box. Place the cursor over the **Polar** button, right-click, and then choose **Settings** or select **Drafting Settings** from the **Tools** menu. The **Polar Tracking** tab can be seen in Figure 3.12. Key elements of the dialog box include the following:

> **Increment angle**—This edit box allows the desired angle to be used with the tracker to be set. Default angles include 5, 10, 15, 18, 22.5, 30, 45, and 90 degrees. Selecting 10 will display a tracking indicator as the cursor is moved in 10° increments.
>
> **Additional angles**—Click the **New** button to see this dialog box, which allows up to 10 additional angles to be added to the default list. If you're drawing an elevation with a roof

at a 6/12 pitch, selecting an angle of 26.5° would aid in the layout of the roof. The setting could be entered as follows:

1. Click the **New** button.

2. Enter **26.5** in the edit box.

3. Click the **OK** button to return the drawing area.

The default angle for tracking will now be 26.5. Settings can be removed from the default list by highlighting the setting to be removed and then clicking the **Delete** button.

Object Snap Tracking Settings—This option selects options for object track settings. Options include orthogonal and polar angle tracking settings.

- **Track orthogonally only**—With this option active, the object tracker displays only horizontal or vertical object snap tracking paths for acquired object snap points when **Object Snap Tracking** is ON.

- **Track using all polar angle settings**—Allows the cursor to track along any polar angle tracking path when **Object Snap Tracking** is ON.

Polar Angle Measurements—Sets the method to be used to measure polar angles. Options include **Absolute** and **Relative**.

- **Absolute**—Displays polar tracking angles based on the current UCS.

- **Relative to last segment**—Displays polar tracking angles on the most recent object.

Options—Clicking this button displays the **Options** dialog box. This dialog box can be used to adjust the color and size of the object snap marker. The dialog box also provides a convenient method of altering the size of the aperture box. The **Options** dialog box will be explored in chapter 4.

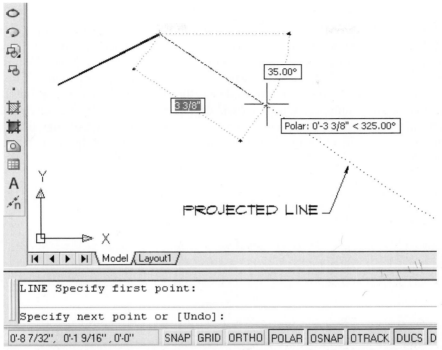

Figure 3.11   *Polar display for tracking a line. The display can be a helpful alternative to entering coordinates at the keyboard. Settings can be adjusted on the **Polar Tracking** tab of the **Drafting Settings** dialog box.*

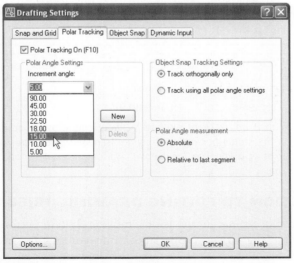

Figure 3.12   *The **Polar Tracking** tab of the **Drafting Settings** dialog box can be used to control the display used with polar tracking.*

OSNAP

Activating Object Snap provides one of the easiest and fastest means of accurately drawing in AutoCAD. With this option **ON** (click OSNAP on the status bar), a marker is placed on the crosshairs as it is moved near an endpoint, midpoint, or intersection of lines. With the marker displayed, the "next point" of a line is placed exactly at the endpoint. Figure 3.13 shows an example of the display as the cursor is moved near the endpoint of an existing line. Rarely will you want to turn this feature off. Chapter 7 will explore methods of altering the default settings of Object Snap. If you want to explore on your own, select **Drafting Settings** from the **Tools** menu and select the **Object Snap** tab to view the options. Use the **Help** menu if you just can't wait until chapter 7.

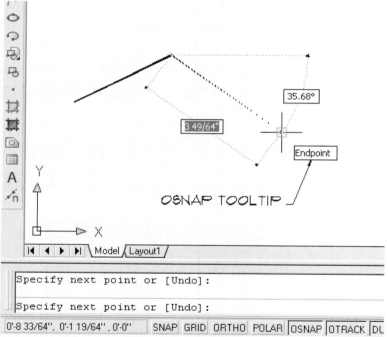

Figure 3.13 *Object Snap is an excellent method for increasing drawing accuracy. With the **OSNAP** button on the status bar active, a square is added to the crosshairs as it moves near the endpoint of a line. The "next point" will automatically snap to the endpoint of the selected line.*

AN INTRODUCTION TO EDITING DRAWING OBJECTS

This section will introduce three methods of removing drawing errors from a drawing file. These methods include the **Erase**, **Oops**, and **Undo** commands. The **Redraw** command will also be introduced as a method of cleaning the drawing display. Chapter 8 will present methods for using these commands more effectively.

ERASE

To remove objects using the **Erase** command, use either of the following methods:

- Click the **Erase** button on the **Modify** toolbar
- Type **E** ENTER
- Select **Erase** from the **Modify** menu.

Each method will display the prompt:

```
Command: E ENTER (Or click the Erase button on the
    Modify toolbar.)
Select objects:
```

As the prompt is displayed, the crosshairs turn into a pick box. The pick box can be used to select objects to be removed. Place the box on the object to be removed and left-click. In Figure 3.14 the column centerline has been selected for removal. Once selected, the line will now appear highlighted. An unlimited number of objects to be erased can be selected. The prompt will continue to show "Select objects" until ENTER is pressed. Pressing ENTER will remove the selected objects and restore the Command prompt, as seen in Figure 3.15.

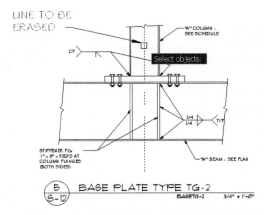

Figure 3.14   *As the **Erase** command is entered, the crosshairs are replaced by a pick box. Placing the pick box on the desired item to be erased and selecting it (the vertical centerline), will select that object for removal. (Courtesy of Van Domelen/Looijenga/ McGarrigle/Knauf Consulting Engineers.)*

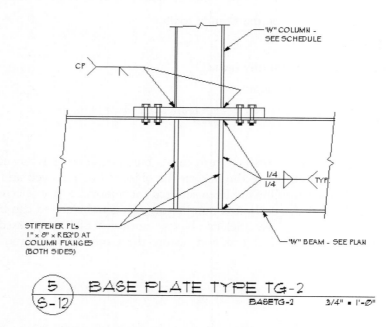

'W' COLUMN -
SEE SCHEDULE

CP

1/4
1/4

TYP.

STIFFENER PL's
1" × 8" × REQ'D AT
COLUMN FLANGES
(BOTH SIDES)

'W' BEAM - SEE PLAN

5
S-12

BASE PLATE TYPE TG-2

BASETG-2   3/4" = 1'-0"

Figure 3.15 *Pressing ENTER or right-clicking will remove the selected object from the drawing screen. In this example, the vertical centerline has been removed. (Courtesy of Van Domelen/Looijenga/McGarrigle/ Knauf Consulting Engineers.)*

Erase allows for the last drawing object to be erased without selecting the object. This can be done using the following sequence:

```
Command: E ENTER (Or click the Erase button on the
     Modify toolbar.)
Select object: L ENTER
Select objects: ENTER
```

As the command line is restored, the last object drawn is eliminated. If you would like to remove the objects drawn with the last drawing sequence type:

```
Command: E ENTER (Or click the Erase button on the
     Modify toolbar.)
Select objects: L ENTER
Select objects: (Select object.)
Select objects (Continue to select objects or press ENTER to end
     sequence.) ENTER
Command:
```

Selecting objects to be erased can be continued indefinitely until the entire drawing is removed, but this process is not the most efficient way to remove multiple objects. Later chapters will provide in-depth coverage of **Erase** and other editing options, as well as more efficient methods of selecting objects to be edited.

OOPS

Occasionally an object is erased by mistake. The **Oops** command will restore the object removed by a previous **Erase** command. The command will restore the last removed object that was erased by typing **OOPS** ENTER at the prompt. Chapter 10 will provide other options for modifying drawing objects.

UNDO

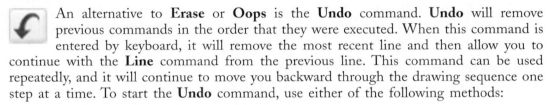

An alternative to **Erase** or **Oops** is the **Undo** command. **Undo** will remove previous commands in the order that they were executed. When this command is entered by keyboard, it will remove the most recent line and then allow you to continue with the **Line** command from the previous line. This command can be used repeatedly, and it will continue to move you backward through the drawing sequence one step at a time. To start the **Undo** command, use either of the following methods:

- Select **Undo** on the **Standard** toolbar
- Type **U** ENTER at the keyboard.

When the toolbar is used to enter the command, simply select the button when you wish to remove the command that has just been executed. Typing **U** ENTER in the middle of a command sequence will remove an unwanted line, without ending the **Line** command. The **Undo** button is inactive during a command process. Selecting the **Undo** button after a command is ended will remove the entire drawings sequence. Right-clicking after a drawing command is completed will allow **UNDO** to be selected from the short cut menu. The menu will reflect the last command that has been executed. If a circle has just been drawn the prompt will read **UNDO CIRCLE**. If you right-click after drawing a line, **UNDO LINE** is displayed at the prompt.

REDO

If you decide that you really do want the line after you have used the **Undo** command, the **Redo** command can be used to correct your change of mind. **Redo** will reverse the effects of **Undo**. To perform a **Redo**, use one of the following methods:

- Click the Erase button on the **Standard** toolbar
- Type **Redo** ENTER at the Command prompt
- Right-click and select **REDO** from the shortcut menu

Selecting the menu arrow between the **Undo** and the **Redo** buttons will display a drop-down menu of previous operations that can be altered using these commands. The **Undo** and **Redo** commands allow you to experiment with a layout by drawing different options, and then using these commands to restore previous options.

AN INTRODUCTION TO PRINTING

Once you've created and edited a drawing to your satisfaction, you'll need some method of passing the contents of the file on to those who will put it to use. This usually means sending the drawing electronically over the Internet to another firm. For many, it also means making a hard copy such as a paper print. No matter how technology advances, for most AutoCAD operators, a paper copy is still a common method for checking drawings. The purpose of this chapter is to help you produce a quick check print of your drawing. This chapter will introduce the process of making a paper print of a drawing in model space using the system printer configured in your Windows platform. Later chapters will provide additional information for plotting in paper space, choosing additional printers, configuring a plotter, plotting layouts, and controlling the many variables associated with plotting drawings using a scale.

EXPLORING THE PLOTTING PROCESS

To start the print process, use one of the following methods:

- Click the **Plot** button on the **Standard** toolbar
- Select **Plot** from the **File** menu
- Type **PLOT** ENTER at the command prompt

Each selection method will produce the **Plot** dialog box shown in Figure 3.16. Notice that the upper left corner shows information about the printer to be used. AutoCAD will automatically use the defaulter printer. Only the key elements needed to make a check print will be explored in this chapter. See Chapter 25 for a detailed discussion on plotting to scale.

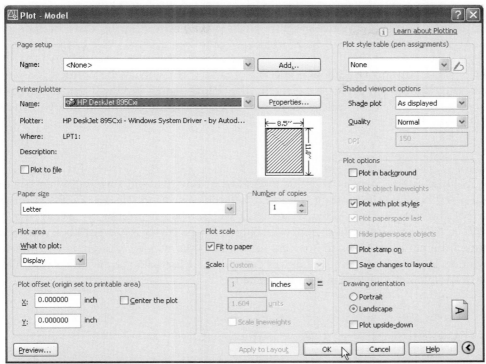

Figure 3.16   *Selecting the **Plot** button on the **Standard** toolbar, selecting **Plot** from the **File** menu, or typing **PLOT** ENTER at the command line will display the **Plot** dialog box. The dialog box can be used to control each aspect of creating a check print.*

Paper Size and Paper Units

The **Plot** dialog box displays the name of the current printing device, the paper size the plot will be made on, and the size of the printable work area. For the current discussion, the default printer is used with a paper size of 8.5 × 11". Additional paper options can be viewed by selecting the arrow by the **Paper size** edit box.

Plot Area

The **Plot area** portion provides options for describing what part of the current drawing will be plotted. Because often the entire drawing does not need to be plotted, four options are given to determine what portion of the drawing will be plotted when working in model space.

Display   The default setting is to plot only the material that is currently displayed on the screen. If you have zoomed into a drawing, only that portion of the drawing that is visible on the screen will be plotted.

Extents   This option will plot the current drawing with the extent of the drawing objects as the maximum that will be displayed in the plot.

Limits   This option will plot all of the drawing objects that lie within the current drawing limits defined during the drawing setup. Even if the entire drawing can't be seen on the screen, the entire drawing will be plotted.

Window   This option will allow a specific area of a drawing to be defined for plotting. When the **Window** option is selected, the **Plot** dialog box will be removed from the screen, allowing you to select what is to be plotted. The command line will show the prompt:

```
Specify window for printing
Specify first corner:
```

AutoCAD is waiting for you to define a window or box that will surround what you would like to plot. To create the plotting window, move the cursor and select a point. Once the first corner of the window is selected, AutoCAD will provide a prompt to specify the opposite corner. Move the cursor to the opposite corner. As the cursor is moved, a box (the window) is formed. Figure 3.17 shows the process for defining the window. Anything that lies within the window is plotted. Once the window is defined, the **Plot** dialog box is redisplayed, allowing you to continue defining the plot to be created.

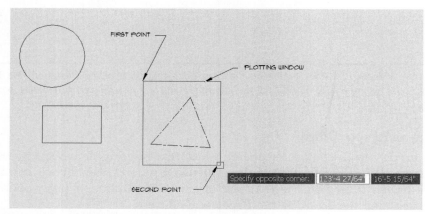

Figure 3.17   *When objects are selected to be plotted using the **Window** option, the size of the window can be set by selecting the window corners with the cursor. Selecting the **Window** option will remove the menu and return the drawing to the screen. Once the corners of the select window are located, the **Plot** dialog box is displayed again so that other print options can be set.*

Drawing Orientation

The **Drawing orientation** portion allows the drawing orientation to be specified. The current orientation in Figure 3.16 shows that the drawing is printed in the normal orientation for construction drawing with the long paper edge horizontal. The drawing is read looking from the bottom to top, and the title block is along the right side of the print. This setting is referred to as **Landscape**. When the print is rotated so that the long edge of the paper is vertical, the print is referred to as a **Portrait** print. Adjust the orientation of the drawing to the paper by selecting the appropriate radio button. The **Plot upside-down** option plots the drawing in reverse of the selected position from the 0,0 origin. If **Landscape** is the selected position, and **Plot Upside Down** is active, the top right corner of the drawing is placed in the 0,0 position rather than the lower left corner of the drawing.

PLOT SCALE

The **Plot scale** portion controls settings for the scale of the plot. Its options include **Fit to Paper** and **Scale**.

Selecting a Scale

With the **Fit to paper** option activated, the selected portion of the drawing is reduced in size to fit on the current printer paper. Unless you'll be plotting a small detail, most architectural drawings can't be printed at full size on a printer. You can, however, print the entire floor plan on 8.5 × 11" material if the scale is reduced. The **Fit to paper** option reduces the selected material to be printed to an appropriate size to fill the current paper. Figure 3.18 shows an example of a detail printed using the **Fit to paper** option.

With the **Fit to paper** option inactive, the **Scale** edit bar is activated. Selecting the arrow by the **Scale** edit box displays the scale menu shown in Figure 3.19. Select the desired scale and the list will close to allow you to continue setting the print parameters. The drawing in Figure 3.20 was printed using the **Window** option of select and by picking a scale of 3/4"=1'–0" from the **Scale** menu. Additional options for setting scale will be addressed in chapter 25.

 Note: Selecting Custom allows you to print a drawing to a scale that you create. Keep in mind that the goal of this chapter is to help you make a quick, non-scaled, check print so that you can evaluate your work. Stick with Fit to paper or one of the common architectural scales until plotting is discussed in detail.

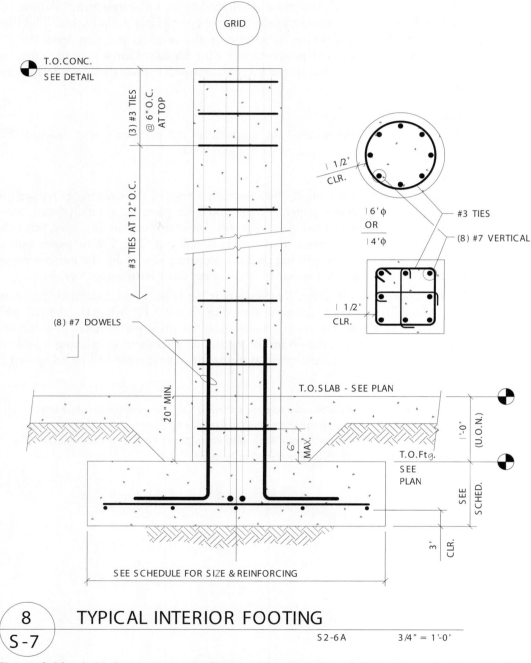

GRID

T.O.CONC.
SEE DETAIL

(3) #3 TIES
@ 6" O.C.
AT TOP

#3 TIES AT 12" O.C.

1 1/2'
CLR.

16'φ
OR
14'φ

#3 TIES

(8) #7 VERTICAL

1 1/2'
CLR.

(8) #7 DOWELS

T.O.SLAB - SEE PLAN

20" MIN.

6" MAX.

T.O.Ftg.
SEE
PLAN

1'-0'
(U.O.N.)

SEE SCHED.

3" CLR.

SEE SCHEDULE FOR SIZE & REINFORCING

8
S-7

TYPICAL INTERIOR FOOTING

S2-6A   3/4" = 1'-0'

Figure 3.18 *A detail printed using the **Fit to paper** option. Most professionals use drawings printed with the **Scaled to Fit** option for check prints prior to plotting the drawing to scale.*

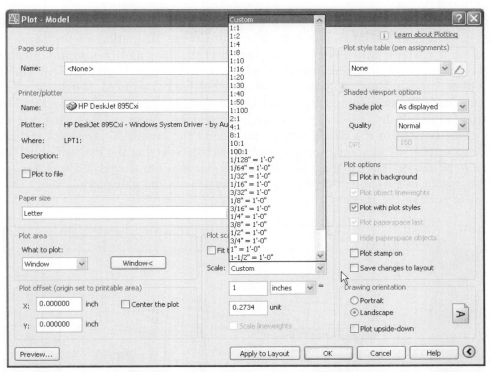

Figure 3.19 *To plot a drawing to a scale, select the arrow beside the edit box to reveal the scale list. Select the desired scale. Once the scale is selected, the list will close to allow you to continue setting the plot parameters.*

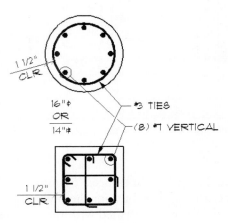

Figure 3.20 *A portion of the detail from Figure 3.18 printed using the window selection method and a plotting scale of 3/4"=1'–0".*

PREVIEWING

Clicking the **Preview** button will produce a display of how the drawing will appear when printed. Figure 3.21 shows an example of a full preview. An outline of the paper size is drawn on the screen, which displays the portion of the drawing to be printed. The cursor is changed to show a magnifying glass with a + and − sign beside it. This represents the **Zoom** command and allows the image to be enlarged or reduced. It will not affect the plotted image. Press and hold the select button to alter the size of the image. Moving the cursor toward the top of the screen enlarges the image, and moving the cursor toward the bottom of the screen reduces the size of the image. A right-click will display a shortcut menu allowing other options to be selected. Selecting the **Pan** option allows the image to be shifted, but the magnification of the image is not altered. Each of the options will be further discussed in chapter 25. To continue with the plot process, select the **Exit** option to remove the menu.

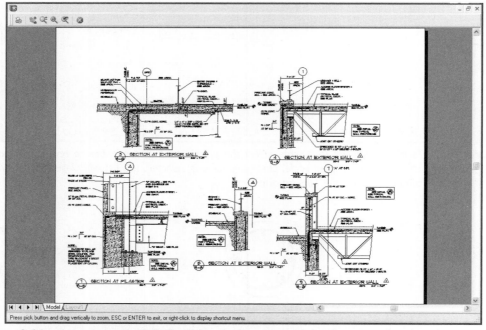

Figure 3.21   *A preview will show the limits of the paper and the drawing to be plotted.*

PRODUCING THE PRINT

Several areas of the **Plot** dialog box have been examined. If you change your mind on the need for a print, select the **Cancel** button; the dialog box will be terminated and the drawing will be redisplayed. If you are satisfied with the plotting parameters that have been established, click the **OK** button. The dialog box will be removed from the screen and the print will be produced.

CHAPTER 3 EXERCISES

Set limits, units, snap, and grid to best suit each project. Save each exercise to a diskette and to a folder on the hard drive or network per your instructor's directions.

1. Draw the following outline using the coordinates that are provided.

   Point of beginning: 1,1

   point b 4.2426<45

   point c 3.1623<342

   point d 2.5<0

   point e 4.0311<120

   point f 4.2720<159

   point g 3.9051<230

   C ENTER

   Save your drawing with a file name of **E-3-1**.

2. Draw the following outline using the relative coordinates that are provided.

   Point of beginning: 2,1

   point b 2.5<N80dE

   point c 3.5<N38d23'15"E

   point d 2.5<S40d20'40"E

   point e 1,1

   point f 1,–1

   point g 4<N

   point h –6,1

   point i 3.5<S39d7'W

   back to the true point of beginning

   Save your drawing with a file name of E-3-2.

3. Draw the following outline using the relative coordinates that are provided. Save the drawing as file **E-3-3**.

   Point of beginning: 9,9

   point b –5,–2.5

   point c 0,–1.5

   point d 4<S39d48'W

point e −1.5,−2

point f 4.5,1.5

point g 1.5<90

point h 2<335

point i 2.5<120

point j 5.5<330

point k 1,1

point l −2.5,1

point m −.5,−.5

point n 3<N61d25'10"W

point o 3.8125<E

@4.5<N to the true point of beginning

4. Use the following relative coordinates to layout the required shape. Save as file **E-3-4.**

Starting point: 7.75, 7.95

1.91<127

−2,0

0,−2

−1.50, 1.75

−.96,0

1.62<137°

.46<227°

1.22<311°

−.92,0

1.62,−.5

−1.11,-4.13

3.5,0

1.62,1.62

1.75<308

3.32<80.73°

back to the point of beginning

5. Use the attached drawing to show a 16" deep steel beam resting on a 3/4" x 12" steel top plate. Support the top plate on a 6" × 6" x 3/8" T.S. column. Provide a 3/4" x 16" steel end plate. Save the drawing with the file **E-3-5**.

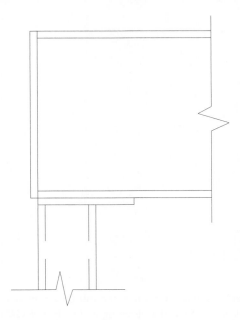

6. Use the attached drawing to draw the roofline of the structure. Assume all hips and valleys are drawn at 45° angles. Save the drawing as file **E-3-6ROOF**.

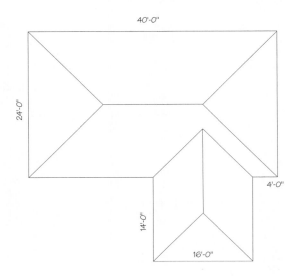

7. Use the attached drawing to draw a section of a one-story footing for a wood floor system. Assume that the bottom of the footing extends 12" into the grade, and the top of the stem wall extends 8" above grade. Save the drawing as file **E-3-7-1STFTG**.

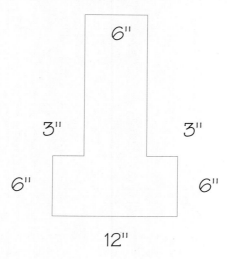

8. Use the attached drawing to draw a section of a one-story footing for an on-grade concrete floor system. Save the drawing as file **E-3-8-CONCFTG**.

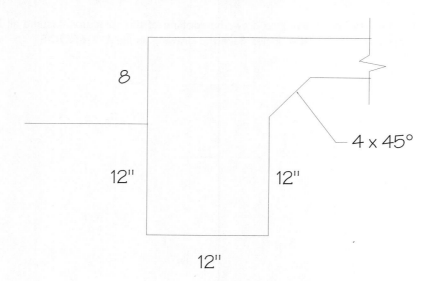

9. Use the attached drawing to complete a section through a 6" wide concrete panel resting on a concrete footing 24" wide × 12" deep. Support the wall panel on 1" deep grout with beveled edges. Show a 3" × 6" pocket in the footing for reinforcing. Save the drawing as file **E-3-9CONCWALL**.

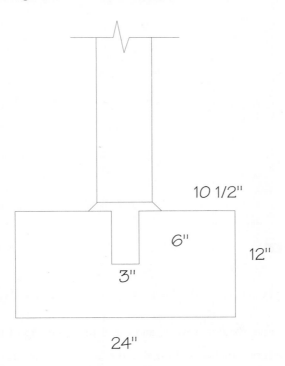

10 1/2"

6"

3"

12"

24"

10. Use the attached drawing to complete a section view of a floor truss intersecting a 6" wide concrete tilt-up wall. Assume the truss to be 24" deep with 2" deep top and bottom. Save the drawing as file **E-3-10TRUSSWALL**.

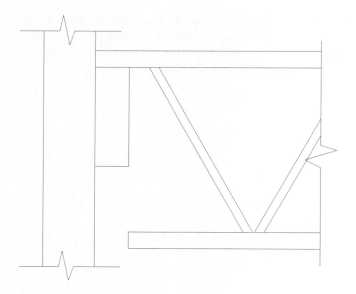

CHAPTER 3 QUIZ

Directions:

Answer the following questions with short complete statements. Type your answers using a word processor.

1. Place your name, chapter number, and the date at the top of the sheet.

2. Type the question number and provide the answer in the form of a statement that includes part of the question. You do not need to write out the entire question.

 Warning: Some of the questions have not been covered in the reading material and will require the use of the help menu. You may also have to do some exploring to answer the questions.

1. Use the drawing below to provide absolute coordinates. Grid=.5", Snap=.25"

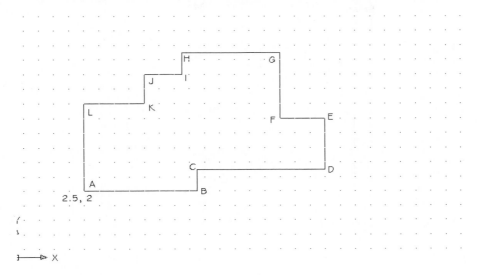

a. 2.5,2	e. ___	i. ___
b. ___	f. ___	j. ___
c. ___	g. ___	k. ___
d. ___	h. ___	l. ___

2. Use the drawing from Problem 1 to provide relative coordinates assuming that **Dynamic Input** is **ON**.

a. 2.5,2	e. ___	i. ___
b. ___	f. ___	j. ___
c. ___	g. ___	k. ___
d. ___	h. ___	l. ___

3. Use the drawing from Problem 1 to provide polar coordinates assuming that **Dynamic Input** is **ON**.

a. 2.5,2	e. ___	i. ___
b. ___	f. ___	j. ___
c. ___	g. ___	k. ___
d. ___	h. ___	l. ___

4. Use the following drawing to provide relative coordinates.

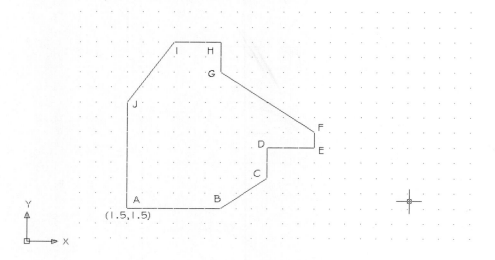

(1.5,1.5)

a. 1.5,1.5   e. ___   i. ___
b. ___   f. ___   j. ___
c. ___   g. ___   k. ___
d. ___   h. ___   l. ___

5. Use the drawing from Problem 4 to provide absolute coordinates.

a. 1.5,1.5   e. ___   i. ___
b. ___   f. ___   j. ___
c. ___   g. ___   k. ___
d. ___   h. ___   l. ___

6. Use the drawing from Problem 4 to provide polar coordinates. Use relative coordinates where insufficient information is available to determine polar entries.

a. 1.5,1.5   e. ___   i. ___
b. ___   f. ___   j. ___
c. ___   g. ___   k. ___
d. ___   h. ___   l. ___

7. What entry method uses the X,Y format to enter coordinates if **Dynamic Input** is **OFF?**

_____

_____

8. What entry method uses the X<# format to enter coordinates?

_____

_____

9. With **Dynamic Input ON**, what entry method uses the X,Y format to enter coordinates?

_____

_____

10. Several entry coordinates are listed below. Circle the coordinates that will not work.

a. 3,4   f. –3,<27   k. 27'6

b. –3,–4   g. @1'–0",<27.5   l. 25',13'6

c. 3.4   h. 27'6<n3615'e   m. 14'<W30N

d. @3,<45   i. 36'<27.5   n. 20'4,S27d36,15E

e. @–3,–5   j. N27d3"27"e,15'   o. 3'6,7'4

11. Can an angled line be drawn with SNAP and GRID both ON?

_____

_____

12. Describe the line that is formed between 2,0 and 4,0.

_____

_____

13. Describe the options for a plot preview.

_____

_____

14. Explain the difference between plotting using the **Display** and **Window** options.

_____

_____

15. What settings must be changed to use absolute, relative, and polar coordinates on the same drawing?

_____

_____

16. You want to plot a drawing at 3/8"=1'–0". How can this be done?

_____

_____

17. List four methods for accessing the **Line** command.

_____

_____

18. What steps are required to erase a line?

_____

_____

19. If a line has been erased by mistake, how can it be restored?

_____

_____

20. What effect does **Erase** have on the cursor?

_____

_____

21. Describe the difference between **Oops** and **Undo**.

_____

_____

22. What effect will typing **C** ENTER have on a line segment?

_____

_____

23. How does the snap setting affect the use of the **Line** command?

_____

_____

24. What type of coordinate entry is best suited for drawing **E-3-6PLAN**?

_____

_____

25. Use the **Help** command and list four methods of accessing the **Layer Properties Manager**.

_____

_____

CHAPTER 4

Working with Drawing Files

This chapter will introduce methods for:

- Creating and saving drawing templates
- Saving user preferences
- Saving drawing files
- Managing files and folders

Commands and major options to be explored in this chapter include:

- Paper space
- Model space
- Viewports
- **Pan**
- **Options**
- **Save**
- **Saveas**
- **Qsave**
- **Savetime**
- **Quit**
- **Exit**
- **Markup**
- **Close**
- **Closeall**
- **Audit**
- **Recover**

This chapter will introduce you to methods for creating a template drawing, explore methods for saving drawing properties for future use, expand your methods for saving and managing drawing files, and explore methods for merging existing drawings.

CREATING AND SAVING DRAWING TEMPLATES

In chapter 2 you were introduced to the templates that AutoCAD provides to help make each new drawing that you start. Most schools and offices have their own templates. Your own template can be created and saved using the **Startup** dialog box, and then used each time a drawing requiring those parameters is needed. Several useful drawing parameters were introduced in chapter 2. Other useful values such as workspace, linetype, lettering size and styles, and dimensioning styles and format will be introduced in the chapters that follow. Necessary values can be added to the drawing template for future use. Once created and saved, they can be opened to form the base for a new drawing file.

USING AN EXISTING TEMPLATE

AutoCAD provides several templates to ease drawing setup. Open AutoCAD using the **Select a Template** option. Scroll through the listing of templates and you'll find two architectural templates. Both templates will provide a drawing screen showing a 24" × 36" sheet of paper with an architectural style border that is suitable for most construction drawings. You'll also notice that most of the other templates are either "named plot styles" or "color dependent." Plot styles alter the way a plotted drawing will appear. Each object in a drawing has a plot style property. These styles can be assigned by color or by named groups. With a change in the plot style, the color, linetype, and lineweight of an object can be altered on the print. This will allow objects within the drawing to be highlighted. The benefits of plotting using assigned styles will be addressed in detail in chapter 25 as plotting is explored. For now, select the **Architectural, English units-Color Dependent Plot Style** template. This will produce the template shown in Figure 4.1. Although the title implies that this template is for architectural drawing, it should not be used without some basic changes. The term *Architectural* is applied for the shape of the title block, rather than the current settings. Three areas of the template should be examined before you complete a drawing. These areas are paper space, model space, and viewport.

Figure 4.1 *The **Architectural** template can be used for a base layout that will meet most of the needs for construction related projects. Display the template by selecting the **Use a Template** option from the **Startup** dialog box and selecting the **Architectural, English units-color dependent** option.*

Paper Space

As the template is displayed, you're working in paper space. Paper space can be identified by the icon in the lower left corner. The **Architectural Title Block** tab is now active, and the **Maximize Viewport** icon has been added to the status bar. The paper is 24" × 36" with a drawing area of approximately 2'–5" × 1'–10". In effect, you're now working on a blank sheet of paper. Move the cursor and you'll notice that it can be moved to any point in the display area. Paper space will allow you to start a line in the drawing area and finish the line at any point outside the drawing border.

 Note: Paper space is a great aid for plotting the drawing to a specified scale, but it should not be used to create a drawing.

Model Space

Model space is the area of the template where the drawing should be completed. Model space allows you to work in "real" space. Select the **Model** tab below the drawing area to switch from paper to model space. This will hide the template and display a blank screen with the model space icon in the lower left corner of the display. With model space active, you'll have a drawing area of approximately 15" × 9". The size will vary depending on the number of command lines displayed and which toolbars you have displayed. The limits of this area can be adjusted to any size at any time. The size of the drawing space is altered using the **Limits** command that was introduced in chapter 2. By changing the size of the space, you can display a large structure on the screen in its entirety.

Viewports

The viewport is a "hole" in paper space that allows the drawing in model space to be viewed in paper space. A viewport allows you to place a 100' long structure on a 36" long sheet of paper. Figure 4.2 shows a representation of how the viewport and the template and model space work together. Switch back to paper space by selecting the **Architectural Title Block** tab. Select the **PAPER** button on the status bar. The **PAPER** button will now change to read **MODEL**, and the drawing will change to display a drawing that appears similar to Figure 4.1, but it has changed to reflect model space inside the template borders. The bold black line represents the edge of the viewport. The UCS icon has changed to reflect model space and has been placed inside the viewport. Move the crosshairs to the border of the template. As soon as the crosshairs touch the edge of the viewport, it changes to an arrow. You're not allowed to draw outside the viewport. You're now working with approximately a 12" × 9" drawing area or viewport. It is a nice looking border, but not many projects will fit into the drawing area. Chapter 23 will introduce methods of altering the viewport size, as well as other needed skills for working with viewports to prepare a drawing for plotting.

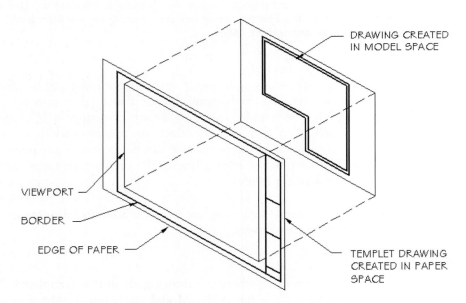

Figure 4.2 *A viewport is provided in each drawing template created in paper space. Additional layouts can be created to allow for different plotting displays.*

Note: As a new AutoCAD operator, you might find the discussion of viewports a bit overwhelming. If nothing else, keep in mind that a template is a great place to store drawing tools. It is like the junk drawer we all have in a desk. All the good stuff goes in the drawer for future use. A template is a base drawing used to store the settings that you use over and over. No one should be setting the precision of units or the drawing limits each time a drawing is started.

The second thing to remember is that you want to work in model space to create objects and paper space to plot them to scale. Each of the preceding chapters has reminded you of this key point. It's important!

CREATING YOUR OWN TEMPLATE

Any drawing can be saved and used as a template. An existing AutoCAD template can be modified and saved with a new name or a template drawing can be created using the methods used to create a drawing file. Open a drawing and establish the drawing units, limits, snap, and grid values. Special consideration will need to be given to the drawing limits as you establish a drawing template. Common paper sizes were introduced in chapter 2 as templates were explored. The size of paper that the completed project is printed or plotted on must be considered as the drawing template is established. Although the drawing limits of model space can be altered at any time throughout the life of the drawing, establishing limits in the viewport that match standard paper sizes will aid in the use of the template. As you enter the template, use the **Viewport** control box to set the paper space viewport to display the desired scale, and set the model space limits to an ample size to display a structure.

SAVING USER PREFERENCES

A drawing template works well for saving drawing settings that must remain constant for each member of the drawing team. There are many features of the drawing environment that do not affect the final project, but that affect the productivity of each team member. These features can be saved as a profile for each team member.

Drawing profiles allow each user to adjust and save the drawing settings to meet their personal preferences. Drawing profiles can be especially helpful if you are working at a computer that you share with several other people. If you like a black background and you share your computer with someone who likes a white background, AutoCAD allows these differences to be stored so that they can be changed quickly. Items that might be included in a profile include the, crosshairs size, display resolution, intervals between automatic saves, text display format, screen colors, and plotter or printer information. Using a profile, you'll be able to adjust, save, and retrieve drawing settings for your next drawing session.

To set a user profile, select **Options** from the **Tools** menu. This will display the **Options** dialog box with ten tabs to choose from. The tab that was used last is displayed by default. Each of these tabs will be explored in detail as commands are introduced throughout the text. The goal of this chapter is to introduce the **Options** dialog box and to help you save a few of your favorite settings. As new features are introduced, they can be added to your profile. Highlights of the **Options** dialog box include the following:

Files—This tab is used to control the directories where AutoCAD will search for support, drivers, menus, and other files. At this point of your AutoCAD career, only a few of the options need to be considered.

Automatic Save File Location—This option specifies the path for the file created each time a file is saved using **Automatic Save** (see the **Open and Save** tab). This is where backup copies are stored. If your machine should crash, this location is the place to look for files that were created using automatic backup system.

Template settings—These options control the template file location, sheet set template file location, and the default template to be used when **QNEW** is used to enter a drawings session. The settings also control the default templates file to be used for creating new sheets (see chapter 24). Use the following procedure to control the template used for **QNEW**:

- Double-click the **Template Settings.**
- Double-click the **Default Template Filename** for **QNEW** listing.
- Double-click the **Browse** button to display a list of available templates.
- Select the name of the desired template *(for most uses this is the* **Architectural, English units-color dependent** *option).*
- Select the **OPEN** button. This will restore the **Files** tab and list the selected template as the default template to be used with **QNEW**.

Temporary Drawing File Location—This setting specifies where temporary files are stored during a drawing session. These files are deleted each time you exit the program.

Display—This tab allows the display of AutoCAD to be customized. Several of the options of this tab have been discussed in previous chapters. When the values are set and saved in a profile, they will automatically be reset when the profile is selected as the active profile. Take time to evaluate each portion of this box. If in doubt, click the **Help** button at the bottom of the tab for an explanation. Items you should adjust on the **Display tab** include:

Window Elements—Two elements in this display are typically **OFF**, but may be of interest to some users. These include:

Display Screen Menu—Activating this button will display a menu on the right side of the drawing area. The **On Screen** menu provides similar functionality to using the drop-down menu and is another option for executing commands.

Use large buttons for Toolbars—Many users prefer using small icons to decrease clutter on the screen, providing more space for other information. Selecting **Use large buttons for Toolbars** will increase the size of the icons for users needing better visual aids.

Colors—Allows the color of each aspect of the display screen to be altered. Selecting this option will display the Color Option dialog box and allow you to set the color of each element in the drawing area. This option will be explored in chapter 5.

Crosshairs size—Allows the size of the crosshairs to be altered by entering a number that represents the percentage of the screen to be filled. See chapter 1 for a review of altering the size.

Display Resolution—The settings in this box affect the quality of the display of objects in the monitor. Do you want to display a circle as a circle or as a series of straight line segments that appear to make a circle? The effects of this area will be discussed in chapter 6. (The default of 1000 is adequate for most drawings.)

Open and Save—This tab will affect how and where drawing files are saved. This tab will be covered in detail later in this and future chapters. Key features include:

SaveAs—This portion of the dialog box can be used to determine the format used when saving your drawings. If you work at multiple workstations this is your opportunity to save files to different drawing formats.

Automatic Save—This option controls the time interval between automatic saves. This option is a great safety feature you can utilize to protect your drawing files. (Ten minutes is the default value and it is adequate for most drawings.)

Create Backup Copy with Each Save—This option toggles the creation of backup files. Typically, having a backup copy is good safety procedure. If you decide to delete backup files, make sure you have made multiple copies of your files on the hard drive and/or other removable media.

File extension for temporary files—The .ac$ extension is used to describe files created by AutoCAD during a work session. These files are the equivalent of scratch paper. If you get in the mood to clean your machine, DO NOT discard these files while you have AutoCAD open. Once all drawing files are closed, these files are normally deleted. If you find some .ac$ files in a directory, they can be safely deleted if AutoCAD is closed.

Plot and Publish—This tab controls all of the options that affect plotting. Components of this tab were explored in chapter 3 and will be discussed in detail in chapter 25.

System—This tab controls how the system will operate. Take time to examine the **General Options** box. These settings control general options that relate to drawing operations. Key elements to consider:

Beep on Error in User Input—In the active setting, AutoCAD will sound an alarm, a soft beep, when an invalid response is given. If you're in the middle of the **Line** command, and AutoCAD is expecting a "to point" to be provided, the alarm will sound if you try and use the **Circle** command. As a new user, keep this activated. When it is deactivated, many new users get frustrated when they execute a command, and the desired results are not achieved. The alarm often serves to bring you back to reality, reminding you to look at the Command prompt so that the desired information can be provided.

User Preferences—This tab affects options that optimize your performance in AutoCAD. One feature that experienced users of AutoCAD might want to explore is the **Right-click Customization** box. Some experienced users are so accustomed to using the right-click button in place of ENTER that they might be willing to forsake the shortcut menus. If you're a new user, leave the menus active. Other options will be addressed in later chapters.

Drafting—Many of the options in this box have been introduced in previous chapters and will be covered in detail in later chapters. Options that you should address as you set your own profile include the following:

Marker—Toggles the display of the AutoSnap marker **ON/OFF**. With the setting **ON**, when the cursor is moved over a snap point, the marker is displayed. Keep this setting activated.

Magnet—Toggles the display of the AutoSnap magnet **ON/OFF**. With the setting **ON**, the magnet is locked onto the nearest snap point. Keep this setting activated.

Display AutoSnap Tooltip—Toggles the display of the AutoSnap tooltip **ON/OFF**. With the setting **ON**, the tooltip is displayed when the AutoSnap magnet locks onto a snap point.

AutoSnap Marker Color—This control can be used to alter the color of the snap box. Select a color that will contrast with the screen color so that the marker can be easily seen.

AutoSnap Marker Size—Displays the object snap location when the cursor moves over or near an object and defines which object snaps are evaluated. The marker shape is dependent on the snap it is marking. This control can be used to alter the size of the snap box. As the size of the box is increased, it gets easier to see. As the drawing complexity increases, a large snap box can be distracting.

Aperture Size—This slide bar can be used to control the size of the aperture box. The aperture box controls the size of the object snap target box.

Drafting Tooltip Settings—This will display options for controlling the display of tooltips including model and layout display controls, as well as controls for colors, size, and transparency.

3D Modeling— This tab provides options for controlling 3-D drawing features that will not be examined in this text.

Selection—These features control how AutoCAD will select drawing objects. This box will be explored in detail in chapter 9. A feature that you might want to set in your initial profile is the pick box size.

Profiles—This tab controls the creation and use of profiles. The **Profiles** tab can be seen in Figure 4.3. The next section of this chapter will help you explore this tab and create your first profile. As you gain experience and discover new settings that you would like to include in the profile, you can update the profile using this tab.

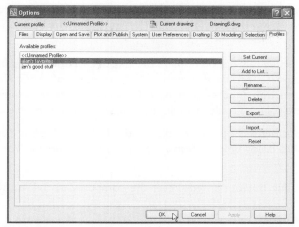

Figure 4.3   *The **Profiles** tab controls the creation and use of user-defined profiles. As you gain experience and discover new settings that you would like to include in the profile, the profile can be updated using this tab.*

CREATING AND SAVING A PROFILE

To create a user profile, select **Options** from the **Tools** menu. Go through each of the tabs and set the options to meet your preferences. Once you have selected the desired settings, follow this procedure:

1. Select the **Profiles** tab to display the dialog box shown in Figure 4.3.

2. Select the **Add to List** option. This will display the **Add Profile** dialog box.

3. Enter a profile name in the **Profile name** edit box. Use a name that will quickly describe the user or drawing type.

4. Provide a short description in the **Description** edit box that will quickly describe the contents.

5. Click the **Apply & Close** button to accept the settings. When the button is clicked, the dialog box is closed and the profile name you provided is added to the **Available profiles** list.

Once more than one profile has been created, you can quickly select the desired profile from the **Profiles** tab by highlighting the listing of the desired profile and then clicking the **Set Current** button. Other buttons in this menu that should be considered include the following:

Rename—This option allows an existing profile to be renamed. Clicking this button will produce the **Change Profile** dialog box. Alter the name as desired and then click the **Apply & Close** button to alter the profile name and close the dialog box.

Delete—This option allows an existing profile to be removed from the **Available Profiles** list. Highlight the profile to be deleted and click the **Delete** button. This will produce a warning to verify that you really want to delete the profile. Selecting the **No** button will terminate the delete process. Selecting the **Yes** button will delete the selected profile.

Export—Selecting this option displays the **Export Profile** dialog box. This dialog box can be used to export a profile as a file with an extension of .ARG. The profile can be saved and then shared with another computer. If you're working on a network, save your profile to your secure folder. Profiles should also be saved to a portable drive or diskettes.

Import—Selecting this option will display the **Import Profile** dialog box. Similar to the **Export Profile** dialog box, the display can be used to import a profile that was saved using **Export**.

Reset—Selecting this option will reset the computer to the AutoCAD default settings.

SAVING DRAWINGS

Chapter 1 introduced the **Save** command, and chapter 3 gave you practice using it. This chapter will help you expand the use of the command. Once you have created your prototype drawing, it will need to be saved. This means saving your drawings in a folder on the hard drive, on a portable drive or diskette, or in your own folder on a network. Check with your instructor to see if saving on the hard disk is allowed. If so, set up a folder so that all of your drawings can easily be accessed. If not, select the letter of the appropriate drive so that your files are saved to the appropriate location.

 Note: You should avoid saving a file to a diskette, except when the drawing session has ended. A file should be saved to the hard drive throughout the drawing session to save time and avoid a possible problem caused by a full diskette.

You might be saving on the hard drive, flash drives, Zip drives, as well as on diskettes. Because working on the hard drive is so much faster, you will be saving to the hard drive during the drawing session and saving on other removable media at the end of the drawing session. Many professionals save on the hard drive and make two backups, which are each stored in different locations. The open drawing file should be saved to the hard drive approximately every 15 minutes. When the drawing session is done, the file should again be saved to update the existing file on the hard drive. The file should then be saved to external, removable media. This is especially true for students who must transport their drawing files between home and school.

No matter where you save your drawing, the most important thing to remember is to save often. Saving every 15 to 20 minutes will help you avoid losing your work if the power supply to your computer is interrupted. As you start saving your drawings, you will have three options: **Qsave**, **Save**, and **Saveas**.

QSAVE

Qsave, or quick save, will probably be your most frequently used method of saving a drawing. Use **Qsave** for saving a drawing that you want to continue working on. If your computer should crash, only work done after the last save will be lost. Select the **Save** button on the **Standard** toolbar or **Save** from the **File** menu to execute the command.

The first time you save the drawing, the **Save Drawing As** dialog box is displayed. During the initial save, selecting the **Save in** edit box arrow will allow you to select a location. When the destination drive has been selected, the contents of the specified drive is listed. Enter the desired file name in the **File name** edit box. The icons on the left side of the window can be used to provide quick access to common folders on your computer. Buzzsaw and FTP icons allow quick access to the Internet for saving files, posting information, and conducting meetings. Use the **Files of type** edit box to control the drawing type to be used. Figure 4.4 shows a list of alternative drawing types that can be used. By default, drawings is saved using the **AutoCAD 2007 drawing** file format. This format is optimized for file compression and for use on a network. Once the type has been selected, click the **Save** button. Guidelines for assigning file names will be given later in the chapter. The .DWG extension is added to the drawing file and it is saved to the designated drive.

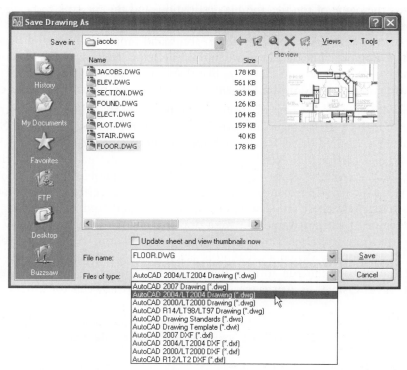

Figure 4.4 *The **Save as type** edit box allows drawings to be saved and be compatible with older versions of AutoCAD.*

The second time the drawing is saved, and all subsequent times the drawing is saved in the drawing session, the save will occur automatically without displaying the **Save Drawing As** dialog box. The name assigned during the initial save sequence is reused.

THE SAVE COMMAND

The **Save** command functions in the same manner as **Qsave**. **Save** will store your drawing in a selected location but still leave it displayed on the screen. This will allow you to continue working on the drawing. To save a drawing, select the **Save** icon from the **Standard** toolbar or type **SAVE** ENTER at the Command prompt.

THE SAVEAS COMMAND

Use **Saveas** to save a drawing with a different name, to save the drawing to a new destination, or to save the drawing as a different file type. The **Saveas** command is similar to the **Qsave** command.

- With **Qsave**, the **Save Drawing As** dialog box is displayed only the first time the drawing is saved.

- With **Saveas**, the dialog box is displayed each time the **Saveas** command is accessed.

Saveas allows the name of the file, the file destination, and the file type to be altered. **Saveas** can be accessed from the **File** menu or from the Command prompt. Select **Save As** from the File menu, or type **SAVEAS** ENTER at the Command prompt. This will produce the **Save Drawing As** dialog box. The default name in the **File name** edit box is the name you selected the last time the drawing was saved. If you like that name, click the **Save** button. If you want to change the name of the drawing file, enter the new name in the **File name** edit box.

Be sure that the correct drive destination is current. Note that the destination is listed in the **Save in** edit box. To change the current drive, use the arrow beside the **Save in** edit box to display the desired drive. Click the **Save** button to save the drawing to the specified drive. The **Saveas** command also allows the file type to be altered. Drawings created using AutoCAD 2007 may also be saved so they will be compatible with previous versions of AutoCAD. These options can be especially useful for students whose home machines are not current with those at school or for architectural and engineering firms with different versions of the software. To save a drawing created using AutoCAD 2007 to be used on a machine running AutoCAD 2000LT, select the AutoCAD 2000/LT2000 Drawing (*.dwg) option. When a drawing is saved using a version other than the default, AutoCAD remembers the file format that was used, and uses the specified format in subsequent **Save**, **Qsave**, or **Autosave** operations that do not call for a dialog box.

Note: You can set the default drawing format used when saving in the **Open and Save** tab of the **Options** dialog box. Save the setting to make it part of your profile.

Reviewing the Save Commands

Open a drawing session, draw a few lines, and then save the file with the name JUNK. As you assign a name, you do not have to type .DWG at the end of the drawing file name.

AutoCAD will add this extension automatically. Once the drawing is saved, add a few more lines to the drawing and save the drawing again. If you save the drawing using the **Save** command, the **Save Drawing As** dialog box will not be displayed; the current file of JUNK.DWG is automatically updated. If you save the drawing using the **Saveas** command, the **Save Drawing As** dialog box is displayed, allowing a new drawing name to be assigned to the drawing. Draw a few more lines and save the file using the **Saveas** command. When the **Save Drawing As** dialog box is displayed, change the drawing name from JUNK to TRASH, using the **File name** edit box. You will now have files named JUNK and TRASH. When you're done with the drawing session, be sure to delete these files using the standard file and folder selection dialog box.

BACKUP FILES

Earlier you discovered that when AutoCAD saves a file it adds an extension of .DWG or .BAK. The .DWG represents a drawing file, and .BAK is the extension for a backup file. When a drawing is terminated, the existing .DWG file becomes the .BAK file and a new .DWG file is created. The .BAK file might not contain all of your recent changes, but it should be only 15 minutes out of date if you've been saving at proper intervals. What a motivator to keep saving your drawings! The .BAK file can be renamed to become a .DWG file using the standard file and folder selection dialog box and then reloaded. If you decide you really don't want .BAK files created, you can deactivate the option by clearing the **Create backup copy with each save** check box on the **Open and Save** tab of the **Options** dialog box.

After a program or system failure, the **Drawing Recovery Manager** will open the next time AutoCAD is started. The manager will display a list of all drawing files that were open when the failure occurred. This will allow backup files and automatically saved versions of the drawing to be retrieved easily without having to use Windows Explorer.

 Note: As a new user, keep the **Create backup copy with each save** check box activated to provide an extra layer of security in case something goes wrong with the master drawings.

AUTOMATIC SAVING

By default, the existing drawing is saved every 10 minutes. You can alter the time between automatic saves by using the **Open and Save** tab of the **Options** dialog box. Use the following procedure to alter the interval between saves:

1. Display the **Options** dialog box by selecting **Options** from the **Tools** menu.

2. Select the **Open and Save** tab.

3. Highlight the current value in the **Minutes between saves** edit box.

4. Enter the desired value.

5. Click the **OK** button.

AutoCAD allows a value between 1 and 120 minutes to be entered. The value entered will start the timer when a change is made to the current drawing file. The commands **Save**, **Saveas**, and **Qsave** will reset the timer. With a value of 20, your drawing file is automatically saved every 20 minutes with a name of ***name filename_a_b_nnnn.sv***. The "a" represents the number of drawing files created in the current AutoCAD session. The "b" represents the number of drawing files created in different sessions of AutoCAD. The "nnnn" is a random number generated by AutoCAD. Files that are automatically saved are deleted when AutoCAD closes a drawing in the normal way. Saved files remain in the event of a crash or power failure. To recover a previous version of your drawing from the automatically saved file, rename the file from a .SV\$ file to a .DWG file. You can disable **Savetime** by entering a value of 0, but this could prove to be a huge mistake if your computer ever loses power.

SAVING A TEMPLATE

Earlier in this chapter you learned how to create a template. Once each of the values for the template is established, the drawing can be saved and reused whenever suitable values are required. Save the template using **Saveas** with the **Drawing Template File** (*.DWT) option selected. Enter the name of the template just as you would do with any other drawing. A name that describes the scale or paper size such as D-ARCH or C-SITE is helpful for sorting templates. Once the name has been entered, click the **Save** button. Provide a brief description of the template, such as *"template for plan views."* This description is displayed whenever you select this template from the **Create New Drawing** dialog box. The final step to save the template is to select the **OK** button. The template can now be reused when a new drawing is opened.

MARKUPS OF DESIGN REVIEW

In addition to DWG and DWT files, AutoCAD allows the creation of DWF (Design Web Format) files. These files allow the drawings files to be sent to clients that do not have AutoCAD. The client can open the DWF file in Autodesk DWF Composer and view, mark up, and make notes on each drawing and then send the DWF file back to the drafter. The drafter then uses the **Markup Set Manager** to review the proposed changes. The **Markup Set Manager** can be accessed by selecting the **Markup Set Manager** icon on the **Standard** toolbar, or by typing **MARKUP** at the command line. The **Markup Set Manager** palette displays the list of markups and allows the CAD user to navigate through the file set, view markups in the DWF, make changes in the DWG file, change the markup status, republish the DWF file, and redistribute updates to the project team. DWF files are highly compressed, which makes them smaller and faster to transmit than the typical CAD drawings. DWF files are not a replacement for DWG files. Data cannot be edited within the DWF file.

LEAVING AUTOCAD

Once a drawing is saved, you might want to leave the drawing and start another project. AutoCAD allows you to exit the program using the **Close** button or by using **Quit**, **Exit**, **Close**, and **Closeall** commands.

The Close Command

The **Close** command can be used to close the current drawings but remain in AutoCAD. The command can be executed by selecting **Close** from the **File** menu or at the Command prompt by typing **CLOSE** ENTER. If there have been no changes, the drawing will be closed automatically but AutoCAD will remain open. If changes have been made since the drawing was last saved, you will be prompted to save or discard the changes.

The Quit and Exit Commands

The **Quit** and **Exit** commands will exit the drawing and close AutoCAD. Execute the **Quit** command by selecting the **Close** button on the AutoCAD title bar or by typing **QUIT** ENTER at the Command prompt. If you attempt to leave a drawing with unsaved changes, you will see the alert box giving you the option to save or exit the drawing session.

- If you click the **Yes** button, AutoCAD will save the changes to the default file.

- Clicking **No** will exit the drawing, close AutoCAD, and return you to other opened programs—any changes made since the last save will be destroyed.

- **Cancel** will terminate the **Quit** command and return you to the drawing.

Exit functions in the same manner as **Quit**. It must be entered from the keyboard by typing **EXIT** ENTER.

The Closeall Command

If you are working in multiple drawings, the **Closeall** command can be used to close all open drawings. The command must be entered from the Command prompt by typing **CLOSEALL** ENTER. If there have been no changes, the drawings will be closed automatically but AutoCAD will remain open. If changes have been made since each drawing was last saved, you will be prompted to save or discard the changes.

MANAGING FILES AND FOLDERS

No office could function if the drafters took their finished drawings and threw them into a room. If a drawing were needed, some poor drafter would have to wade in and hunt through thousands of sheets of paper to find the specific project. Architectural and engineering offices typically use well-organized filing systems to keep track of each project for quick, easy retrieval.

FILES

No matter what method is used to save a drawing, information that is to be kept on your disk must be stored in some orderly manner to allow quick, easy retrieval. Disks are sorted in files and folders. Your drawings are stored as files and will need names to distinguish them from other files. File names for AutoCAD can contain up to 255 characters. Other guidelines for naming files include the following:

- Layer names can contain letters, digits, spaces, and special characters such as $ (dollars), - (hyphen), and _ (underscore).

- Special characters used by Microsoft, Windows, or AutoCAD cannot be used. Do not use the following symbols:

* - (asterisk)	< > - (less-than and greater-than symbols)
" - (back quote)	? - (question mark)
: - (colon)	" " - (quotation marks)
, - (comma)	; - (semicolon)
= - (equal sign)	| - (vertical bar)
/ \ - (forward and back slashes)	

FOLDERS

Files should be organized in folders to provide efficient storage. Folders are the equivalent of dividers in a file cabinet drawer. By default, your drawings are stored in the *Acad2008* folder. Information about files such as file size, time of creation, and the time of last file revision is automatically kept as files are stored. Folders can be created using the **Standard File** and **Folder Selection** dialog box to organize information based on client names, class requirements, or any other similar criteria. Folder naming should follow the same guidelines used to name drawing files. Names should be used that will quickly define the contents such as DRAWINGS, MAPS, REPORTS, or CLIENTS. Folders can be placed inside other folders to better organize information. Subfolders can be created and named using the same guidelines used to create a folder.

OFFICE PROCEDURE

Many small residential offices keep work for each client in separate folders or disks and label the disk by the client name. Drawings are then saved by contents such as floor, foundation, elevation, sections, specs, or site. Some offices assign a combination of numbers and letters to name each project. Numbers are usually assigned to represent the year the project is started, as well as a job number with letters representing the type of drawing. For example, 0853fl would represent the floor plan for the fifty-third project started in 2008. A few offices will save an entire house plan in one file. The drawings are separated by layers, which will be discussed in chapter 5. The drawback to saving large amounts of information in one file is that more time is needed to load and process the information. Later chapters will introduce **External Reference** as a method of saving drawing files and methods of posting drawings on the Internet.

U.S. National CAD Standards Guidelines for Naming Files

Because it is rare that only one person in an office will use a file, most architectural and engineering offices use file names based on the American Institute of Architects (AIA) guidelines. These standards have been combined by the National Institute of Building Sciences (NIBS) to create the U.S. National CAD Standards. These guidelines provide a uniform file naming system that is recognized by the various consultants that work on each project. The suggested method for identifying drawing sheets is based on a four level system that features a combination of letters and characters. A typical file name should include a **discipline designator**, followed by a hyphen, followed by a **sheet type designator**, followed by a **sheet sequence number**. A typical name would resemble **AA-NNN**, where each A represents an alphabetical character and each N represents a numerical character. On very simple projects, this can be reduced to **A-N,** where **A** represents the first **A** and **N** represents the final **N** of the complete specification.

The first character (A) is used to represent the **discipline designator**. Common discipline designators and the order of each sheet within a drawing set include:

G	General
H	Hazardous materials
V	Survey / Mapping
B	Geotechnical
W	Civil Works
C	Civil
L	Landscape
S	Structural
A	Architectural
I	Interiors
Q	Equipment
F	Fire Protection
P	Plumbing
D	Process
M	Mechanical
E	Electrical
T	Telecommunications
R	Resources
X	Other Disciplines
Z	Contractor / Shop Drawings
O	Operations

Each letter represents the type of firm that will provide the drawings. Architectural projects typically include drawings supplied by the G, C, L, S, A, P, M teams for even the simplest projects.

A second character may be used on complex projects to provide further description of the contents. Examples of letters that might be used for drawings originating with the Architectural team include:

> **AS**—Site plan
>
> **AD**—Architectural demolition (protection and removal)
>
> **AE**—Architectural elements (general architectural)

AI—Architectural interiors

AF—Architectural finishes

AG—Architectural graphics.

A hyphen always follows the designator whether one or two characters are used.

The first number in the next designator is used to define the **sheet type**. Numbers that are used for the first digit include:

0—General—symbols, abbreviations, notes, and location maps

1—Plans—These sheets may also contain closely related schedules.

2—Exterior elevations

3—Sections and wall sections

4—Large-scale views including floor plans, elevations, stair sections

5—Details

6—Schedules and diagrams

7—User defined—for material that does not fall into other categories.

8—User defined

9—3D views—isometrics, perspectives, and photographs

On small or simple projects, different types of drawings can be combined on the same sheet. If multiple types of drawings are combined on one sheet, the sheet contents must be clearly identified in the title block.

The last two numbers are used to identify the **sheet sequence number** where 01 is used to identify the first sheet of each series of drawings of the same discipline and sheet type. Numbers are assigned sequentially from 01 through 99. Using the complete code, a designator of AD-103 would represent:

A—a drawing originated by the architectural team

D—demolition

1—plan view

03—Page 3 (the third floor level, with the first floor labeled 01, and the second floor 02).

Final notations that may be part of a file name, such as AA-NNN-??, include:

R-1 where **R** represents a minor revision.

X-1 where **X** represents major changes in the drawing.

A- where **A** represents the first of a phase of drawings followed by **B**, **C**, etc.

The number 1 may be replaced with sequential numbers to represent additional revisions.

STORAGE PROBLEMS

One of the great thrills for a new CAD operator is seeing the message, "DISK FULL." It's not a serious problem, but it is discouraging. As a drawing is started, AutoCAD examines the disk to find a suitable storage space. If you open a drawing for a work session from a diskette, because AutoCAD creates space for several files, there will not be enough space to create the intended drawing. In the middle of a command you'll find that a DISK FULL error is displayed. Your drawing will be terminated, but all work up to that point will be saved.

Occasionally a full disk will cause FATAL ERROR to flash across the screen just before your machine destroys two hours worth of drawing. Sure, you know that you should save every 15 minutes, but sometimes you've got to learn the hard way. The unfortunate aspect of a fatal error is that it might disrupt or destroy other files on your diskette. You'll find that you only have a FATAL ERROR when you're 99% done with a project.

 One Last Reminder: As you work with a drawing file, save the drawing to the hard drive. Only save to a diskette when you are done with the drawing session and are closing the file. Working from a diskette will slow the drawing session and can lead to storage problems.

Correcting File Errors

No matter how carefully you handle files, they will occasionally become corrupted. AutoCAD provides the **Audit** and **Recovery** commands to attempt to make corrupted drawing files useable, as well as the **Drawing Recovery Manager**. Each is located in **Drawing Utilities** on the **File** menu. The **Purge** command can be used to clean unneeded information from the drawing base and will be introduced in chapter 15.

Examining a Drawing File with Audit   The **Audit** command can be used to examine the current drawing file for errors and to correct any errors that exist. These errors are not spelling mistakes or drawing errors, but errors that are occasionally made by the software. This command will detect errors, generate extensive descriptions of the problems, and recommend actions to correct the errors. Use **Audit** by selecting **Audit** from **Drawing Utilities** on the **File** menu or by typing **AUDIT** ENTER at the Command prompt. Each method will produce the following prompt:

```
Fix any errors detected? [Yes/No] <N>
```

Responding with an **N** ENTER will produce a report of the file and list the errors but will not repair the errors.

Responding **Y** ENTER at the prompt will produce a report of the file and list the errors.

With the **Auditctl** system variable set to 1, an Audit report file is created using the file name with .ADT as the extension. This file is placed in the directory containing the original drawing file and lists all of the functions required to fix file errors.

Recovering a Drawing   If errors are discovered that cannot be repaired by **Audit**, the **Recover** command can be used to try and salvage the file. Access **Recover** by selecting

Recover from **Drawing Utilities** on the **File** menu. AutoCAD will automatically attempt to repair damaged drawings as they are opened. If the drawing can't be opened, a message will be displayed that the drawing must be recovered. Once the file to be recovered has been selected, AutoCAD will begin a series of diagnostic procedures. When the process is complete, a listing of problems and the corrective measures taken by the program is displayed. Generally, the recovery process will open a damaged drawing. If the **Recover** command fails, it's time to get out the backup file.

Drawing Recovery Manager The Drawing Recovery Manager palette allows backup files or automatically saved versions of drawings to be recovered. After a program or system failure, the Drawing Recovery Manager opens the next time AutoCAD is started. A list of all drawing files that were open when the failure occurred is displayed. Each file can be previewed and opened allowing you to choose the files that should be saved.

CHAPTER 4 EXERCISES

1. Start a new drawing. You will be using this exercise as a base for a drawing of a small subdivision. The drawing will be drawn at a scale of 1" = 30' and plotted on D-size paper. North will be at the top of the paper (twelve o'clock). Set all applicable defaults to two-place decimals.

   What unit setting should be used for this drawing? _____

   Set the limits at the proper value. What option number will be required?
   _____

   Many of the property lines are described with angles measured in degrees, minutes, and seconds. Set the angle measurement system in the appropriate system. What default number was selected? _____

2. Start a new drawing. Draw a square, a rectangle, and a triangle. Save the drawing using the name **E-4-2**.

3. Start a new drawing called **EXER4-3** that has the same drawings contained on **E-4-2**. Add another circle and a rectangle. Save as **E-4-3**.

4. Edit drawing **E-4-3** by adding any figure that you would like. Save file as **E-4-4**.

5. Get a directory listing of your portable storage medium; list the following:

   The contents

   The time your first exercise was finished

   The date of your last exercise

   The number of bytes used

   The number of bytes free

6. Open the **ARCHITECTURAL.DWT** template.

What is the current snap resolution? X ___ Y ___

What is the current grid setting? X ___ Y ___

What layer will new lines be added to? _____

You will be using this exercise to set up a drawing for a plan view of a room that will be drawn at 1/4" = 1'–0" and plotted on D-size paper. Two-place decimal angles are desired.

Set the units at the proper value. What option number will be required? _____ What was the default? _____

The smallest fraction to be worked with will be 1/4". What denominator should be used? _____

Give the current system for measuring angles. _____

Will the current setting allow for two-place decimals? _____

In which direction will angles be drawn? _____

What are the current drawing limits? _____

What limits will be needed to achieve the listed parameters of this exercise? _____

What is the current snap value? _____

Set the snap value to 6".

What is the current grid default value? _____

Set the grid to 12".

List two other methods of listing the 12" value.

CHAPTER 4 QUIZ

DIRECTIONS

Answer the following questions with short complete statements. Type your answers using a word processor.

1. Place your name, chapter number, and the date at the top of the sheet.

2. Type the question number and provide the answer in the form of a statement that includes part of the question. You do not need to write out the entire question.

Warning: Some of the questions have not been covered in the reading material and will require the use of the help menu. You may also have to do some exploring to answer the questions.

1. You're working on the fifth level floor plan of a multilevel project. Suggest possible names that can be used to save the file.

_____

_____

2. What command will store the drawing on the hard disk but leave the drawing on the screen?

_____

_____

3. How many characters can be used for naming a file in AutoCAD 2008?

_____

_____

4. What command will store the drawing on a disk and terminate the drawing?

_____

_____

5. List three different methods for accessing the **Save** command.

_____

_____

6. You have a drawing file labeled **SECTION**. You draw a roof detail and start to save it as **SECTION**. What will AutoCAD respond and how will your drawings be affected?

_____

_____

7. Give four forms of information that are listed when the dialog box is used to save a drawing.

   _____

   _____

8. List four common extension names that can be assigned to a drawing file.

   _____

   _____

9. What does **Quit** do differently from **Save**?

   _____

   _____

10. What is a fatal error?

    _____

    _____

11. Give two reasons for backing up a drawing in multiple locations.

    _____

    _____

12. List two reasons why you might receive an alert message while saving a drawing.

    _____

    _____

13. You've opened an existing drawing named **FLOOR.DWG**, made some changes, and now want to save the drawing while keeping **FLOOR.DWG** intact. How can this be done?

    _____

    _____

14. You've saved a file named **LOST.DWG** somewhere on your hard drive, but you're not sure where. How can you find the file?

_____

_____

15. Explain the difference between **Save** and **Saveas**.

_____

_____

Drawing Organization

INTRODUCTION

This chapter will introduce:

- Common linetypes used on construction drawings
- Methods of controlling linetype and lineweight
- Methods of controlling colors that are assigned to drawing objects
- The use of layers for controlling objects, linetypes, lineweights and colors

Commands to be introduced include:

- **Linetype**
- **Properties**
- **Lineweight**
- **Ltscale** (Linetype scale)
- **Celtscale** (Current entity linetype scale)
- **Color**
- **Layer**
- **Layer Previous**
- **Laymch** (Layer match)
- **Laycur** (Change to current layer)
- **Layiso** (Layer isolate)
- **Layuniso** (Layer unisolate)
- **Copy to Layer** (Copy object to new layer)
- **Laywalk** (Layer walk)
- **Layfrz** (Layer Freeze)
- **Layoff** (Layer off)
- **Laylck** (Layer lock)

- **Layulk** (Layer unlock)

- **Laymrg** (Layer merge)

- **Match Properties**

- **Properties**

- **Qselect** (Quick select)

Now that you're able to set up drawing units and limits, basic drawing components can be explored. The main component of the drawings you will be working with will be lines. This chapter will introduce you to various linetypes used on construction drawings and to methods for assigning and controlling varied linetypes. As linetypes are added to a drawing, a method of managing these linetypes must also be considered. Color can be assigned to lines to help distinguish types of building components. The **Linetype, Color,** and **Lineweight** commands will be introduced in this chapter. Once you have mastered these skills, **Layers** will be introduced to provide an efficient method of utilizing a variety of colors, linetypes and lineweights to represent drawing objects.

AN INTRODUCTION TO LINETYPES

Varied linetypes are used in every type of drafting and consulting firm involved in the construction process. This chapter will introduce common linetypes used to make construction drawings. It will explain how to load linetypes to be used in a drawing session and how to alter and control linetypes.

COMMON LINETYPES IN CONSTRUCTION DRAWINGS

Although each office may have its own standard, guidelines established by the American National Standards Institute (ANSI) and the National CAD Standards published by the National Institute of Building Sciences are used throughout much of the construction industry to ensure drawing uniformity. Methods of altering line width will be introduced in this chapter and discussed later in chapter 25 when the drawing is plotted. Common linetypes include object, hidden, center, cutting plane, section, break, phantom, extension, dimension, and leader. Examples of most of these types of lines can be seen in Figure 5.1.

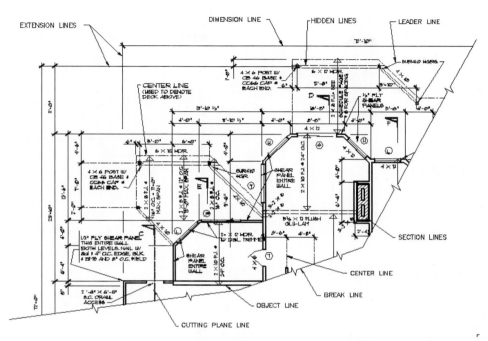

Figure 5.1 *Common linetypes used in construction drawings. Courtesy Residential Designs*

Object Lines

Object lines are continuous lines used to describe the shape of an object or to show changes in the surface of an object. Object lines can be thick or thin depending on how they are being used. Common line widths include:

- Object lines are 0.6 mm when representing walls or other key features.
- Object lines range in width from 0.0 to 0.3 when representing doors, windows, cabinets, or appliances.

Figure 5.2 shows both uses of object lines in a detail showing the change of floor elevations.

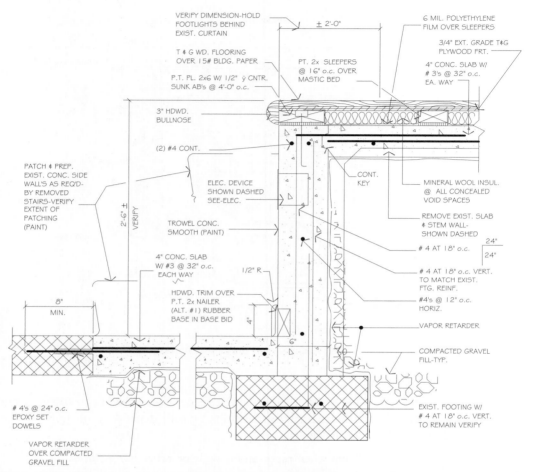

VERIFY DIMENSION-HOLD
FOOTLIGHTS BEHIND
EXIST. CURTAIN

± 2'-0"

6 MIL. POLYETHYLENE
FILM OVER SLEEPERS

3/4" EXT. GRADE T&G
PLYWOOD FRT.

T & G WD. FLOORING
OVER 15# BLDG. PAPER

PT. 2x SLEEPERS
@ 16" o.c. OVER
MASTIC BED

4" CONC. SLAB W/
3's @ 32" o.c.
EA. WAY

P.T. PL. 2x6 W/ 1/2" ÿ CNTR.
SUNK AB's @ 4'-0" o.c.

3" HDWD.
BULLNOSE

(2) #4 CONT.

CONT.
KEY

MINERAL WOOL INSUL.
@ ALL CONCEALED
VOID SPACES

PATCH & PREP.
EXIST. CONC. SIDE
WALL'S AS REQ'D-
BY REMOVED
STAIRS-VERIFY
EXTENT OF
PATCHING
(PAINT)

ELEC. DEVICE
SHOWN DASHED
SEE-ELEC.

REMOVE EXIST. SLAB
& STEM WALL-
SHOWN DASHED

2'-6" ± VERIFY

TROWEL CONC.
SMOOTH (PAINT)

4 AT 18" o.c.

24"
24"

4" CONC. SLAB
W/ #3 @ 32" o.c.
EACH WAY

1/2" R

4 AT 18" o.c. VERT.
TO MATCH EXIST.
FTG. REINF.

8"
MIN.

HDWD. TRIM OVER
P.T. 2x NAILER
(ALT. #1) RUBBER
BASE IN BASE BID

4"

#4's @ 12" o.c.
HORIZ.

VAPOR RETARDER

6"

COMPACTED GRAVEL
FILL-TYP.

4's @ 24" o.c.
EPOXY SET
DOWELS

EXIST. FOOTING W/
4 AT 18" o.c. VERT.
TO REMAIN VERIFY

VAPOR RETARDER
OVER COMPACTED
GRAVEL FILL

Figure 5.2   *Object lines are continuous lines used to describe the shape of an object or to show
changes in the surface of an object. Courtesy Architects Barrentine, Bates & Lee, A.I.A.*

 Note: AutoCAD will automatically control the spacing for each of the following linetypes
that require a pattern. Specific line sizes are given only to aid you in adjusting the default
segment length. Altering the segment length will be described later in this chapter.

Hidden Lines

Dashed or *hidden lines* are thin lines used to represent a surface or object that is hidden
from view. They can also be used to provide visible contrast with other features represented
by object lines. Figure 5.3 shows an example of hidden lines. Common hidden linetype
traits include:

- ANSI specifies that the lines should be 1/8" long with 1/16" space between line segments.

- A line width of 0.25 mm is recommended by NIBS for hidden lines.

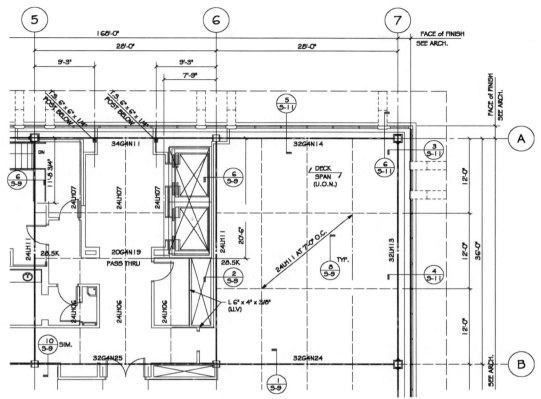

Figure 5.3   *Dashed or hidden lines are thin lines used to represent a surface or object that is hidden from view. On this plan, hidden lines are used to represent the roof outline and the location of major and supporting beams. Courtesy Van Domelen/Looijenga/McGarrigle/ Knauf Consulting Engineers*

Centerlines

A *centerline* is composed of thin lines that create a long-short-long pattern. Key components of a centerline include:

- The short line should be 1/8" long.
- The long segment must be between 3/4" and 1" long.
- The space between line segments should be 1/16".
- Centerlines should have a width of 0.25 mm.

Centerlines are used to locate the center axis of circular features such as drilled holes, bolts, columns, and piers. Figure 5.4 shows the use of centerlines on a foundation plan.

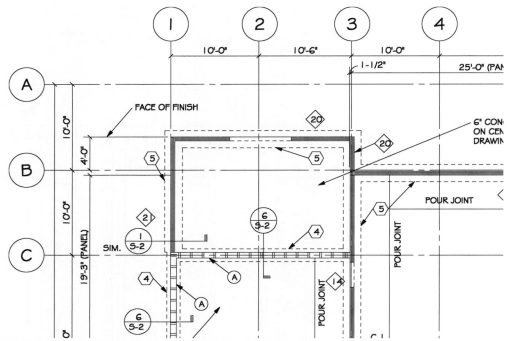

Figure 5.4 *A centerline is composed of thin lines that create a long-short-long pattern. Centerlines are used on this drawing to represent an imaginary grid that is used as an aid in reading the drawing. Courtesy Van Domelen/Looijenga/McGarrigle/Knauf Consulting Engineers.*

Cutting Plane Lines

A *cutting plane line* is placed on the floor or framing plan to indicate the location of a section and the direction of sight when viewing the section. Cutting planes are represented by a long-short-short-long line pattern. Key components of a cutting plane include:

- Cutting plane lines should be drawn with a width of 0.70 mm.
- Long line segments should be between 3/4" and 1" long.
- Short lines should be 1/8" long.
- The space between lines should be 1/16".
- Terminate the line with arrows to indicate the viewing direction.
- Place a letter at each end of the line using 1/4" high text to indicate the view or a detail reference bubble.

Figure 5.5 shows examples of cutting planes and line terminators.

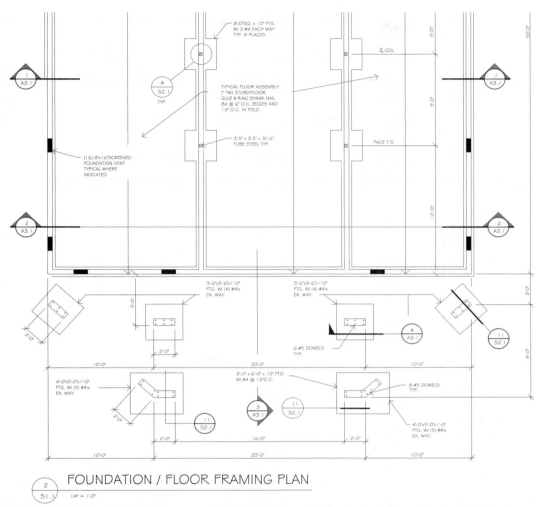

2 / S1.1   FOUNDATION / FLOOR FRAMING PLAN   1/4" = 1'-0"

Figure 5.5   *A cutting plane line is used to indicate the location of a section and the direction of sight when viewing the section. The line is thicker than object lines and must terminate with arrows to indicate the viewing direction or a detail reference bubble. Courtesy Scott R. Beck Architect*

Section Lines

When an object has been cut to reveal its cross section, a pattern is placed to indicate the portion of the material that has been cut by the cutting plane. In their simplest form, *section lines* are:

- Thin parallel lines used to indicate what portion of the object has been cut by the cutting plane.

- Usually drawn at a 45° angle but should not be parallel or perpendicular to any portion of the object that has been sectioned.
  - If an angle other than 45° is used, it should be between 15° and 75°.
- If two adjacent materials have been sectioned, section lines should be placed at opposing angles (see Figure 5.6).
  - Spacing between section lines can be altered, depending on the size of the materials to be sectioned.
  - A spacing of 1/8" should be provided between section lines.
  - Section lines should be placed using the **Hatch** command (see chapter 13).

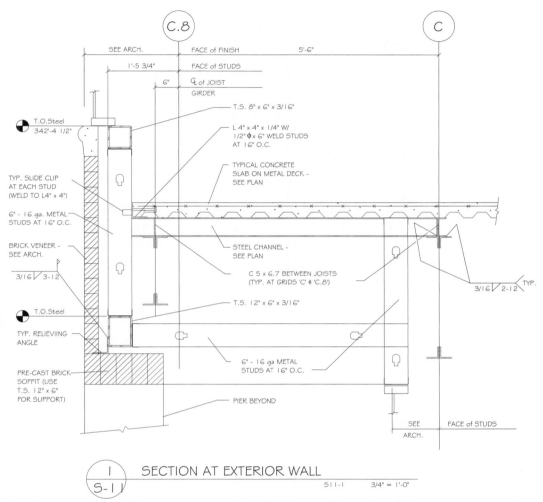

SECTION AT EXTERIOR WALL

Figure 5.6 *Section lines are thin parallel lines drawn at a 45° angle to indicate what portion of the object has been cut by the cutting plane. Section lines are generally used only to represent masonry in plan or sectioned views. Courtesy Van Domelen/Looijenga/McGarrigle/ Knauf Consulting Engineers*

Section lines are generally used only to represent masonry in plan or sectioned views. Most materials have their own special pattern. Figure 5.7 shows common patterns used to indicate that an object has been sectioned. AutoCAD refers to these patterns as hatch patterns. Chapter 13 will discuss representing sectioned material, guidelines for placement of section lines and hatch patterns, and methods of using the **Hatch** command.

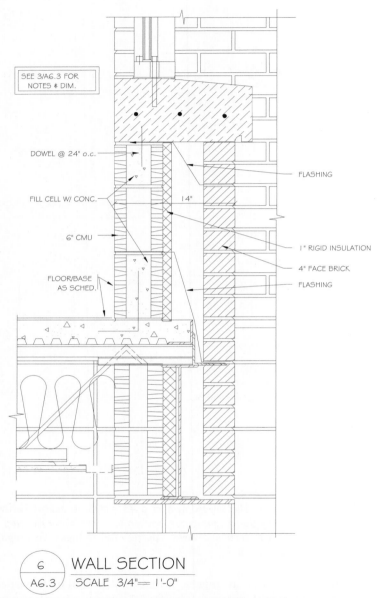

SEE 3/A6.3 FOR
NOTES & DIM.

DOWEL @ 24" o.c.

FILL CELL W/ CONC.

6" CMU

FLOOR/BASE
AS SCHED.

14"

FLASHING

1" RIGID INSULATION

4" FACE BRICK

FLASHING

6 / WALL SECTION
A6.3 / SCALE 3/4" = 1'-0"

Figure 5.7 *Common patterns used to indicate that an object has been sectioned. Most construction materials have their own special pattern. AutoCAD creates these patterns through the **Hatch** command. Courtesy G. Williamson Archer A.I.A., Archer & Archer P.A.*

Break Lines

Break lines are used to remove unimportant portions of an object from a drawing so that it will fit into a specific space. If you consider a 10' long steel column that supports a beam and rests on a concrete footing, the column is the same between the beam and the bottom plate. If you need to draw a 10' steel column at full scale in a 5' space, break lines can be used to remove 5' of the column. Three common break lines used on construction drawings including:

- Short break line are thick jagged lines placed where material has been removed (see Figure 5.8 at A).

- Long break lines are thin lines with a zigzag shape or inverted S shape inserted into the line at intervals (see Figure 5.8 at B).

- Cylindrical break lines are thin lines resembling a backward S (see Figure 5.8 at C).

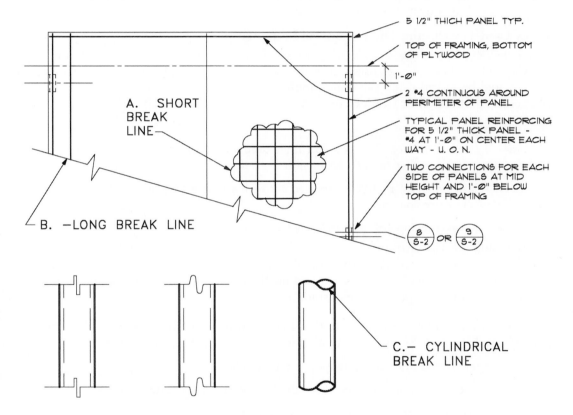

COMMON   BREAK LINES

Figure 5.8   *Common break lines of construction drawings: a short break line (A) is a thick jagged line placed where material has been removed; a long break line (B) is a thin line with a zigzag shape or inverted S shape inserted into the line at uniform intervals; a cylindrical break line (C) is a thin line resembling a backward S.*

Phantom Lines

Phantom lines are thin lines in a long-short-short-long pattern that are used to show motion, or to show an alternative position of a moving part. They can also be used in place of centerlines to represent upper levels or projections of a structure, or on site plans to represent easements or utilities. The linetype is represented by:

- Long line segments that are between 3/4"–1" long. Segments should remain a constant length within each drawing.
- Short line segments should be 1/8" long.
- The space between line segments should be 1/16".

Lines Used For Placing Dimensions and Annotation

Three types of lines are associated with referencing dimensions and annotation to a drawing. These lines are extension, dimension and leader lines. Extension and dimension lines should be placed using the dimensioning tools of AutoCAD. Leader lines should be placed using the **Quick Leader** command. Each command will be introduced in later chapters.

Extension Lines   *Extension lines* are thin lines used to relate dimension text to a specific surface of a part (see Figure 5.9). Extension lines should be placed so that:

- A space of 1/16" to 1/8" is provided between the object being described and the start of the extension line.
- They extend line 1/8" past the dimension line.

An extension line can cross other extension lines, object lines, center, hidden, and any other line, with one exception. An extension line can *never* cross a dimension line. *Never!* Chapters 16 and 17 will introduce other guidelines for placing extension lines using the various commands for placing dimensions.

Dimension Lines   *Dimension lines* are thin lines used to relate dimension text to a specific part of an object (see Figure 5.9). Dimension lines should be placed so that:

- They extend from one extension line to another.
- A line terminator is placed where the line intersects the extension line.
- The terminator is usually a tick mark or a solid arrowhead.

Dimension text is typically 1/8" high and should be placed to meet the following guidelines:

- Dimension text should be placed centered over the dimension line.
- The space between the dimension text and the dimension line should be equal to half the height of the dimension text.

Leader Lines   *Leader lines* are thin lines used to relate a dimension or note to a specific portion of a drawing. The following guidelines should be used to place leader lines:

- An arrowhead should be used at the object end of the leader line.

- A horizontal line approximately 1/8" long should be used at the note end of the leader line.

- Leader lines can be placed at any angle, but angles between 15° and 75° are most typical.

    - Vertical and horizontal leader lines should never be used.

    - The arrow end of the leader line should touch the edge of the part it describes.

- Leader lines should extend far enough away from the object it is describing so that the attached note is a minimum of 3/4" from the outer surface of the object. See Figure 5.9 and 5.10.

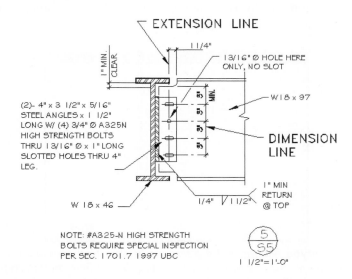

Figure 5.9   *Dimension lines are thin lines used to relate dimension text to a specific part of an object.*

COMMON LEADER LINES

Figure 5.10   *Leader lines are thin lines used to relate a dimension or note to a specific portion of a drawing. A terminator should be used at the object end of the leader line. Leader lines should extend a minimum of 3/4" from the object being described.*

CONTROLLING LINETYPES

AutoCAD contains many of the linetypes found in the construction industry in the Standard linetype library. To access these linetypes, use one of the following methods:

- Select **Linetype** from the **Format** menu.

- Type **LT** ENTER at the Command prompt.

- Select **Other** from the **Linetype Control** box on the **Properties** toolbar.

Each method will produce the **Linetype Manager** dialog box shown in Figure 5.11. Clicking the **Load** button will display the **Load or Reload Linetypes** dialog box that shows a portion of the available linetypes. See Figure 5.12. As you work with linetypes, it is important to remember that you can't use a linetype unless it has been loaded into your drawing. The linetypes shown in Figure 5.12 exist in a library file stored in AutoCAD. They will be loaded automatically as a function of the **Linetype Manager** dialog box when the **Linetype** command is used.

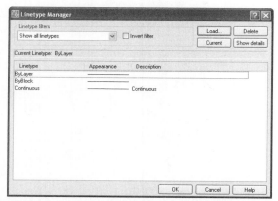

Figure 5.11   *The **Linetype Manager** dialog box can be used to change and load linetypes. The **Linetype** manager is accessed by selecting **Other** from the **Linetype Control** edit menu on the **Properties** toolbar, or selecting **Linetype** from the **Format** menu.*

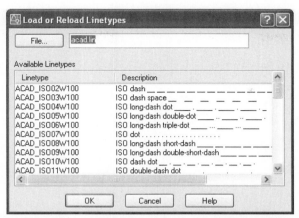

Figure 5.12   *Selecting the **Load** button will display the **Load or Reload Linetypes** dialog box that displays a list of AutoCAD linetypes. Use the scroll bar to see the balance of the list.*

LOADING LINETYPES INTO A DRAWING

As you display the **Linetype Manager** dialog box, you'll notice that three listings are displayed: ByLayer, ByBlock, and Continuous. Additional linetypes are loaded by using the **Load** button in the **Linetype Manager** dialog box. Use the following procedure to load the center, dashed, hidden, and phantom linetypes.

1. Display the **Linetype Manager** dialog box by selecting **Linetype** from the **Format** menu. Click the **Load** button to display the **Load or Reload Linetypes** dialog box shown in Figure 5.12.

2. Scroll through the listing and select the name, appearance, or description of the line to be loaded into the drawing. To load more than one linetype, press and hold CTRL while selecting the names of the desired linetypes with the select button of the mouse.

3. When you're through selecting the linetypes to be loaded into a drawing, click the **OK** button.

The dialog box will be removed, and the selected linetypes will be displayed in the **Linetype** window of the **Linetype Manager** dialog box (see Figure 5.13). The lines are also shown in the **Linetype Control** box on the **Properties** toolbar when the display arrow is selected. Figure 5.14 shows the linetypes currently loaded into the drawing base.

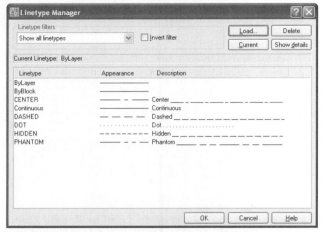

Figure 5.13 *Once a linetype has been selected to load, it will be displayed in the Linetype list.*

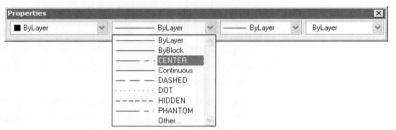

Figure 5.14 *The linetypes currently loaded into the drawing base can be found in the **Linetype Control** box on the **Properties** toolbar.*

MAKING A LINETYPE CURRENT

All you've done so far is to load linetypes into the drawing. To use the linetype, you must select one of the loaded linetypes as the current linetype. The following process will allow you to alter the lines that are used to create a drawing. Later in this chapter you'll be introduced to methods of assigning linetypes to a specific layer. Most users find linetypes easier to control when they are associated to a specific layer rather than scattered throughout one layer.

By default, you've been drawing with a continuous black line. Use the following procedure to alter the current linetype:

1. Select the linetype that will become the current linetype in the **Linetype Manager** dialog box. Select **Center** for now.

2. Click the **Current** button.

3. Click the **OK** button.

This process will make the selected linetype the current linetype to be used, close the dialog box, and return you to the drawing screen. Notice the name and pattern of the selected line will now be displayed in the **Linetype Control** box on the **Properties** toolbar. See Figure 5.15. The **Linetype Control** box can also be used to alter the linetype used to create objects. Click the arrow to display a list of current linetypes. Select the name of the desired linetype to be used. This will close the list and make the selected linetype the current linetype.

Figure 5.15   *The current linetype is displayed on the **Properties** toolbar. The **Linetype Control** box can be used to alter the linetypes that will be drawn.*

To display the linetypes that are loaded into a drawing, use one of the following methods:

- Select the **Properties** button on the **Standard** toolbar.
- Select **Properties** from the **Modify** menu.
- Type **PROPERTIES** ENTER at the Command prompt.

Each method will produce the **Properties** tool palette. Selecting the current linetype will display a list arrow similar to Figure 5.16a. Selecting the arrow will display a listing of the current linetypes similar to Figure 5.16b.

Note: New users of AutoCAD often expect the **Linetype** command to work miracles. Once a linetype has been made the current linetype, it will affect only drawing objects that are added after the linetype is changed. Objects drawn prior to the selection will not be affected. Objects drawn prior to the selection can be transformed to a different linetype using **Match Properties**, **Properties**, or the **Change** command.

Figure 5.16a *The linetypes that are loaded into a drawing can be displayed by selecting the* **Properties** *button on the* **Standard** *toolbar or by selecting* **Properties** *from the* **Modify** *menu.*

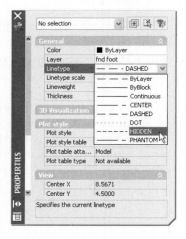

Figure 5.16b *Current linetypes can be displayed or altered by selecting the* **Linetype** *arrow.*

ALTERING THE LINETYPE SCALE

Enter the **Line** command and draw several lines. The new linetype should now be reflected on the drawing screen. If the line still appears as a continuous line, check the **Properties** toolbar to verify that you did select the current button. If the name and linetype show the desired linetype, most likely the linetype does not appear altered because you've altered the drawing limits from the original 12" × 9" screen. A centerline drawn in a 12" x 9" drawing area will appear as a centerline until you zoom in on the line. If the same centerline is used in a drawing area that is 50' wide, the centerline will be displayed as a continuous line.

To have the centerline appear as a centerline requires that the scale of the line be altered. Use the following steps to alter the linetype scale.

1. Redisplay the **Linetype Manager** dialog box and click the **Show Details** button. This will produce a dialog box similar to Figure 5.17.

2. Alter the **Global scale factor**. If you're working on a template set for 1/4"=1'–0" plotting, change the **Global scale factor** to 48.

3. Click the **OK** button.

As you return to the drawing screen, lines should now reflect the selected pattern.

 Note: Although it will be meaningless now, open your drawing template and type **PSLTSCALE** ENTER at the Command prompt. Change the default value from 1 to 0. This will ensure proper display of linetypes as drawings are plotted in the future. This process will need to be repeated for each layout that is created within a drawing.

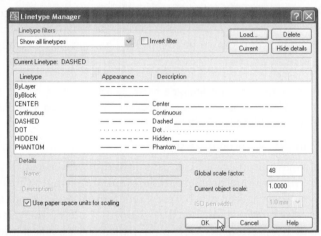

Figure 5.17   *Selecting the **Show Details** button of the **Linetype Manager** dialog box will display controls for linetype scaling. If you've altered the limits of the drawing screen from the 12" x 9", a line drawn with a centerline may appear to be drawn as a continuous line.*

You can also alter the segment length of each pattern by typing **LTS** ENTER at the Command prompt. (LTS represents linetype scale.) No matter how the value is entered, by default AutoCAD thinks of the selected length as one unit. This unit could be in inches, feet, millimeters, kilometers, or any other unit of measurement. Altering the value will allow you to alter the selected unit length. The distances specified in the linetype definition will be multiplied by the **Ltscale** value to produce a new length for all patterns. The command sequence is as follows:

```
Command: LTS ENTER
Enter new linetype scale factor <1.00>: 2 ENTER
```

If a value of 2 is entered, all line segments and spaces will be doubled in size. The effect of **Ltscale** on line segments can be seen in Figure 5.18.

ORIGINAL
LTSCALE ⎯ ⎯⎯⎯ ⎯ ⎯⎯⎯⎯

LTSCALE = 2
⎯⎯⎯ ⎯⎯ ⎯⎯⎯⎯

Figure 5.18 *What appears as a centerline on a 12" × 9" screen will appear as a continuous line when the limits are changed to 100' × 75'. The **Ltscale** command allows the length of line segments to be altered so that the pattern can be seen.*

The linetype scale is determined by the size the drawing will be when plotted. Using a linetype scale factor of 2 will make the line pattern twice as big as the line pattern created by a scale of 1. For a drawing to be plotted at 1/4"=1'–0", a dashed line that should be 1/8" when plotted, would need to be 3" long when drawn at full scale. This can be determined by:

- Multiply .125 × 24 (the desired height × scale factor).
- You can also determine the linetype scale factor by using half of the scale factor. The drawing scale factor is always the reciprocal of the drawing scale. Using a drawing scale of 1/4"=1'–0" would equal .25"=12".
- Divide 12 by .25 will produce a scale factor of 48 with a linetype scale value of 24.

Keep in mind that when the linetype scale is adjusted, it will affect ALL linetypes within the drawing. Usually this is not a problem, unless you are trying to create one unique line. Typically, the linetype scale factor can be preset as template drawings are set up. In most offices, template drawings are established based on the size of the paper or the scale the finished drawing will be plotted at. Common plotting scales and their respective scale factors include the following:

ARCHITECTURAL VALUES		ENGINEERING VALUES	
Drawing scale	LTSCALE	Drawing scale	LTSCALE
1"=1'–0"	6	1"=1'–0"	6
3/4"=1'–0"	8	1"=10'	60
1/2"=1'–0"	12	1"=100'	600
3/8"=1'–0"	16	1"=20'	120
1/4"=1'–0"	24	1"=200'	1200
3/16"=1'–0"	32	1"=30'	180

ARCHITECTURAL VALUES		ENGINEERING VALUES	
Drawing scale	LTSCALE	Drawing scale	LTSCALE
1/8"=1'–0"	48	1"=40'	240
3/32"=1'–0"	64	1"=50'	300
1/16"=1'–0"	96	1"=60'	360

CHANGING INDIVIDUAL LINE SEGMENT SIZES

Using **Ltscale** changes the values of *all* of the lines in a drawing. Using the **Celtscale** command (current entity linetype scale) allows the scale of existing lines to remain unaltered, but changes the scale of future lines. Typing **CELTSCALE** ENTER at the Command prompt will produce the prompt:

```
Enter new value for CELTSCALE <1.0000>:
```

The value that is assigned to Celtscale is used in combination with the value assigned to the linetype scale. The **Celtscale** acts as a multiplier of the linetype scale, producing a net scale effect on lines being drawn. With a linetype scale value of 2, and a celtscale of .25, the net scale factor would be $2 \times .25 = .50$. Figure 15.19 shows the effects of using **Celtscale** to control the current line width.

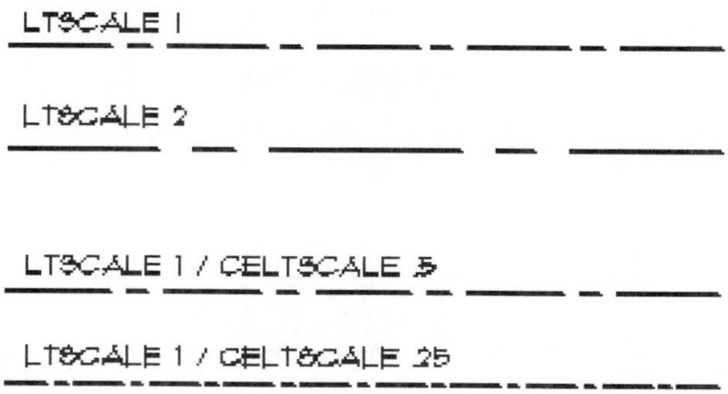

Figure 5.19   *The effects of **Ltscale** and **Celtscale** on line length.*

CONTROLLING LINEWEIGHTS

Varied line widths will help you and the print reader keep track of various components of a complex drawing. As you assign lineweight to drawings, remember that you're using thick and thin lines to add contrast, not represent thickness of an object. Chapter 11 will introduce an alternative method of using a thickness to accurately represent objects.

EXPLORING LINEWEIGHTS

Lineweights can be assigned to an object by selecting:

- **Lineweights** from the **Format** menu
- Typing **LW** ENTER at the Command prompt

Each method will produce the **Lineweight Settings** dialog box. Once you've changed some basic settings, lineweights can be altered using the **Lineweight Control** box on the **Properties** toolbar. For now, display the **Lineweight Settings** dialog box by selecting **Lineweights** from the **Format** menu. This will produce a display similar to Figure 5.20. Use the following NIBS guidelines for determining lineweight.

mm	Description	Use
0.18	Fine	Material indications, surface marks, hatch lines and patterns
0.25	thin	1/8" (3mm) annotation, setback, and grid lines
0.35	Medium	5/32" - 3/8" (4-10 mm) annotation, object lines, property lines
0.50	wide	7/32" - 3/8" (4-10 mm) annotation, edges of interior and exterior elevations, profiling cut lines
0.70	x-wide	1/2 - 1" (13-25 mm) annotation, match lines, borders
1.00	xx-wide	Major titles underlining and separating portions of designs
1.40	xxx-wide	Border sheet outlines and cover sheet line work
2.00	xxxx-wide	Border sheet outlines and cover sheet line work

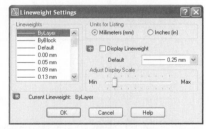

Figure 5.20 *The **Lineweight Settings** dialog box can be used to control drawing lineweights. Access the dialog box by selecting **Lineweights** from the **Format** menu or by typing **LW** ENTER at the Command prompt.*

ASSIGNING LINEWEIGHTS

The **Lineweight Settings** dialog box can be used to set the units for listing line width, the default lineweight, the display scale, and as a display of the current settings. As you display the dialog box, notice that the **Display Lineweight** check box is inactive. In its current state, you can draw using varied lineweights, but the width will be reflected only on a print and not in the drawing display. For most users, this should be the first adjustment you make. Select the **Display Lineweight** check box to make it active, and to display varied lineweights on the screen. You may be more comfortable working in inches rather than metric units. Make this change by choosing the **Inches** radio button. To assign a lineweight, scroll down the list of lineweights, find the desired lineweight, and click the **OK** button. As you use the **Line** command, any line added to the drawing will now reflect the current linetype and lineweight.

Once the unit and display values have been set, you can alter lineweights by selecting the **Lineweight Control** arrow on the **Properties** toolbar shown in Figure 5.21. Lineweights can also be toggled **ON/OFF** by clicking **LWT**, the **lineweight display control** button on the status bar.

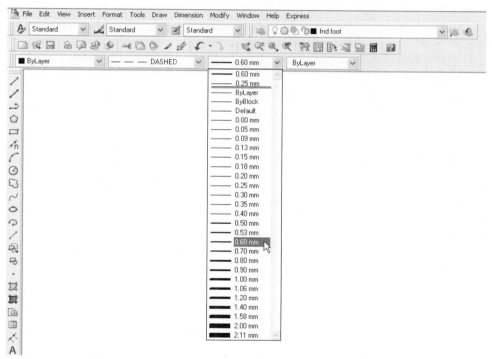

Figure 5.21 *Once the unit and display settings have been set, lineweights can also be altered by selecting **Lineweight Control** on the **Properties** toolbar.*

CONTROLLING COLOR

Up to this point, all of your drawings have been done with black lines on a white background, or white lines on a black background. The choice of screen background can be altered by the following procedure:

1. Select **Options** from the **Tools** menu. This will display the **Options** dialog box.

2. Select the **Display** tab in the **Options** dialog box.

3. Click the **Colors** button from the **Window Elements** portion of the **Display** tab. This will produce the **Drawing Window Colors** dialog box shown in Figure 5.22.

4. The **Color** edit box allows the color of each of the drawing window display elements to be altered. To change the color of the drawing area, highlight the **2D model space** option.

5. To change the color, select the arrow by the edit box and then select desired color tile.

6. Click the **Apply & Close** button to close the **Drawing Window Colors** dialog box.

7. Click the **OK** button in the **Options** dialog box. The dialog box will close and the screen will display the selected background color.

8. Alter other display features as desired, and then be sure to update your profile.

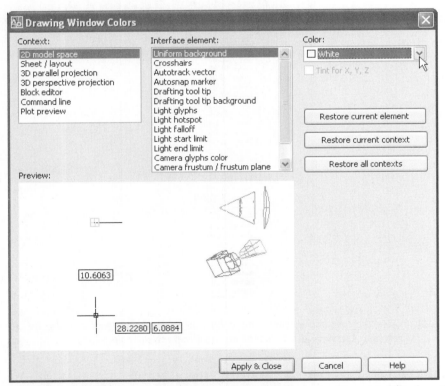

Figure 5.22 *The **Colors** button on the **Display** tab will produce the **Drawing Window Colors** dialog box. Select the desired workspace to be altered (2D model space), the element to be altered (background), and then select the desired color from the color menu.*

One of seven base colors can be selected, or the color palette can be displayed by selecting **Select colors** from the menu. The **Select Color** palette allows the color of drawing objects to be altered using one of 255 colors. A name or an AutoCAD Color Index (ACI) number ranging from 1 through 255 can identify the color for objects.

COLOR NUMBERS AND NAMES

In addition to changing the display features, each drawing object or layer can be assigned a color using the **Color Control** display. The seven basic colors used by AutoCAD are listed in Figure 5.23. Although only seven colors are shown in the **Color Control** display, AutoCAD identifies colors using numbers from 1 through 255, with the selection based on the color displayed in the color palette. Multiple objects or layers can have the same color number.

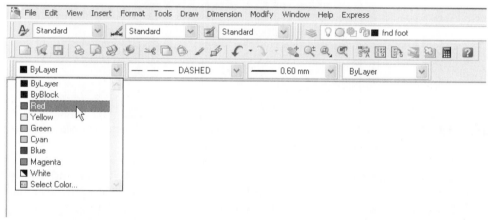

Figure 5.23   *The eight major colors of AutoCAD. The default color is number 7. Depending on the background color, number 7 will produce either a black or white line.*

SETTING OBJECT COLOR

The **Color** command can be used to assign colors to an entire layer or to individual objects on a layer. Setting colors for layers is the best method and will be explained as you proceed through the **Layer** options later in this chapter. The easiest method for assigning color to an object is to select the **Color** button from the **Properties** toolbar. You can select the color tile, the color name, or the down arrow. Each method will display a list similar to Figure 5.23. Selecting one of the listed colors will close the list and return the drawing display. Now if a line is drawn, it will be drawn using the selected color. Using the current settings, the line would be a red line drawn with a dashed linetype.

Select the **Color** button again to redisplay the color list. Notice the last option is **Select Color**. Selecting this option will display a menu similar to Figure 5.24. The **Select Color** dialog box can also be displayed by selecting **Color** from the **Format** menu or by typing **COLOR** ENTER at the Command prompt. The **Select Color** dialog box allows shades of the standard colors to be selected. Notice the current color is listed as red. Selecting any of the color tiles will alter the current color display tile and list the color by number.

Clicking the **OK** button will close the display and make that the current color. Using the **Line** command now will create lines using the selected color.

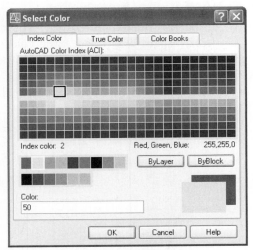

Figure 5.24 *Selecting **Other** from the **Color** list displays the color palette and allows a wide variety of color tints to be selected for drawing objects.*

Notice that the **Select Color** dialog box contains buttons for **ByLayer** and **ByBlock**. If **ByLayer** is accepted as the default, the color will be assigned once each layer has been defined. Most offices set colors using the **ByLayer** option rather than assigning colors to individual objects. The **ByBlock** option will be discussed in chapter 18.

True Color Tab

Selecting the **True Color** tab of the **Select Color** dialog box produces a display similar to Figure 5.25. The **True Color** tab allows over sixteen million colors to be displayed. When specifying true colors, you can use either an RGB (red, green, blue) or HSL (hue, saturation, luminance) color model. With the RGB color model, you can specify the red, green, and blue components of the color; with the HSL color model, you can specify properties of colors including the hue, saturation, and luminance aspects of the color. Figure 5.25 shows an example of the **True Color** tab with the **HSL** color model active. Figure 5.26 shows an example of the **True Color** tab with the **RGB** color model active.

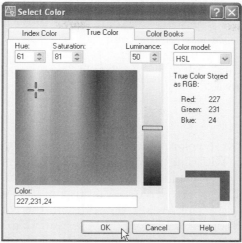

Figure 5.25   *Specify hue, saturation, and luminance on the **True Color** tab with the **HSL** color model active.*

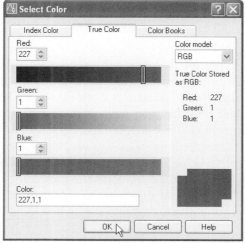

Figure 5.26   *Specify red, green, and blue on the **True Color** tab with the **RGB** color model active.*

The HSL Color Model   Hue describes the specific wavelength of light for a specific color. To specify a hue, move the crosshairs at the top of the image box from side to side or change the value in the **Hue** edit box. Valid hue values are from 0 to 360 degrees. *Saturation* specifies the purity of a color. High saturation causes a color to look more pure while low saturation causes a color to look washed-out. To specify color saturation, move the crosshairs (from top to bottom) over the color spectrum or specify a value in the **Saturation** box. Valid saturation values are from 0 to 100%. Notice that the saturation of red changes in the image box from red to a brownish color as the saturation is altered. *Luminance* specifies the brightness of a color. To specify color luminance, adjust the bar on the color slider or specify a value in the **Luminance** box. Valid luminance values are from 0 to 100%. A value of 0% represents the color black, 100% represents white, and 50% represents the optimal brightness for the color.

 Note: Changing the value for the hue, saturation or luminance will also affect the display of RGB colors.

RGB Color Model   The RGB color model in Figure 5.26 uses the colors of red, green, and blue to form other colors. The values specified for each component represent the intensity of the red, green, and blue components. The combination of these values can be manipulated to create a wide range of colors. Altering the **Red**, **Green**, and **Blue** slide bars specify the components of a color. Adjust the slider on the color bars or specify a value from 1 to 255 in one or more of the color edit boxes. In addition to the three-color slide bars, the **Color** edit box is used to specify the RGB color value. This option is updated when changes are made to HSL or RGB options. You can also edit the RGB value directly using the format of 000,000,000 where each group of numbers represents the value for red, green, and blue.

Color Books

This method of selecting colors for objects specifies colors using third-party color books or user-defined color books. AutoCAD includes several standard Pantone color books. You can also import other color books such as the DIC color guide or RAL color sets. To load a color book, use the **Color Book Locations** option in the **Options** dialog box, on the **Files** tab. Once a color book is loaded, you can select colors from the color book and apply them to objects in your drawings. Once a color book is selected, the **Color Books** tab will display the name of the selected color book. You can select a color book from the **Color Book** drop-down list of all the color books that are found in the **Color Book Locations** specified in the **Options** dialog box. To navigate through color book pages, select an area on the color slider or use the up and down arrows. The corresponding colors and color names are displayed by page as you navigate through the color book.

CONTROLLING LAYERS

 Other than the first few days of learning AutoCAD, you should *NEVER* create drawings without the use of layers. Layers allow:

- Different transparent levels to be stacked above each other so that information can be sorted by type.

- Items such as the walls of a structure to be separated from the text or dimensions.

- Drawing information can also be sorted based on content, such as placing all doors on one layer, and all windows on another.

- Information to be sorted by linetype, lineweight, and color.

- Once assigned to a layer, information can be temporarily removed from the drawing display by making one or more layers invisible.

Figure 5.27 shows the use of layers to display a valve assembly with the annotation hidden from display. Figure 5.28 shows the completed drawing with all required information.

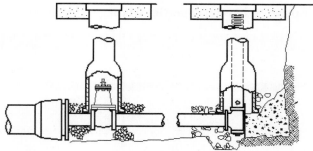

Figure 5.27 *The use of layers is similar to placing information on several sheets of paper that are stacked above each other.. This drawing shows the basic drawing with no supplemental information provided. Courtesy Department of Environmental Services, City of Gresham, Oregon.*

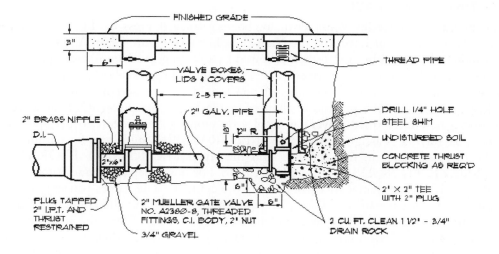

Figure 5.28 *Supplemental information added to the base drawing. Courtesy Department of Environmental Services, City of Gresham, Oregon.*

STARTING THE LAYER COMMAND

To start the **Layer** command, use one of the following methods:

- Select the **Layer Properties Manager** button on the **Layers** toolbar.
- Select **Layer** from the **Format** menu.
- Type **LA** ENTER at the Command prompt.

Each method will display the **Layer Properties Manager** dialog box shown in Figure 5.29.

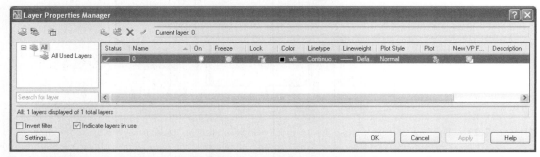

Figure 5.29   *Click the **Layer Properties Manager** button on the **Layer** toolbar or type **LA** ENTER tol display the **Layer Properties Manager** dialog box. The box is used to create and control the properties of each drawing layer.*

CREATING A NEW LAYER

Once the **Layer Properties Manager** dialog box is displayed, press ENTER. This will add the highlighted name of Layer1 to the **Name** edit box (see Figure 5.30). A new layer name can also be selected with a right-click followed by selecting the **New Layer** option. With **Layer1** still highlighted, a new layer name can be created by keyboard. Click the select button to accept the new layer name. If you would like to add multiple layers, press ENTER, which adds the previously changed name to the list. Press ENTER as second time and a new layer titled **Layer1** is displayed, allowing the process to be repeated. In Figure 5.31 **Layer1** is assigned the name of FLOR WALL. Guidelines for naming layers will be discussed later in this chapter. Add the layers FLOR ANNO, FLOR DOOR, FLOR GLZE, FLOR APPL, and FLOR CABS. For now, keep the names short, simple, and descriptive, using names that will identify the contents. The names you've just entered describe layers suitable for the floor plan template (FLOR) and the contents of specific layers (annotation, doors, windows, appliances, and cabinets). Once the name has been supplied, click the **OK** button to save the layer and return to the drawing area. You have created new layers, but the current layer is still 0. This can be confirmed by looking in the **Layer Edit** box that displays the name and status of the current layer.

RECEIPT

No. 696416

DATE 6-18-2010

RECEIVED FROM Stroud, Alexandra N.

Twenty and no/100 ——————— $ 20

○ FOR RENT

○ FOR co-pay

ACCOUNT		○ CASH	○ MONEY ORDER
PAYMENT	copay/cash	○ CHECK	FROM _____ TO _____
BAL. DUE		○ CREDIT CARD	BY _____

DOLLARS

Adams 2701

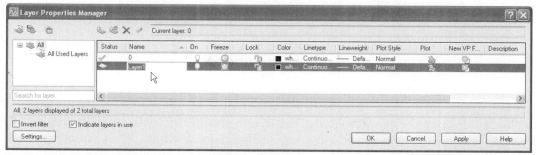

Figure 5.30 *Press ENTER, select the **New Layer** button, or right-click and select **New Layer** from the shortcut menu to display a new layer titled **Layer1**.*

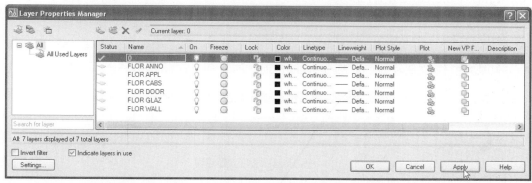

Figure 5.31 *The name LAYER 1 in Figure 5.30 was altered to reflect the desired layer name (FLOR WALL) and additional layer names were added. Guidelines for naming layers will be introduced later in this chapter.*

MAKING A LAYER CURRENT

Close the **Layer Properties Manager** dialog box and reenter the drawing area and add a few lines. The lines will be the same linetype and color as those drawn before you created the FLOR WALL layer. Redisplay the **Layer Properties Manager**. The dialog box displays 0, FLOR WALL, and the other layers that you created. The blue box highlights the 0 layer, indicating that it is the current layer. Use the following steps to make FLOR WALL the current layer:

1. Double-click the title FLOR WALL.
2. Click the **Apply** button.
3. Click **OK** to return to the drawing area.

Clicking the **OK** button will make the selected layer the current layer, close the **Layer Properties Manager** dialog box, and return you to the drawing area.

Note: You can also make a layer current by displaying the **Layer Properties Manager** and double-clicking the name of the layer to be made current, or by right-clicking and selecting **Set current** from the shortcut menu.

Notice that just to the right of the **Layers Properties Manager** button on the **Layers** toolbar is an edit box that describes the current layer status (see Figure 5.32). You'll see icons that describe the current status of the layers, with FLOR WALL listed as the current layer. Clicking the down arrow will create a display of all of the layers contained in the drawing, similar to Figure 5.33. This is a very convenient method for altering the current layer. To alter the current layer, select the down arrow, and then select the name of the layer you want as the current layer.

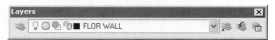

Figure 5.32   *The **Layer** box on the **Layers** toolbar describes the properties of the current layer.*

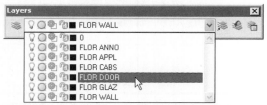

Figure 5.33   *Selecting the edit arrow displays the status of each drawing layer.*

ALTERING LAYER QUALITIES

Redisplay the **Layer Properties Manager** dialog box, and double-click the FLOR APPL layer. FLOR APPL has the same color, linetype, and lineweight qualities as the other layers. Layers become powerful drawing tools when varied linetypes, lineweights, and colors are assigned to specific layers. Creating a layer with thin, blue, hidden lines for appliances can be a great way of easily separating them from walls that are drawn with thick, continuous red lines.

Assigning Linetypes

The default linetype for each layer is a continuous line. By using the **Linetype** option of the **Layer Properties Manager,** all lines drawn on a layer can be assigned a selected linetype. Use the following steps to alter the linetype of the FLOR APPL layer:

1. Display the **Layer Properties Manager** dialog box.

2. Select the current linetype for the layer to be altered (FLOR APPL). This will produce the **Select Linetype** dialog box shown in Figure 5.34a. Only linetypes that have been loaded into the drawing will be displayed. Load the desired linetypes using

methods presented earlier in this chapter. (Load CENTER, DASHED, HIDDEN, and PHANTOM if you have not previously loaded these linetypes.)

3. Select the desired linetype to be used. For this example, HIDDEN will be assigned to the layer.

4. Click the **OK** button to close the **Select Linetype** dialog box.

5. Click **APPLY** followed by selecting **OK** to close the **Layer Properties Manager** dialog box and return to the drawing area.

Selecting one of the linetypes will use that linetype for all drawing objects created using that layer.

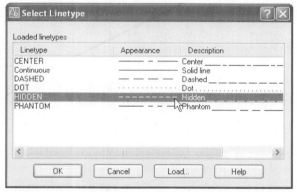

Figure 5.34a *Selecting the current linetype for the layer will produce the **Select Linetype** dialog box. Only linetypes that have been loaded into the drawing will be displayed. You can load additional linetypes into the drawing by selecting the **Load** button and following the guidelines presented early in this chapter.*

Altering Layer Color

Selecting the color display will produce the **Select Color** dialog box shown in Figure 5.24. The same tools that are available to assign color to one line can be used to assign color to an entire layer. Objects and layers can be assigned any color you desire, although most offices have a company standard. As you work through this chapter assign the following colors:

FLOR APPL	blue	FLOR GLAZ	green
FLOR CABS	160	FLOR WALL	red
FLOR DOOR	7	FLOR ANNO	232
0	20		

Altering Layer Lineweight

Earlier in this chapter you learned how to assign line width for individual lines. A similar process can be used to assign lineweight to an entire layer. Selecting the lineweight display for FLOR WALL will display the **Lineweight** dialog box similar to Figure 5.34b. Select the desired lineweight, and then select the **OK** button to close the **Lineweight** menu. The lineweight can also be assigned by double-clicking the desired value. Once the desired qualities of a layer have been assigned, select the **Set current** option if you want to use that layer when you return to the drawing editor. Select the **Apply** button to apply any alterations when the drawing area is restored. Select the **OK** button to close the **Layer Properties Manager** and return to the drawing screen.

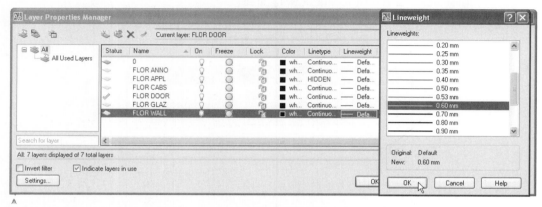

Figure 5.34b   *The lineweight of a layer can be adjusted by selecting the current lineweight display for that line. The **Lineweight** dialog box allows the line width to be altered. Select the desired lineweight and press the **OK** button in the Lineweight menu. Adjust any other options and then press the **Apply** button followed by the **OK** button to return the to the drawing area.*

 Note: Assigning colors, linetypes, and lineweights will affect all objects drawn on the layer that is altered, even objects drawn prior to making the changes.

CONTROLLING THE VISIBILITY OF LAYERS

Three options are displayed by each layer name. These options are used to control layer visibility. Each setting can be made from the **Layer Properties Manager** dialog box or from the **Layer Properties** edit display on the Layers toolbar.

Setting Layers On/Off

All of the new layers you have created up to this point are set to **ON**. If you were to switch to a layer such as FLOR DOOR, and then draw, you would be able to see what is drawn. In a multilevel structure it is sometimes helpful to view the walls of several different levels to check alignment, but typically, you will want to have only one level of a structure displayed at a time. If a layer is set to **OFF**, objects drawn on that layer will be invisible. The information is retained, but not displayed or plotted. The **Lightbulb** button is a toggle for the **ON/OFF** setting of a layer. For the drawing shown in Figure 5.27, the ANNO layer was toggled **OFF**, and for Figure 5.28, the ANNO layer was toggled **ON**.

Layer Off   Use the following steps to turn layers off:

- Display the **Layer Properties Manager** or the **Layer Control** edit box.
- Select the **ON/OFF** button of the desired layer (the lightbulb, with yellow representing **ON** and blue representing **OFF)**.
- Click the **OK** button (press ENTER if the edit box was used).

The dialog box will be removed or the edit box will be removed, and any information stored on an **OFF** layer will be removed from the screen. If you attempt to set the current layer to be turned **OFF** you will be given the following prompt:

> **Cannot freeze current layer**

If you accept the current layer OFF, the current layer will be made invisible. Now if you draw objects, they will be drawn but not displayed. This situation occurs more by accident than by design. If you find yourself drawing a line but nothing is displayed, check the listing to determine if the current layer has been accidentally switched **OFF**.

Layer On   Layers that have been made invisible by using the **OFF** option can be restored with the **ON** option. Display the **Layer Properties Manager** dialog box or the Layers edit box and highlight the desired layers to be set to **ON**. Select the blue lightbulb to turn it **ON** and click the **OK** button to restore the selected layers, remove the dialog box, and display the information on the selected layers.

Thawing and Freezing Layers

The second option, a sun, can also be used to toggle the visibility of a layer. If you have set layers to **OFF**, you removed information from the display screen but not from the drawing base. The information merely became invisible. When AutoCAD sifts through the information in your drawings, it looks at all of the information on all of the layers whether they are **ON** or **OFF**. Using the **Freeze/Thaw** option, you will be able to control the extent of information scanned by AutoCAD as it processes information. Layers that are frozen are not processed during certain functions of AutoCAD, reducing the time required to complete the task. On small drawings, this may only be a millisecond, but this will be a factor in complex drawing files.

Freeze   Layers are frozen and thawed using the same methods (different icons) used to set layers ON/OFF. Display the **Layer Properties Manager** dialog box, select the layer that you would like to freeze, and select the **Thaw** (sun) button, showing the Thaw mode. It will be changed to a snowflake, showing the Freeze mode. Clicking the **OK** button will activate the selection, restore the drawing screen, and remove the objects on the selected layer from the display.

Thaw   The **Thaw** option allows frozen layers to be thawed. To thaw a layer that is frozen, display the **Layer Properties Manager** dialog box and select the **Freeze** (snowflake) button of the layer to be thawed. As the snowflake is selected, it will change to the **Sun** button, indicating that the layer has been thawed. Clicking the **OK** button will activate the selection, restore the drawing screen, and display the material on the layer that is now thawed.

Safeguarding Layers

The final pair of options is **Lock** and **Unlock**. As the name implies, a lock is used to protect something. AutoCAD allows you to lock layers so that they cannot be edited accidentally. Information on a locked layer is still visible; it's just protected. To lock a layer, highlight the desired layer and select the **Lock** icon. Clicking the **OK** button will activate the lock the selection from editing, and restore the drawing screen.

To unlock a layer, highlight the desired layer and select the **Unlock** button, and click the **OK** button. This process will reverse the consequences of the Lock. Remember that locking a layer is not a security device, but a method of protecting information on a layer from careless editing. Locking a layer will allow you to view objects on the layer, but objects on a locked layer will not be included in a selection set of an edit command. You can continue to draw on a locked layer and objects can still be selected for OSNAP locations, thawed or frozen, or altered using the **Linetype** or **Color** button.

MANAGING LAYERS

The **Layer Properties Manager** and the Layer Panel in the Dashboard offer many options to create and manage layers efficiently. Chapter 23 will offer additional methods that affect the display of layers in viewports. Using layer filters allows you to sort layers by property or name. With layer groups, you can organize layers into categories and quickly apply property changes to all layers in the group. Utilizing the **Layer Properties Manager** allows you to:

- Create filters.
- Organize layers into groups.
- Save a layer state.

Creating Filters

Filters provide the ability to display certain layer names in the listing box, based on a specific quality. Layers can be filtered based on layer name, color, lineweight, contents, locked or unlocked status, freeze or thaw status, and plot styles. There are several other filter options that will not be considered until viewports and xrefs are explored in later chapters. Filters are controlled in the **Layer Properties Manager** using the **Named layer filters** edit box. The left side of the **Layer Properties Manager** dialog box contains a tree view of all the groups and filters. This is where the names of group filters will be displayed once the filters have been established. The right side of the display shows the names of the layers that will be in a selected group. Because no filters have been created yet, only the **All** option is currently displayed. If **All** is selected in the tree view, the list view displays all layers currently in the file. See Figure 5.35.

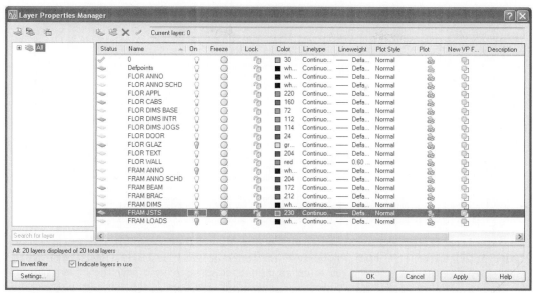

Figure 5.35 *Selecting the **New Property Filter** button, allows layers properties to be defined to create a filter.*

Creating a layer properties filter allows layers to be grouped based on similar properties such as names, colors, or status. For instance, all layers that contain the name FLOR as part of the title, all green layers, or all frozen layers can be displayed as a group. This provides an efficient and accurate method of modifying more than one layer at a time. Use the following steps to create a filter based on all layers with the properties ON and BLACK:

1. On the **Layers** toolbar, click the **Layer Properties Manager** icon.

2. In the **Layer Properties Manager** dialog box, click on the **New Property Filter** button to display the **Layer Filter Properties** dialog box as shown in Figure 5.36.

3. You can use one or more properties to define a filter. Properties include status, name, on or off, frozen or thawed, color, linetype, lineweight, and plot style. For this example ON and Black were used.

4. Click the **OK** button. This will close the **Layer Filter Properties** dialog box and return to the **Layer Properties Manager**. The new filter group is displayed in the tree view (left side). Clicking on the filter in the tree view will display the layers within the filter group in the list view (right side).

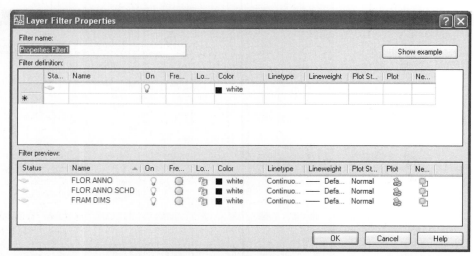

Figure 5.36 *Selecting the **New Property Filter** button accesses the **Layer Filter Properties** dialog box, allowing you to define the properties to create a filter. The current filter is displaying all layers that are ON and BLACK compared to the entire list of layers displayed in Figure 5.35.*

Organizing Layers into Groups

In addition to adding filters to layers by property, AutoCAD allows you to create group filters. Group filters allow you to manually arrange layers into groups. This is especially helpful when multiple drawings are contained in one drawing file. All layers for the floor, framing, electrical, and foundation plans can be stored in individual groups, allowing quick display of each layer group. Figure 5.35 shows the display of the layers assigned to the floor and framing drawings. Use the following steps to create a group filter for the floor and framing layers:

1. In the **Layer Properties Manager** dialog box, click on the **New Group Filter** button. A new group filter is created and displayed in the tree view.

2. Type in the name of your new group filter. Some examples include furniture, plumbing, electrical, and dimensions. In the tree view, click **All** to display all the layers in the drawing.

3. In the list view, hold down CTRL and select the layers you want to include in the new filter group.

4. Drag the layers into the new group filter in the tree view. as shown in Figure 5.37.

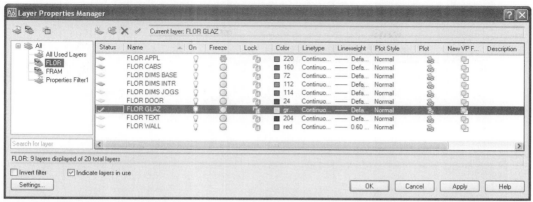

Figure 5.37   *Select the desired layers and drag them into the name of your new group filter. Once the group is made, selecting the FLOR listing from the tree now displays all of the layers associated with the floor plan. Selecting FRAM from the tree will display all layers assigned to the framing plan.*

Saving a Layer State

You can save the current layer settings in a drawing and then restore them later. This dialog box also allows layer states to be imported or exported to a different file. This may be useful when working on different disciplines of the project or on different stages of a drawing. Use the following steps to save a layer state:

1. In the **Layer Properties Manager**, select the **Layer States Manager** button. This will display the **Layer States Manager** dialog box shown in Figure 5.38.

2. Select the **New** button.

3. In the **New Layer State Name to Save** dialog box, enter the name and click **OK**.

4. In the **Layer States Manager** dialog box, click **Select All.**

5. Select **Close** and return to the **Layer Properties Manager**.

Restoring a Layer State

1. In the **Layer States Manager** dialog box, highlight the name of the group.

2. Select **Restore**.

3. Select **Apply** in the **Layer Properties Manager**. The changes are displayed immediately after you select the **Apply** button.

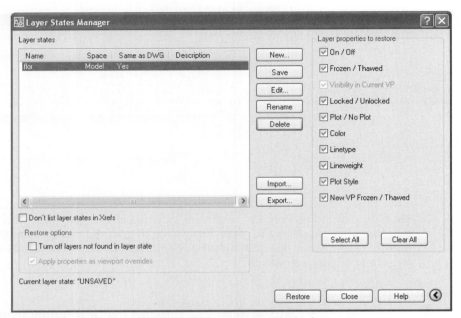

Figure 5.38 *Selecting **Select All** saves all the settings with the layer state.*

NAMING LAYERS

As your drawing ability increases, it is crucial to fully understand the importance of using layers. Layers are used like separate sheets of paper to separate information within a drawing file. The base components of a floor plan can be placed on one layer, and the text, dimensions, framing information, and plumbing can each be placed on separate layers to provide the information needed by various subcontractors. To help you control information, consideration must be given to layer names, controlling linetypes and colors by layer, as well as methods of displaying and hiding layer information.

STANDARDIZED LAYER NAMES

The U.S. National CAD Standards are incorporated into the *AIA CAD Layer Guidelines* that are used by many architectural firms. Copies of the standard can be obtained from the NIBS at http://www.nibs.org. The guideline recommends a name comprised of a Discipline Designator and a Major Group name. A Minor Groups name and a Status name can also be added to the layer name, depending on the complexity of the project. Because name components are common to both file and layer names, care must be taken to avoid creating project-specific references within a layer name. Project designations should not be included in folder names. Use the guidelines presented in chapter 4 for naming project folders and the following guidelines for naming layers.

Discipline Code

The discipline code was introduced in chapter 4. It is a character code that is used to identify the originator of the drawing. The discipline code will be the same for a model file, sheet file, or layer name. The code allows instant recognition of the originator of the drawing or layer. The code is one character that represents one of the following 16 disciplines:

Discipline Codes for the Drawing Originator			
A	Architectural	M	Mechanical
C	Civil	P	Plumbing
E	Electrical	Q	Equipment
F	Fire Protection	R	Resource
G	General	S	Structural
H	Hazardous Materials	T	Telecommunications
I	Interiors	X	Other Disciplines
L	Landscape	Z	Contractor/Shop Drawings

Major Group Name

The major group code is a four-character code that identifies a building component specific to the defined layer. Major group layer codes are divided into the major groups of architectural, civil, electrical, fire protection, general, hazardous, interior, landscape, mechanical, plumbing, equipment, resource, structural, and telecommunication. Codes such as ANNO (annotation), EQIP (equipment), FLOR (floor), GLAZ (glazing), and WALL (walls) are examples of major group codes that are associated with the architectural layers. A complete listing of major codes for each group can be found in the U.S. National CAD Standard.

Minor Group Name

The minor group code is an optional four-letter code that can be used to define subgroups to the major group. The code A–FLOR (architectural–floor) might include minor group codes for OTLN (outline), LEVL (level changes), STRS (stair treads or escalators), EVTR (elevators), or PFIX (plumbing fixtures). A layer name of A FLOR IDEN would contain room names, numbers, and other related titles or tags. A complete listing of minor group codes specific to each discipline is listed in the national standard. Figure 5.39 shows a listing of common layer names and their contents.

ARCHITECTURAL LAYER NAMES

ANNOTATION

A-ANNO-DIMS	DIMENSIONS
A-ANNO-KEYN	KEY NOTES
A-ANNO-LEGN	LEGENDS & SCHEDULES
A-ANNO-NOTE	NOTES
A-ANNO-NPLT	NON-PLOTTING INFORMATION
A-ANNO-NRTH	NORTH ARROW
A-ANNO-RBVS	REVISIONS
A-ANNO-REDL	REDLINE
A-ANNO-SYMB	SYMBOLS
A-ANNO-TEXT	TEXT
A-ANNO-TTLB	BORDER & TITLE BLOCK

FLOOR PLAN

A-FLOR-EVTR	ELEVATOR CAR & EQUIPMENT
A-FLOOR-FIXT	FLOOR INFORMATION
A-FLOR-FIXD	FIXED EQUIPMENT
A-FLOR-HRAL	HANDRAILS
A-FLOR-NICN	EQUIPMENT NOT IN CONTRACT
A-FLOR-OVHD	OVERHEAD ITEMS (SKYLIGHTS)
A-FLOR-PFIX	PLUMBING FIXTURES
A-FLOR-STRS	STAIR TREADS
A-FLOR-TPTN	TOILET PARTITIONS

ROOF PLAN

A-ROOF	ROOF
A-ROOF-OTLN	ROOF OUTLINE
A-ROOF-LEVL	ROOF LEVEL CHANGES
A-ROOF-PATT	ROOF PATTERNS

ELEVATIONS

A-ELEV	ELEVATIONS
A-ELEV-FNSH	FINISHES & TRIM
A-ELEV-IDEN	COMPONENT IDENT. NUMBERS
A-ELEV-OTLN	BUILDING OUTLINES
A-ELEV-PATT	TEXTURES & HATCH PATTERNS

SECTIONS

A-SECT	SECTIONS
A-SECT-IDEN	COMPONENT IDENT. NUMBERS
A-SECT-MBND	MATERIAL BEYOND SECTION CUT
A-SECT-MCUT	MATERIAL CUT BY SECTION PLANE
A-SECT-PATT	TEXTURES AND HATCH PATTERNS

CIVIL LAYER NAMES

ELEVATIONS

C-BLDG	PROPOSED BUILDING FOOTPRINT
C-PKNG	PARKING LOTS
C-PKNG-CARS	AUTOMOBILES
C-PKNG-DRAN	DRAINAGE INDICATORS
C-PROP	PROP. LINES & BENCHMARKS
C-PROP-BRNG	BEARINGS & DISTANCE LABELS
C-PROP-CONS	CONSTRUCTION CONTROLS
C-PROP-ESMT	EASEMENTS, RIGHTS- OF -WAY
C-ROAD	ROADWAYS
C-ROAD-CNTR	CENTER LINES
C-ROAD-CURB	CURBS
C-SSWR	SANITARY SEWERS
C-STRM	STORM DRAIN / CATCH BASINS
C-STRM-UNDR	UNDERGROUND DRAIN. PIPES
C-TOPO	PROPOSED CONTOURS & ELEV..
C-TOPO-BORE	TEST BORINGS
C-TOPO-RTWL	RETAINING WALL

STRUCTURAL LAYER NAMES

GENERAL LAYERS

S-ABLT	ANCHOR BOLTS
S-BEAM	BEAMS
S-COLS	COLUMNS
S-DECK	STRUCTURAL FLOOR DECKING
S-GRID	COLUMN GRIDS
S-GRID-DIMS	COLUMN GRID DIMENSIONS
S-GRID-EXTR	COLUMN GRIDS OUTSIDE OF BUILD.
S-GRID-IDEN	COLUMN GRID TAGS
S-JOIS	JOIST
S-METL	MISCELLANEOUS METAL
S-WALL	STRUCT. BEARING / SHEAR WALLS

FOUNDATION /SLAB PLAN LAYERS

S-FNDN	FOUNDATION
S-FNDN-PILE	PILES/ DRILLED PIERS
S-FNDN-RBAR	FOUNDATION REINFORCING
S-SLAB	SLAB
S-SLAB-EDGE	EDGE OF SLAB
S-SLAB-REBAR	REINFORCING
S-SLAB-JOIN	SLAB CONTROL JOINTS

Figure 5.39 *Common layer names and their contents based on the National CAD Standard Layer Guidelines.*

Status Code

The status code is an optional single-character code that can be used to define the status of either a major or minor group. The code is used to specify the phase of construction. The layer names for the walls of a floor plan (A WALL FULL) could be further described using one of the following status codes:

E	Existing to remain		N	New work
D	Existing to demolish		T	Temporary work
F	Future work		X	Not in contract
M	Items to be moved		1-9	Phase numbers

RENAMING LAYERS

The current name of a layer can be altered to better define the use of the layer at any time throughout the life of a drawing. Only the 0 layer and XREF layers can't be renamed. (XREF layers are layers that are created on a drawing that is attached to the current drawing, and they will be explained in later chapters.) Rename layers inside the **Layer Properties Manager** using the following steps:

- Highlight the name of the layer to be renamed.
- Right-click to display the shortcut menu.
- Select **Rename Layer**. As it is selected the name will be highlighted and a box will be placed around the original name with a blinking cursor.
- Press BACKSPACE while the name is highlighted to remove the existing name and allow a new name to be entered.
- Select **OK** to accept the new name, remove the dialog box and restore the drawing area.

DELETING LAYERS

An unused layer can be deleted from the drawing base at any time throughout the life of the drawing—with the exception of the 0 layer or an XREF layer. Layers that are referenced to a BLOCK and a layer named DEFPOINTS cannot be deleted, even if the layer is empty. XREF, DEFPOINTS, and BLOCK will be explored in later chapters. Delete an unused layer from inside the **Layer Properties Manager** using the following steps:

- Select the name of the layer to be removed.
- Select the **Delete Layer** button (the red X below the title bar) or right-click and select the **Delete** from the shortcut menu.
- An **X** will be placed in front of the layer to be removed, but the dialog box will remain open. Selecting the **APPLY** button will remove the selected layer.

NON-PLOTTING LAYERS

The final layer control to be considered in this chapter is a toggle between the **Plotting** and **Non-plotting** settings. Generally layers should be set to **Plotting**. Layers such as **JUNK** that contain notes to yourself can be set to **Non-plotting**. Information on a non-plotting layer will be displayed on the drawing screen but not on the print. This is an excellent method of making and keeping track of notes about the project. Chapter 25 will explore each of the layer control options in greater depth.

 Note: While you're still thinking about layers, enter your template and create a layer titled **VPORT**. Assign the layer a color of **RED** and make it a **NON-PLOTTING** layer. The importance of this layer will be explored as future chapters are explored.

ALTERING LINETYPE, COLOR, AND LAYERS

AutoCAD allows individual properties within a layer to be altered, as well as the properties of the entire layer to be changed. The **Layer Control** button on the **Layers** toolbar can be used to alter the qualities of an entire layer. Individual objects on a layer can be changed using the **Make Object's Layer Current** button, or by using the **Match Properties** or **Properties** toolbar, the **Properties** palette, and **Qselect** command.

LAYER CONTROL MENU

The **Layer Control** menu shown in Figure 5.40 can be used to quickly alter the visibility of an existing layer. Display the menu by selecting the **Layer Control** arrow on the right of the text box on the **Layers** toolbar. This menu can be used to toggle **ON/OFF**, **Thaw/Freeze**, **Lock/Unlock**, and to set the current layer.

- Adjust the visibility of layers by selecting the **ON/OFF** button or the **Thaw/Freeze** button. As the button for each layer is selected, the layer will be displayed or removed, depending on the setting of the button.

- Lock or unlock layers by selecting the button of the desired layer.

- Alter the current layer by selecting the name of the layer that you want to make current. As soon as the layer name is selected, the menu will be closed, the layer will be changed, and the drawing area will be restored.

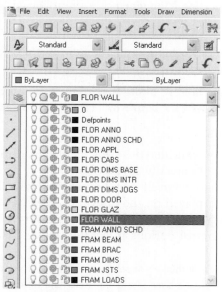

Figure 5.40 *Access the **Layer Control** menu by selecting the **Layer Control** edit arrow on the **Layers** toolbar. The menu can be used to toggle **ON/OFF**, **Thaw/Freeze**, and **Lock/Unlock**.*

MAKING AN OBJECT'S LAYER CURRENT

 The **Make Object's Layer Current** button is located at the right end of the **Layers** toolbar. The button can be used to alter the current layer. Selecting the button displays the following prompt at the Command prompt:

```
Select object whose layer will become current:
```

The cursor will be changed to a pick box as the program waits for you to select a drawing object. Selecting a drawing object will make the layer that contains the selected object the current layer. The edit box will then display the name of the current layer, and drawing can continue.

ALTERING LAYERS USING LAYER PREVIOUS

The **Layer Previous** command can be used to undo changes made to layer settings. The command will affect settings such as **On/OFF Freeze, Thaw, Lock, Color, Linetype, and Lineweight**. **Layer Previous** will not undo the renamed layers, deleted layers, or added layers. Execute the command by selecting the **Layer Previous** button on the **Layers** toolbar.

USING THE LAYERS II TOOLBAR

The Layer II toolbar displays ten additional methods for controlling layers. Several of these methods produce fairly advanced results and will only be introduced now. They'll be covered in depth in later chapters. Display the tool bar by right-clicking the empty space beside an existing docked toolbar, select ACAD, and then select Layers II. This will display the toolbar shown in Figure 5.41.

Figure 5.41   *The Layer II tool bar offers several options for controlling objects and their current layer. Access the toolbar by right-clicking in the space beside any docked toolbar, select ACAD, and then select **Layers II** from the menu.*

Each command is also available by selecting **Layer Tools** from the **Format** menu or by keyboard using the specified Command prompt. The following commands are listed in the order their icon is displayed in the default setting of the toolbar. These commands include:

Layer Match

The **Laymch** command can be used to change the layer of a selected object to match the destination layer. This command will remove an object from its current layer and place it on a new layer of your choosing. Once the icon is selected from the toolbar, the following prompt is displayed at the Command prompt:

Select objects to be changed: *(Select the desired object that will be changed to a new layer)*

Continue selecting objects as needed. When the selection set is complete, press the ENTER key. This will produce the following prompt:

Select object on destination layer or [*Name*]:

This is your opportunity to select the new layer in which the objects will be placed. Select an object on the desired destination layers, and the original object will be moved to the new layer.

Change to Current Layer

The **Laycur** command changes the layer of selected objects to the current layer. Once the icon is selected from the toolbar, the following prompt will be displayed:

Select objects to be changed to the current layer:

Select objects to be changed to the current layer and press ENTER when you are finished.

Layer Isolate

The **Layiso** command isolates the layer of selected objects so that all other layers are turned off. This command provides a very quick method of displaying the information for one specific layer. Once the icon is selected from the toolbar, the following prompt will be displayed:

Select objects on the layer(s) to be isolated or [Settings]:

Select an object and select ENTER to leave the layer of the selected object ON. All other layers will now be OFF, and their objects will be removed from the display. Layers can be restored individually using the **Layer Properties Manager**.

Select the **Settings** option by typing **S** ENTER at the Command prompt and the following prompt will be displayed:

In paper space viewport use [Vpfreeze/Off] <Vpfreeze>:

Remember that a viewport is the "hole" in the template that allows the drawing in model space to be viewed for plotting. These options determine whether layers are frozen in viewports; they will be explored in more detail in later chapters.

Layer Unisolate

The **Layuniso** command restores the layer display to the state just before the **Layiso** command was executed. Changes to layer settings after **Layiso** is used are retained when you enter the **Layuniso** command. This allows you to freeze layers, add objects to the drawing base, and then restore the balance of the drawings using **Layuniso**. If **Layiso** was not used, **Layuniso** does not restore any layers.

Copy Object to New Layer

The **Copytolayer** command copies existing objects to a new layer and places the copy in a new location. This command can be helpful when you want to copy an item such as a skylight on a floor plan to the roof plan. Once the icon is selected from the toolbar, the following prompt will be displayed:

```
Select objects to copy: (Select objects as desired and then
     press ENTER when the selection is complete)
```

Once you press ENTER, a prompt will be displayed requesting the destination layer to be specified. The destination layer is specified by selecting any object on the proposed layer. The prompt reads:

```
Select object on destination layer or [Name] <Name>: (Select an
     object that is on the layer to which the original object
     will be transferred)
```

Once the object to be copied and the proposed layer it will be placed on have been selected, a prompt will be given to specify a base point. A base point is an end or center point that can be selected using **Osnap**. The base point is similar to a handle on a briefcase. It is the convenient method for moving an object from one point to another. Future chapters will introduce other methods of selecting "handles." Once the base point is selected, the following prompt will be given to specify the new location for the base point:

```
Specify base point or [Displacement/eXit] <eXit>: (Select an end,
     mid, or center point of the object to be copied)
Specify second point of displacement or <use first point as
     displacement>: (Select the new location for the base point)
```

Once the displacement point is selected, the original object will be copied to its new location on the specified layer, and the command will be closed.

Layer Walk

Selecting the **Laywalk** icon will display a **LayerWalk** display similar to Figure 5.42. The box provides a display of the objects on the layers that you select in the **Layer** list. Currently all layers in the drawing are displayed, and the number of layers in the drawing is displayed in the dialog box title. Specific layers can be selected, and layers that are not selected will be removed from the display. You can change the current layer state when you exit, save layer states, and purge layers that are not referenced.

This command will be very helpful in the future when plotting a sheet containing multiple drawings. The **LayerWalk** dialog box can be used in a paper space viewport to select layers to turn on and thaw in the layer table and the current viewport. Any layer that is not selected in the **Layer** list is frozen in the current viewport. You can change the display of one viewport without altering the display of another viewport.

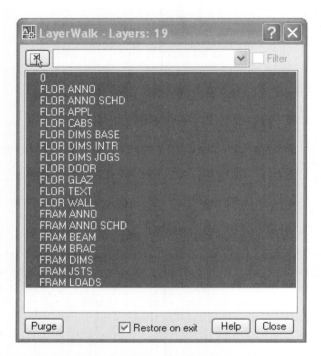

Figure 5.42 *The **Laywalk** command provides an alternative method for choosing what layers will be displayed. Its full power will be examined as viewports and plotting are explored*

Layer Freeze

The **Layfrz** command freezes the layer of selected objects. Once the icon is selected from the toolbar, the following prompt is displayed:

```
Select an object on the layer to be frozen or [Settings/Undo]:
```

By selecting an object, all objects on the same layer as the object will be frozen. Selecting the **Setting** option will produce the following prompt:

```
Enter setting type for [Viewports/Block selection]:
```

The results of selecting the **Viewport** or **Blocks** options will be explored in future chapters.

Layer Off

The **Layoff** command turns the layer of the selected object off. The following prompt is displayed as the command is entered:

```
Select an object on the layer to be turned off or
    [Settings/Undo]:
```

As objects are selected, the layer of those objects is turned to the OFF position. The prompt will continue to be displayed until the ENTER key is pressed to end the command. If the Setting option is selected, options will be displayed for controlling layers in paperspace viewports, or for controlling layers of selected objects. Each setting will be explored in later chapters.

Layer Lock

The **Laylck** command locks the layers of selected objects. As the command is selected, the following prompt will be displayed:

```
Select an object on the layer to be locked:
```

Selecting an object will automatically lock that layer so that it can't be edited until it is unlocked. A prompt will be displayed showing the name of the layer that has just been locked, and the command line is returned. As you move the cursor over objects on a locked layer, a lock will be displayed beside the cursor to remind you of the setting.

Layer Unlock

The **Layulk** command unlocks the layer of selected objects. As the command is selected, the following prompt will be displayed:

```
Select an object on the layer to be unlocked:
```

Selecting an object will automatically unlock that layer so that it can be edited. A prompt will be displayed showing the name of the layer that has just been unlocked, and the command line is returned.

Layer Merge

Although not on the layer toolbar, the **Laymrg** command merges selected layers onto a destination layer. The command can be started by selecting **Layer Merge** from **Layer Tools** listing of the **Format** menu or by typing **LAYMRG** ENTER at the Command prompt. Each will produce the following prompt:

```
Select object on layer to merge or [Name]: (Select an object to
     represent the layer to be merged to another layer)
```

The prompt will continue to be displayed to allow additional layers to be selected for the merge. Press the ENTER key when the selection set is complete. The name of the selected layer will be displayed, and a prompt to select the destination layer will be displayed.

```
Select object on target layer or [Name]: (Select the layer to
     receive the selected material)
```

Once a destination layer has been selected, a prompt will be displayed reminding you that you're about to do something serious.

```
******** WARNING ********
You are about to merge layer "FLOR DIMS INTR" into layer
     "FLOR CABS".
Do you wish to continue? [Yes/No] <No>:
```

Selecting **Y** ENTER will complete the layer merge and restore the Command prompt for your next command. Selecting **N** ENTER will terminate the command, with no changes in the existing layer status. When in doubt, select **Y** ENTER. If you're unhappy with the results, use the UNDO command.

ALTERING THE DISPLAY WITH MATCH PROPERTIES

The properties of one layer object can be transferred to another object by the use of the **Match Properties** command. The command will copy the properties of a selected object to additional objects, including the layer, the linetype, and the color. The command is started using one of the following methods:

- Select the **Match Properties** button on the **Standard** toolbar.
- Type **MA** ENTER at the Command prompt.
- Select **Match Properties** from the **Modify** menu.

Each method will produce the following prompt:

```
Select Source Object: (Select the object with the properties that
    you would like to copy.)
Select destination object(s) or (Select the object that you would
    like to change.)
Select destination object(s) or ENTER
```

Instead of selecting the object that you would like to change, typing **S** ENTER for Settings at the second prompt will produce the **Property Settings** dialog box shown in Figure 5.43. The box can be used to specify the qualities that will be copied to selected objects. One or more of the properties can be assigned including the **Color, Layer, Linetype, Linetype Scale, Lineweight,** and **Thickness**—as well as several other qualities that will be discussed in later chapters. Once the qualities have been selected, click the **OK** button and resume the command sequence by selecting the object to receive the new properties.

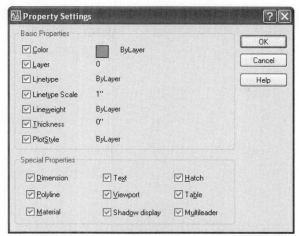

Figure 5.43   *Once a source object has been selected, typing **S** ENTER for **Settings** at the Command prompt will display the **Property Settings** box. The dialog box allows specific properties of a layer to be altered when using **Match Properties** without affecting other layer qualities.*

CHANGING OBJECT PROPERTIES

The **Properties** command can be used to control the properties of existing objects. To start the command, use one of the following methods:

- Select the **Properties** button on the **Standard** toolbar.
- Type **MO** ENTER at the Command prompt.
- Select **Properties** from the **Modify** menu.

Each method will display the **Properties** palette similar to Figure 5.44A, showing the qualities of the current layer. Object properties can be listed in alphabetical order or categorized by groups depending on which tab is selected. The palette allows drawing properties, including the Color, Layer, Linetype, Thickness, and Linetype scale, to be altered. To alter a property for the selected object, choose either the name or value of the desired property. Each property will display a specific listing. Figure 5.44B shows the list for altering lineweights. Once the desired properties are altered, select the **Close** button to remove the palette and return to the drawing.

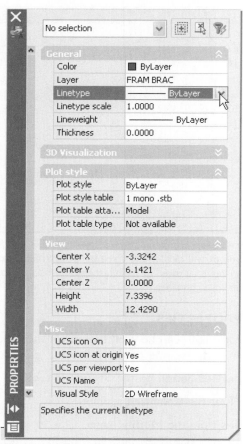

Figure 5.44a *The **Properties** palette can be used to control the properties of existing objects.*

Figure 5.44b *To alter a property for the current object, choose either the name or value of the desired property. Each property will display a specific listing. Note that each area of the palette except the **General** portion has been collapsed.*

CHANGING OBJECTS WITH QSELECT

The **Qselect** command can be used to create selection sets for altering the drawing objects. You can create selection sets to either include or exclude objects matching the specified object type and criteria. For instance, **Qselect** allows you to make a selection set made up of all objects that are green, or a selection set could be made to include everything except green objects. To start Qselect, use one of the following methods:

- Right-click **Quick Select** from the shortcut menu.
- Select the **Quick Select** button in the **Properties** palette.
- Select **Quick Select** from the **Tools** menu.
- Type **QSELECT** ENTER at the Command prompt.

Each method will produce the **Quick Select** dialog box shown in Figure 5.45. The dialog box contains eight key areas for adjusting the selection set:

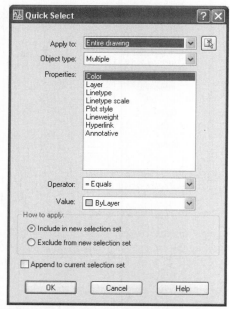

Figure 5.45 *Display the* **Quick Select** *dialog box by right-clicking and choosing* **Quick Select** *from the shortcut menu, or by selecting the* **Quick Select** *button from the* **Properties** *palette.*

Apply to—If a selection set has been defined and is current, it will be selected automatically in the **Apply to** box. If no selection set has been defined, by default the only option is to apply changes to the entire drawing. Choosing the **Select Objects** button creates selection options.

Select Objects—The **Select Objects** button is located beside the **Apply to** edit box. Selecting the button will remove the **Quick Select** dialog box temporarily so that items can be selected to apply the selection filters to.

Object type—This box specifies the object type for the filter. Until a selection set is created, the **Object type** box displays a list of each shape included in the drawing, similar to Figure 5.46. Once a selection set is created, the list will include only the object types contained in the selection set.

Figure 5.46 *The **Object type** box will only display a list of each shape included in the drawing. Once a selection set is created, the list will include only the object types contained in the selection set.*

Properties—The **Properties** list shows all searchable properties for the selected object to be used with the filter. The **Properties** display will vary depending on the object selected in the **Object type** edit box. If a circle is selected in the **Object type** edit box, a listing similar to Figure 5.47 will be displayed.

Figure 5.47 *The **Properties** display provides a list of all searchable properties for the selected object to be used with the filter. The display will vary depending on the object selected in the **Object type** edit box and the operator and value selected.*

Operator—The **Operator** box controls the filter range and is dependent on the selected property. Operator options include: = Equals, < > Not Equal, > Greater than, < Less than and Select all (see Figure 5.46).

Value—The **Value** edit box allows a value for the filter to be entered. For instance, if the property is a circle, a value of 24" could be used to include all circles equal to, not equal to, greater than, or less than 24" to define the selection set.

How to Apply—The **How to Apply** area contains radio buttons for including or excluding objects from the selection set.

- When **Include in new selection set** is active, a new selection set will be created composed of objects that match the filtering criteria.

- When **Exclude from new selection set** is active, a selection set will be created composed of objects that do not match the filtering criteria.

Append to current selection set—This check box determines if the selection set replaces or appends the current selection set.

Creating a Selection Set Using Qselect

A selection set for editing all green objects can be created using the following steps:

1. Display the **Quick Select** dialog box using one of the methods described earlier.

2. Using the **Apply to** edit box, select **Entire drawing**.

3. Select **Multiple** for the **Object type**.

4. Select **Color** in the **Properties** edit box.

5. Select **=Equals** in the **Operato**r edit box.

6. Use the down arrow beside the **Value** edit box, and select **Green**.

7. Select **Include in new selection set** in the **How to apply** area.

8. Click the **OK** button to close the box.

AutoCAD will close the **Quick Select** dialog box and provide grips on each green object in the entire drawing. The Command prompt will show the number of items in the selection set, similar to Figure 5.48.

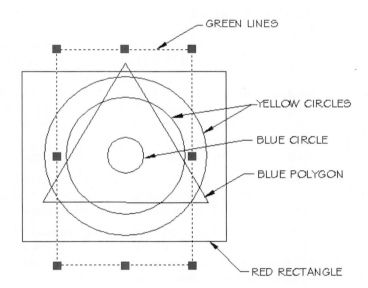

Figure 5.48   *Qselect can be used to create a selection set made up of all objects of a specific property description. In this example, all objects that are green have been selected for editing.*

Excluding Objects from the Selection Set

A selection set excluding objects of a specific property can be created. Using the objects in Figure 5.48, a selection set excluding all circles larger than 24" could be created using the following sequence:

1. Display the **Quick Select** dialog box using one of the methods described earlier.

2. Using the **Apply to** edit box, choose **Current Selection**.

3. Select **Circle** for the **Object type**.

4. Select **Diameter** in the **Properties** edit box.

5. Select **>Greater Than** in the **Operator** edit box.

6. Enter a value of 24.

7. Select **Exclude from new selection set** in the **How to apply** area.

8. Click the **OK** button to close the box.

AutoCAD will close the **Quick Select** dialog box and remove from the selection set all circles with a diameter greater than 24".

CHAPTER 5 EXERCISES

Open the Architectural drawing template, set the **Ltscale**, and set the scale values appropriate to 1/4"=1'–0". Set the units to Architectural, with 1/16" fractional value. Establish layers that would be suitable for a one-level residential floor plan. Load CENTER, DASHED, HIDDEN and PHANTOM linetypes. Include the following: walls, dimensions, architectural notes, structural text, joist, beams, furniture, appliances, planting, plumbing, cabinets, section tags, doors and windows, decks at each level, electrical fixtures, and electrical wiring. No provisions need to be made for other drawings that typically are grouped with a floor plan. Assign colors to each layer. Assign linetypes and weights appropriate to the layer contents. Save the drawing as ARCHBASE. This will become the basis for future floor plan drawings that are to be plotted at 1/4"=1'–0".

Set limits, units, snap, and grid to best suit each project. Use appropriate linetypes and lineweights that best suit each project. Dimensions, hatch patterns, and text do not need to be placed at this time. Save the following exercises in safe locations.

1. A parcel of land is to be developed. Use the following metes and bounds supplied by the surveyor and draw the site plan and the indicated utility easement. Set limits to 200',150'. Use a phantom line to represent the property line and use a centerline to represent the easement. Select different colors and layers for each linetype.

   SITE. Beginning at a point to be described as the northwesterly corner, 100.00' due south, thence 37.52' N89°37'E, thence 96.85' N58°40'52"E, thence 57.25' N3°50'40"W, and then back to the true point of beginning.

   EASEMENT. The easement is approximately an inverted L-shaped easement. The long leg of the "L' is 10' wide; the short leg is 15' wide. Layout the easement, beginning at a point that lies on the southerly property line, 20'–0" from the southwesterly corner of the property. Thence northerly 70'–0" along the centerline of said easement to a point at N7°32'15"W, thence a 15' wide easement lying along a centerline extending to the westerly property line at N50°55'04"W. Save the drawing with the file name **E5-1 SITE**.

2. Create a template drawing containing appropriate values for an architectural drawing. Include settings for each of the options discussed in previous chapters. Set limits and paper sizes assuming 24" × 36" plotting paper. Load linetypes and lineweights for lines that might be necessary on a plan view. Create a layer titled **FLOR WALL**. Save the file as a **PLAN TEMPLATE** drawing.

3. Use the drawing on the next page to complete a side view of a steel beam resting on a 6"×12"×3/4" steel plate supported on a steel column. Assume the beam to be 18" deep, and show the thickness of each plate as 7/8" thick. Provide a 10" long support plate, and an 18" high stiffener plate at the end of the beam. The support column is a TS 6" × 6" × 3/8". Provide a 4" × 4" × 3/8" TS brace @ 45° and a 4" × 12" × 12" × 3/8" gusset plate. Provide appropriate break lines where needed. Save the drawing as **E-5-3**.

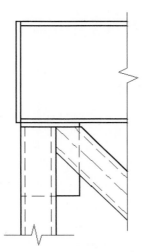

4. Use the following drawing to complete a side view of a steel beam resting on a steel plate supported on a steel column. Assume the beam to be 16" deep, and show the thickness of each plate as 5/8" thick. Provide a 14" long support plate. The support column is a TS 6" × 6" × 3/8" with a TS 6" × 6" × 3/8" chevron at a 45° angle. Provide centerlines for each TS. Provide appropriate break lines where needed. Save the drawing as **E-5-4**.

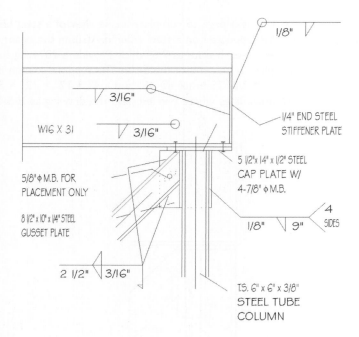

1/8"

3/16"

1/4" END STEEL
STIFFENER PLATE

WI6 X 31

3/16"

5/8" φ M.B. FOR
PLACEMENT ONLY

5 1/2"x 14" x 1/2" STEEL
CAP PLATE W/
4-7/8" φ M.B.

8 1/2" x 10" x 1/4" STEEL
GUSSET PLATE

1/8"   9"   4
                SIDES

2 1/2"   3/16"

T.S. 6" x 6" x 3/8"
STEEL TUBE
COLUMN

5. Use the following drawing to complete a side view of a steel column resting on a steel plate that is supported on 1" mortar over a 24" ×24" concrete pedestal. Assume the thickness of the plate as 5/8" and show 3/4" × 14" long anchor bolts 1 1/2" from each edge of the plate. Provide a 14"×10"×3/8" support plate. The support column is a TS 6" × 6" × 3/8" with a TS 6" × 6" × 3/8" chevron at a 45° angle. Provide centerlines for each TS. Provide appropriate break lines where needed. Save the drawing as **E-5-5**.

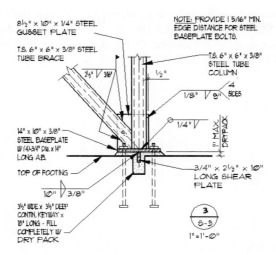

8½" x 10" x 1/4" STEEL
GUSSET PLATE

T.S. 6" x 6" x 3/8" STEEL
TUBE BRACE

NOTE: PROVIDE 1 5/16" MIN.
EDGE DISTANCE FOR STEEL
BASEPLATE BOLTS.

T.S. 6" x 6' x 3/8"
STEEL TUBE
COLUMN

14" x 10" x 3/8"
STEEL BASEPLATE
W/ (4) 3/4" DIa. x 14"
LONG A.B.

TOP OF FOOTING

3½" WIDE x 3½" DEEP
CONTIN. KEYWAY x
18" LONG - FILL
COMPLETELY W/
DRY PACK

3/4" x 2½" x 10"
LONG SHEAR
PLATE

3
S-3

1"=1'-0"

6. Use the following drawing to draw a partial foundation plan and place a detail
marker. Show the outline of the footing with thin dashed lines and show the outline
of the slab with a thick, continuous line. Draw a side view for a concrete slab
foundation. Save the drawing as **E-5-6**.

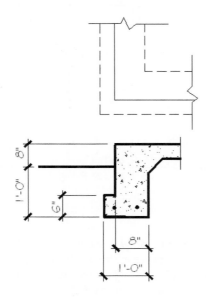

7. Use the drawing from Exercise 6 to obtain the needed sizes to draw a partial foundation plan and place a detail marker. Show the outline of the footing with thin dashed lines and show the outline of the 8"wide stem wall with thick, continuous lines. Save the drawing as **E-5-7**.

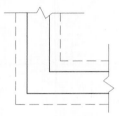

8. Use the following drawing to draw a site plan. Create your own north arrow. Save the drawing as **E-5-8**.

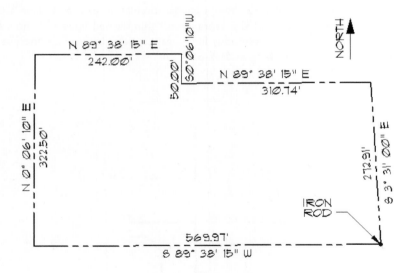

9. Use the following drawing to draw a section of a wall constructed of 8" concrete blocks. Use thin lines to represent the block. Show a #5 rebar centered in the wall using a thick dashed line. The individual blocks and hatch pattern do not need to be shown at this time. Save the drawing as **E-5-9**.

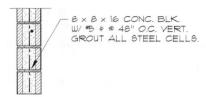

8 × 8 × 16 CONC. BLK.
W/ #5 ∅ @ 48" O.C. VERT.
GROUT ALL STEEL CELLS.

10. Use the following drawing to draw a 24' wide × 26' high concrete wall panel with a 10' × 12' door and a 3' × 7' door. Place each door 2' from the panel edge. Use thin lines to represent the wall outline and openings. Show the finish floor 12" above the bottom of the panel. Show a portion of the panel broken out to reveal #5 @ 12" o.c. horizontal and #5 @ 15" o.c. vertical. In addition to the standard reinforcing, draw lines to represent #5 rebar in the following locations:

- 2" in from each edge of the panel

- 2" clear from each edge of each opening

- Extend the vertical bars to within 2" of the top and bottom of the panel.

- Extend the horizontal door steel 12" past each opening.

Save the drawings as **E-5-10**.

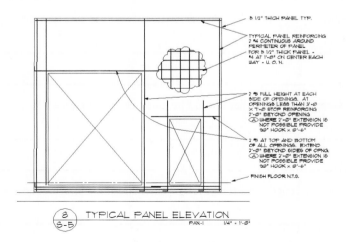

TYPICAL PANEL ELEVATION

CHAPTER 5 QUIZ

DIRECTIONS

Answer the following questions with short complete statements. Type your answers using a word processor.

1. Place your name, chapter number, and the date at the top of the sheet.

2. Type the question number and provide the answer in the form of a statement.

 Warning: Some of the questions have not been covered in the reading material and will require the use of the help menu. You may also have to do some exploring to answer the questions.

1. List the methods for accessing linetype. Which is your favorite method? Which is your least favorite method? Explain each answer.

2. List two possible uses for an object line.

3. Describe the length and thickness of each line used in a centerline pattern.

4. List examples of lines that should be thick on a floor plan.

5. Describe the length and thickness of each line used in a hidden line pattern.

6. Describe the use of a section cutting plane line.

_____

_____

7. A note must be placed to describe a feature. Name and describe the type of line to be used.

_____

_____

8. A length of 24'–6" must be placed to describe the length of a room. What kind of line is used to reference the text to the drawing? In addition to this line, what else is needed to complete the correct placement?

_____

_____

9. Sketch how long and short break lines should be drawn.

_____

_____

10. List nine types of lines often found on construction drawings.

_____

_____

11. Describe the differences between a section line and a section cutting plane line.

_____

_____

12. A portion of a beam needs to be removed so that it will fit in a drawing. Name and describe the line used where the portion of the object has been removed.

_____

_____

13. List four objects that are affected by a change in linetype.

_____

_____

14. What command will allow the line pattern to be altered?

_____

_____

15. What would be the line scale value for a line that is to be plotted at a scale of 1" = 20'?

_____

_____

16. List four options for controlling the visibility of a layer.

_____

_____

17. You've just loaded five linetypes, and have entered the **Line** command. Only continuous lines are being drawn. List two possible problems.

_____

_____

18. What is the difference between assigning a color to an object or a layer?

_____

_____

19. List four options for entering the **Layer** command.

_____

_____

20. What is the current layer name, and how can a new layer be named?

_____

_____

21. What are the limitations to assigning layer names?

_____

_____

22. A friend needs help to load five linetypes into a drawing. How can this be done quickly?

_____

_____

23. Centerline has just been made the current linetype. You return to the drawing and nothing is different. What's happened and why did it happen?

_____

_____

24. You've used the **Layer Properties Manager** dialog box to make centerline the current linetype. You return to the drawing. What's happened and why did it happen?

_____

_____

25. Use the **Help** menu and explore how the global scale factor will affect linetypes.

_____

_____

26. You've used **Lineweight Control** on the **Properties** toolbar to alter the lineweight. You return to the drawing area and draw several lines, but no change has been made to the lineweight. Why is there no change, and what must be done to see varied lineweights?

_____

_____

27. List the suggested lineweight for thick, moderate, and thin lines.

_____

_____

28. You want to draw one red line. How can it be done?

_____

_____

29. You want all of your lines to be red. What is the best way to do this?

_____

_____

30. When describing the state of a layer, what qualities are considered?

_____

_____

31. List the four groups of information to be used in layer names recommended by the USNCS.

_____

_____

32. What general aspects of **Properties** are listed if a line is selected?

_____

_____

33. You are working on a drawing containing several different linetypes and would like to find out their names. What command and option should be used?

_____

_____

34. List the preferred linetype scale for the following scales.

Full size 3" = 1'–0"   1" = 1'–0"
3/4" = 1'–0"   3/8" = 1'–0"
1/4" = 1'–0"   1/8" = 1'–0"
1/16" = 1'–0"   1" = 10'–0"
1" = 20'–0"   1" = 60'–0"

_____

_____

35. Describe the difference between the effects of **Ltscale** and **Celtscale**.

36. A layer has a title of **S-ANNO-SYMB**. Who created the drawing and what do you think it would contain?

37. You're about to start the plans for a one-level residence that will require a site plan, floor plan, framing plan, electrical plan, foundation plan, exterior elevations, and sections. How many drawing files do you expect to use, and how will you create layers for these files? (It's not in this chapter, you've got to search through your cranium.

38. What is the advantage of using **Qselect**?

39. How can the **Properties** dialog box be accessed and what can be done with it?

40. What are the uses for the include/exclude buttons of **Qselect**?

CHAPTER 6

Drawing Geometric Shapes

INTRODUCTION

Lines and points may be the base of all drawings, but it's difficult to imagine a large engineering or architectural project without curved features. This chapter will introduce drawing methods for creating common geometric shapes, including circles, arcs, ellipses, polygons, and rectangles.

Commands to be introduced include:

- **Dragmode**
- **Circle**
- **Arc**
- **Donut**
- **Fill**
- **Regen**
- **Ellipse**
- **Ellipse Arc**
- **Polygon**
- **Rectangle**

CIRCLE

Circles are found in a variety of projects throughout the construction world. Round windows, concrete piers and columns, steel tubing, and conduit lines—circular features will be prevalent throughout your drawings. Figure 6.1 uses circles to represent concrete piers for a structure. To start the **Circle** command, use one of the following methods:

- Select the **Circle** button on the **Draw** toolbar.
- Type **C** ENTER at the Command prompt.
- Select **Circle** from the **Draw** menu (as seen in Figure 6.2).

Note: Each of the options shown in brackets [3p/2p/Ttr] is displayed when the **Dynamic Display** edit arrow is selected during the command input. Options are automatically displayed when the command is selected from the drop-down menu.

Entering the **Circle** command from the toolbar or keyboard will display the prompt:

```
Specify center point for circle or [3P/2P/Ttr (tan tan radius)]:
```

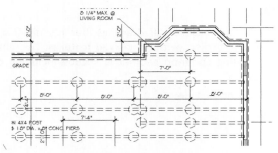

Figure 6.1 *Circles are used throughout architecture to represent many building components.*

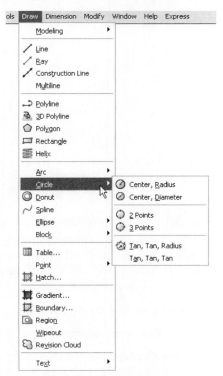

Figure 6.2 *To draw a circle, select the **Circle** button on the **Draw** toolbar, type **C** ENTER at the Command prompt, or select **Circle** from the **Draw** menu.*

AutoCAD is waiting for you to select a center point for the origin point of the circle. Points to define the circle can be selected on the screen using the mouse or by entering coordinates. Once the center point is selected, you're prompted for the edge of a circle. The process can be seen in Figure 6.3.

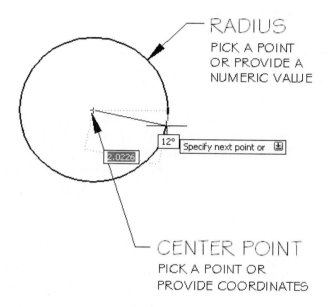

Figure 6.3 *Control the size of a circle by entering a value for the radius or the diameter.*

Circles can be drawn by entering seven different combinations of information about the circle. Six of the circles are listed in the **Circle** submenu shown in Figure 6.2. The seventh, the **Donut** command on the **Draw** menu, produces a circle with varied line width and will be discussed later in this chapter.

Notice that once the center point is selected, a circle will appear on the screen, but the size will be altered as the mouse is moved. AutoCAD refers to this as drag or rubber band. Moving the mouse and changing the circle size is called dragging the circle into position. This drag image is helpful in determining the final locations of lines and other geometric shapes. If you find it distracting, type **DRAGMODE** ENTER at the Command prompt and select **OFF** to disable it.

Note: As you draw circular features, you might want to review the **Display** tab of the **Options** dialog box. AutoCAD constructs circles using a series of straight lines. The **Arc and Circle Smoothness** setting can be used to control the smoothness of the display of arcs, circles, and ellipses. The default value is 1000, with a range of 1 through 20,000. The higher the value setting, the smoother the display. A lower value setting will display circular features with a series of straight lines. The true shape of objects will be displayed on plots, regardless of the resolution.

CENTER AND RADIUS

The default method of drawing a circle is to enter the center point and a radius. As the **Circle** command is started, select the center point with the pointing device or type coordinates at the Command prompt. In Figure 6.4 three circles have been drawn, and a fourth has been dragged into position. Notice that the rubber-band line extends from the chosen center point to the radius indicated by the crosshairs. Move the crosshairs to change the size of the circle. Type a numeric value and press ENTER to draw a circle with the desired radius. To draw a 24" diameter circle, the sequence is as follows:

```
Command: Click the Circle button on the Draw toolbar (Or type C
    ENTER)
Specify center point for circle or [3P/2P/Ttr (tan tan radius)]
    (Select center point.)
Specify radius of circle or [Diameter] <0'-0">:12 ENTER
```

 Note: To draw another circle, continue the command by pressing ENTER or SHIFT, or right-click and then select **Repeat CIRCLE** from the shortcut menu.

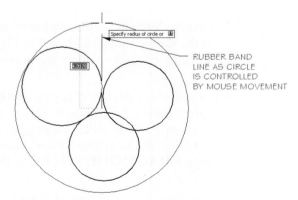

Figure 6.4 *Rather than entering a specific size, use the rubber-band line to see the effect on the circle to be drawn.*

CENTER AND DIAMETER

Sizes of circular features such as piers, pipes, and windows are typically given as a diameter. At the CIRCLE prompt, type **D ENTER** to switch the prompt from radius to diameter. When entered by toolbar or keyboard, the command sequence is as follows:

```
Command: Click the Circle button on the Draw toolbar (Or type C
    ENTER)>
Specify center point for circle or [3P/2P/Ttr (tan tan radius)]
    (Select center point.)
Specify radius of circle or [Diameter] <0'-0">: D ENTER
Specify diameter of circle <2'-0">: (Enter a diameter value or
    move the crosshairs to the desired location.)
```

The command sequence is similar when **Center, Diameter** is selected from the **Circle** submenu. AutoCAD simplifies the process with the menu so that the **Diameter** option is automatic.

Once the circle center is selected, the circle will be dragged across the screen but will not line up with the crosshairs. Figure 6.5 shows an example of a circle being created using the diameter. Notice that the length of the line extending from the center point is the diameter of the circle to be drawn. The line will disappear once the diameter is selected. The diameter entered will become the default setting for the next circle drawn.

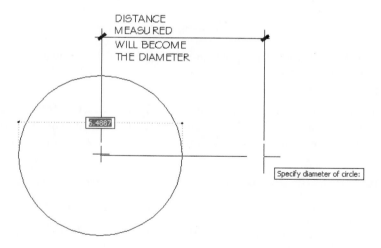

Figure 6.5 *With the Diameter option, the distance that the cursor is moved from the center point will become the diameter of the circle.*

TWO-POINT CIRCLES

Drawing a two-point circle can be useful if the diameter is known, but the center point is not. Draw the two-point circle by selecting a point and then selecting a point on the opposite side of the circle, as seen in Figure 6.6. The points can be selected with the crosshairs, or coordinates can be entered. The command sequence is as follows:

```
Command: Click the Circle button on the Draw toolbar (Or type C
    ENTER)
Specify center point for circle or [3P/2P/Ttr (tan tan radius)]
    2P ENTER
Specify first end point of circle's diameter: (Enter coordinates
    or s elect a point.)
Specify second end point of circle's diameter: (Enter coordinates
    or select a point.)
```

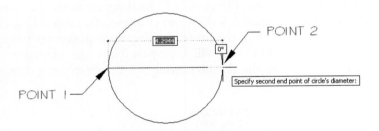

Figure 6.6 *A circle can be drawn by specifying the endpoints of the diameter.*

THREE-POINT CIRCLES

If certain points of a circle are known, a circle can be drawn by entering any three points. Figure 6.7 shows an example of a three-point circle. The points can be selected with the crosshairs, or coordinates can be entered. The process for drawing a three-point circle is as follows:

```
Command: Click the Circle button on the Draw toolbar (Or type C
    ENTER.)
Specify center point for circle or [3P/2P/Ttr (tan tan radius)]:
    3P ENTER
Specify first point on circle: (Enter coordinates or select a
    point.)
Specify second point on circle: (Enter coordinates or select a
    point.)
Specify third point on circle: (Enter coordinates or select a
    point.)
```

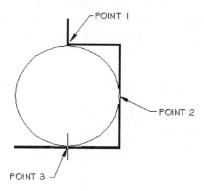

Figure 6.7   *The size of a circle can be determined by specifying three tangent points.*

TANGENT, TANGENT, RADIUS

An object is tangent if it intersects another object at one and only one point. A circle can be drawn tangent to existing features by selecting two features that the circle will be tangent to and then specifying a radius. This can be seen in Figure 6.8. The command sequence is as follows:

```
Command: Click the Circle button on the Draw toolbar (Or type
    C ENTER.)
Specify center point for circle or [3P/2P/Ttr (tan tan radius)]:
    TTR ENTER
Specify point on object for first tangent of circle: (Select a
    line or circle.)
Specify point on object for second tangent of circle: (Select a
    line or circle.)
Specify radius of circle <2'-0">: (Enter a numeric value or
    select a point.) ENTER
```

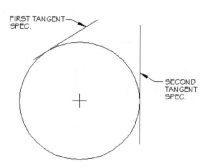

Figure 6.8   *A circle can be drawn by specifying two tangent surfaces and a radius.*

In Figure 6.9 each line was selected as a tangent point and a radius was entered. Because the radius is smaller than the distance between the lines, the first line does not touch the circle. If the line were extended, it would be tangent to circle.

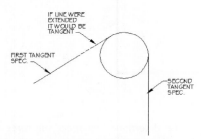

Figure 6.9 *If a radius is selected that is not large enough for the specified circle to touch each line, a circle is drawn tangent to one line and tangent to where the other line would be if it were extended.*

TANGENT, TANGENT, TANGENT

This option works well for drawing a circle that is tangent to three objects. This option can only be accessed by selecting **Tan, Tan, Tan** from the **Circle** submenu. The command sequence is as follows:

```
Select Tan, Tan, Tan from Circle on the Draw menu.
Specify first point on circle: _ tan to (Select the first object
    the circle is to be tangent to.)
Specify second point on circle: _ tan to (Select the second
    object the circle is to be tangent to.)
Specify third point on circle: _ tan to (Select the third object
    the circle is to be tangent to.)
```

Figure 6.10 shows the development of a circle tangent to three lines.

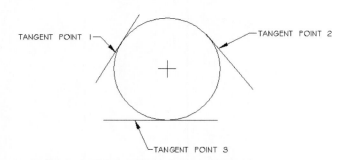

Figure 6.10 *Tan, Tan, Tan can be used to draw circles that are tangent to three objects.*

CIRCLE@

This option allows a circle to be drawn using the last endpoint of the last line drawn for the center point of a new circle. The option is best used for drawing concentric circles associated with pipes and circular columns. The command sequence is as follows:

```
Command: Click the Circle button on the Draw toolbar (Or type
    C ENTER).
Specify center point for circle or [3P/2P/Ttr (tan tan radius)]:
    @ ENTER
Specify radius of circle or [Diameter] <1.000>: (Specify desired
    diameter.)
```

DRAWING ARCS

Curved components are found throughout architectural drawings. To start the **Arc** command, use one of the following methods:

- Select the **Arc** button on the **Draw** toolbar.

  Typing **A** ENTER at the Command prompt.

  Select **Arc** from the **Draw** menu.

The menu can be seen in Figure 6.11. Eleven different options are available for drawing arcs, depending on what information is available or personal preference. Arc points for each option can be selected by mouse or by entering coordinates. The default setting for drawing an arc is the three-point arc. Selecting **Arc** on the toolbar will produce the prompt:

```
Specify start point of arc or [Center]:
```

 Note: When the *Arc* command is started from the *Draw* menu and the method is specified, prompts are automatically provided to match the option selected. The following examples assume the use of the toolbar or keyboard methods.

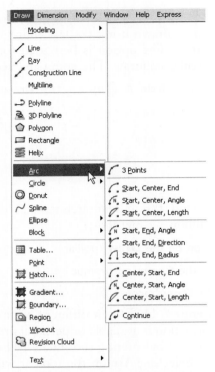

Figure 6.11 *Arcs are found throughout many structures. The Arc menu will aid in selecting the type of arc to draw.*

THREE-POINT ARCS

Drawing a three-point arc is similar to drawing a 3P circle and can be seen in Figure 6.12. The first and third points are the arc endpoints; the second point is any point on the arc. The command sequence for a three-point arc is as follows:

```
Command: Click the Arc button on the Draw toolbar (Or type
    A ENTER.)
Specify start point of arc or [Center]: (Select any point.)
Specify second point of arc or [Center/End]: (Select second
    point.)
Specify end point of arc: (Select the arc's endpoints.)
```

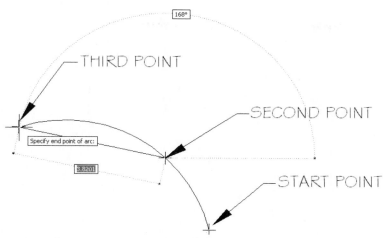

Figure 6.12   *Drawing a three-point arc is similar to drawing a three-point circle.*

Once the second point is selected, the arc will drag across the screen with the crosshairs. The arc will continue to change size until the endpoint is selected.

START, CENTER, END ARCS

When the ends and center of an arch are known, the Start, Center, End option can be used. The command sequence is as follows:

```
Command: Click the Arc button on the Draw toolbar (Or type
    A ENTER.)
Specify start point of arc or [Center]: (Select any point or
    enter coordinate.)
Specify second point of arc or [Center/End]: C ENTER
Specify center point of arc: (Select a center point.)
Specify end point of arc [Angle/chord Length]: (Select the arc's
    endpoint or enter coordinates.)
```

As the last point is entered, the arc will be drawn and the Command prompt will be returned. The process can be seen in Figure 6.13. Notice that as the last point is entered, the arc cannot end at the point entered. The endpoint that was specified is used only to determine the angle at which the arc will end. The actual radius was determined by the start and center points.

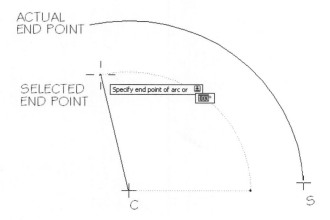

Figure 6.13 *The selected endpoint is used to determine the angle of the specified circle. The actual size of the circle is determined by the distance from the start to the center point.*

START, CENTER, ANGLE ARCS

On site plans for road and utility layouts, it's often necessary to draw an arc using lengths or angles. Drawing an arc using the start, center point, and the included angle is ideal for this type of work. If a positive value is used for the angle, the angle will be measured in a counterclockwise direction. If a negative value is used, the angle will be measured in a clockwise direction. The command sequence for a Start, Center, Angle arc is as follows:

```
Command: Click the Arc button on the Draw toolbar (Or type
    A ENTER)
Specify start point of arc or [Center]: (Select any point.)
Specify second point of arc or [Center/End]: C ENTER
Specify center point of arc: (Select center point of arc.)
Specify end point of arc or [Angle/chord Length]: A ENTER
Specify included angle: 30 ENTER
```

When the desired angle is entered, the arc will be drawn and the next command can be started.

START, CENTER, LENGTH ARCS

A chord is the straight line connecting the endpoints of an arc. Subdivision maps and drawings relating to land often include arcs that require use of the chord length. These arcs are drawn by specifying the start and center points and then supplying the chord length. A positive or negative value must be entered with the length. Figure 6.14 shows the difference between a chord of +3" or –3". The command sequence for a Start, Center, Length arc is as follows:

```
Command: Click the Arc button on the Draw toolbar (Or type
    A ENTER.)
```

```
Specify start point of arc or [Center]: (Select start point or
    enter coordinates.)
Specify second point of arc or [Center/End]: C ENTER
Specify center point of arc: (Select any point.)
Specify end point of arc [Angle/chord Length]: L ENTER
Specify length of chord: 3 ENTER
```

When a length is specified, the arc will be drawn and the command line will return.

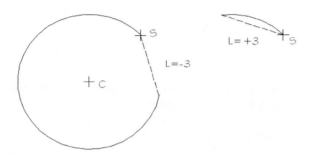

Figure 6.14   *Entering a positive or negative chord length determines which portion of the arc will be drawn using the Start, Center, Length arc option.*

START, END, ANGLE ARCS

An arc drawn using Start, End, and included Angle is drawn counterclockwise from the start point to the endpoint if a positive angle value is used. If a negative angle is used, the angle will be drawn clockwise. The results of coordinate entry can be seen in Figure 6.15. The command sequence is as follows:

```
Command: Click the Arc button on the Draw toolbar (Or type
    A ENTER).
Specify start point of arc or [Center]: (Select start point.)
Specify second point of arc or [Center/End]: E ENTER
Specify end point of arc: (Select the desired endpoint.)
Specify center point of arc or [Angle/Direction/Radius]: A ENTER
Specify Included angle: (Select desired angle.) 90 ENTER
```

When an angle is specified, the arc will be drawn and the command line will return.

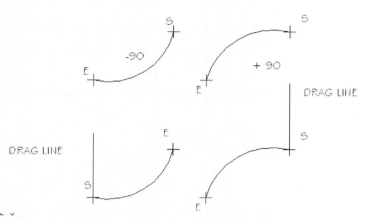

Figure 6.15 *Entering a positive or negative angle determines which direction an arc will be drawn using the Start, End, Angle arc option.*

START, END, DIRECTION ARCS

The Start, End, Direction option allows the direction of the arc to be selected. Major or minor arcs can be drawn in any orientation to the start point. Effects can be seen in Figure 6.16. The command sequence is as follows:

```
Command: Click the Arc button on the Draw toolbar (Or click
    A ENTER).
Specify start point of arc or [Center]: (Select start point.)
Specify second point of arc or [Center/End]: E ENTER
Specify end point of arc: (Select the desired endpoint.)
Specify center point of arc or [Angle/Direction/Radius]: D ENTER
Specify tangent direction for the start point of arc: (Select
    desired direction.)
```

When a distance is specified or a point is selected, the arc will be drawn and the command line will return.

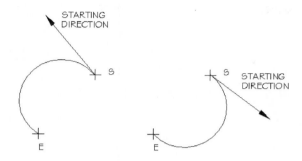

Figure 6.16   *The Start, End, Direction arc option allows the direction of the arc to be selected.*

START, END, RADIUS ARCS

Start, End, Radius arcs can only be drawn counterclockwise. Because the variables can produce different options, the arc is drawn counterclockwise from the Start point. Entering a positive or negative value for the radius will determine if the major or minor arc will be drawn. The effects of the value can be seen in Figure 6.17. The command sequence is as follows:

```
Command: Click the Arc button on the Draw toolbar (Or type
    A ENTER)
Specify start point of arc or [Center]: (Select start point.)
Specify second point of arc or [Center/End]: E ENTER
Specify end point of arc: (Select the desired endpoint.)
Specify center point of arc or [Angle/Direction/Radius]: R ENTER
Specify radius of arc: (Enter the desired radius or select a
    point.)
```

When a radius is specified or a point is selected, the arc will be drawn and the command line will return.

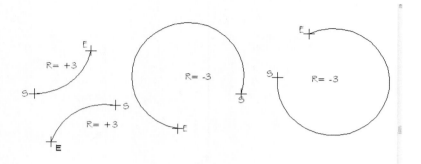

Figure 6.17 *Entering a positive radius determines which portion of the arc will be drawn using the Start, End, Radius arc option. The direction will always be counterclockwise.*

CENTER, START, END ARCS

The Center, Start, End option is similar to the Start, Center, End option. The effects of each command can be seen in Figure 6.18. The command sequence is as follows:

```
Command: Click the Arc button on the Draw toolbar (Or type
    A ENTER)
Specify start point of arc or [Center]: C ENTER
Specify center point of arc: (Select center point.)
Specify start point of arc: (Select the start point.)
Specify end point of arc or [Angle/chord Length]: (Select the
    desired endpoint.)
```

When the end point is selected, the arc will be drawn and the command line will return.

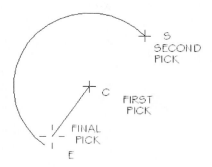

Figure 6.18　*The Center, Start, End arc option is similar to the Start, Center, End option.*

CENTER, START, ANGLE ARCS

The Center, Start, Angle option is similar to the Start, Center, Angle option. The effects of each command can be seen in Figure 6.19. The command sequence for this arc is as follows:

```
Command: Click the Arc button on the Draw toolbar (Or type
    A ENTER).
Specify start point of arc or [Center]: C ENTER
Specify center point of arc: (Select center point or enter
    coordinates.)
Specify start point of arc: (Select the start point or enter
    coordinates.)
Specify end point of arc or [Angle/chord Length]: A ENTER
Specify included angle: (Select the desired angle.)
```

When an angle is specified or a point is selected, the arc will be drawn and the command line will return.

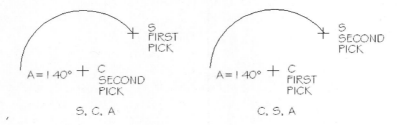

Figure 6.19 *The Center, Start, Angle arc option produces results similar to the Start, Center, Angle option.*

CENTER, START, LENGTH OF CHORD ARCS

The Center, Start, Length of chord option is similar to the Start, Center, Length option. The effects of each command can be seen in Figure 6.20. The command sequence is as follows:

```
Command: Click the Arc button on the Draw toolbar (Or type
    A ENTER).
Specify start point of arc or [Center]: C ENTER
Specify center point of arc: (Select center point.)
Specify start point of arc: (Select the start point.)
Specify end point of arc or [Angle/chord Length]: L ENTER
Specify length of chord: (Select desired length.)
```

When a length is specified or a point is selected, the arc will be drawn and the command line will return.

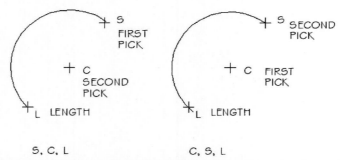

Figure 6.20 *The Center, Start, Length arc option produces results similar to the Start, Center, Length arc option.*

CONTINUE

An arc can be continued from a previous arc by pressing ENTER or SPACEBAR, or by selecting **Repeat ARC** from the shortcut menu in lieu of selecting the first arc entry point. When continued from a previous arc, the current arc's start point and direction are taken from the endpoint and ending direction of the last arc.

The **Continue** option also can be used to extend a line tangent to the endpoint of an arc. Figure 6.21 shows an example of continuous arcs drawn from a straight line. After drawing Line 1 and starting the **Arc** command, enter was selected when prompted for the start of the first, second and third arcs. Line 2 was drawn tangent to arc 3 because the Continue option was used. To place a line that continues from an arc but is not tangent to the arc will require the arc endpoint to be selected as the line "next point," using the methods described in chapter 8.

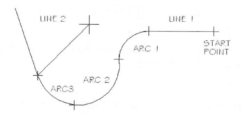

Figure 6.21   *Continue can be used to extend an arc tangent to the last line segment.*

DRAWING DONUTS

A donut in AutoCAD is a filled circle or ring. The ends of the steel shown in Figure 6.22 were drawn using the **Donut** command. To start the command, use either of the following methods:

- Type **DO** ENTER at the Command prompt.

- Select **Donut** from the **Draw** menu.

The command sequence to create a donut is as follows:

```
Command: DO ENTER
Specify inside diameter of donut <0'-0">: 0 ENTER
Specify outside diameter of donut <0'-1">:.75 ENTER
Specify center of donut or <exit>: (Select the desired location
     for the donut.)
Specify center of donut or <exit>: (The command will continue
     indefinitely; press ENTER to end the command.)
```

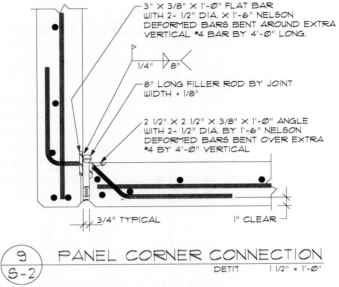

3" X 3/8" X 1'-∅" FLAT BAR
WITH 2- 1/2" DIA. X 1'-6" NELSON
DEFORMED BARS BENT AROUND EXTRA
VERTICAL #4 BAR BY 4'-∅" LONG.

1/4" 8"

8" LONG FILLER ROD BY JOINT
WIDTH + 1/8"

2 1/2" X 2 1/2" X 3/8" X 1'-∅" ANGLE
WITH 2- 1/2" DIA. BY 1'-6" NELSON
DEFORMED BARS BENT OVER EXTRA
#4 BY 4'-∅" VERTICAL

3/4" TYPICAL   1" CLEAR

9
S-2   PANEL CORNER CONNECTION
                DETIT   1 1/2" = 1'-∅"

Figure 6.22 *Donuts can be drawn by selecting Donut from the Draw menu or by typing DO ENTER at the Command prompt.*

Three different options are available for drawing donuts: a ring, a filled circle that is solid in color, or a circle filled with lines. Each can be seen in Figure 6.23. The fill material is determined by the status of the **Fill** command.

- Type **FILL** ENTER at the Command prompt to allow for a choice to be made for **Donut**.

- If **OFF** is entered, filled objects will resemble the bottom circle in Figure 6.23.

- With **Fill ON**, filled objects will resemble the middle circle in Figure 6.23.

 Note: *Fill* will affect several of the commands you will soon be using. In addition to donuts, *Fill* will affect the appearance of polylines, arrows created with *Leader*, and the display of hatch patterns. Unless you're trying to reproduce hollow objects, leave *Fill* in the **ON** setting.

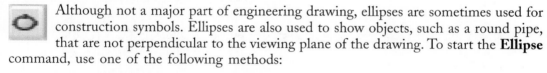

INTERIOR ⏀= .5 / EXTERIOR ⏀ = 1
FILL = ON

INTERIOR ⏀= ∅ / EXTERIOR ⏀ = 1
FILL = ON

INTERIOR ⏀= .5 / EXTERIOR ⏀ = 1
FILL = OFF

Figure 6.23   *Three options are available for the display of a donut.*

The effect of **Fill** is not seen immediately. To see the effect of **Fill,** enter the command, alter the setting, and press ENTER. No change shows until you type **REGEN** ENTER at the Command prompt. Now the effect of **Fill** can be seen.

USING THE ELLIPSE COMMAND

Although not a major part of engineering drawing, ellipses are sometimes used for construction symbols. Ellipses are also used to show objects, such as a round pipe, that are not perpendicular to the viewing plane of the drawing. To start the **Ellipse** command, use one of the following methods:

- Select the **Ellipse** button on the **Draw** toolbar.

- Type **EL** ENTER at the Command prompt.

- Select **Ellipse** from the **Draw** menu.

Figure 6.24 shows the major and minor axes that will be referred to in laying out an ellipse. The endpoints of each axis can be specified by selecting points with the cursor or by entering their coordinates. The command sequence is as follows:

```
Command: Click the Ellipse button on the Draw toolbar (Or type
   EL ENTER)
Specify axis endpoint of ellipse or [Arc/Center] (Select a point
   for the major axis endpoint.)
Specify other endpoint of axis: (Select a point for the opposite
   major axis endpoint.)
Specify distance to other axis or [Rotation]: (Select a point
   for the minor axis endpoint.)
```

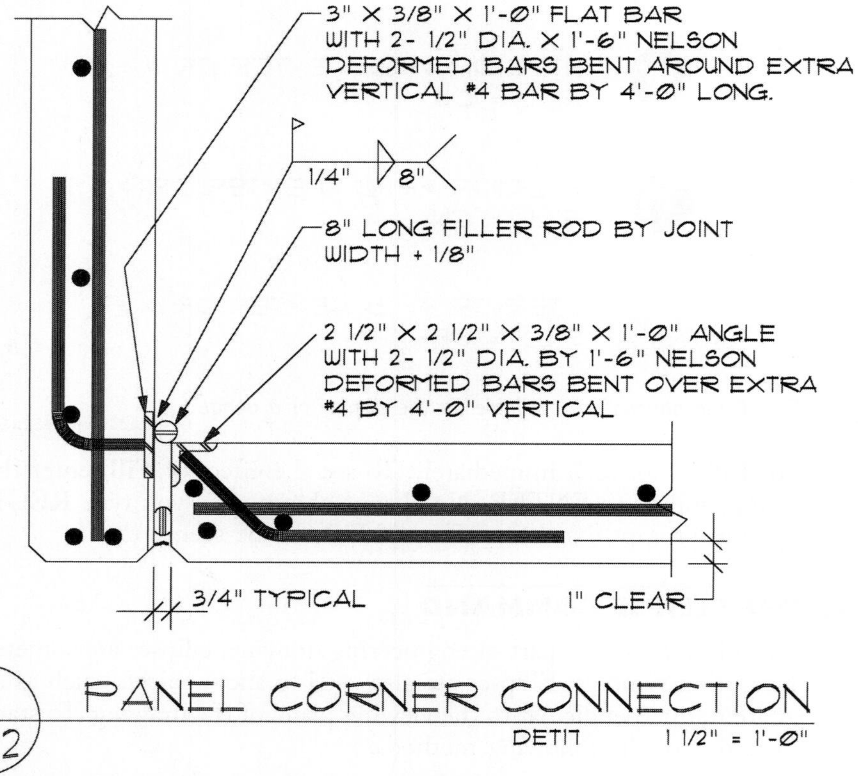
3" X 3/8" X 1'-Ø" FLAT BAR
WITH 2- 1/2" DIA. X 1'-6" NELSON
DEFORMED BARS BENT AROUND EXTRA
VERTICAL #4 BAR BY 4'-Ø" LONG.
1/4"
8"
8" LONG FILLER ROD BY JOINT
WIDTH + 1/8"
2 1/2" X 2 1/2" X 3/8" X 1'-Ø" ANGLE
WITH 2- 1/2" DIA. BY 1'-6" NELSON
DEFORMED BARS BENT OVER EXTRA
#4 BY 4'-Ø" VERTICAL
3/4" TYPICAL
1" CLEAR
9
S-2
PANEL CORNER CONNECTION
DETI7
1 1/2" = 1'-Ø"

Figure 6.24 *An ellipse is defined by the major and minor axis.*

An ellipse is now shown on the screen between the two selected points. The process can be seen in Figure 6.25. AutoCAD is waiting for you to enter the axis or rotation. If the angle is not known, the crosshairs can be moved until the ellipse represents the desired shape.

If you know the ellipse represents a pipe that is at a 60° angle to the viewing plane, type **R** ENTER. The command sequence is as follows:

```
Command: Click the Ellipse button on the Draw toolbar (Or type
     EL ENTER)
Specify axis endpoint of ellipse or [Arc/Center]: (Select a point
     for the major axis endpoint.)
Specify other endpoint of axis: (Select a point for the opposite
     major axis endpoint.)
Specify distance to other axis or [Rotation]: R ENTER
Specify rotation around major axis: (Enter desired angle of
     rotation.) 60 ENTER
```

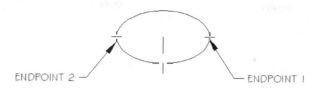

ENDPOINT 2 —⏜   ⏜— ENDPOINT 1

Figure 6.25   *Once the two endpoints are specified, the ellipse will change as the cursor is moved.*

Figure 6.26 shows the effect of rotating a circular pipe at 10° increments. An ellipse with a 90° rotation cannot be drawn using the **Ellipse** command because it would produce a single straight line.

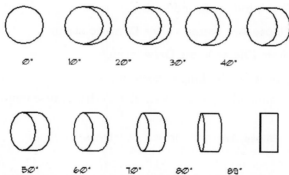

0°   10°   20°   30°   40°

50°   60°   70°   80°   89°

Figure 6.26   *The effects of rotating a circular object in 10° increments.*

An ellipse also can be drawn by locating the center point and then specifying an endpoint for each axis. This process can be seen in Figure 6.27. The command sequence is as follows:

```
Command: Click the Ellipse button on the Draw toolbar (Or type
    EL ENTER)
Specify axis endpoint of ellipse or [Arc/Center]: C ENTER
Specify center of ellipse: (Select a point for the major axis
    center point.)
Specify endpoint of axis: (Select a point for the endpoint of
    major axis.)
Specify distance to other axis or [Rotation]: (Select a point
    for the minor axis endpoint.)
```

When the second axis endpoint is entered, the Command prompt is returned and the ellipse is drawn on the screen.

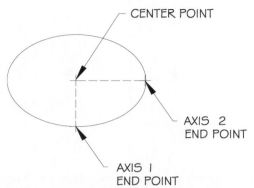

Figure 6.27 *An ellipse can be drawn by locating the center point followed by an endpoint for the major and minor axis.*

DRAWING ARCS WITH THE ELLIPSE ARC COMMAND

The **Ellipse Arc** command can be used to draw an arc based on an ellipse. To start the command, use one of the following methods:

- Select the **Ellipse Arc** button on the **Draw** toolbar.
- Select **Arc** from **Ellipse** in the Draw menu.
- Use the Arc mode of the **Ellipse** command.

The first half of the command sequence is similar to the **Ellipse** command. The command sequence is as follows:

```
Click the Ellipse Arc button on the Draw toolbar.
Specify axis endpoint of elliptical arc or [Arc/Center]: A ENTER
Specify axis endpoint of elliptical [Center]: (Select a point for
    the major axis endpoint.)
Specify other endpoint of axis: (Select a point for the opposite
    major axis endpoint.)
```

An ellipse is now shown on the screen between the two selected points. The next portion of the command sequence allows a portion of the ellipse to be selected for display as an arc. The display will be altered depending on how the selection points are entered.

If the points are entered in a left to right motion, a portion of the ellipse will be removed between the selection points.

If the selection points are entered from right to left, only the portion of the arc between the selection points will be displayed, with the balance of the arc removed.

The second half of the command sequence is as follows:

```
Specify distance to other axis or [Rotation]: (Select a point or
    enter coordinates for the minor axis endpoint.)
Specify start angle or [Parameter]: (Select a start point for
    the arc.)
```

```
Specify end angle or [Parameter/Included angle]: (Select an end
     point for the arc.)
Command:
```

The complete process can be seen in Figure 6.28.

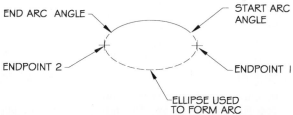

Figure 6.28 *An elliptical arc is drawn between the two selected points depending on how the selection points are entered. If the points are entered with a left-to-right motion, a portion of the ellipse will be removed between the points. If the selection points are entered from right to left, only the portion of the arc between the points will be displayed.*

DRAWING POLYGONS

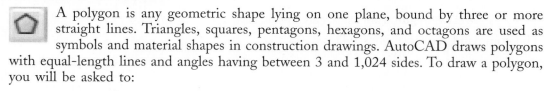

A polygon is any geometric shape lying on one plane, bound by three or more straight lines. Triangles, squares, pentagons, hexagons, and octagons are used as symbols and material shapes in construction drawings. AutoCAD draws polygons with equal-length lines and angles having between 3 and 1,024 sides. To draw a polygon, you will be asked to:

- Choose the number of sides.

- Choose whether the object is to be inscribed in or circumscribed around a circle.

  - When the polygon is *inscribed* in a circle, the circle diameter provides the distance across the polygon points.

  - When the polygon is *circumscribed* around a circle, the circle diameter provides the distance across the flat surfaces of the polygon.

The effects of a triangle inscribed in and circumscribed around a 2.5" diameter circle can be seen in Figure 6.29.

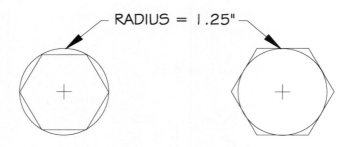

RADIUS = 1.25"

INSCRIBED　　　　　　　CIRCUMSCRIBED

Figure 6.29　*Polygons are drawn based on their relationship to a circle. Once the number of sides has been specified, you will be asked to specify whether the polygon will be placed inside or outside the circle.*

To start the **Polygon** command, use one of the following methods:

- Select the **Polygon** button on the **Draw** toolbar.

- Type **POL** ENTER at the Command prompt.

- Select **Polygon** from the **Draw** menu.

The command sequence is as follows:

```
Select the Polygon button on the Draw toolbar (Or enter POL ENTER
     at the Command prompt.)
polygon Enter number of sides <4>: (Select the desired number of
     sides.) 8 ENTER
Specify center of polygon or [Edge]: (Select a point for the
     center point.)
Enter an option [Inscribed in circle/Circumscribed about circle]
     <I>: (Accept I or select C depending on your need and press
     ENTER) I ENTER
Specify radius of circle: (Enter desired radius.) 2.75 ENTER
```

The result of the sequence can be seen in Figure 6.30. The setting of ORTHO will affect how the polygon will be placed.

- When ORTHO is **ON**, the edges of the polygon will be parallel or perpendicular to the drawing screen.

- When ORTHO is **OFF**, the edges of the polygon will rotate based on the movement of the mouse.

To activate ORTHO, click the **ORTHO** button on the status bar or press F8.

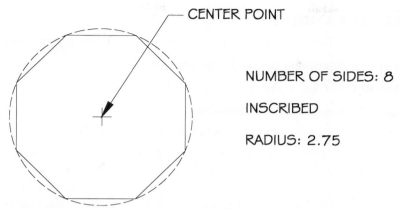

Figure 6.30   *The result of drawing an inscribed octagon with a 2.75" radius.*

A polygon can also be drawn based on the location of one edge. Once the number of sides has been provided, you will be asked for an edge or center point. When **E ENTER** is typed at the prompt, a polygon similar to the one in Figure 6.31 can be drawn.

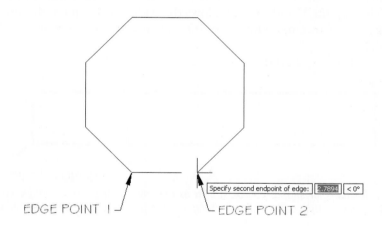

Figure 6.31   *A polygon also can be drawn by specifying the points that define one edge.*

As the first point is entered, a polygon will be displayed that will vary in size as the mouse is moved to the second point. As the second point is entered, the Command prompt is returned and the polygon is drawn. The setting of ORTHO will affect the placement of a polygon using the **Edge** option.

DRAWING RECTANGLES

To access the Rectangle command, use one of the following methods:

- Select the **Rectangle** button on the **Draw** toolbar.
- Type **REC** ENTER at the Command prompt.
- Select **Rectangle** from the **Draw** menu.

The command sequence for drawing a rectangle is as follows:

```
Select the Rectangle button on the Draw toolbar (Or enter
    REC ENTER at the Command prompt.)
_rectang
Specify first corner point or [Chamfer/Elevation/Fillet/
    Thickness/Width]: (Select a corner of the rectangle with the
    mouse.)
Specify other corner point or [Area/Dimensions/Rotation]: (Select
    the opposite corner of the rectangle.)
```

As the cursor is moved between the first and second corner, a rectangle will be displayed on the screen allowing the exact size to be determined. The sequence can be seen in Figure 6.32. Selecting either the Chamfer or Fillet option allows the corners of the rectangle to be altered. Selecting the Width option allows the width of the line to be altered. The other options affect only a rectangle drawn in 3D and will not be discussed in this chapter.

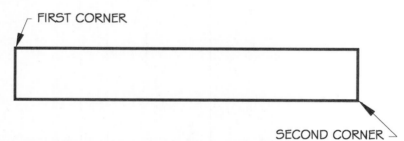

Figure 6.32 *Rectangular shapes can be drawn as one object using the Rectangle command on the Draw toolbar and the Draw menu, or by typing REC ENTER at the command line.*

CHAMFER

A chamfer is an angled or mitered corner. Drawing a rectangle with the **Chamfer** option active will place a chamfer on each corner of the rectangle. Once the **Chamfer** option is selected, prompts will be given for the first and second chamfer distances. Figure 6.33 shows examples of equal and unequal chamfer values. The command sequence to draw a rectangle with an equal chamfer is as follows:

```
Select the Rectangle button on the Draw toolbar (Or enter
    REC ENTER at the Command prompt.)
_rectang
Specify first corner point or [Chamfer/Elevation/Fillet/
    Thickness/Width]: C ENTER
```

Select first chamfer distance for rectangles <0'-0">: *(Enter the desired size.)* **.5** ENTER
Second chamfer distance for rectangles <0'-0 1/2">: ENTER
Specify first corner point or [Chamfer/Elevation/Fillet/ Thickness/Width]: *(Select a corner of the rectangle.)*
Specify other corner point or [Dimensions]: *(Select the opposite corner of the rectangle or enter a dimension to define the length.)*

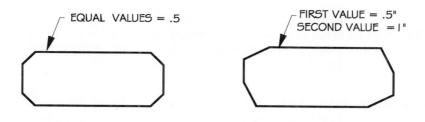

EQUAL DISTANCES   UNEQUAL DISTANCES

Figure 6.33   *The Chamfer option of the Rectangle command can be used to provide a chamfer at each corner of a rectangle. The rectangle on the left shows equal values for the chamfers while the rectangle on the right shows unequal values.*

FILLET

A fillet is a rounded corner. Drawing a rectangle with the **Fillet** option active will place a fillet on each corner of the rectangle. Once the **Fillet** option is selected, a prompt will be given to provide a radius for the fillet. Figure 6.34 shows an example of a rectangle with filleted corners. The command sequence to draw a rectangle with a fillet is as follows:

Select the **Rectangle** button on the **Draw** toolbar *(Or enter **REC** ENTER at the Command prompt.)*
_rectang
Specify first corner point or [Chamfer/Elevation/Fillet/ Thickness/ Width]: **F** ENTER
Select fillet radius for rectangles <0'-0 1/2">: *(Enter the desired size.)* ENTER
Specify first corner point or [Chamfer/Elevation/Fillet/ Thickness/Width]: *(Specify a corner of the rectangle.)*
Specify other corner point or [Area/Dimensions/Rotation]: *(Select the opposite corner of the rectangle.)*

The current radius value is 0'–0 1/2". The value that was provided for the chamfer will become the default value for all future chamfers and fillets until a new value is provided.

 Tip: Experiment to determine the results of entering a negative value.

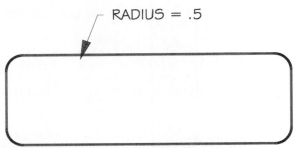

RADIUS = .5

Figure 6.34 *The Fillet option of the Rectangle command can be used to provide a fillet at each corner of a rectangle. The value used for the radius will remain constant for future use until it is altered.*

WIDTH

Selecting the **Width** option allows the line thickness to be altered. Because the fillet option was just used, setting the thickness for a rectangle will result in a rectangle with 1/2" radius corners to be drawn with a thick line. To draw a rectangle with square corners, enter the **Rectangle** command and alter the value for **Fillet** to 0, and then set the thickness, or vice versa. The command sequence is as follows:

```
Select the Rectangle button on the Draw toolbar (Or enter
    REC ENTER at the Command prompt.)
_rectang
Specify first corner point or [Chamfer/Elevation/Fillet/
    Thickness/Width]: W ENTER
Specify line width for rectangles <0'-0">: .25 ENTER
Specify first corner point or [Chamfer/Elevation/Fillet/
    Thickness/Width]: F ENTER
Select fillet radius for rectangles <0'-0 1/2">: 0 ENTER (Enter
    the desired size.)
Specify first corner point or [Chamfer/Elevation/Fillet/
    Thickness/Width]: (Specify a corner of the rectangle.)
Specify other corner point or [Dimensions]: (Select the opposite
    corner of the rectangle.)
```

The width will remain the current width for all future rectangles until the value is altered. Methods of providing thickness to other entities will be discussed as polylines are introduced.

AREA

This option can be used to create a rectangle using the area and either a length or a width. If the Chamfer or Fillet option is active, the area includes the effect of the chamfers or fillets on the corners of the rectangle.

```
Select the Rectangle button on the Draw toolbar (Or enter
    REC ENTER at the Command prompt.)
_rectang
Specify first corner point or [Chamfer/Elevation/Fillet/
    Thickness/Width]: (Specify a corner of the rectangle.)
Specify other corner point or [Area/Dimensions/Rotation]: A ENTER
```

```
Enter area of rectangle in current units <100>: (Enter a
    positive number) 50 ENTER
Calculate rectangle dimensions based on [Length/Width] <Length>:
    (Enter L or W) L ENTER
Enter Rectangle length <10>: (Enter a positive number) 24 ENTER
```

DIMENSIONS

This option will create a rectangle using length and width values. The command sequence to create a 10" × 13" rectangle is:

```
Select the Rectangle button on the Draw toolbar (Or enter
    REC ENTER at the Command prompt.)
_rectang
Specify first corner point or [Chamfer/Elevation/Fillet/
    Thickness/Width]: (Specify a corner of the rectangle.)
Specify other corner point or [Area/Dimensions/Rotation]:
    D ENTER
Specify length for rectangles <0.0000>: (Enter a positive
    number.) 10 ENTER
Specify width for rectangles <0.0000>: (Enter a positive number.)
    13 ENTER
Specify other corner point or [Area/Dimensions/Rotation]: (Move
    the cursor to display one of four possible locations for the
    rectangle and click.)
```

ROTATION

This option creates a rectangle at a specified rotation angle. The command sequence to locate a 12" ×15" rectangle rotated 15° from horizontal is:

```
Select the Rectangle button on the Draw toolbar (Or enter
    REC ENTER at the Command prompt.)
_rectang
Specify first corner point or [Chamfer/Elevation/Fillet/
    Thickness/Width]: (Specify a corner of the rectangle.)
Specify other corner point or [Area/Dimensions/Rotation]: R ENTER
Specify rotation angle or [Points] <0>: (Specify an angle by
    entering a value) 15 ENTER
Specify other corner point or [Area/Dimensions/Rotation]: D ENTER
Specify length for rectangles <0.0000>: (Enter a positive
    number.) 12 ENTER
Specify width for rectangles <0.0000>: (Enter a positive number.)
    15 ENTER
Specify other corner point or [Area/Dimensions/Rotation]: (Move
    the cursor to display one of four possible locations for the
    rectangle and click.)
```

CHAPTER 6 EXERCISES

Save the following exercises using the indicated file names.

1. Draw a 3" diameter circle, a circle with a 1" radius, and a 1" diameter circle that is tangent to two 1" long perpendicular lines. Make the smallest circle green with continuous lines. Make the largest circle red with hidden lines. Give one of the lines a width of .065". Save the drawing as **E-6-1**.

2. Draw two parallel lines that are 1" long and 1" apart, and provide an arc at each end of the lines to form a slotted hole. Draw these objects on a layer named SLOTS, using dashed blue lines, with a thin lineweight. Save the drawing as **E-6-2**.

3. Draw a 4" long line. Use the continuous option to place a 1" diameter circle at each end of the line. Save the drawing as **E-6-3**.

4. Draw a box formed by three 1" long lines. Draw an arc to close the polygon. Save as drawing **E-6-4**.

5. Draw an arc with a start point that is 2" to the right of the endpoint with an included angle of 232°. Save as drawing **E-6-5**.

6. Draw an arc with an endpoint that is @2,2 from the center point with an included angle of 45°. Save as drawing **E-6-6**.

7. Draw an ellipse with a 3" long axis at a 42° rotation. Save as drawing **E-6-7**.

8. Draw an ellipse with a center point that is 2.577 inches from the end of the major axis and 1" from the minor axis. Draw a four-sided polygon using the center of the ellipse as the center of the polygon. Make two edges of the square touch the ellipse. Save as drawing **E-6-8.**

9. Draw the following polygons:

3 sides inscribed	2" diameter
3 sides circumscribed	2" diameter
4 sides inscribed	.5" radius
4 sides circumscribed	.5" radius
6 sides inscribed	.75" radius
6 sides circumscribed	.75" radius
8 sides inscribed	1.25" radius
8 sides circumscribed	1.25" radius

Save the drawing as E-6-9.

10. Draw a donut with an inside diameter of .25", outside diameter of 75", and **Fill** set at **ON**. Draw another donut with an inside diameter of .0" and an outside diameter of .375" with **Fill** set at **OFF**. Draw a donut with an inside diameter of 0 and an outside diameter of .5" with **Fill** set to ON. Save the drawing as **E-6-10**.

11. Draw a rectangle with filleted corners using a 2" radius. Save the drawing as **E-6-11**.

12. Draw a rectangle with chamfered corners using a 1" distance for both settings. Save the drawing as **E-6-12**.

13. Draw a rectangle with chamfered corners using a 1.5" distance for the first distance and a 2.5 value for the second setting. Assign a line width of .065 and save the drawing as **E-6-13**.

14. Start a new drawing and draw at least three connected combinations of lines and arcs. Save the drawing as **E-6-14**.

15. Draw two perpendicular lines that are 2" long and form a corner. Draw a 3" diameter circle that is tangent to each of the lines. Draw three lines that are not connected and are neither parallel nor perpendicular to each other. Draw a circle that is tangent to all three line segments. Save this drawing as **E-6-15**.

CHAPTER 6 QUIZ

DIRECTIONS

Answer the following questions with short complete statements. Type your answers using a word processor.

1. Place your name, chapter number, and the date at the top of the sheet.

2. Type the question number and provide the answer.

 Warning: Some of the questions have not been covered in the reading material and will require the use of the help menu. You may also have to do some exploring to answer the questions.

1. What is the maximum number of sides that can be drawn for a polygon?

2. What are various points that are use to create arcs?

3. What is the default method for drawing arcs from the toolbar?

4. What is the command sequence to draw three continuous arcs?

_____

_____

5. What command determines whether the interior of a donut will be black or hatched with lines?

_____

_____

6. What is the difference between inscribed and circumscribed, and what are the effects of each on a polygon?

_____

_____

7. What four components can be used to construct an ellipse?

_____

_____

8. When would TTR be used for drawing a circle?

_____

_____

9. What are the options available for drawing circles?

_____

_____

10. What command controls the rubber-band effect of geometric shapes?

_____

_____

11. What options must be entered before a donut can be drawn?

_____

_____

12. What is the largest angle rotation that can be used with an ellipse?

_____

_____

13. What shape would a polygon with 500 sides appear as?

_____

_____

14. How does ORTHO affect the drawing of a polygon?

_____

_____

15. Use the **Help** menu and determine the use of Elevation as it relates to a rectangle.

_____

_____

Controlling Drawing Accuracy

INTRODUCTION

This chapter will introduce methods of controlling drawing accuracy by:

- Controlling the aperture box
- Controlling drawing settings using the **Drafting Settings** dialog box
- Effectively using Object Snap options

Commands and options introduced in this chapter include:

- **Aperture**
- **Dsettings**
- **OSNAP**

In previous chapters, you were introduced to using the tools on the status bar to make the crosshairs move at specific intervals. As you begin to combine geometric shapes into drawings, it is often helpful to join the shapes at exact locations. Joining objects by eye is difficult and should rarely be done. Objects that appear to be tangent might overlap when the **Zoom** command is used. Perfect intersections can be achieved by using the Object Snap modes that were introduced in chapter 3.

CONTROLLING THE APERTURE BOX

The normal cursor for the drawing area is the crosshairs surrounded by a box. If you enter a drawing command such as **Line**, the pick box is removed from the crosshairs as you select the first and next points. With OSNAP activated, an aperture box is added to the crosshairs, as seen in Figure 7.1. Any object that lies within the target or aperture box is subject to the drawing or editing command being used. Although objects can be added or deleted from within the aperture box, that is time consuming. To obtain higher accuracy in object selection, you can adjust the size of the aperture box.

Note: The aperture box will not be displayed unless the ***Display AutoSnap aperture box*** checkbox is activated. Even if it is not displayed, an invisible aperture box will still function as you draw. The aperture box is displayed on the ***Drafting*** tab of the ***Options*** dialog box. To display the dialog box, select ***Options*** from the ***Tools*** menu.

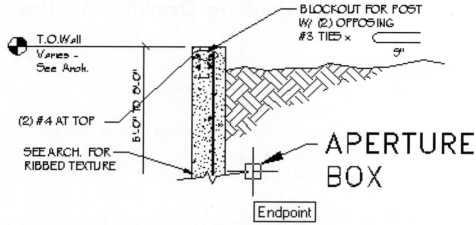

Figure 7.1 *As a drawing command is started with Object Snap activated, an aperture box is added to the crosshairs. The box is used for selecting existing drawing objects to be used as a selection point for a new line, circle, arc, or ellipse. (Courtesy Van Domelen/Looijenga/ McGarrigle/Knauf Consulting Engineers.)*

ALTERING THE SIZE

The size of the aperture box can be adjusted using the **Drafting** tab of the **Options** dialog box. This will produce the dialog box shown in Figure 7.2. Select the slide bar in the **Aperture Size** area and hold the select button down. Dragging the bar to the left will decrease the size. Moving the bar to the right will increase the size. The size will be indicated in the **Aperture Size** window display.

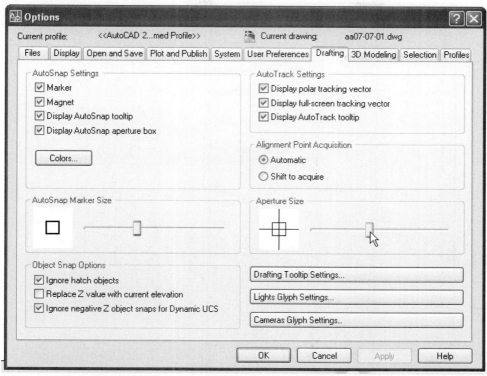

Figure 7.2   *The **Options** dialog box is accessed by selecting **Options** from the **Tools** menu. The dialog box can be used to adjust the aperture box, the marker size and color, Auto Snap settings, and AutoTrack settings.*

ALTERING THE APERTURE COLOR

In addition to adjusting the size of the box, you can adjust the aperture color using the **Display** tab of the **Options** dialog box. Select the **Color** button. This will display the **Drawing Window Colors** dialog box shown in Figure 7.3. Alter colors by using one of the following methods:

- Select the location from the **Context** menu.

- Select the element to be changed from the **Interface** menu.

- Choose a color that will offer good contrast to the screen color for easy viewing.

Once the desired color has been selected, click the **Apply & Close** button to change the crosshairs and box to reflect the selected color.

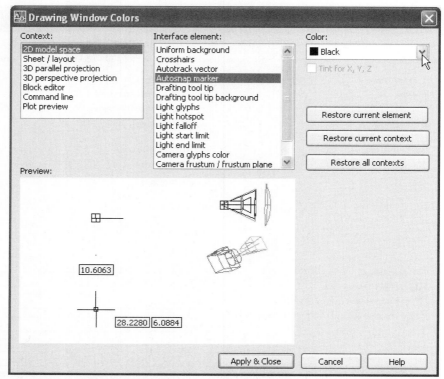

Figure 7.3 *The **Drawing Window Colors** dialog box can be used to adjust the marker color and all other display features. Access the dialog box by choosing **Options** from **the Tools** menu, selecting the **Display** tab, and clicking the **Color** button.*

CONTROLLING DRAWING SETTINGS

To be used effectively, Object Snap settings need to be understood and put to use. Major areas to be considered include the AutoSnap settings and AutoTrack settings. Each can be set using the **Options** dialog box shown in Figure 7.2.

AUTOSNAP SETTINGS

The AutoSnap settings control the settings for each Object Snap mode. Settings include the marker, magnet, and tooltip.

Marker

As the cursor moves close to an object while a drawing command is being executed, an Object Snap mode is displayed. The marker will vary based on what portion of an object the cursor is adjacent to. AutoCAD uses a different geometric shape to represent each Object Snap mode. Specific markers will be introduced later in this chapter. With the **Marker** checkbox activated, each time the cursor is moved to a point controlled by Object Snap, the appropriate Object Snap marker will be displayed. For example, when you use the **Line** command, each time the cursor is moved toward the end of a line, a square **Endpoint** marker is added to the cursor to indicate that the endpoint will be selected. If you deactivate this setting, the marker box will not be displayed, but the current Object Snaps will be unaffected. AutoCAD has it **ON** by default for a reason. It is usually very helpful for new users to be reminded of the current settings.

Altering the Marker Size   The size of the Object Snap marker can be adjusted using the **Drafting** tab of the **Options** dialog box. Select the slider bar in the **AutoSnap Marker Size** area and hold the select button down.

- Dragging the bar to the left will decrease the marker size.

- Moving the bar to the right will increase the marker size.

The size will be indicated in the **AutoSnap Marker Size** window display.

Altering the Marker Color   Selecting the arrow in the **Color** field will display a menu of the seven basic drawing colors. Selecting a color name will automatically change the marker to that color and close the color menu. Choose a color that will offer good contrast to the screen color for easy viewing.

Magnet

The **Magnet** checkbox is a toggle for the **ON/OFF** setting. By default, AutoCAD magnet is **ON**. As you've used a drawing command, the cursor often moved toward the end of a line. That is the effect of having the magnet **ON**. The magnet automatically moves the cursor to lock it onto the nearest Object Snap point as the cursor is moved. Keeping the magnet on will increase your drawing speed.

Tooltip

The **Display AutoSnap tooltip** checkbox is a toggle for the ON/OFF setting. It is **ON** by default. Each time the cursor is moved near an active Object Snap mode, the Object Snap mode to be used will be displayed in a tooltip below the cursor. Figure 7.4 shows the tooltip for the **Endpoint** Object Snap mode. On simple drawings, the tooltip might not seem important. On complex drawings with several Object Snap modes active, the tooltip can be helpful in determining what Object Snap mode will be used. Once you recognize each of the Object Snap markers, you might find the tooltips unnecessary.

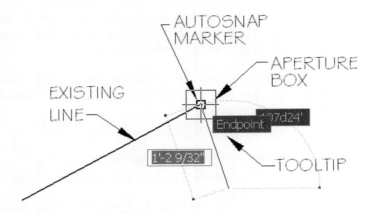

Figure 7.4 *Each time the cursor is moved near an active Object Snap mode, a tooltip below the cursor shows the Object Snap mode to be used.*

AUTOTRACK SETTINGS

The AutoTrack settings control the settings that can enhance each Object Snap mode. Settings include the polar tracking vector, full-screen tracking vector, and AutoTrack tooltip displays. Each option is an **ON/OFF** toggle switch. By default, each is set to **ON**.

Polar Tracking Vector

If **Polar Tracking** is **ON**, the **Display polar tracking vector** checkbox, when activated, sets Polar Tracking so that lines can be drawn along angles relative to the drawing "first" or "next" points. Angle settings must be 90° divisors, such as 15°, 30°, and 45°. Tracking increments are set on the **Polar Tracking** tab of the **Drafting Settings** dialog box. Review chapter 3 and see Figure 3.12. You can access the dialog box by using one of the following methods:

- Move the cursor over the OTRACK button, right-click, and select Settings from the pop-up menu.

- Select **Drafting Settings** from the **Tools** menu.

- Type **OS** ENTER at the Command prompt.

With this option **ON**, tracking angles will be displayed similar to Figure 7.5. The polar tracking display can be toggled **ON/OFF** using the **POLAR** button on the status bar.

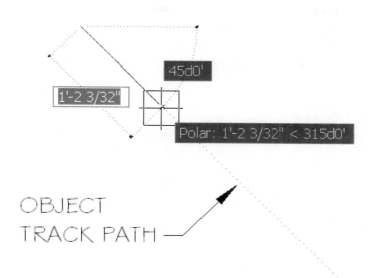

Figure 7.5 *If the Polar Tracking option is* **ON**, *tracking angles will displayed below the cursor.*

Full-Screen Tracking Vector

The **Display full-screen tracking vector** checkbox controls the display of tracking vectors. Tracking vectors are construction lines that can be used as a base, enabling objects to be drawn at a specific angle (see Figure 7.5). They can also be used to draw objects at a specific relationship to other objects. In their default setting, tracking vectors extend to the edge of the drawing display area. Alignment paths can be shortened or eliminated. When this checkbox is not selected, the path is displayed only from the Object Snap point to the cursor. Controls for setting Object Snap Tracking are found on the **Polar Tracking** tab of the **Drafting Settings** dialog box.

Autotrack Tooltip

The AutoTrack tooltip is similar to the AutoSnap tooltip. The **Display AutoTrack tooltip** checkbox controls the display of the tracking coordinates.

OBJECT SNAPS

Object snaps will save time and greatly improve your accuracy. Object Snap modes can be used to select an exact start or endpoint for connecting lines, rather than hoping you're close. Object snaps (Osnaps) allow a line or other object to be snapped to a specific point. For instance, object snaps will allow a line to be started exactly at the midpoint of an existing line with absolutely no guessing. Osnap has two methods of operation: single use and running.

- The running Osnaps allow modes to be set and used throughout the drawing session.

- The single use of Osnaps allows you to select an Osnap mode to be used once for a specific command.

In either method, once the Osnap command is activated, the marker box is used to select the object to snap to. Move the crosshairs so that the desired object lies within the marker box, and then click the select button. If more than one object lies within the marker box, the closest object will be selected.

OBJECT SNAP MODES

Object Snap tools include ENDpoint, MIDpoint, INTersection, EXTension, APParent Intersection, CENter, NODe, QUAdrant, INSertion, PERpendicular, TANgent, NEArest, PARallel, and NONe. By default, the Endpoint, Midpoint, Center, Intersection, and Extension modes of Osnap are active. In the active setting, these modes can be used for drawing lines or arcs that connect at certain points to existing lines, circles, arcs, or ellipses. Object Snap modes also can be used any time a point needs to be specified, and for **Copy**, **Move**, and **Insert** commands, which will be discussed in later chapters.

Running Object Snaps can be controlled using the **Drafting Settings** dialog box. The dialog box can be accessed by using one of the following methods:

- Select **Drafting Settings** from the **Tools** menu.

- Right-click the **Osnap** button on the status bar and then selecting **Settings**.

- Type **OS** ENTER at the Command prompt.

Each method will produce a dialog box similar to Figure 7.6. Options are toggled **ON/OFF** by selecting the check box.

- The **Select All** option can be used to toggle all of the Osnap settings **ON**.

  - With options **ON**, the cursor will snap to the nearest Endpoint, Midpoint or other active Osnap.

- The **Clear All** option can be used to toggle all Osnap settings **OFF**.

On simple drawings, all options can be toggled **ON** without causing problems. On a complex drawing with all options activated, an incorrect selection may be made. The cursor may snap to a midpoint rather than to an intersection that is near the midpoint. A temporary override of continuous Osnaps can be achieved by selecting the **Snap to None** button on the **Object Snap** toolbar, or by typing **NON** ENTER at the prompt for an object snap. Deactivate continuous Osnap by clicking the **OSNAP** button on the status bar or by pressing F3. As a general rule, only activate the options that apply to the needs of the current drawing.

Figure 7.6 *Select **Drafting Settings** from the **Tools** menu and then select the **Object Snap** tab to access the controls for running Osnaps. Options with a check in the selection box are active. Inactive options can be selected at any time based on selection needs.*

Figure 7.7 shows a listing of each Object Snap mode found on the **Object Snap** toolbar. The toolbar allows a specific mode to be selected for a single use. Notice that the first three letters of some of the modes are written in capital letters. Typing these letters at the start point or endpoint prompt will also activate the desired Object Snap mode for single use. Whether you choose to use running or single use, the options function in the same manner.

Note: Assuming that you will choose running Osnaps, setting the desired options is a matter of simply using the **Object Snap** tab of the **Drafting Settings** dialog box. Use the tab to set options related to the objects needed for the current project. The options will be explained in the following section, and will assume the use of the toolbar for single Osnaps. This method will allow you to activate options as needed to complete your project. Remember that running Osnaps can be adjusted at any point in the drawing setup.

The following discussion will assume that the default settings of Endpoint, Midpoint, Center, Intersection, and Extension modes of Osnap are active.

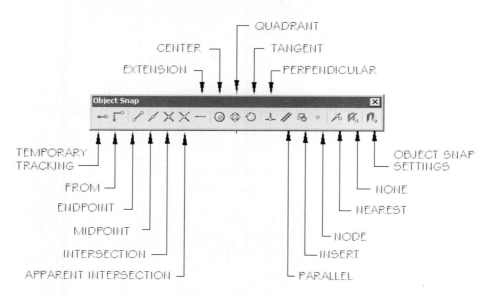

Figure 7.7 *The **Object Snap** toolbar and the marker box for each mode. The toolbar allows a specific mode to be selected for single use.*

ENDPOINT

The Endpoint mode allows an arc, line, or circle to be drawn to or from the endpoint of a previous line or arc by snapping to the selected endpoint. By default Endpoint is active. If you've accidentally made it inactive, return to the Object Snap tab and activate this option. To add a line that extends between the endpoints of two existing lines, the command sequence would be as follows:

```
Command: L ENTER (Or click the Line button.)
Specify first point: (Move the cursor near the end of the line
    that will be the source of the new line.)
```

The cursor does not have to touch the end of the line, but it must be between the midpoint and endpoint so the correct end of the line is selected. As the cursor touches a line, the square marker box is displayed around the nearest endpoint. (See Figure 7.8.) Click the select button, and the new line segment will be connected to the selected endpoint. Repeat the process again at the prompt for the "next point" to connect the new line to the end of another existing line. The new line will automatically end at the selected end of the existing line segment as shown in Figure 7.9.

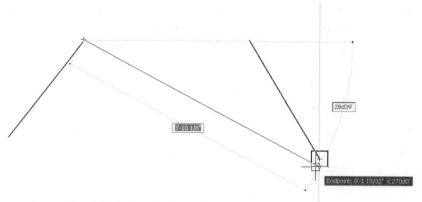

Figure 7.8   *When the **Endpoint** mode is active,, a marker (the square surrounding the aperture box) is placed on the endpoint of the selected line to indicate the selected "first" or "next" point. Notice that the aperture box does not need to touch the endpoint; it must only be near it to activate the object snap mode.*

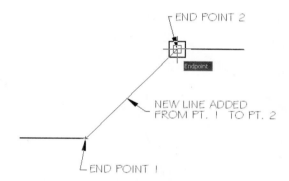

Figure 7.9   *To draw a line from the endpoint of one line to the endpoint of another, click the **Endpoint** button or type **END** ENTER when the "first point" and the "next point" prompts of the **Line** command are displayed.*

If the Endpoint option is inactive, the endpoint can also be selected from the Object Snap toolbar or at the keyboard, by typing **END** ENTER as the first point of the desired command. Move the marker box and select the desired line. To draw a line from the endpoint of an existing line to the endpoint of a second line, use the following command sequence:

```
Command: L ENTER (Or click the Line button.)
Specify first point: (Click the Endpoint button or type END ENTER)
    (Select endpoint of the desired line)
Specify next point or [undo]: (Click the Endpoint button or type
    END ENTER.)
Specify next point or [undo]: ENTER
```

Note: If you're thinking, "This is extra work," you're right. By default, AutoCAD automatically draws lines from endpoint to endpoint. However, the process will work wonders as different Object Snap modes are entered.

MIDPOINT

Midpoint allows the selection of a line or arc at its midpoint. The **Midpoint** button can be selected in response to the "first point" prompt. To add a line that extends between the midpoints of two existing lines, the command sequence would be as follows:

```
Command: L ENTER (Or click the Line button.)
Specify first point:
```

Move the cursor near the line that you would like the new line to connect to. As the cursor touches a line, a triangular marker is displayed at the midpoint. Click the select button, and the new line segment will be connected to the selected midpoint of the existing line. The command sequence can be seen in Figure 7.10.

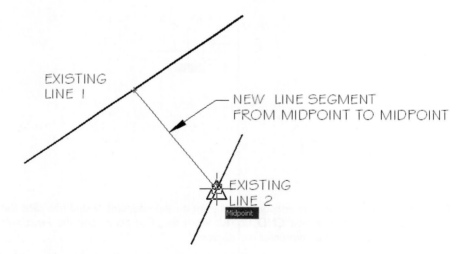

Figure 7.10 *The **Midpoint** mode can be used to draw a line from the midpoint of one line or arc to the midpoint of another.*

MID BETWEEN TWO POINTS

Although not on the Object Snap tab or on the Object Snap toolbar, **Mid between 2 Points** is available to place a **Start** or next point in the middle of two points. This option is available on the **Temporary Override** menu. Activate the menu by selecting SHIFT + right-click, and the select **Mid between 2 Points** from the flyout menu. Picking a point and entering **M2P** ENTER at the command line will also access this option. Use the following command sequence to draw a line from the midpoint between two points:

```
Command: L ENTER (Or click the Line button.)
Specify first point: M2P ENTER
First point of mid: (Select the desired first point.)
Second point of mid: (Select the desired second point.)
Specify next point or [Undo]: (Select desired line endpoint.)
    ENTER
```

The command sequence can be seen in Figure 7.11.

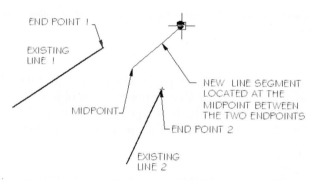

Figure 7.11   *The **Midpoint between two points** mode can be used to draw a line from the midpoint of two points. Endpoints or other Object Snaps can be used to pick the two points that will be used to locate the new line.*

INTERSECTION

The Intersection mode allows a line or arc to be extended to or from the intersection of any lines, arcs, or circles. The command sequence is similar to other Object Snap command entries and can be seen in Figure 7.12. By default, this option is **ON**. To draw a line from the intersection of two lines, move the cursor so the desired intersection lies within the target box. This will cause the cursor to snap to the intersection, allowing the next line to be drawn based on the selected intersection.

Single use of the Intersection mode is accessed by selecting the **Intersection** button on the **Object Snap** toolbar, or by typing **INT** ENTER as a first point or next point; the desired intersection can then be snapped to.

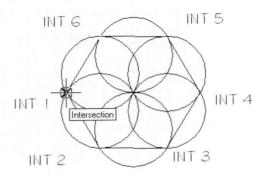

Figure 7.12 *The **Intersection** mode of Object Snap is used to draw a line from the intersection of each circle.*

EXTENSION

The Extension mode of Object Snap will use a point that lies along an extended path of a line as a First, Next, or Center point for an object. By default, this option is **ON**. This mode has two methods of operation. The first method will extend a line from an existing line. (See Figure 7.13.) This method can be used to lengthen a line.

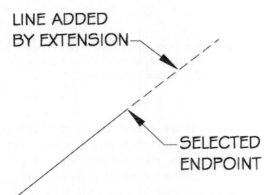

Figure 7.13 *The **Extension** mode of Object Snap will use a point that lies along an extension path of a line as a "next" point. When the endpoint of an existing line is selected for the extension point, the existing line will be lengthened along the projected path to the "next" point.*

The second method will use a point that lies on the path as the line is extended from the existing line. This method will work well to place the endpoint of a new line on a projected path created from an existing line. (See Figure 7.14.) Place the start point for the new point at any location. To place the "next" point on a projection from an existing line, move the cursor over, but do not select the existing endpoint. Moving the cursor over the endpoint will activate the **Extension** mode and show a projection from the existing line.

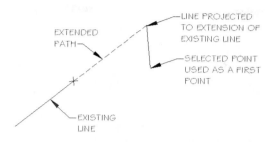

Figure 7.14 *Passing the EXT marker over the endpoint of an existing line, but not selecting it as an endpoint, will cause that point to be a projection point for the extension path. Extension will use a point that lies on the path from the existing line as the endpoint for the new line.*

APPARENT INTERSECTION

The Apparent Intersection mode finds the apparent intersection where two lines would intersect if one or both were extended. To place a line from the apparent intersection of two lines, activate the **Apparent Intersection** button on the **Object Snap** tab of the **Drafting Settings** dialog box. To place the line, move the cursor over, but do not select the existing endpoint to be "extended." Moving the cursor over the endpoint will activate the tracking mode and show a projection to the second line. Move the cursor along the tracking line until the **Apparent Intersection** marker is displayed at the apparent intersection. This point can now be selected for the first point. The process can be seen in Figure 7.15.

> **Note:** Make sure tooltips are activated and that *Intersection* is displayed in the tip. With several Osnap options active, the cursor may snap to *Nearest* or *Midpoint* instead of the desired *Intersection* option.

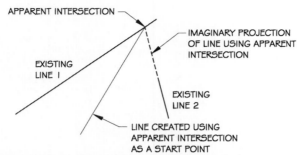

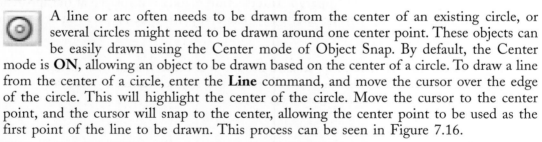

Figure 7.15   The **Apparent Intersection** mode of Object Snap can be used to extend a line between two lines that would intersect. The line was started by using the **Endpoint** mode to select the end of line 2. Then the **Apparent Intersect** button was selected, or **APP** ENTER was typed for the "to" point. A marker is automatically placed on line 1 where the lines would intersect. The second endpoint of the original line must be selected to display the new line segment.

The command can be started for single use by selecting the appropriate icon from the **Object Snap** toolbar or by typing **APP** ENTER at the "first" or "next" prompt. The prompt will read "_appint of" as the first point is selected. The new line will be placed where the selected lines would meet. Chapter 10 will present two alternatives to this option.

CENTER

A line or arc often needs to be drawn from the center of an existing circle, or several circles might need to be drawn around one center point. These objects can be easily drawn using the Center mode of Object Snap. By default, the Center mode is **ON**, allowing an object to be drawn based on the center of a circle. To draw a line from the center of a circle, enter the **Line** command, and move the cursor over the edge of the circle. This will highlight the center of the circle. Move the cursor to the center point, and the cursor will snap to the center, allowing the center point to be used as the first point of the line to be drawn. This process can be seen in Figure 7.16.

With the option **OFF**, a line or arc can be extended from the center point of a circle by selecting the **Center** button on the **Object Snap** toolbar. The Center mode can also be selected by keyboard by typing **CEN** ENTER at the prompt for the first point.

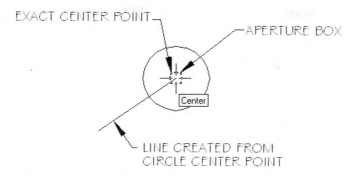

EXACT CENTER POINT

APERTURE BOX

Center

LINE CREATED FROM
CIRCLE CENTER POINT

Figure 7.16 *To extend a line from the center of a circle, enter the **Line** command. When prompted for a "first" point, move the cursor over the edge of the circle. AutoCAD will automatically use the center point of that circle as the "first" point for a line.*

QUADRANT

The Quadrant mode allows a line or arc to be snapped to the 0°, 90°, 180°, and 270° positions of a circle or arc. With this option active on the **Object Snap** tab, when prompted for a **first** or **next** point, the cursor will automatically snap to the nearest quadrant as the cursor touches the edge of a circle. Figure 7.17 shows the process for snapping to a quadrant. With the option inactive, single use of the Quadrant mode can be started by selecting the **Quadrant** button on the **Object Snap** toolbar or by typing **QUA** ENTER as the "first" or "next" point.

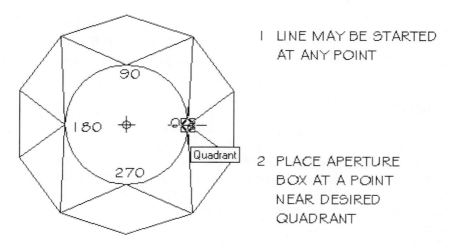

90

180

270

Quadrant

1 LINE MAY BE STARTED
   AT ANY POINT

2 PLACE APERTURE
   BOX AT A POINT
   NEAR DESIRED
   QUADRANT

Figure 7.17 *The **Quadrant** mode of Object Snap can be used to attach a line or arc to the quadrants of an existing circle.*

TANGENT

The Tangent mode allows a line to be snapped tangent to a circle, arc, ellipse, elliptical arc, or spline. With this option active on the **Object Snap** tab, when prompted for a next point, move the cursor to the desired circle. A tracking line will automatically be extended from the start point to a tangent point on the selected circle. Before selecting the selected endpoint on the circle, verify that the tooltip indicates **Tangent** and not **Quadrant**. Figure 7.18 shows an object created with tangent lines.

The Tangent option can be used in the middle of a command sequence by selecting the **Tangent** button on the **Object Snap** toolbar or by typing **TAN** ENTER at the "first" or "next" point prompt.

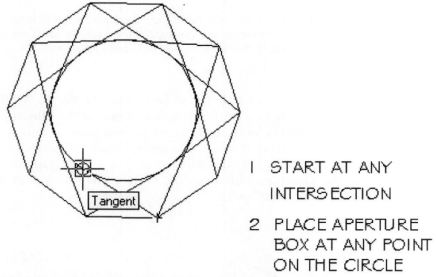

I START AT ANY
    INTERSECTION
2 PLACE APERTURE
    BOX AT ANY POINT
    ON THE CIRCLE

Figure 7.18 *The tangent option allows a line to be placed tangent to a circle, arc, ellipse, elliptical arc, or spline.*

PERPENDICULAR

Perpendicular lines are a basis for most construction drawings. The Perpendicular mode is used to draw a line that is perpendicular to an existing line. With this option active, select a first point, and move the cursor to the target line. As the projected line approaches a perpendicular position to the existing line, the perpendicular marker is added to the cursor. Press the select button to create the line. The process can be seen in Figure 7.19. The option can also be used to draw lines perpendicular to an arc, circle, ellipse, elliptical arc, multiline, ray, region, solid, xline, or spline. Each of these features will be introduced in future chapters.

A single use of the option can be activated by selecting the **Perpendicular** button on the **Object Snap** toolbar or by typing **PER** ENTER at the Command prompt when prompted for a "first" or "next" point.

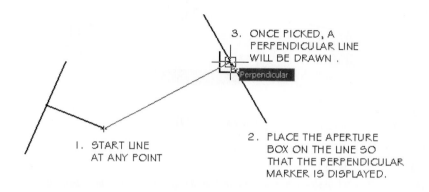

3. ONCE PICKED, A PERPENDICULAR LINE WILL BE DRAWN.

Perpendicular

1. START LINE AT ANY POINT

2. PLACE THE APERTURE BOX ON THE LINE SO THAT THE PERPENDICULAR MARKER IS DISPLAYED.

Figure 7.19 *The **Perpendicular** mode can be used to project a line so that it will be at a 90° angle to an existing line, arc, circle, ellipse, elliptical arc, multiline, ray, region, solid, xline, or spline. Each of these features will be introduced in future chapters.*

PARALLEL

The **Parallel** mode draws a line parallel to an existing line. Unlike other modes, Parallel is specified after the first point is selected. After the first point of the new line is specified, move the cursor over the original line. This will place the parallel marker on the line, indicating that AutoCAD will draw the new line parallel to the selected line. Now move the cursor to establish the location of the proposed line. When the path of the new line is parallel to the original line segment, an alignment path is displayed, which you can then use to create the parallel object. Figure 7.20 shows the use of the Parallel mode. Chapter 9 will explore an alternative method of placing parallel lines using the **Offset** command.

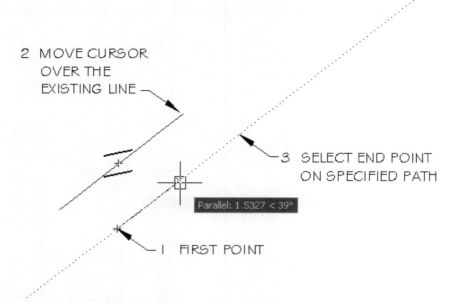

Figure 7.20   *The Parallel mode will draw a line parallel to an existing line.* **Parallel** *is specified after the first point is selected. Move the cursor over, but do not select the line that is to be the basis for the second line. Once the first line has been identified, specify the next point for the new parallel line.*

INSERTION

The Insertion mode will snap to the insertion point of a shape, text, attribute, or block. This might seem like meaningless stuff now, but combined with the information in later chapters, it will allow information to be inserted at an exact point. This mode can be left inactive or active as needed by selecting the **Insertion** button from the **Osnap** toolbar or by typing **INS** ENTER at the command line at either the first or next point as needed.

NODE

The Node mode can be used to snap to a point, allowing lines or arcs to be extended from or to predetermined points. These points typically include dimension definition points and dimension text origins. Methods of drawing a point will be introduced in chapter 12. This mode can be left inactive and started as needed by selecting **Node** from the **Osnap** toolbar or by typing **NOD** ENTER at the command line at either the first or next point.

NEAREST

The Nearest mode can be used to snap to the circle, line, arc, ellipse, or elliptical arc that is nearest to the cursor. This mode can be left inactive since it may accidentally be substituted for the Midpoint or Intersection options. Activate the mode as needed by selecting the **Nearest** icon from the **Osnap** toolbar or at the command line by typing **NEA** ENTER at either the first or next point.

FROM

The From mode can be used to establish a point for a line or other drawing objects based on a known base point. The base point can be located by selecting a point with the cursor, or by providing polar or relative coordinates for the point. The From mode could be useful on site-related drawings where objects are to be located based on a specific point. This mode can be left inactive and accessed as needed by selecting the **From** icon from the **Osnap** toolbar or at the command line by typing **FRO** ENTER. Figure 7.21 shows the use of the From mode for locating a circle from a specific point. The command sequence is as follows:

```
Command: Click the Circle button (Or type C ENTER).
Specify center point or [3P/2P/Ttr (tan tan radius)]: (Click the
    From button or type FRO ENTER.)
From Base point: (Click the Endpoint button or type END ENTER).
    (Select the desired line endpoint)
<offset>: 1.5 ENTER (Select location or provide coordinates from
    base point.)
Specify radius of circle or [Diameter]: (Pick radius point or
    provide coordinates.)
```

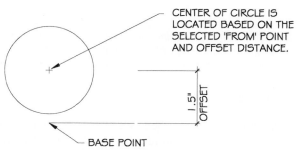

CENTER OF CIRCLE IS LOCATED BASED ON THE SELECTED 'FROM' POINT AND OFFSET DISTANCE.

1.5" OFFSET

BASE POINT

Figure 7.21 *The From mode of Object Snap allows a line, circle, or arc to be drawn from a base point by selecting the **From** button or by typing **FRO** ENTER for one of the selection points for the drawing object.*

TEMPORARY TRACKING

The Temporary Tracking Point mode can be used to place a first or next point that lies along an implied projection of a line. Figure 7.22 uses the mode to draw a line from the center of a rectangle in a manner similar to the Apparent Intersection mode. By selecting the midpoint on the bottom side of the rectangle, and then selecting the midpoint of one of the sides, you can draw a line from the imaginary intersection of these two lines. This mode can be left inactive and accessed as needed by selecting the **Temporary Tracking Point** button on the **Object Snap** toolbar or by typing **TK** ENTER at the Command prompt. The command sequence to draw a line from the center of a rectangle is as follows:

```
Command: Click the Line button (Or type L ENTER)
Specify first point: (Click the Tracking button or type TK ENTER.)
Specify first point: (Click the Midpoint button or type MID
    ENTER.)
Select the bottom edge of the rectangle
Specify first point: (Click the Tracking button or type TK ENTER.)
Specify first point: (Click the Midpoint button or type MID
    ENTER.)
Select the edge of the rectangle. Once the second point is
    selected, crossing lines will be extended from each midpoint
    locating the center of the rectangle.
Track point: (Press the select button to start the line.)
Specify next point or [Undo] (Select the desired endpoint of the
    line.)
```

As the tracking sequence is ended, the rubber-band line will now be centered in the rectangle and the prompt for a "To" point is given. The command sequence can be seen in Figure 7.22.

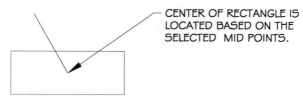

CENTER OF RECTANGLE IS LOCATED BASED ON THE SELECTED MID POINTS.

Figure 7.22 *Tracking can be used to determine the center point of a rectangle. After selecting the Tracking mode, select the midpoint of the bottom and then select the midpoint of one side of the rectangle. A line or arc can now be projected from the center of the rectangle.*

CHAPTER 7 EXERCISES

1. Draw an equilateral triangle inscribed in a 4" diameter circle. Use Object Snap Midpoint to form a second triangle inside the first. Draw a third triangle inside the second. Create six different layers, each with a different linetype and color. Assign each of the existing shapes to a different layer. Draw a different geometric shape on each of the remaining layers. Use continuous OSNAP. Save the drawing as **E-7-1**.

2. Draw the base to a chimney in plan view that is 60" × 32" with 8" wide walls. Save the drawing as **E-7-2**.

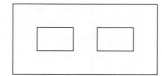

3. Use the drawing below to draw a 1" diameter circle, a 2.25" diameter circle, a four-sided polygon, and a triangle. Complete the drawing using the CEN, MID, INT, and TAN modes. Save the drawing as **E-7-3**.

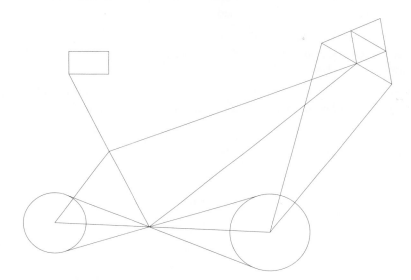

4. Set the limits and units to draw a 100' × 75' lot. Assume north to be at the top of the page. The north and south property lines will be 100' long. Five trees are located on the property. Each is located from the southwest corner with coordinates and diameters listed as follows:

	Coordinates	Trunk diameter	Branch diameter
a.	15',12'-9"	9" diameter	6' diameter
b.	30',17'	13" diameter	7.5' diameter
c.	42'-6", 19'-3"	19" diameter	9.5' diameter
d.	50', 28'-2"	15" diameter	8.5' diameter
e.	62'-0", 5'-3"	26" diameter	14' diameter

Locate each tree with a point. Use a circle to represent the trunk and branch diameters of each tree. Use the appropriate Object Snap modes to draw the diameter of the trunk and the branch structure around each tree center. Draw a line from the southwest property corner through the center of the trees to the southeast corner. Draw another line from the southeast property corner to the northern limits and tangent to the branch diameter of each tree and ending at the southwest corner.

Create layers with these specifications:

PROPLINE	red	Centerx2	property line
TREES	Green	Dashed	tree shape
T-LAYOUT	7	Dot2	point/baseline

Set the **Ltscale** appropriate for future plotting at a scale of 1"=10'–0". Save the drawing as **E-7-4**.

5. Use the drawing below to draw a 4' x 4' window with a half-round window above. Draw all window dividers as 1/2" wide. Save the drawing as **E-7-5**.

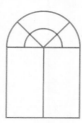

6. Using the drawing below as a guide, open the **ARCHBASE** template and draw a 32" x 21" double sink. Show the outline of the sink with rounded corners using a 2" radius. Divide the sink into two equal portions. Draw a 2" diameter circle in the center of the right portion of the sink. Save the drawing as **E-7-6SNK**.

7. Using the drawing below as a guide, open the **ARCHBASE** template and draw a rectangle to represent a 60" × 32" tub/shower. Draw a rectangle with 3" radius rounded corners at one end and a full radius at the other end. Draw a triangle at the squared end to represent a showerhead. Save the drawing as **E-7-7TUB**.

8. Using the drawing below as a guide, open the **ARCHBASE** template and draw a 48" wide × 24" deep, 0' clearance fireplace. Draw the face of the fireplace as 8" wide on each side, with the interior edges 20" long, set at a 15° angle. Save the drawing as **E-7-8FIRE**.

9. Open the **ARCHBASE** template and draw a 42" × 36" rectangle using continuous lines. Draw diagonal lines from opposite corners using hidden lines and show a 3" diameter circle at the intersection of the interior crossing lines. Save the drawing as **E-7-9SHOW**.

10. Draw a plan view of a 10' long x 24" wide bathroom vanity. The vanity is to have a 30" wide knee space in the center that should be represented by hidden lines. Place an 18" × 16" oval sink centered on each side of the knee space. Save the drawing as **E-7-10VAN**.

CHAPTER 7 QUIZ

DIRECTIONS

Answer the following questions with short complete statements. Type your answers using a word processor.

1. Place your name, chapter number, and the date at the top of the sheet.

2. Type the question number and provide the answer.

 Warning: Some of the questions have not been covered in the reading material and will require the use of the help menu. You may also have to do some exploring to answer the questions.

1. What command is used to change the size of the aperture box?

_____

_____

2. What is the size range for the aperture box?

_____

_____

3. Type the command sequence to provide running snap to an endpoint.

_____

_____

4. List three Object Snap modes for connecting a line to a circle.

_____

_____

5. What other OSNAP mode is similar to temporary tracking? Explain your answer.

_____

_____

6. List three methods to stop continuous OSNAP.

_____

_____

7. What steps are needed to draw a line perpendicular to an existing line?

_____

_____

8. List the letters that are used to activate the Object Snap mode used to connect geometric shapes.

_____

_____

9. How can you get the dialog box for Object Snaps to be displayed?

_____

_____

10. With the Quadrant mode of Object Snap active, at what degree will lines be attached to a circle?

_____

_____

Drawing Display Options

INTRODUCTION

In the last chapter you were introduced to drawing accurately by snapping to specific points of an object. Drawing accuracy can also be improved by changing the size of the drawing view and the drawing resolution. You were introduced to the **Zoom** and **Pan** commands in chapter 2. This chapter will explore

- Options of the **Zoom** command

- Other commands that can be used to alter or move through the display

Commands to be explored in this chapter include:

- **Zoom**
- **Pan**
- **View**
- **Redraw**
- **Regen**
- **Regenall**
- **Viewres**

ZOOM

Similar to the zoom lens of a camera, AutoCAD's **Zoom** command has a potential zoom ratio of 10 trillion to 1. "Zooming in" on a drawing can magnify a portion of a drawing to allow for better visual quality. Zoom can be used in the middle of another command sequence, which can greatly improve accuracy on detailed drawings. As the apparent size of the object is increased, the area of the drawing that can be seen is reduced. Figures 8.1 and 8.2 compare the change in display. The opposite also holds true: when you reduce the apparent size of an object, more of the drawing can be seen. It is important to keep in mind that the actual size of the object is not changing, only the magnification.

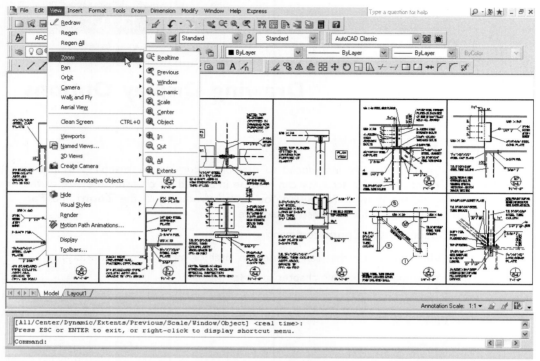

Figure 8.1 As the drawing limits are enlarged, many features of the drawing might not be legible. (Courtesy Kenneth D. Smith Architects & Associates, Inc.)

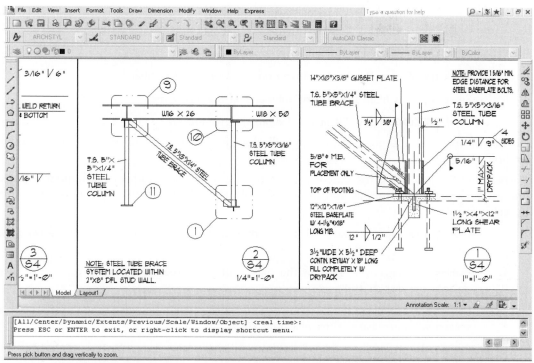

Figure 8.2   *The **Zoom** command alters the position of a drawing to be viewed, giving the sense that the portion of the drawing being viewed has been enlarged. (Courtesy Kenneth D. Smith Architects & Associates, Inc.)*

ACCESSING THE COMMAND

The Zoom command can be accessed from the **Zoom** toolbar and by selecting one of three different buttons on the **Standard** toolbar, **Zoom Realtime**, **Zoom Window** and **Zoom Previous**. You can also access the **Zoom** command from the **View** menu shown in Figure 8.1, by the right-click shortcut menu, or by keyboard. Typing **Z** ENTER at the Command prompt will start the command sequence. Each will produce the prompt:

```
Command: Click the Zoom Window button (Or type Z ENTER).
Specify corner of window, enter a scale (nX or nXP) or [All,
    Center, Dynamic, Extents, Previous, Scale, Window, Object]
    <real time>:
```

Each of the options listed at the Dynamic Display menu or Command prompt and in the menu can be selected from the **Zoom Flyout** menu on the **Standard** toolbar.

ZOOM REALTIME

The Realtime option of the **Zoom** command alters the magnification as you move the cursor. At a workstation equipped with a wheel mouse, the **Zoom** command can be executed using the wheel. Rolling the wheel forward will increase the magnification, and rolling the wheel backward will decrease the magnification. The **Zoom Realtime** option can also be executed using one of the following methods:

- Right-click the shortcut menu.
- Click the **Zoom Realtime** button on the **Standard** toolbar.
- Select options from the **Zoom** submenu of the **View** menu.
- Type **RTZOOM** ENTER at the Command prompt.

As the command is activated, the cursor will change to a magnifying glass with a + and – symbol.

Normally when you move the cursor around on the screen nothing special happens. Once you enter **Zoom Realtime**, as you press and hold the select button, the following results are achieved:

- Moving the cursor toward the top of the screen, the magnification of the drawings is enlarged.
- Placing the magnifying glass in the middle of the screen and moving to the top will enlarge the display 100%.
- Pressing and holding the select button and moving from the middle to the bottom of the screen will decrease the magnification 100%.
- Placing the magnifying glass at the bottom of the screen and moving to the top will enlarge the display 200%.
- Moving from the top to the bottom of the screen will decrease the magnification 200%.

Each command can be repeated indefinitely until the desired magnification is achieved. **Zoom** is not limited by the edge of the screen. During a zoom, you can drag your cursor at the edge of the monitor and continue to zoom.

 Note: If you have trouble making **Realtime Zoom** alter the drawing size, remember that you must enter the command, and you must hold down the select button as the cursor is moved.

The command is ended by pressing ESC, SPACEBAR, ENTER, or right-clicking to display the shortcut menu and selecting **Exit**. Selecting **Exit** will end **Zoom** and return the Command prompt. The menu also allows other viewing options to be selected. By alternating between the **Pan** and the **Zoom** commands, you can enlarge a drawing while keeping the desired portion on the screen. Each of these options will be discussed throughout the balance of this chapter.

ZOOM WINDOW

The Window option of **Zoom** is a common option for enlarging a portion of a drawing. This option lets you select the area you wish to view by providing two opposite corner locations that will form the viewing window. Select the size of the view box by using the select button to select locations on the screen. By default, the command is waiting for you to enter two points that will form the corners of a rectangle. The command sequence is as follows:

```
Command: Click the Zoom Window button (Or type Z ENTER).
Specify corner of window, enter a scale (nX or nXP) or [All,
     Center, Dynamic, Extents, Previous, Scale, Window, Object]
     <real time>: (Pick a corner.)
Specify opposite corner: (Pick a corner.)
```

This process can be seen in Figure 8.3a. Select the first corner of the zoom display by moving the cursor to the desired location. Once a corner is selected, the crosshairs will switch to a box that is enlarged or reduced as the mouse is moved across the drawing area. As the second point is selected, AutoCAD will smoothly zoom in to magnify the view of the selected objects. Objects within the view box will be redisplayed as the new display. Figure 8.3b shows the results of the selection made in Figure 8.3a.

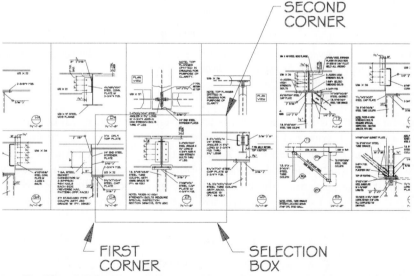

Figure 8.3a *The **Zoom Window** option allows the display area to be selected by specifying opposite corners of the "window." (Courtesy Kenneth D. Smith Architects & Associates, Inc.)*

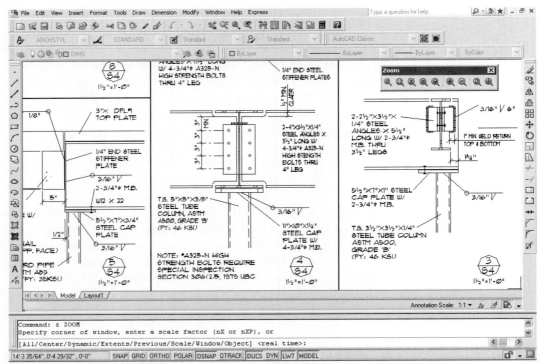

Figure 8.3b *The **Zoom** window created in Figure 8.3a now displays the selected area. (Courtesy Kenneth D. Smith Architects & Associates, Inc.)*

TRANSPARENT ZOOM

One of the best features of **Zoom** is that it is transparent, meaning it can be used in the middle of another command. If you start to draw a line and realize that you're straining to see the line, the screen image can be magnified using **Zoom** in the middle of a **Line** command sequence. Simply select the button of the desired **Zoom** option or enter the command by keyboard. To use a transparent command from the keyboard, type an ' (apostrophe) prior to the command. The command sequence for using the window option of **Zoom** in the middle of the **Line** command would be

```
Command: Click the Line button (Or type L ENTER).
Specify first point: (Select first line endpoint.)
Specify next point or [Undo]: 'Z ENTER
Specify corner of window, enter a scale (nX or nXP) or [All,
    Center, Dynamic, Extents, Previous, Scale, Window, Object]
    <real time>: (Pick a corner.)
Specify opposite corner: (Select opposite corner to be enlarged.)
Resuming LINE command.
Specify next point or [Undo]: (Select desired next point.)
Specify next point or [Undo]: ENTER
```

Zoom can be used at any point during a command when a prompt is given. In the **Line** command, the '**Zoom** command could have been started at the prompt for "first point" or at the "next point" prompt. Start the **Line**, **Circle**, or **Arc** command and then use '**Zoom** to become familiar with this option of zoom. Your eyes and back will love you as you zoom more and squint less.

ZOOM ALL

 The **Zoom All** option will zoom to show the entire drawing. **Zoom All** should be used if you are zoomed in to a small area of a drawing, such as Figure 8.2. The command can be started from the **Zoom** toolbar, or at the command line:

- Clicking the **Zoom All** button will automatically return all of the drawing to the screen so that the display resembles Figure 8.1.

- The command is executed at the command line using the follow sequence:

```
Command: Z ENTER
Specify corner of window, enter a scale (nX or nXP) or [All,
    Center, Dynamic, Extents, Previous, Scale, Window, Object]
    <real time>: A ENTER
```

ZOOM CENTER

The **Zoom Center** option allows a zoom to be specified by its desired center point. Once a center point is selected, a prompt will be given for the height of the display. If the current default is maintained, the drawing is redisplayed based on the new center point, but the magnification is not changed. If a smaller value for the height is selected, the display magnification will be increased. If a greater value for the height is selected, the display magnification will be reduced. The command sequence is as follows:

```
Command: Z ENTER (Or click the Zoom Center button.)
Specify corner of window, enter a scale (nX or nXP) or [All,
    Center, Dynamic, Extents, Previous, Scale, Window, Object]
    <real time >:C ENTER
Specify center point: (Select a point or enter coordinates.)
Enter magnification or height < 4'-4">:10 ENTER
```

ZOOM EXTENTS

The Zoom Extents option displays the entire drawing based on the size of the drawing producing the largest possible display of all the objects. Zoom Extents is helpful if the extent of the drawing exceeds the drawing limits because the option allows the entire drawing to be examined. The command can be started from the **Zoom** toolbar, or at the command line.

- Clicking the **Zoom Extents** button will automatically alter the display.

- The command is executed at the command line using the follow sequence:

```
Command: Z ENTER
Specify corner of window, enter a scale (nX or nXP) or [All,
    Center, Dynamic, Extents, Previous, Scale, Window, Object]
    <real time>:E ENTER
```

ZOOM PREVIOUS

The **Previous** option allows you to go backward, so that the previous drawing display can be seen. A maximum of 10 previous views can be restored using the Previous option. The command can be started from the Standard toolbar or at the command line.

- Clicking the Zoom Previous button will automatically alter the display.

- The command sequence for **Zoom Previous** at the command line is as follows:

```
Command: Z ENTER
Specify corner of window, enter a scale (nX or nXP) or [All,
    Center, Dynamic, Extents, Previous, Scale, Window, Object]
    <real time>:P ENTER
```

ZOOM SCALE

The Scale option of the **Zoom** command alters the display to a specified factor. The scale can be set to be relative to the full view or to a current view.

Scale Relative to Full View

When you enter a number for the scale factor, the entire drawing will be affected. A scale factor of 1 will display the current size. A scale factor of 3 will make objects appear three times as large. By entering a number smaller than 1, you will decrease the size of the object. If you enter .5, the object will appear half as big as the full display. The command sequence is as follows:

```
Command: Click the Zoom Scale button (Or type Z ENTER)
Specify corner of window, enter a scale (nX or nXP) or [All,
    Center, Dynamic, Extents, Previous, Scale, Window, Object]
    <real time>:S ENTER
Enter scale factor (nX or nXP):.5 ENTER
```

Scale Relative to Current View

If you enter a numeric value followed by an "X," the scale of the zoom will be relative to the current view, rather than the entire drawing. The command sequence is as follows:

```
Command: Click the Zoom Scale button (Or type Z ENTER).
Specify corner of window, enter a scale (nX or nXP) or [All,
    Center, Dynamic, Extents, Previous, Scale, Window, Object]
    <real time>: S ENTER
Enter scale factor (nX or nXP): 1.5X ENTER
```

Scale Relative to Paper Space Units

AutoCAD thinks of space as model space and paper space. **Zoom XP** can be used to scale each space relative to paper space units. This will be helpful, for instance, in printing drawings on the same sheet that are drawn at different scales. This topic will be covered in more detail in later chapters.

ZOOM IN

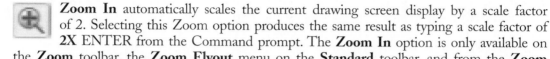

Zoom In automatically scales the current drawing screen display by a scale factor of 2. Selecting this Zoom option produces the same result as typing a scale factor of **2X** ENTER from the Command prompt. The **Zoom In** option is only available on the **Zoom** toolbar, the **Zoom Flyout** menu on the **Standard** toolbar, and from the **Zoom** submenu of the **View** menu. Once selected, the zoom will be automatically executed.

ZOOM OUT

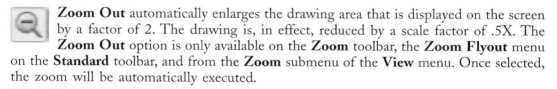

Zoom Out automatically enlarges the drawing area that is displayed on the screen by a factor of 2. The drawing is, in effect, reduced by a scale factor of .5X. The **Zoom Out** option is only available on the **Zoom** toolbar, the **Zoom Flyout** menu on the **Standard** toolbar, and from the **Zoom** submenu of the **View** menu. Once selected, the zoom will be automatically executed.

PAN

You can best visualize the **Pan** command by thinking of your drawing as a large sheet of paper that is rolled up. As you pan across your drawing, it's just as if you were unrolling the drawing to view a different portion. The drawing magnification is not changed, just the portion of the drawing that is displayed. This can be helpful when you've zoomed into a drawing, similar to the one in Figure 8.2, and you are looking at a specific detail. Rather than having to **Zoom All** and then **Zoom Window** to a new detail, **Pan** can be used to move to the desired detail. To start the **Pan** command:

- Click the **Pan Realtime** button on the **Standard** toolbar.
- Type **P** ENTER at the Command prompt.
- Select **Pan** from the **View** menu.
- Select **Pan** from the shortcut menu.

PAN REALTIME

The **Pan Realtime** option is similar to **Zoom Realtime**. The easiest method of starting the command is to click the **Pan Realtime** button on the **Standard** toolbar. If you're using a wheel mouse, **Pan Realtime** is executed by pressing and holding the wheel down, and then moving the mouse. As the command is activated, the cursor will change to a hand cursor. If the command line display is **ON**, the following prompt is displayed:

```
Press ESC or ENTER to exit, or right-click to display
       shortcut menu.
```

No prompt is given at the Dynamic Display if the command line prompt is removed from the drawing. Move the cursor (the hand) to the desired pick point and then press and hold the select button. This point now becomes the displacement point of the **Pan** command. As the cursor is moved while the select button is being held, the screen display is altered. **Pan** can be repeated indefinitely until the desired display is achieved. **Pan** is not limited by the edge of the screen. During a pan, you can drag your cursor at the edge of the monitor and continue to pan. End the command by pressing ESC or ENTER. Figures 8.4a and 8.4b compare the effects of using the **Pan** command.

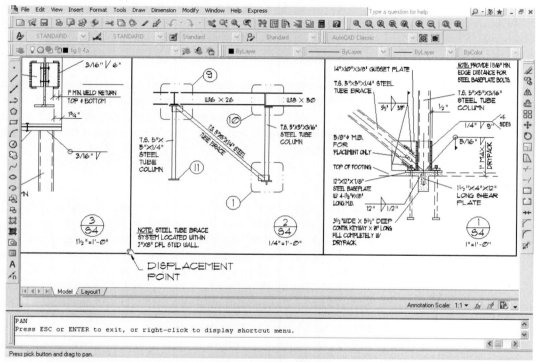

Figure 8.4a *Pan allows a drawing to be "scrolled" across the screen without the drawing magnification being changed. (Courtesy Kenneth D. Smith Architects & Associates, Inc.)*

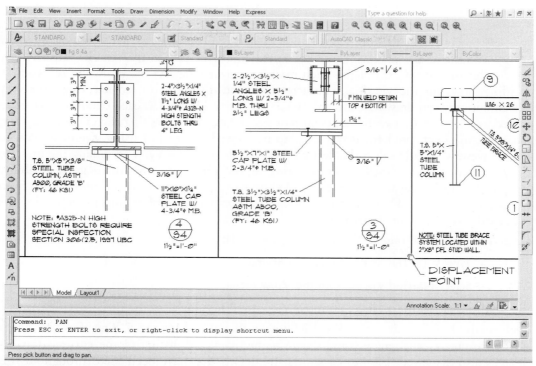

Figure 8.4b   *The effects of the **Pan** that was started in Figure 8.4a. (Courtesy Kenneth D. Smith Architects & Associates, Inc.)*

Combining Realtime Zoom and Pan

AutoCAD allows consecutive zoom and pan operations to be combined into one operation if the **Combine zoom and pan** button found on the **User Preferences** tab of the **Options** dialog box is toggled ON. By entering either **Pan Realtime** or **Zoom Realtime** using the appropriate button on the **Standard** toolbar or the shortcut menu, you can easily toggle between the two options and have a convenient method of moving between different portions of a large drawing. To experiment, try this:

1. Open a drawing and draw several lines and circles.

2. Click the **Zoom Realtime** button and use the select button to enlarge the drawing.

3. Now click the **Pan Realtime** button, and the cursor will change to the hand and allow the enlarged view to be panned.

Remember that when you right-click, a shortcut menu will be displayed that allows you to toggle between the **Zoom** and **Pan** commands.

VIEW

An alternative to **Zoom** and **Pan** is the use of predefined drawing views. The **View** command allows a display to be named and saved so that they can be retrieved. By default, AutoCAD begins each drawing as one view. As the limits are set, you are deciding how big the view is. The screen can be divided into smaller views for easy movement between drawings. On a drawing similar to Figure 8.1, each detail box could be named as a view so that each detail could be revised easily. By naming each detail as a view, you have, in effect, predefined the areas to be viewed. Instead of using **Zoom** or **Pan** to select a view, you could select the desired predefined view. The **View** command might seem like a waste of time on small exercises. However, the **View** command offers great flexibility for quickly viewing areas of a large, detailed drawing if you need to constantly move from one place to another. Dividing the drawing space of your template into predefined views will greatly reduce your drawing time.

To define views, use one of the following methods:

- Type **V** ENTER at the Command prompt.
- Select **Named Views** from the **View** menu.
- Select **Named Views** from the **View** toolbar.

Each method will produce the **View Manager** shown in Figure 8.5. The **View Manager** is used to create, set, rename, and delete named views. A listing of the current drawing views is also displayed. Figure 8.6 shows the details that have been used throughout this chapter and the view names that will be used.

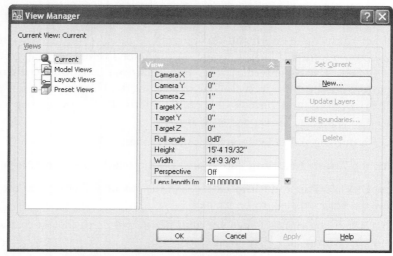

Figure 8.5 *The **View** command offers great flexibility for quickly viewing areas of a large, detailed drawing. The **View Manager** can be accessed by selecting **Named Views** from the **View** menu, by typing **V** ENTER at the Command prompt, or by selecting the **Named Views** button on the View toolbar.*

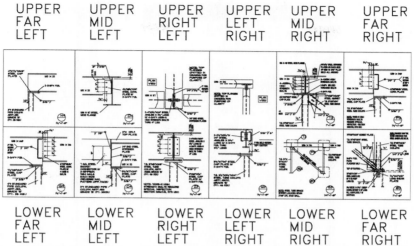

Figure 8.6 *The sheet of the details that have been used throughout this chapter and the view names that will be used.*

CREATING AND SAVING A NAMED VIEW

The following steps can be used to create a named view:

1. Click the **New** button from the **View Manager**. This will display the **New View** dialog box show in Figure 8.7.

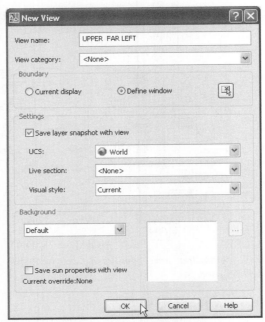

Figure 8.7 *Clicking the **New** button in the **View Manager** produces this display. The **New View** dialog box is used to assign drawing view names and to select the contents of the view.*

2. Assign a view name in the **View name** edit box (***upper far left***).

3. Choose the display mode.

- Choose the **Current display** radio button if the view on the screen is the view that you would like to have saved.

- Choose the **Define window** radio button if you need to select the objects to be included in the view.

4. If you selected the **Define window** button, use the **Define View Window** button to define the view. Selecting this button will remove the dialog box and restore the drawing area. Select the objects that will make up the view. Once the objects for the view are selected, the **New View** dialog box will be restored.

5. Select **OK**. The **View** dialog box will be restored, and the name of the selected view will be displayed.

Figure 8.8 shows the **View** dialog box with a partial listing of the twelve views described in Figure 8.6. The **New** option selects and saves the current screen display using a name that you supply. View titles can be up to 255 characters long, following the same guidelines used to name drawing files. Views are often named by their location in the job or by their specific function. If four rows of details will be drawn, each row could be given a letter and each column of details could be given a number to produce an easy grid to work with. Each room of a floor plan can also be turned into a view for easy movement through a project.

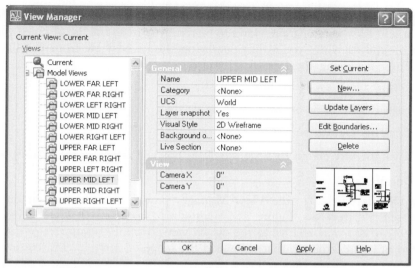

Figure 8.8 *The named views reflected in Figure 8.6 can be seen in the View Manager.*

Altering the Current View

The screen display can easily be altered between named views. To make the LOWER FAR LEFT view of Figure 8.6 the current view, use the following steps:

1. Type **V** ENTER at the Command prompt to display the **View Manager.**

2. Highlight the name of the view that is to become the current view.

3. Click the **Set Current** button.

4. Click the **OK** button. The dialog box will be closed and the LOWER FAR LEFT view will now occupy the screen display area.

Renaming a Named View

The name of a view can be altered in the View dialog box. Use the following steps to change the existing name of a view:

1. Display the **View Manager**.

2. Highlight the name of the view to be renamed. This will produce a list of information to the view in the **General** display.

3. Highlight the view name in the **Name** edit box in the **General** box.

4. Enter the desired name of the view.

5. Click the **OK** button to accept the new name, close the dialog box, and return to the drawing area.

Delete

This option will allow one or more views to be removed from the list of saved views. You're not removing the drawing, you're just removing the saved view from the list of named views. Delete views using the following procedure:

1. Display the **View** dialog box.

2. Highlight the name of the view to be deleted.

3. Click the **Delete** button.

4. Click the **OK** button to close the dialog box and return to the drawing area.

Viewing Details

The **View Details** portion of the **View Manager** provides information on the width and height of a selected view. See Figure 8.8. Highlighting a named view will provide the view name, the area of the view relative to the overall viewport, the coordinates of the view target location, and other information that is beyond the scope of entry-level discussion.

REFRESHING THE SCREEN

This chapter has introduced methods of controlling the drawing display using the **Zoom**, **Pan**, and **View** commands. Each command alters the view that is placed in the drawing area. The final aspect of controlling the drawing view is adjusting the quality of the view that is displayed. The view quality can be controlled using redraw, regeneration, and view resolution commands.

REDRAW

The **Redraw** command removes the current screen display momentarily and then replaces the drawing. As the drawing is replaced, blips are removed from the drawing editor. Lines that might have been partially erased from view while other objects were edited will also be restored. To execute the command:

- Type **R** ENTER at the Command prompt.

- Select **Redraw** from the **View** menu.

Redraw can be used while another command is in progress if it is entered at any non-text prompt. Typing **R** ENTER in the middle of a command sequence where text is expected will confuse AutoCAD and frustrate you with a misspelled word. On drawings similar in size to the exercises that you've been doing, **Redraw** will seem almost instantaneous. On larger drawings, such as a subdivision map or floor plan, **Redraw** may take a full nanosecond.

REGEN

A regeneration or **Regen** clears the drawing screen and then redisplays the entire drawing. AutoCAD recalculates the current screen display with 14-place accuracy, keeping track of each pixel in the X and Y directions. If you have zoomed in on a specific part of a drawing, **Regen** is regenerating even the part of the drawing that cannot be seen. To start a screen regeneration:

- Type **REGEN** ENTER at the Command prompt.

- Select **Regen** from the **View** menu.

REGENALL

In later chapters you'll learn how to divide the drawing screen into different viewports so that specific areas of the drawing can be viewed quickly. When multiple viewports are created, the **Regen** command will only affect the current drawing viewport. All other viewports will remain unaffected. The **Regenall** command will regenerate all of the views, so that blips in each view will be removed, and reindex the drawing database to increase the accuracy of object selection. To start the command, use one of the following methods:

- Select **Regen All** from the **View** menu.

- Type **REA** ENTER at the Command prompt.

REGENAUTO

As you create drawings, certain qualities of drawing objects may be altered as you change the zoom aspect. AutoCAD will automatically regenerate the drawing to restore the settings. On most drawings this may take a fraction of a nanosecond, or slightly longer on larger drawings. AutoCAD allows the regeneration mode to be set. AutoCAD allows a prompt to be given to remind you that the drawing is about to be regenerated. **Regenauto** can be entered only by keyboard. The command sequence is as follows:

```
Command: REGENAUTO ENTER
Enter mode [ON/OFF] <ON>: (Enter On or Off, or press ENTER.)
```

If the **ON** option is retained, AutoCAD will perform an automatic regeneration when needed while altering the screen display in commands such as **Zoom**, **Pan** and **View Restore**. When **OFF** is the current mode and a regeneration is needed as the screen is altered, AutoCAD will display a warning that it is about to regenerate.

VIEW RESOLUTION

The speed of each **Zoom** or **Pan** can be enhanced if a **Redraw** is used, rather than a **Regen**. The speed of all view changes also can be influenced by the view resolution. Resolution refers to the amount of detail that is represented when arcs and circles are drawn. The higher the resolution, the more lines that are used to represent an object and the smoother the arc or circle will appear. The resolution of arcs and circles can be controlled from the **Options** dialog box or by keyboard. To control the resolution from the dialog box, select **Options** from the **Tools** menu and then select the **Display** tab. The **Arc and circle smoothness** option in the **Display Resolution** box can be used to alter the smoothness of an arc. The **Segments in a polyline curve** option will be considered in later chapters. The other two options in this box affect 3D drawings and will not be considered in this text.

Arc and Circle Smoothness

The **Arc and circle smoothness** option controls the smoothness of circles. A value from 1 through 20,000 can be provided in the edit box in place of the default value of 1000. With a low number such as 25 or 50, circles will appear as a series of flat lines. As the value increases, the smoothness of the circle or arc is increased. Regeneration time on large drawings will slow slightly as the value is increased.

CHAPTER 8 EXERCISES

1. Open drawing **E-6-9**. Use the **Zoom** command to make the view containing the triangles the current view. Save this view as **TRIANGLES**. Zoom into each of the remaining shapes, and make each shape a separate named view. Save the entire drawing as **E-8-1**.

2. Using drawing **E-8-1** as a base, make the view containing the squares the current view. Draw a square in the center of one of the existing squares using Midpoint Object Snap to form a diagonal square. Repeat this process four additional times, using **Zoom** as required. Inside the final square, draw a circle that is tangent to the smallest square. Save this view with the name **TINY** and save the entire exercise as **E-8-2**.

3. Load **E-8-1** and enlarge the drawing view named **TRIANGLE** to twice its current size. Save this view as **VIEW2X**. Zoom out so that the edges of the largest triangle touch the edge of the screen. Name this view with the name of the command required to perform the zoom. Set the drawing so that screen markers will be produced and save the drawing as **E-8-3**.

4. Load drawing **E-8-3** and use it as a base for this exercise. Change the limits of this drawing to 36,18 and set the screen display so that the total content of the limits will be displayed. The original view is now displayed. Use **Zoom** and **Pan** to make the object's original view fill the screen. Save the drawing as **E-8-4**.

5. Start a new drawing and set the limits to 96',144' with a grid of 24" and a snap of 6". Set units to measure in architectural, with fractions set at 16. Set angles to be measured in degrees/minutes/seconds with two-place accuracy. Set the view resolution to 150. Divide the drawing into the following seven views: **All**, **Upper Left**, **Upper Cen**, **Upper Right**, **Low Right**, **Low Cen**, and **Low Left**. Set **All** as the current view. Save the drawing as a drawing template. This drawing can be used as a future base for drawings done at 1/4"=1'–0" scale. Save the drawing as **E-8-5**.

6. Open a file of your choice and establish a minimum of three named views. Provide names that quickly describe the area captured in the viewport. Save the file as **E-8-6**.

7. Open a file of your choice and use **Zoom** to view a small portion of the drawing. Save the enlarged view as **E-8-7**.

CHAPTER 8 QUIZ

DIRECTIONS

Answer the following questions with short complete statements. Type your answers using a word processor.

1. Place your name, chapter number, and the date at the top of the sheet.

2. Type the question number and provide the answer.

 Warning: Some of the questions have not been covered in the reading material and will require the use of the help menu. You may also have to do some exploring to answer the questions.

1. What command controls the use of screen blips, and how should it be set to provide markers?

2. Compare and explain the difference in **Viewers** settings between 50 and 500 and the default value.

3. Explain the difference between **Regen** and **Redraw**.

4. Describe the difference between **Zoom All** and **Zoom Extents**.

5. What commands will be needed to make a drawing exactly twice as large as the current screen display?

6. What option is used to scale space relative to paper space units?

_____

_____

7. Describe the meaning of "displacement point" and "second point."

_____

_____

8. List the steps to enlarge a portion of a drawing using the **Zoom** command.

_____

_____

9. List six options for accessing the **Zoom** command.

_____

_____

10. Describe the process that allows the **Zoom** command to be used in the middle of another command.

_____

_____

Basic Methods of Selecting and Modifying Drawing Objects

INTRODUCTION

By now you've explored the basic components of a drawing. This chapter will explore methods for

- Selecting objects for editing, including the pick box, Window, Crossing window, automatic selection mode, Wpolygon, Cpolygon, Fence, All, and Last

- Modifying the selection process using Noun/Verb, Shift to Add, Press and Drag, Implied Window, and Object grouping

- Modifying drawings by removing objects

- Modifying drawings by adding objects

Commands explored in this chapter include:

- **Pickbox**

- **Erase**

- **Copy**

- **Mirror**

- **Mirrtext** (Mirror text)

- **Offset**

- **Array**

Each command will reproduce all or parts of a drawing. A second group of editing commands for refining drawings and two new methods of sorting objects will be explored in the next chapter.

SELECTING OBJECTS FOR EDITING

AutoCAD provides several options for selecting objects. Individual objects can be selected using a pick box. Groups of objects can be selected for editing using one of the following options: Window, Crossing, Auto, WPolygon, CPolygon, Fence, All, Last, Add, Remove, or Multiple. As different methods of selecting objects for editing are discussed, each selection method will be introduced using the Erase command, although other commands can be used. These object selection methods can be used any time an object selection needs to be made within an editing command sequence. Clicking a button on a toolbar or entering the command at the keyboard are generally the fastest methods. No matter which method is used to access the command, a prompt will be displayed asking you to

Select objects:

Each of the following selection methods can be entered at the Select objects prompt for any of the editing commands.

SELECTING OBJECTS USING A PICK BOX

As you've edited your drawings to this point, you've selected objects to be edited one at a time. One or more objects have been selected using the pick box and clicking the select button. AutoCAD highlights each object that the cursor moves over to aid in the selection process. As the prompt for the edit command is entered, the crosshairs will change to an object selection target or pick box and the object to be edited can be selected. Any object that is included in the pick box will be affected by the command. If the size of the pick box is altered, it will affect the accuracy of your selections. As the size of the box is enlarged, information that is not desired in the selection set might be added. With the box too small, you might have trouble selecting information.

The size of the box can be altered to enlarge or reduce the size of the selection area. Adjust the size of the pick box by selecting **Options** from the **Tools** menu and then using the **Pickbox size** slide bar on the **Selection** tab.

- Move the slide bar to the right to enlarge the size of the pick box.
- Move the slide bar to the left to decrease the size of the pick box.

Selecting the Object

You can select objects to be edited by choosing any point on the object. Place the pick box on top of the object and click the select button. When you select objects with a thickness, it is important to place the pick box so that it touches an edge and not the center of the object. You also need to be careful not to select objects at their intersections with other objects.

AUTO SELECTION MODE

Rather than selecting objects to edit one at a time, you can select a group of objects by surrounding them completely in a window. If the select button is clicked, the window method of selecting objects will automatically be used. The window can be created before or after an editing command is selected. Since you've explored Erase already, enter a drawing, draw a few lines, and then click the ERASE button. When prompted for the object to be erased, click the select button on the mouse. This will turn the pick box into the corner of the selection box. As the cursor is moved, a shaded, semi-transparent window will be formed. Notice that the area inside of the window is highlighted to accent the objects that will form the selection set. Any items that the window encloses will be selected. When you're satisfied with the size of the selection box, click the select button to accept the objects. To remove the selected objects from the drawing, press ENTER. Figures 9.1a through 9.1c show the process of erasing objects using automatic selection. The command process is as follows:

Command: **E** ENTER *(Or click the **Erase** button.)*
Select objects: *(Click the select button to identify the first corner of the selection window.)*
Specify opposite corner: *(The window will drag across the screen until a second point is selected.)*
Select objects: **(Continue to add or remove objects to the selection set, or accept the selection by pressing** ENTER) ENTER

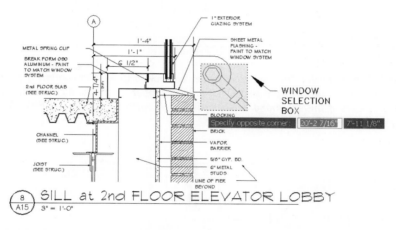

Figure 9.1a   *Use automatic selection or the Window option to form a selection set for the **Erase** command. Objects to be edited will be highlighted to ensure accuracy in the selection set. (Courtesy Peck, Smiley, Ettlin Architects.)*

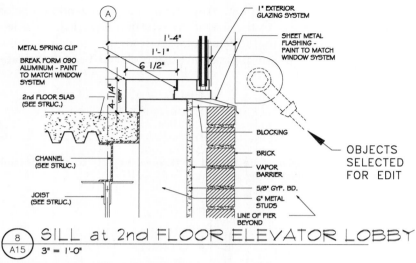

Figure 9.1b *Results of the Window selection set are now displayed to confirm the choice of objects to be included in the selection set. The number of objects included in the set is displayed if the command line is activated. (Courtesy Peck, Smiley, Ettlin Architects.)*

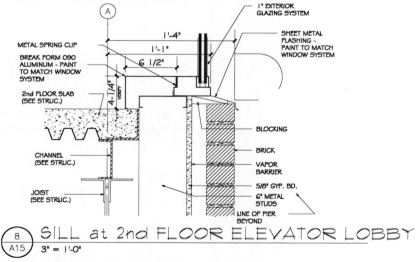

Figure 9.1c *The effect of the Window option used with the **Erase** command. (Courtesy Peck, Smiley, Ettlin Architects.)*

Implied Window Selection

The direction in which the cursor is moved as the selection window is defined will affect the selection process in two ways.

- If the cursor is moved from left to right when picking the selection set is picked, only objects completely enclosed in the window will be erased.

- When the selection window is created by moving from right to left, any object that is inside the window or that the window crosses will be affected by the edit command.

The results of implied window selection can be seen in Figure 9.2. Figures 9.3a and 9.3b show the use of a window created from right to left.

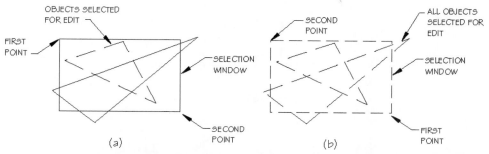

Figure 9.2 *If the cursor is moved from left to right, only objects completely enclosed in the window will be in the selection set. When the selection window is created by moving from right to left, any object that is in the window or that the window crosses will be affected by the edit command.*

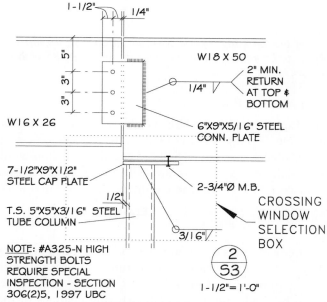

Figure 9.3a *Creating the selection window by moving from right to left is referred to as a **Crossing** window. The option is used here to remove the column, support plate, and related information. Any object touched by the window will be edited. (Courtesy Kenneth D. Smith Architect & Associates, Inc.)*

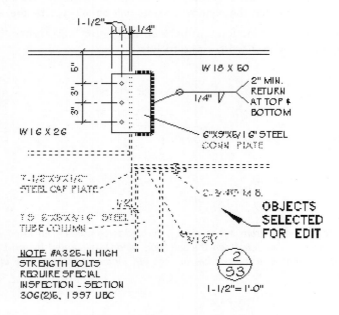

Figure 9.3b *Results of the **Crossing** window selection set are now displayed to confirm the choice of objects to be included in the selection set. (Courtesy Kenneth D. Smith Architect & Associates, Inc.)*

Selecting with a Window or Box

The same results accomplished with the left to right auto selection method can be achieved with the Window mode. Typing **W** ENTER at the Select objects prompt starts this selection method. Any object that lies entirely inside the selection window will be edited. Any object that is partially inside the window will not be edited as shown in Figure 9.2. The command sequence to use the **Erase** command with a window is as follows:

```
Command: E ENTER (Or click the Erase button.)
Select objects: W ENTER
```

The command continues in the same manner as with implied window selection. When prompted to select objects, you now have an opportunity to alter the selection set by adding additional objects using another window, or by using the Add or Remove selection method. If no changes need to be made, press ENTER. Making changes to the selection set will be introduced later in this chapter. Typing **B** ENTER will select objects for editing using a box similar to the one generated by the window options.

Selecting with a Crossing Window

The same results accomplished with the right to left auto selection method can be achieved with the Crossing window mode. Typing **C** ENTER at the Select objects prompt starts this selection method. Choosing an object to edit with the Crossing option produces the same results as the automatic Window option when the cursor is moved from right to left. See Figure 9.2. Any objects that lie entirely in the window will be edited. Any objects that are partially inside the window will also be edited.

 Note: To help keep track of the differences between window and crossing window, the crossing window is constructed using a dashed line, and a continuous line represents the Window option window.

SELECTION OPTIONS

In addition to selecting single objects or objects with windows, AutoCAD offers several other selection methods. Enter the Erase command, and when prompted to select objects, type **?** ENTER. The selected following menu options will be displayed:

```
Window/Last/Crossing/BOX/ALL/Fence/WPolygon/CPolygon/Group/Add/
    Remove/Multiple/Previous
/Undo/AUto/SIngle/SUbobject/Object
```

Selecting the Last Object

The **Last** option allows the most recently created object to be selected for editing.

```
Command: E ENTER (Or click the Erase button.)
Select options: L ENTER
Select options: ENTER
```

Selecting All Drawing Objects

When the All option is used, all objects that are current will be affected. You've learned how to freeze certain layers so that they cannot be seen. All objects except those on frozen or locked layers will be edited.

```
Command: E ENTER (Or click the Erase button.)
Select objects: ALL ENTER (All items will be highlighted.)
Select objects: ENTER (All items will be removed.)
```

SELECTING OBJECTS USING A FENCE

This option selects objects using a series of line segments. As seen in Figure 9.4, the selection fence edits only items that intersect or cross the fence. Fence points can be selected with a pointing device or by entering coordinates. The command sequence is as follows:

```
Command: E ENTER (Or click the Erase button.)
Select options: F ENTER
Specify first fence point: (Enter a point.)
Specify next fence point or [Undo]: (Enter a point.)
Specify next fence point [Undo]: (Continue until object set is
    complete.) ENTER
Select objects: ENTER (Pressing ENTER will execute the editing
    command.)
```

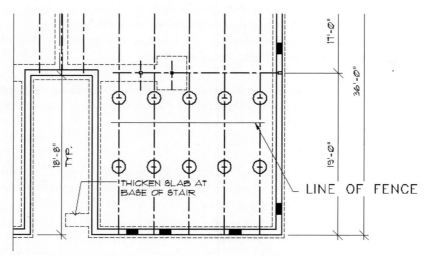

Figure 9.4 *The **Fence** option allows the use of straight lines to include objects in the selection set. Any objects that are touched by the fence will be included in the selection set. (Courtesy of Scott R. Beck, Architect.)*

Selecting Objects with the WPolygon Option

The WPolygon option is useful for selecting an irregularly shaped group of objects to edit. The option is similar to the Window option, but you are allowed to select points to form a polygon around objects to be edited. The command sequence is started once the editing option has been entered and you are prompted to select points to form the polygon. You can select points to define the polygon by using a pointing device as seen in Figure 9.4 or enter coordinates. Press ENTER to close the polygon. All objects lying totally inside the polygon will be edited. The process is completed using the following steps:

```
Command: E ENTER (Or click the Erase button—any editing command can
      be used.)
Select objects: WP ENTER
First polygon point: (Enter a point.)
Specify endpoint of line or [Undo]: (Select a point.)
Specify endpoint of line or [Undo]: (Select a point.)
Specify endpoint of line or [Undo]: (Select a point.)
Specify endpoint of line or [Undo]: (Continue until desired
      objects are selected.) ENTER
```

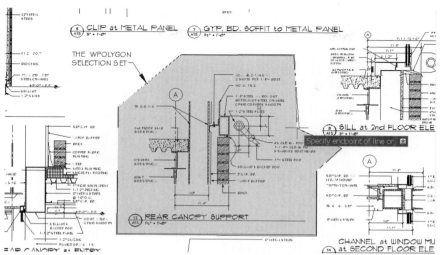

Figure 9.5   *The WPolygon option can be used as an alternative to the Closing option to select objects in an irregularly shaped group. Points for the polygon can be entered by selecting points or entering coordinates. (Courtesy Peck, Smiley, Ettlin Architects.)*

SELECTING OBJECTS WITH THE CPOLYGON OPTION

Objects can also be edited by typing **CP** ENTER at the Select objects prompt. This option will be similar to the effects of combining Crossing and WPolygon, in that it will edit all objects that are within or are crossed by the polygon. The command sequence is as follows:

```
Command: E ENTER (Or click the Erase button.)
Select objects: CP ENTER
First polygon point: (Select a point.)
Specify endpoint of line or [Undo]: (Select a point.)
Specify endpoint of line or [Undo]: (Continue until object set
     is complete) ENTER
```

Groups

AutoCAD allows multiple objects to be associated with each other in named groups. For instance, all of the symbols used in a bathroom on a floor plan can be linked together to form the BATH SYMBOLS group. Placing objects together in groups allows them to be edited as one object. Future chapters will explore the use of the **Group** option.

ADDING OBJECTS TO THE SELECTION SET

When selection sets are defined, you might not have selected all objects intended for editing. Typing **A** ENTER, for Add, at the Select objects prompt will allow for additional objects to be added to the selection set. Once Add is entered, objects can be added to the selection set individually or by any other selection method. When you are satisfied with the selection set, press ENTER to edit the drawing. Objects are added to a selection set using the following steps:

```
Command: E ENTER (Or click the Erase button.)
Select objects: (Click the select button to form an automatic
    window. Any selection method can be used.)
Specify opposite corner: (Specify second point.)
Select objects: A ENTER
Select objects: C ENTER (Any selection method can be used.)
Specify first corner: (Select first corner.)
Specify opposite corner: (Specify second point.)
Select objects: ENTER
```

REMOVING OBJECTS FROM THE SELECTION SET

This option will allow for objects to be removed from the selection set. Once objects are selected for editing, typing **R** ENTER will allow objects to be taken away from the selection set. Objects can be removed from the selection set individually or with any other selection method. When you are satisfied with the selection set, press ENTER to edit the drawing. Figure 9.6 shows material removed from the selection set of Figure 9.3b. The command sequence is as follows:

```
Command: E ENTER (Or click the Erase button.)
Select objects: (Click the select button to form an automatic
    window. Any selection method can be used.)
Specify opposite corner: (Specify second point.)
Select objects: R ENTER
Select objects: (Select objects to be removed using any selection
    method.)
Remove objects: ENTER
```

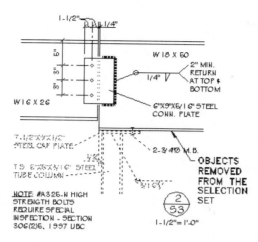

Figure 9.6   *If you would like to include more objects in the set, type **A** ENTER at the Select objects prompt. Typing **R** ENTER at the prompt will remove objects from the set. To accept the set, press ENTER. (Courtesy Kenneth D. Smith Architect & Associates, Inc.)*

Multiple

This option selects multiple points without highlighting the objects, which decreases the amount of time to complete the selection process for complex objects. The **Multiple** method also selects two intersecting objects if the intersection point is specified twice.

Previous

This option is used for editing groups of objects. It selects the most recent selection set for editing. The **Previous** selection set is cleared by operations that delete objects from the drawing. The program keeps track of whether each selection set was specified in model space or paper space. The Previous selection set is ignored if you switch spaces.

DEFINING THE SELECTION MODE

The **Selection** tab of the **Options** dialog box (choose **Options** from the **Tools** menu) offers six modes that will affect the formation of the selection set. An examination of this tab will reveal that six settings are available. Any combination of selection methods can be used.

NOUN/VERB SELECTION

Up to this point, as you have edited an object, you've used a process AutoCAD refers to as Verb/Noun. You've selected the verb (**Erase**) and then selected the noun (object to be edited). With the **Noun/Verb selection** check box active, AutoCAD allows you to reverse the process. The object can be selected first and then you can use the editing command. With **Noun/Verb selection** active, any object can be selected merely by choosing it with the pick box. Open a drawing and select a line. Notice the line is highlighted and a box appears in the middle and at each end of the line, but the Command prompt has not changed. Typing **E ENTER** will erase the selected line.

The Noun/Verb selection can be used with the following commands to edit objects:

Align	Copy	List	Rotate
Array	Dview	Mirror	Scale
Block	Erase	Move	Stretch
Change	Explode	Properties	WblockChprop

Commands that require objects to be selected before the edit command can be executed include the following:

Break	Divide	Fillet	Offset
Chamfer	Extend	Measure	Trim

USING SHIFT TO ADD SELECTION

This selection method controls how objects are added to an existing selection set. Start a new drawing and draw a few lines. Toggle **Noun/Verb selection** to **OFF** (no check) and **Use Shift to Add to selection ON**. Select **OK** and return to the drawing. Select an object to erase and it will be highlighted. Select a second object, and the first object no longer will be highlighted. To retain the first object, hold the shift key down while the second object is selected. Both objects will now be edited. If **Use Shift to Add to selection** is not enabled, objects are added to the selection by selecting them individually or using a selection window. To remove objects from the selection set, press SHIFT while selecting the object.

PRESS AND DRAG SELECTION

This mode determines the method used for drawing a selection window. With the **Press and drag** option active, the selection window can be made with one button selection rather than two. Pressing and holding the select button and moving the cursor diagonally creates the selection window. Releasing the select button completes the window when it is the size you want. When this option is not selected, the selection box is automatically created, but a second select is required to locate the opposite corner of the box.

IMPLIED WINDOWING

With the **Implied windowing** option in the **ON** mode, a selection window is automatically created when the Select object prompt is displayed. Clicking the select button will automatically select the first corner point. A prompt will then be given requesting the other corner. If you draw the window from left to right, the window selects objects that lie entirely within the window. If the window is drawn from right to left, all objects in and touching the window will be selected. With the option off, the window must be set for each edit. In the **OFF** position, clicking the select button does nothing. A mode such as Window or Crossing needs to be entered to start the creation of a window.

OBJECT GROUPING

The **Object grouping** option determines if grouped objects will be recognized as individual objects or as a group. In the default setting of **ON**, grouped objects function as a group. When **OFF**, grouped objects can be edited individually. Grouping of objects will be discussed in later chapters.

ASSOCIATIVE HATCH

The **Associative hatch** option affects how the hatch patterns are edited and will be discussed in chapter 13. You may wish to activate this option while exploring the program's features.

MODIFYING A DRAWING USING ERASE

 The **Erase** command allows you to remove unwanted objects (mistakes) from a drawing. The command has been used throughout this chapter as selection methods were introduced.

Objects to be erased can be selected by picking single objects, or by using any of the methods just discussed. The selected objects are highlighted, while AutoCAD waits for you to select other objects. When the selection process is complete, press ENTER to remove the object.

SALTERING DRAWINGS BY COPYING OBJECTS

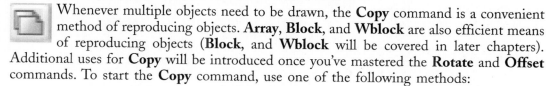 Whenever multiple objects need to be drawn, the **Copy** command is a convenient method of reproducing objects. **Array**, **Block**, and **Wblock** are also efficient means of reproducing objects (**Block**, and **Wblock** will be covered in later chapters). Additional uses for **Copy** will be introduced once you've mastered the **Rotate** and **Offset** commands. To start the **Copy** command, use one of the following methods:

- Click the **Copy** button on the **Modify** toolbar.
- Select **Copy** from the short-cut menu.
- Type **CP** ENTER at the Command prompt.
- Select **Copy** from the **Modify** menu.

Figure 9.7 shows an example of a drawing created using the **Copy** command. The drawing on the left was drawn first and copied to produce the drawing on the right. Once copied, the text was edited slightly to produce the detail. The command sequence is as follows:

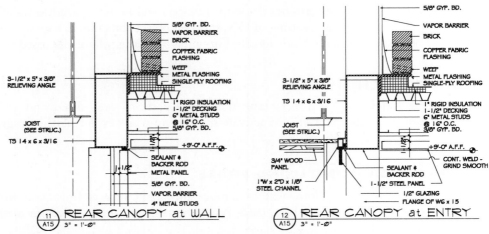

Figure 9.7 *The **Copy** command can be used to duplicate common features. The command can be selected from the **Modify** toolbar, typed, selected from the **Modify** drop-down menu, or selected from the short-cut menu. (Courtesy Peck, Smiley, Ettlin Architects.)*

```
Command: Click the Copy button (Or type CP ENTER.)
Select objects: (Select objects by desired selection method.)
Specify first corner: (Select window corner.)
Specify opposite corner: (Select opposite window corner.)
Select objects: ENTER
Specify base point or [displacement/mOde]<Displacement>: (Select
    base point or enter coordinates of object to be moved.)
Specify second point or < use first point as displacement>
    (Select a point to relocate the base point to.)
Specify second point or [Exit/Undo]: ENTER
```

The objects in the selection set will be copied, and the prompt will be displayed to continue the process indefinitely. When you've made the desired number of copies, press ENTER to end the command. The copy process can be seen in Figures (9.8a and 9.8d).

The mode setting controls whether the command will automatically repeat. Selecting this option will display the following prompt:

```
Enter a copy mode option [Single/Multiple] <Multiple>:
    Enter S or M
```

With the default setting, multiple copies of the selection set can be made. If for some reason you decide you don't like the convenience of making multiple copies, change the setting to Single.

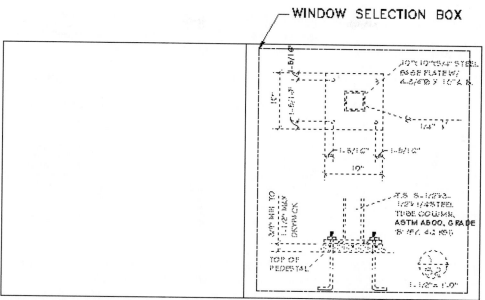

Figure 9.8a   *Start the **Copy** command by defining the selection set. (Courtesy Kenneth D. Smith Architect & Associates, Inc.)*

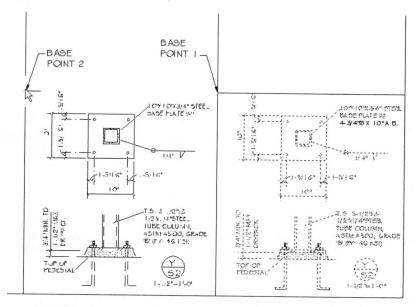

Figure 9.8b   *The second step of the **Copy** command is to specify a base point and then specify a new location for the base point. (Courtesy Kenneth D. Smith Architect & Associates, Inc.)*

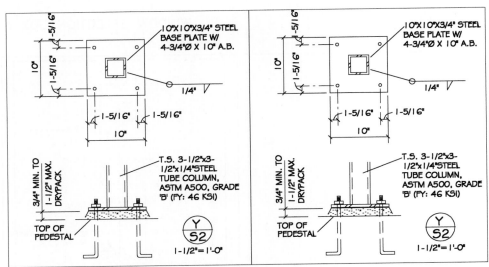

Figure 9.8c *Once ENTER is pressed for the new location, the objects in the selection set will be duplicated. (Courtesy Kenneth D. Smith Architect & Associates, Inc.)*

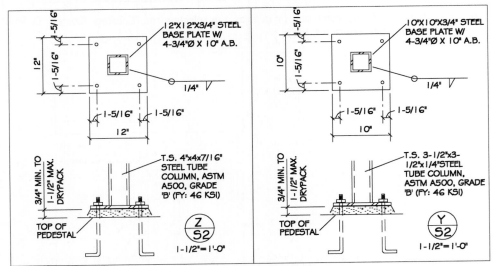

Figure 9.8d *Once copied, the detail can be edited to provide information about a similar column-plate intersection. Notice that the size of the plate, column, and the bolt locations have been edited. (Courtesy Kenneth D. Smith Architect & Associates, Inc.)*

CREATING SYMMETRICAL OBJECTS WITH MIRROR

Objects in a drawing often need to be reversed. Something as simple as reversing a door swing, or as complex as flipping a floor plan for an apartment building plan, (Figure 9.9) can be reversed using the **Mirror** command. The command also can be used any time symmetrical objects such as a fireplace, bay windows, sinks, convenience outlets, toilets, tubs, or double doors need to be represented. When half the object is drawn, the other half can be created using the **Mirror** command. To start the **Mirror** command, use one of the following methods:

- Click the **Mirror** button on the **Modify** toolbar.

- Type **MI** ENTER at the Command prompt.

- Select **Mirror** from the **Modify** menu.

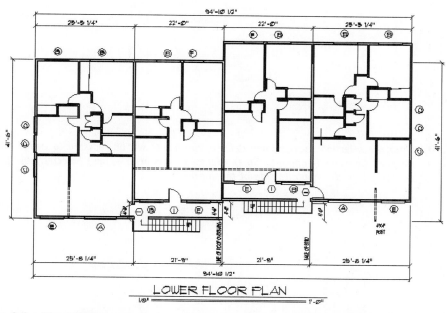

LOWER FLOOR PLAN

Figure 9.9 *The **Mirror** command can be used to flip existing objects. The command can be selected on the **Modify** toolbar, from the **Modify** menu, or entered by keyboard. (Courtesy Kenneth D. Smith Architect & Associates, Inc.)*

Once objects to be mirrored (flipped) are selected, you will be asked to describe a "mirror line" by selecting each endpoint. The process can be seen in Figure 9.10. Think of the mirror line as a fold line between the old and the new objects to be mirrored. As you pick the first point of the mirror, drag the selection set into position and alter it as the cursor is moved. The use of ORTHO and Object Snaps can greatly aid in the placement of the mirrored image. The new placement can be seen in Figure 9.11.

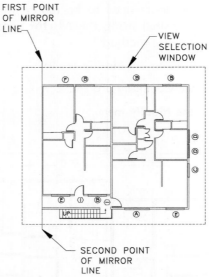

Figure 9.10 *Once the selection set has been defined, two points will be required to define the mirror line. (Courtesy Kenneth D. Smith Architect & Associates, Inc.) Selected objects will be duplicated on the opposite side of the mirror line.*

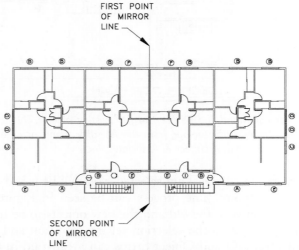

Figure 9.11 *As the second point is entered, the selection set will be mirrored into position. (Courtesy Kenneth D. Smith Architect & Associates, Inc.)*

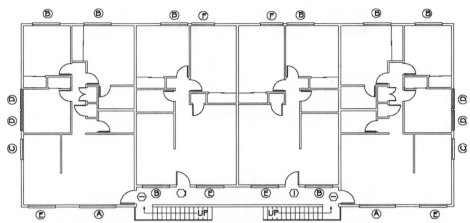

Figure 9.12   *Before the **Mirror** command is complete, an option will be displayed to "Erase source objects?" If you respond with **N** ENTER, the selection set will be mirrored and the original objects will be retained. (Courtesy Kenneth D. Smith Architect & Associates, Inc.)*

You'll be given the option of keeping or discarding the original object prior to making the **Mirror** operation permanent. Your choice will depend on the use of the drawing. If the object is being completed by **Mirror**, don't delete the old object. The final result will resemble Figure 9.12. The command sequence is as follows:

```
Command: Click the Mirror button (Or type MI ENTER.)
Select objects: (Choose objects using the desired selection
    method or combination of methods. Select window corner.)
Specify opposite corner: (Select opposite window corner.)
Select objects: ENTER
Specify first point of mirror line: (Select a point or enter
    coordinates.)
Specify second point of mirror line: (Select a point or enter
    coordinates.)
Erase source objects? [Yes/NO] <N>: ENTER
```

If the object is being flipped, you will want to remove the old object, as in Figure 9.13. The Command prompts would be similar to those just noted, until the last line is reached.

```
Erase source objects? [Yes/NO] <N>: Y ENTER
```

Typing **Y** ENTER will delete the original object and display the mirrored copy of the original object.

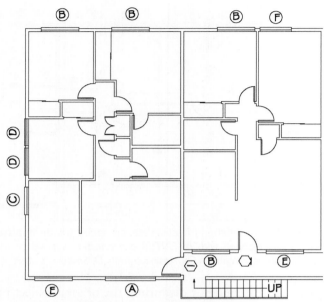

Figure 9.13 *If you respond **Y** ENTER to "Erase source objects?" the selection set will be mirrored and the original objects will be removed. (Courtesy Kenneth D. Smith Architect & Associates, Inc.)*

Note: One of the drawbacks of the **Mirror** command is how it affects text. Although you will not add text to a drawing for a few more chapters, an additional feature of **Mirror** will be helpful soon. By default, AutoCAD will not reverse text (see Figure 9.14). If you share a workstation with others, you may find this setting has been altered. To mirror the drawing without flipping the text, the **Mirrtext** value should be set to zero. With **Mirrtext** set to zero the drawing will be flopped, but the text will remain readable, as seen in Figure 9.15. This will be covered again as text is explored, but you might want to set the value in your drawing template profile no, so that when you do add text, you won't need the **Undo** or **Oops** command. The command sequence is as follows:

```
Command: MIRRTEXT ENTER
New value for MIRRTEXT <1>: 0 ENTER
```

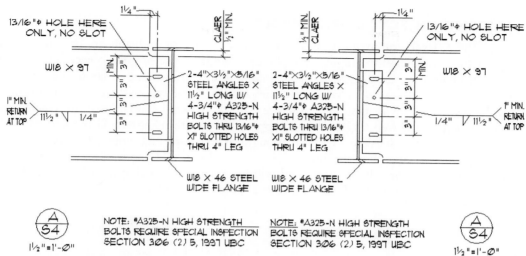

Figure 9.14 *The default setting of the* **Mirror** *command also will flop the drawing but not the text. (Courtesy Kenneth D. Smith Architect & Associates, Inc.)*

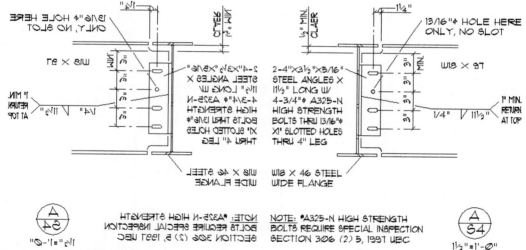

Figure 9.15 *If the* **Mirrtext** *variable is set to 1, text will be reversed. (Courtesy Kenneth D. Smith Architect & Associates, Inc.)*

CREATING OBJECTS USING OFFSET

The **Offset** command allows a line, circle, arc, or polygon to be copied and relocated parallel to and at a specific distance from the original object. The uses of this command are almost endless. A circle can be drawn and offset to represent the inner and outer surfaces of a conduit or sewer pipe. Starting with a horizontal and a vertical line, an entire rectangular structure can be drawn by offsetting lines and then cleaned up using other editing commands presented in later chapters. Using **Offset**, lines can be drawn in exact locations without thought as to their exact length. Get the location right without worrying that the line might be too long, and then edit the length with **Fillet, Stretch**, or **Trim**. Each command will be introduced in chapter 10. Figures 9.16a and 9.16b show the use of **Offset** to layout a simple floor plan.

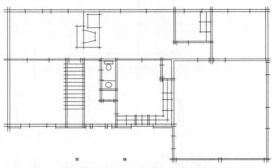

Figure 9.16a *With one horizontal and one vertical line, the **Offset** command can be used for drawing parallel lines at a specified distance apart to create a floor plan.*

Figure 9.16b *Once the basic lines have been put in the correct position using the **Offset** command, other editing commands can be used to finish the drawing.*

To start the command process, use one of the following methods:

- Click the **Offset** button on the **Modify** toolbar.

- Type **O** ENTER at the Command prompt.

- Select **Offset** from the **Modify** menu.

The command sequence to offset a line 18 from the original line is as follows:

```
Command: O ENTER (Or click the Offset button.)
Specify offset distance or [Through/Erase/Layer] <Through>:
    I'6 ENTER
```

A distance can be entered by providing a numeric value or by typing **T** ENTER. Choosing the **Erase** option will remove the original object from the drawing once the offset is complete. Choosing **Layer** allows you to indicate if the object to be offset will be placed on the current layer, or on the layer of the source object. When the distance is entered, a new prompt will be displayed that prompts

```
Select the object to offset or [Exit/Undo] <Exit>:
Lines, arcs, circles, ellipses, xlines, rays, and polylines can
    be offset (the last three objects will be introduced in
    later chapters).
```

Once selected, the object will be highlighted and the prompt will read

```
Specify point on side to offset [Exit/Multiple/Undo] <Exit>:
```

Move the cursor slightly to the desired side and click the select button to offset the object and redisplay the prompt:

```
Select the object to offset or [Exit/Undo] <Exit>:
```

If the Multiple option is selected, the process will continue indefinitely with the offset distance that was entered originally remaining as the default. Terminate the **Offset** command by pressing ENTER. Once the command is terminated, a new default distance can be entered or a new command sequence can be started. The command sequence can be seen in Figure 9.17.

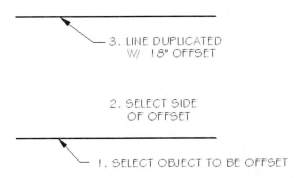

3. LINE DUPLICATED
W/ 18" OFFSET

2. SELECT SIDE
OF OFFSET

1. SELECT OBJECT TO BE OFFSET

Figure 9.17 *Complete the* **Offset** *command by selecting the distance to be offset, the object to be offset, and the side of the original on which the offset will occur.*

OFFSET OPTIONS

The offset command has several options that increase the effectiveness of the command, including Through, Erase, Layer, Exit, Multiple and Undo.

Through

Typing **T** ENTER at the original prompt allows a point to be entered with a pointing device, and the object to be offset will pass through the indicated point. The command sequence is as follows:

```
Command: O ENTER (Or click the Offset button.)
Specify offset distance or [Through/Erase/Layer]<1'-6>: T ENTER
Select the object to offset or [Exit/Undo]: (Select an object.)
Specify through point: (Move cursor to the desired point and
    select.)
Select the object to offset or [Exit/Undo]: ENTER
```

The Through option of the **Offset** command can be seen in Figure 9.18.

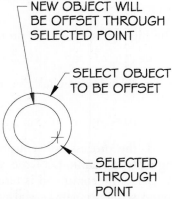

NEW OBJECT WILL
BE OFFSET THROUGH
SELECTED POINT

SELECT OBJECT
TO BE OFFSET

SELECTED
THROUGH
POINT

Figure 9.18 *The Through option of **Offset** allows the offset distance to be specified using the pointing device.*

Erase

Using the **Erase** option will eliminate the source object after is has been offset. When using this option the prompt will read:

```
Erase source object after offsetting [Yes/No]<current>: (Enter Y
    or N) ENTER
```

Layer

This option determines if objects that are offset will be placed on the current layer or on the layer of the source object. When using the **Layer** option, the prompt will read:

```
Enter layer option for offset objects [Current/Source]<current>:
    (Enter an option.)
```

Exit

Pressing ENTER when this option is the default value will end the **Offset** command.

Multiple

The **Multiple** option mode repeats the offset operation using the current offset distance as additional through points are selected. Once a distance and side have been selected, the **Multiple** option will continue to prompt for new locations.

Undo

The **Undo** option removes the most recent offset and allows the command sequence to continue. The prompt reads:

```
Select object to offset or [Exit/Undo] <Exit>: U ENTER
Select object to offset or [Exit/Undo] <Exit> (Select new object,
    exit the command, or continue to undo the previous offsets.)
    ENTER
```

PRACTICAL USES FOR OFFSET

Open your architectural template. Now draw intersecting horizontal and vertical lines. Refer to the floor plan in Figure 9.16b. The **Offset** command can be useful in the layout of a floor plan. Assume the two lines that have been drawn represent the outer edges of the walls in the upper right corner of the plan shown in Figure 9.16b. Offset each line 6 in toward the inside of the house. Because the room in Figure 9.16b is 12'–4 wide, offset the vertical line you just created a distance of 12'–4 to the left. Offset this same line 4 to represent the wall thickness. Offset this line 8'–2, and then again 4. This will provide the wall on the left side of the Utility room.

Now go back to the inner horizontal line that you drew. Offset this line 11'–4 toward the bottom of the drawing. Offset this line 6 toward the bottom of the drawing. These two lines will represent the wall between the shop and the garage. To place the wall between the utility room and the hall, offset the inner garage wall 42. Offset this new line 4. The entire process is shown in Figure 9.19. Save this crude drawing as **FLOOR10**.

Figure 9.19 *By drawing two perpendicular lines and using the **Offset** command, you can easily draw a floor plan.*

ARRANGING MULTIPLE OBJECTS WITH ARRAY

Until now, to reproduce a drawing element you've used the **Copy** command. The **Array** command will also allow for multiple copies of an object or group of objects to be reproduced—such as the beam, post, and columns of a post-and-beam foundation. The command also provides several added features not available with **Copy**. **Array** reproduces objects in rectangular or circular (polar) patterns, allows the object to be rotated as it is reproduced, and allows for easy control of the spacing of the object during the **Array** operation. Figures 9.20a and 9.20b show examples of a rectangular and a polar array. To start the command, use one of the following methods:

- Click the **Array** button on the **Modify** toolbar.
- Type **AR** ENTER at the Command prompt.
- Select **Array** from the **Modify** menu.

Each method will produce the **Array** dialog box shown in Figure 9.21. Once the dialog box has been accessed, the type of array pattern can be selected.

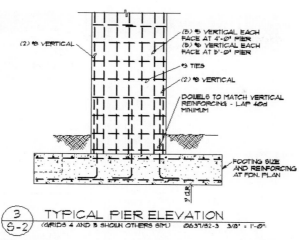

Figure 9.20a   *The steel reinforcement was placed using the **Array** command. (Courtesy Van Domelen/Looijenga/McGarrigle/Knauf Consulting Engineers.)*

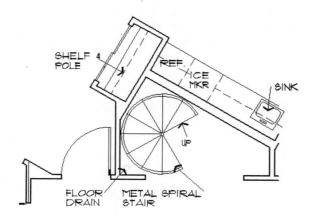

Figure 9.20b   *The treads of this stair were placed using the Polar option of the Array command. (Courtesy Piercy & Barclay Designers, Inc.)*

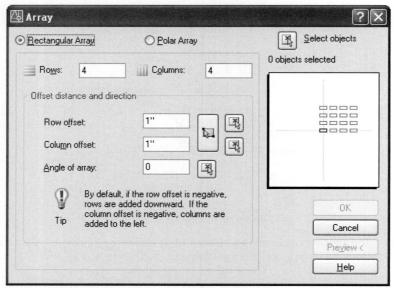

Figure 9.21 *The **Array** dialog box is used to recreate objects in rectangular and circular patterns. Access the display dialog box by selecting the Array button on the modify toolbar, by typing **AR** ENTER at the Command prompt or by selecting **Array** from the **Modify** drop-down menu.*

RECTANGULAR ARRAY

A rectangular array will prove suitable for projects that are arranged in rows or columns. Selecting the **Rectangular Array** button starts the process. Six additional specifications must be provided to perform a rectangular array including: selecting the object to be arrayed, the number of rows and columns, the distance between row and column offsets, and the array angle. The elements can be seen in Figure 9.22.

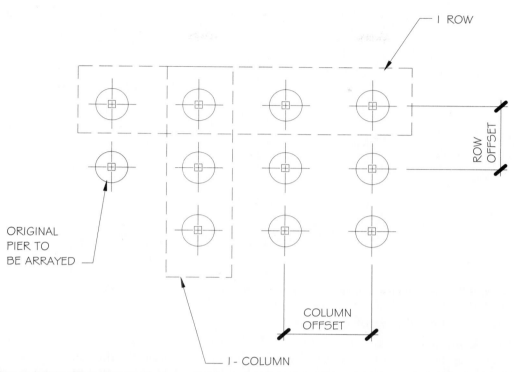

Figure 9.22   *The number of rows and columns and the spacing of each must be specified to use a rectangular array.*

Selecting the Object to Array

The **Select objects** button removes the **Array** dialog box and allows objects to be selected that will form the array pattern. Drawing objects can be selected using any of the selection methods introduced earlier. When the selection set is complete, press ENTER or SPACEBAR to redisplay the dialog box.

Specifying the Number of Rows

In the **Rows** edit box, specify the quantity of horizontal rows to be created. Think of rows as the seats in a theater. Any whole number can be entered for the value. Figure 9.23 shows an example of an array with two, three, and four rows.

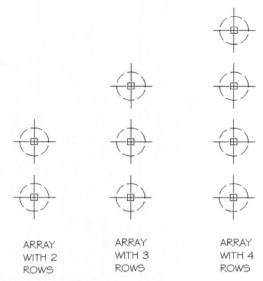

ARRAY
WITH 2
ROWS

ARRAY
WITH 3
ROWS

ARRAY
WITH 4
ROWS

Figure 9.23 *Altering the number of rows of an array.*

Specifying the Number of Columns

In the **Columns** edit box, specify the number of vertical columns to be used. Any whole number can be entered for the value. If the value of 1 was used for the row value, a value other than 1 must be used for the column value. Figure 9.24 shows the effect of an array using two, three, and four columns.

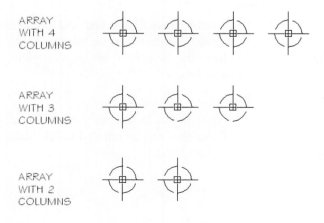

ARRAY
WITH 4
COLUMNS

ARRAY
WITH 3
COLUMNS

ARRAY
WITH 2
COLUMNS

Figure 9.24 *Altering the number of columns in an array.*

Specifying the Row Offset

The **Row offset** edit box allows the offset distance to be set for rows. The offset distance includes the size of the object to be arrayed. If you are arranging 16 wide chairs in rows, and would like 16 between rows, the spacing would require a 32 unit cell. This is shown in Figure 9.25. You can set the distance between rows by selecting the **Pick Row Offset** button and then selecting a point with the cursor, or by selecting the **Pick Both Offset** button and then selecting the unit cell as two opposite points of a rectangle. The rectangle will provide information for the row spacing. If more than one column is to be arrayed, the rectangle will provide information for both the row and the column distances.

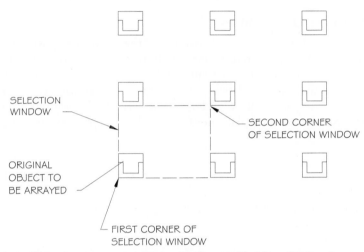

Figure 9.25 *The offsets for rows and columns can be specified by selecting the two opposite corners of a window.*

Specifying the Column Offset

The **Column offset** edit box for a rectangular array provides the distance between columns. The same guidelines for spacing rows apply to spacing column offsets. Figure 9.26 shows the results of an array pattern with three rows and eight columns.

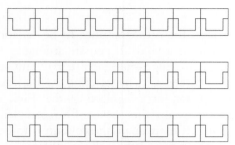

Figure 9.26 *The effect of row and column spacing in an array of three rows and eight columns.*

Positive and Negative Offset Values

As values are provided for rows and columns, they can be positive, negative, or a combination of both. Using positive values for both will array the objects up and to the right of the original object, as seen in Figure 9.27. If a negative value is used for the row distance, new objects will be added below the original. If negative values are used for the column distance, new objects will be added to the left of the original object. Figure 9.28 shows the effects on the pattern placement of combining positive and negative values.

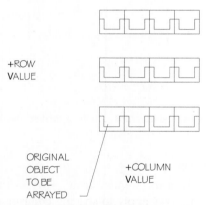

Figure 9.27 *An array based on entering two positive distances.*

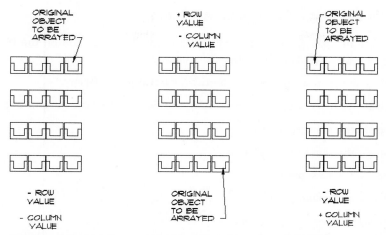

Figure 9.28   *An array pattern will vary based on the values entered. The pattern on the left is made using two negative values. The center option uses positive row and negative column values. The pattern on the right uses a negative row and a positive column value.*

Specifying an Angle for Rectangular Arrays

The array pattern is normally based on a horizontal or vertical baseline. Entering a numeric value in the **Angle of array** edit box will rotate the baseline angle. Selecting the **Pick Angle of Array** button will also allow the angle array to be altered. Selecting this button will prompt for two lines to be specified. The angle of the line drawn will become the angle for the array pattern. Figure 9.29 shows the effect of using an angled array.

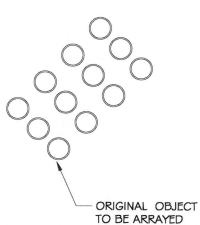

ORIGINAL OBJECT
TO BE ARRAYED

Figure 9.29   *A rotated array can be drawn by selecting the **Pick Angle of Array** button.*

Summary of Steps in the Rectangular Array Process

1. Display the **Array** dialog box by selecting the **Array** button on the **Modify** toolbar or by typing **AR** ENTER at the Command prompt.

2. Select the method of array to be used, **Rectangular Array**.

3. Select the objects to be arrayed using the **Select objects** button.

4. Specify the number of rows.

5. Specify the number of columns.

6. Specify the row offset.

7. Specify the column offset.

8. Specify the array angle if required.

9. Preview the array by selecting the **Preview** button. If the pattern is okay, press the **Accept** button. If the pattern is not what you expected, select the **Modify** button, and alter the pattern as needed.

10. Select the **Accept** button to accept the array as specified. If you decide not to complete the pattern, select the Cancel button.

POLAR ARRAY

A polar array is a circular layout of objects based on a center point. Selecting the **Polar Array** button starts the process. Once **Polar Array** is selected, the dialog box will be altered as shown in Figure 9.30. Six additional specifications must be provided to perform a polar array. These include selecting the center point of the array pattern, the method of array, the rotation of objects, and the object base point.

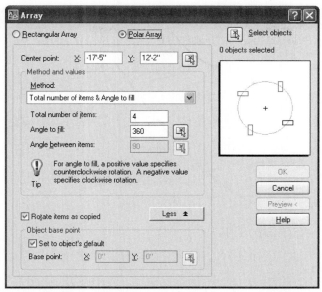

Figure 9.30 *Selecting the **Polar Array** button of the **Array** dialog box will alter the display of the dialog box, allowing objects to be created in a circular pattern.*

Selecting the Center Point

The center point is the point that the selected objects will be arrayed around. The center point can be selected by entering coordinates in the **Center point** X and Y edit boxes or by selecting the **Pick Center Point** button. Selecting the **Pick Center Point** button will temporarily remove the dialog box to allow a center point to be specified.

Selecting the Method of Array

Three methods of completing a polar array pattern are available. Each method can be selected from the **Method** edit box. Options include the following:

- Total number of items & Angle to fill
- Total number of items & Angle between items
- Angle to fill & Angle between items

The option selected will determine which of the three prompts in the **Method and values** area will be active.

Number of Items—The **Total number of items** edit box allows you to specify how many items, including the original, will be displayed around the center point.

Angle to Fill—The **Angle to fill** edit box controls the limits of the display around the center point. The default of 360° will locate the specified number of items around a full circle. A number less than 360° will array the indicated number of objects around a selected center point in a specified degree range. The pattern will start with the original object and spread the balance of the pattern from its location. A positive value will produce a counterclockwise rotation. A negative value will cause a clockwise rotation. Examples of each pattern can be seen in Figure 9.31.

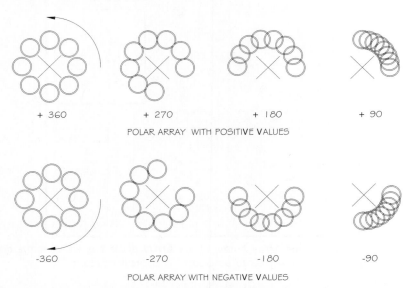

Figure 9.31 *The effect of adjusting the **Angle to fill** edit box for a polar array.*

Angle Between Items—The **Angle between items** edit box can be used to specify the angle between items. The number of items to be rotated, combined with the angle distance between each item, is often used to locate bolts or other similar features for drawing connectors. Materials also can be arrayed over a specified degree range and intervals, such as showing pipes that intersect a cooling unit over a 75° area at each 15°. Figure 9.32 shows an example of each type of placement.

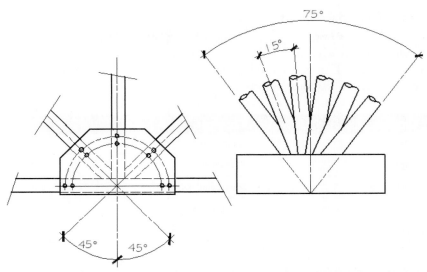

Figure 9.32   *In the example on the left, the pattern was completed by specifying the number of items to be rotated and the **Angle between items.** The example on the right, the **Angle to fill** and the **Angle** between items is used to define the pattern.*

Rotating Arrayed Objects

Selecting the **More** button will display this option. When active, this option allows objects to be arrayed and the object should be rotated as it is arrayed. The default is yes, which will rotate objects as they are arrayed around the center point. If the **Rotate items as copied** check box is inactive, the objects will not be rotated and will appear similar to the right side of Figure 9.33.

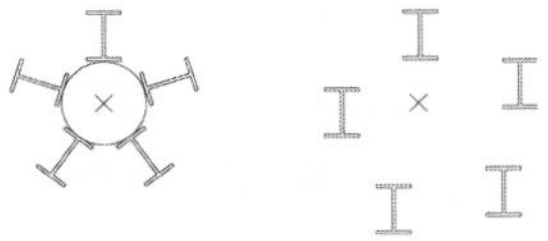

Figure 9.33   *Objects can be rotated as they are arrayed if the **Rotate objects as copied** check box is activated.*

Selecting an Object Base Point

The **Object base point** option specifies a new base point relative to the selection set. The base point will remain a constant distance from the center point of the array and the objects to be arrayed. The point used depends on the type of the object to be arrayed. Default base points are listed as follows:

Object type	Default base point
Arc, circle, ellipse	Center point
Polygon, rectangle	First corner
Donut, line, polyline, ray	Start point
Block, paragraph text, single-line text	Insertion point
Construction lines	Midpoint
Region	Grip point

Although some of these terms are unfamiliar, they will be introduced in later chapters.

Summary of Steps in the Polar Array Process

The entire process for a polar array of 50 segments around a 250° arc is as follows:

1. Display the **Array** dialog box by clicking the **Array** button on the **Modify** toolbar or by typing **AR** ENTER at the Command prompt.

2. Select the method of array to be used, **Polar Array**.

3. Select the objects to be arrayed using the **Select objects** button.

4. Select the center point of the array pattern.

5. Select the method. For this example, the **Total number of items & Angle to fill** option was used. Notice that the **Angle between items** edit box is inactive.

6. Enter 270 in the **Total number of items** edit box.

7. Enter 250 in the **Angle to fill** edit box.

8. Keep the **Rotate items as copied** check box active.

9. Preview the array by selecting the **Preview** button.

10. Select the **Accept** button to accept the array as specified. If you decide not to complete the pattern, select the **Cancel** button.

CHAPTER 9 EXERCISES

1. Start a new drawing and draw a horizontal line. Draw polygons with three, four, five, six, and eight sides with the centerline of each polygon on the horizontal line. Draw all polygons inscribed in a circle with a .5" radius. Make copies of the polygons with three, five, and eight sides and place them directly below their counterparts. Make copies of the four-sided and six-sided polygons and place them in a third row directly below their counterparts. Mirror the entire drawing with a horizontal mirror line, and delete the old objects. Save the drawing as **E-9-1**.

2. Open the existing file **E-7-5**. Copy the window so that a total of three pairs of windows are drawn exactly 4" apart. Align the bottoms of the windows. Save the drawing as **E-9-2**.

3. Draw half of a W 8" × 28" (I-shaped) beam. The beam has a total height of 8.28". The top and bottom flanges are .25" thick, with a total width of 5.25". The web is a total of .28" thick. Use the **Mirror** command to complete the drawing. Save the drawing as **E-9-3BM**.

4. Using the attached drawing as a guide, draw an exterior swinging door that is 36" wide with a 12" wide sidelight. Use the **Mirror** command to draw a pair of doors with double sidelights. Save the drawing as **E-9-4-DOOR**.

5. Using the following figure as a guide, draw a steel connector plate. The plate is 15 1/2" x 10"h. Draw the bolts as 3/4" diameter hexagons, 1 1/2" down and 1 1/2" in from the top and ends. Bolts are 3" O.C. Each end of the plate is symmetrical. Draw a 2 1/2" x 8" strap, centered in the bottom of the plate. Provide two bolts with similar spacing as the plate. Save the drawing as **E-9-5**.

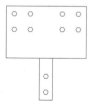

6. Use the drawing below as a guide and lay out the outline of the apartment floor plan. Draw the exterior walls as 6" wide. Use this unit plan and lay out four different building layouts with a minimum of four units per building. Save the drawing as **APARTFLR**.

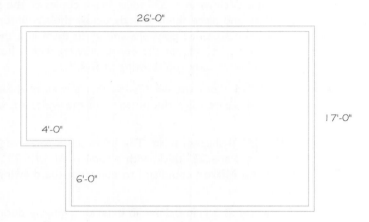

26'-0"

17'-0"

4'-0"

6'-0"

7. Draw a post-and-beam foundation plan for a 24' x 16' structure. Exterior walls will be 8" thick. Piers are 18" diameter at 8' O.C. along a 4" wide beam. Repeat the beam and piers at 4' O.C. Save this exercise as **PB-FND**.

8. Start a new drawing that can be used as a base for shingle siding in elevation. Draw several lines approximately 36" long, at 10" O.C. Draw shingles similar to the sketch with widths that vary from 4" through 12'. Copy this row of shingles to form a minimum of five rows of shingles. Stagger each row with the rows above and below it, so that the shingle seams do not line up. Save the drawings as **SHINGLE**.

9. Start a new drawing and set the limits as needed to draw a 4" square centered in a 15" diameter circle. Draw an 8" and a 12" diameter circle with the same center point as the 15" circle. Make three additional copies of this object and move each so that the four 15" circles are tangent. Save the drawing as **E-9-9**.

10. Open drawing **E-9.2**. Your client thinks he would like a grid in each half but isn't really sure. Copy the drawing and draw a grid in one of the windows. Provide one vertical and three horizontal dividers. Mirror the grid into the other. Save the drawing as **E-9-10**.

11. Start a new drawing and draw a 60' wide x 44' deep structure. Use 8" wide walls. No doors or windows need to be shown. Create a layer to represent the suspended ceiling and draw the plan for a suspended ceiling using 24" x 48" panels. Make a second copy of the plan and show the panels laid out in the opposite direction. Save the drawing as **E-9-11**.

12. Use the base drawing started in Exercise 9.11 to draw a roof framing plan. Freeze the suspended ceiling information and create a new layer to represent roof trusses at 32" O.C. Use the attached drawing as a guide to show a portion of the roof with 4' x 8' plywood over the trusses. Save the drawing as **E-9-12**.

13. Draw a column that is 10'–0" tall and 12" wide. Show a 20" x 4" and a 16" x 4" corbel 4" down from the top and 4" up from the bottom of the column. Save the drawing as **E-9-13**.

14. A circular structure is to be built on a hillside. Use the drawing on the next page as a guide and draw the pier plan. Save the drawing as **E-9-14**.

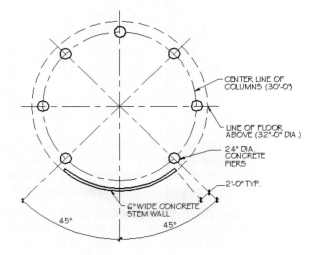

CENTER LINE OF
COLUMNS (30'-0")

LINE OF FLOOR
ABOVE (32"-0" DIA.)

24" DIA.
CONCRETE
PIERS

2'-0" TYP.

6"WIDE CONCRETE
STEM WALL

45°   45°

Figure 9.38

15. Draw a plan view of a 40' diameter water reservoir. The reservoir will be supported on (12) 6" diameter × 3/8" steel columns. The columns will be equally spaced, on a 36' diameter circle centered on an 18" diameter 3 3/4" thick vertical inlet pipe in the center of the reservoir. Assume north to be at the top of the drawing, and locate a 12" diameter 3 3/8" thick vertical outlet pipe at the outer edge of the tank, at 37° west of true north. Save the drawing as **E-9-15**.

CHAPTER 9 QUIZ

DIRECTIONS

Answer the following questions with short complete statements. Type your answers using a word processor.

1. Place your name, chapter number, and the date at the top of the sheet.

2. Type the question number and provide the answer.

Warning: Some of the questions have not been covered in the reading material and will require the use of the help menu. You may also have to do some exploring to answer the questions.

I. Describe the process used to draw a wall that is 6" thick.

2. Several copies of an object need to be made. What option is available so that the **Copy** command does not have to be repeated?

_____

_____

3. You have selected several objects for editing when you realize you have accidentally selected an object that you do not want changed. What can you do so that you do not have to start the edit sequence over again?

_____

_____

4. How will **Undo** affect objects that have just been moved?

_____

_____

5. List three methods suitable for selecting irregularly shaped objects for editing.

_____

_____

6. When the Fence option is used, what objects will it select for editing?

_____

_____

7. What editing mode will allow you to select objects and then decide how to edit them?

_____

_____

8. What process will display the dialog box that controls selection methods?

_____

_____

9. List the commands in sequence required to complete Exercise 9.

_____

_____

10. Explain when the **Through** option of **Offset** could be used.

_____

_____

11. Start the **Erase** command using each method and describe the advantages or disadvantages.

_____

_____

12. Experiment with the **Mirror** command. Describe the effects of ORTHO on the mirror line.

_____

_____

13. Mirror is being used to complete a symmetrical object. What options are available and how do they affect the mirror sequence?

_____

_____

14. Where is **Erase** listed in a menu?

_____

_____

15. Describe how to change the size of the pick box.

_____

_____

16. What is the effect of Base point on the **Copy** command, and how should it be selected?

_____

_____

17. List ways that the **Copy** command can be used to draw a floor plan.

_____

_____

18. What is the benefit of having so many different methods of selecting objects for editing?

_____

_____

19. What are the drawbacks to enlarging the pick box to twice its normal size, or decreasing it to half its normal size?

_____

_____

20. Use the **Help** menu to explain the difference between the normal setting for an editing command and the Single option.

_____

_____

21. Explain the difference between Noun/Verb and Verb/Noun selection.

_____

_____

22. Explain the meaning and the effect of the following letters for selection groups: W, C, F, WP, CP, R, A, and P.

_____

_____

23. What effect does Press and drag have on selection?

_____

_____

24. Use the **Help** menu to list the effects of the **Pickadd**, **Pickauto**, and Pickdrag variables.

_____

_____

25. How can Object Snap be used with editing commands?

_____

_____

26. Explain the difference between the two types of array patterns.

_____

_____

27. Twelve steel columns need to be drawn 22' on center in one horizontal line. What will the array sequence be?

_____

_____

28. Three horizontal lines of 10 steel columns need to be drawn 18' on center. The horizontal lines of columns will be 32' apart. What will the array sequence be?

_____

_____

29. Use the drawing for Exercise 9.14 and give the command sequence to array the supports.

_____

_____

30. Using an illustration as an example, define a unit cell.

_____

_____

31. How could an array pattern for objects aligned at 45° from horizontal be drawn?

_____

_____

32. Ten rows and three columns are to be arrayed below and to the left of the original object at 60" O.C. How can this be done?

_____

_____

33. Describe how to produce a clockwise rotation.

_____

_____

34. Give the combination of alternatives for describing a polar array.

_____

_____

35. If a positive column distance and a negative row distance are entered, where will the objects be arrayed in relation to the original object?

_____

_____

Modifying the Position and Size of Drawing Objects

INTRODUCTION

In chapter 9 you were introduced to editing commands that can affect an entire drawing. This chapter will explore:

- Commands that can be used to modify part of an object by altering the position or size of the object
- Commands that can be used to edit objects by altering their length
- Methods to select objects for editing using grips
- Methods for selecting objects for multiple edits
- Methods for naming object groups

Commands explored in this chapter include:

- **Move**
- **Align**
- **Rotate**
- **Scale**
- **Stretch**
- **Lengthen**
- **Trim**
- **Extend**
- **Break**
- **Join**
- **Chamfer**
- **Fillet**

- **Select**

- **Group**

Mastering these commands will greatly increase your drawing ability. All of the commands, with the exception of **Align**, **Select** and **Group**, can be selected on the **Modify** toolbar, entered by keyboard or selected from the **Modify** menu. Each command can also be entered from the shortcut menu, by selecting an object to edit and then right-clicking to select the editing command.

ALTERING THE POSITION OR SIZE OF OBJECTS

The **Modify** menu offers five options for modifying a drawing by changing the position of an object or by changing the size of the object. These commands include **Move**, **Rotate**, **Scale**, **Stretch**, and **Lengthen**. The **Align** command, found in the **3D Operation** listing of the **Modify cascading** menu, can also be used to change the position of a 2D object.

MOVE

The **Move** command allows one or more objects to be selected and moved from their current location to a new one. To access the command, use one of the following methods:

- Click the **Move** button on the **Modify** toolbar.

- Type **M** ENTER at the Command prompt.

- Select **Move** from the **Modify** menu.

The command sequence for **Move** is as follows:

```
Click the Move button (Or type M ENTER.)
Select objects: (Use any selection method.)
Select objects: (Continue to select objects or select ENTER to
    end) ENTER
Specify base point or [Displacement] <Displacement>:(Select a
    point or enter coordinates.)
Specify second point or <Use first point as displacement >:
    (Select a point or enter coordinates.)
```

The process can be seen in Figures 10.1a through 10.1c.

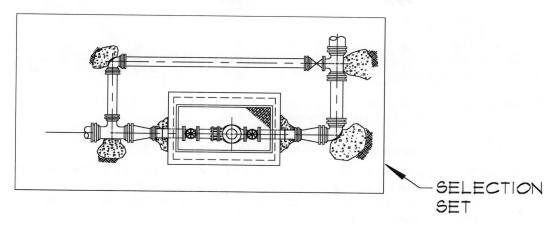

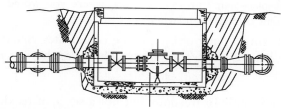

Figure 10.1a   *The first step in the process for moving a drawing is to select the objects to be moved. (Courtesy Lee Engineering, Inc.)*

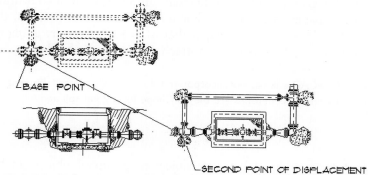

Figure 10.1b   *The second step is to select the base point of displacement and the new location for this point. (Courtesy Lee Engineering, Inc.)*

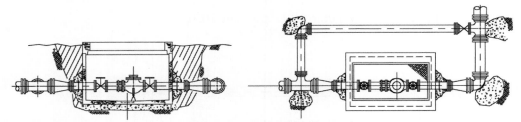

Figure 10.1c *Finally, the second point of displacement is selected and the Move process is complete. (Courtesy Lee Engineering, Inc.)*

- Use any selection method to select the objects to be moved.

- The "base point of displacement" prompt is asking for a reference point for moving the selection set. Any point may be selected, but it may help you to visualize the results of moving by selecting a point in the drawing, such as an intersection, midpoint or endpoint. This is an excellent time to make use of OSNAP.

- The "second point of displacement" prompt is asking where you would like the new location for the base point of displacement to be.

Each point can be entered by selecting a point or by entering coordinates. SNAP, ORTHO, OSNAP, and Zoom can be very helpful for aligning related objects as they are moved.

Practical Use of Move

Move should give you an immense sense of freedom, but not a sense of recklessness. You still need to plan your drawings, but the **Move** command offers you great flexibility in drawing arrangement. Space between views can be altered as the needs of the drawing change. On the other hand, you have to keep in mind that each of your drawings will need to fit on a sheet of paper of a set size. For most architectural projects, this means your drawing will need to fit on a sheet of paper that is 36 × 24. If you're going to try and place ten details on this sheet, there is only so much space where the details can be moved.

Note: AutoCAD still has not incorporated the **Cram** command. Planning and wise use of the editing commands are the only hope for a good solution.

ROTATE

The **Rotate** command allows an object or group of objects to be selected and rotated around a base point to change their orientation. To start the command, use one of the following methods:

- Click the **Rotate** button on the **Modify** toolbar.

- Type **RO** ENTER at the Command prompt.

- Select **Rotate** from the **Modify** menu.

- Select the object to be rotated, right-click, and then select **Rotate**.

Figure 10.2 shows an example of a window that was drawn for a horizontal position and then rotated 90°.

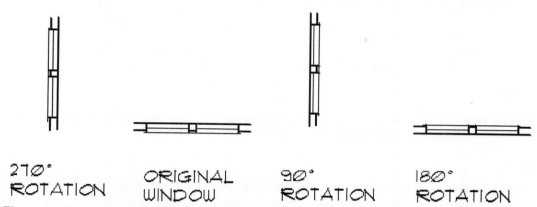

Figure 10.2   *The Rotate command allows objects to be rotated into a new position.*

To complete the command, you will need to select objects to be rotated, a base point to rotate the object around, and the desired degree of rotation.

• Enter a positive rotation angle to rotate the selected object in a counterclockwise direction.

• Enter a negative rotation angle to produce a clockwise rotation.

The base point can be anywhere on the drawing screen, but selecting one corner of the object as the base point can ease visualization of the rotation. The command sequence is as follows:

```
Click the Rotate button (Or type RO ENTER.)
Current positive angle in UCS: ANGDIR=counterclockwise
    ANGBASE=0.00
Select objects: (Use any selection method.)
Select objects: ENTER
Specify base point: (Select any point such as an endpoint or
    midpoint.)
Specify rotation angle or [Copy/Reference]<0>: (Enter angle or
    drag into position.) 90 ENTER
```

The angle that is entered will become the default angle for the current editing session.

Instead of entering a precise angle, you can drag the selected objects into the new position as the cursor is moved. If ORTHO is toggled **ON**, the objects will be snapped at 90° intervals. With ORTHO toggled **OFF**, the objects can be dragged into any desired angle. When an angle is entered, objects can be rotated to exact locations based on the current location. The process can be seen in Figure 10.3.

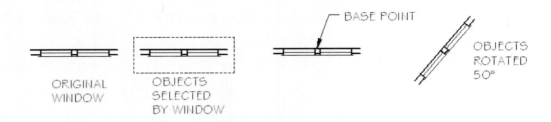

Figure 10.3 *Objects are rotated by defining the selection set, specifying a base point for the rotation, and providing the rotation angle.*

Copy

Typing **C** ENTER at the Rotation angle prompt allows a copy of the selection set to be created prior to rotation. When this option is selected, the original objects will be copied and left in the original position. The original objects will be left intact and a copy will be rotated to the desired angle. The **Move** command can then be used to place the new copy in the desired position. Figure 10.4 shows the use of **Rotate Copy** to create three additional windows at the desired orientation. The **Move** command was then used to place the windows in the desired location. The command sequence to rotate the windows with the **Copy** option is:

```
Click the Rotate button (Or type RO ENTER.)
Current positive angle in UCS: ANGDIR=counterclockwise
    ANGBASE=0.00
Select objects: (Use any selection method.)
Select objects: (Continue to select objects, and press ENTER when
    complete) ENTER
Specify base point: (Select any point such as an endpoint or
    midpoint.)
Specify rotation angle or [Copy/Reference]<90>:C ENTER
Rotating a copy of the selected objects:
Specify rotation angle or [Copy/Reference] <90>: ENTER
Specify rotation angle or [Copy/Reference]: (Enter angle or drag
    into position.)
```

The command was repeated to rotate and make copies of the object with rotation angles of 180° and 270°. The **Move** command was then used to place them in the proper location.

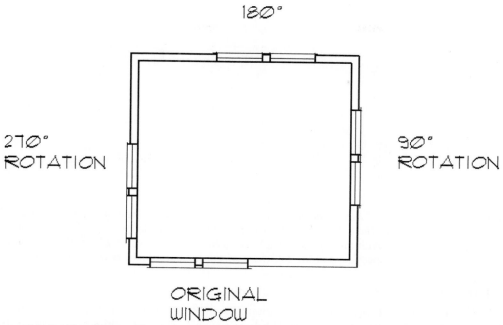

Figure 10.4 *The **Copy** option of the **Rotate** command will make a copy of and rotate the original objects. The original set of objects is left intact, and a new set is copied and rotated to the desired angle. The windows in this structure were rotated and then placed in the desired position using the **Move** command. Future commands will introduce faster and more efficient methods to achieve the same results.*

Using the Reference Option

Objects also can be rotated based on using absolute rotation angles. Figure 10.5 shows the top chord of a truss drawn at 22 1/2° for a 5/12 pitch. The owners change their minds and desire a 6/12 pitch (27 1/2°). The Reference option will place the roof exactly where it should be. The option requires the existing angle as well as the new angle to complete the command sequence. The option uses the following command sequence:

```
Click the Rotate button (Or type RO ENTER.)
Current positive angle in UCS: ANGDIR=counterclockwise ANGBASE=0
Select objects: (Use any selection method.)
Select objects: ENTER
Specify base point: (Select any point.)
Specify rotation angle or [Copy/Reference]<90>: R ENTER
Specify the reference angle <0>: 22.5 ENTER
Specify the new angle or [Points] 27.5 ENTER
```

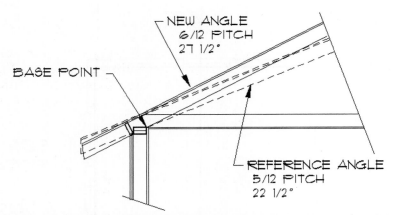

Figure 10.5 *The pitch of a roof can be altered by use of the **Reference** option of the **Rotate** command. The selection set (shown in dashed lines) is defined, and the base point is specified. With the reference angle specified (existing angle of 22.5°), the new angle of 27.5° is specified, and the top chord of the truss is rotated to the correct angle.*

ALIGN

The **Align** command provides a combination of the **Move** and **Rotate** commands. Although primarily a command for a 3D drawing, it works well when an object or group of objects need to be moved and rotated. To start the command:

- Type **AL** ENTER at the Command prompt

- In the **Modify** menu, select **3D Operation**, and then select **Align**.

  The command will prompt you for source points and destination points.

- Source points are used to define points on the original object in its original position.

- A Destination point represents the location where the source point will be placed.

- The command can be used in two ways: If two pairs of points are provided, the object can be moved, scaled, and rotated.

- If three pairs of points are provided, the selected object can be moved and rotated.

The following command sequence can be used to move, scale, and rotate a kitchen from one area of a floor plan to another.

```
Command: AL ENTER
Select objects: (Select object using any selection method.)
Select objects: ENTER
Specify first source point: (Select point with mouse.)
Specify first destination point: (Select point with mouse.)
Specify second source point: (Select point with mouse or enter
    coordinates.)
Specify second destination point: (Select point with mouse or
    enter coordinates.)
```

```
Specify third destination point: ENTER
Scale objects based on alignment points? [Yes/No] <No>: ENTER
```

When ENTER is pressed, the object will be moved. The selection process can be seen in Figure 10.6. The result of the command can be seen in Figure 10.7. If **Y** ENTER is typed at the last prompt, the objects being rotated will be scaled to match the distance indicated by the first and second destination points. Figure 10.8 shows the process of scaling the objects to meet the alignment points.

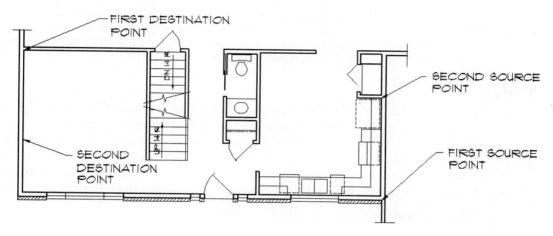

Figure 10.6   *The **Align** command allows portions of a drawing to be moved and rotated. The command will prompt you to provide source points and then the new location of those points.*

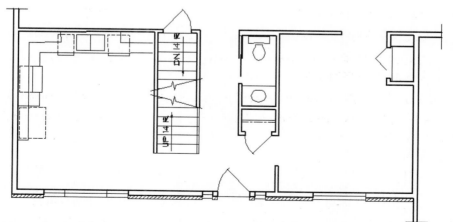

Figure 10.7   *By using the destination points shown in Figure 10.6, the kitchen was moved and rotated to a new position with the **Align** command.*

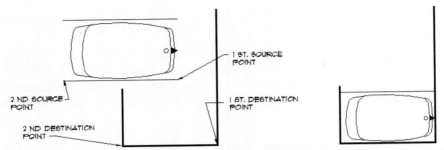

Figure 10.8 *If you choose to scale objects based on alignment points, an object or group of objects can be moved and scaled to match the selected destination points. The example on the right shows the 6' long tub is scaled to fit in a 5' space.*

SCALE

Up to this point, your drawings have been drawn at full scale. If you needed to draw a square parcel of land that is 100' × 100', you set the limits and units accordingly, zoomed out to the limits, and draw the parcel of land. The whole process was done at real scale with the size of the zoom used to control the viewing screen. The **Scale** command allows you to change the size of an existing object or an entire drawing, as seen in Figure 10.9. To start the command, use one of the following methods:

- Click the **Scale** button on the **Modify** toolbar.
- Type **SC** ENTER at the Command prompt.
- Select **Scale** from the **Modify** menu.

The command can be completed using the following process:

```
Click the Scale button (Or type SC ENTER.)
Select objects: (Select objects using any selection method.)
Select objects: ENTER
Specify base point: (Select base point.)
Specify scale factor or [Copy/Reference] <0'-1">: .5 ENTER
```

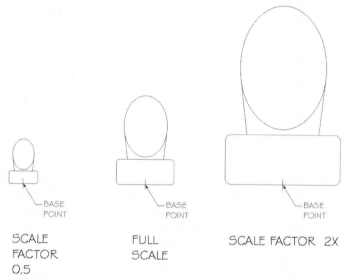

BASE   BASE   BASE
POINT   POINT   POINT

SCALE   FULL   SCALE FACTOR 2X
FACTOR   SCALE
0.5

Figure 10.9   *The Scale command allows the size of objects to be altered.*

Scale Factor

The default for entering the effect of scaling is to enter a numeric value. To enlarge an object to twice its existing size, type **2** ENTER at the prompt. Using whole numbers will enlarge the selected objects by that factor. Typing a fraction, such as **.5** ENTER, will reduce the selected object to half its original size. The number that is entered will become the default value for the balance of the current editing session.

Copy

Selecting the **Copy** option will copy and scale the new object, and leave the original objects at their current size and location.

Reference Factor

This option will allow an object to be enlarged to an absolute length, rather than a relative scale. For instance, a tub that is 5' long can be scaled to 6' long. Rather than figuring the proportion of enlargement, type **R** ENTER at the Command prompt. The command can be completed using the following process:

```
Click the Scale button (Or type SC ENTER.)
Select objects: (Select objects using any selection method.)
Select object: ENTER
Specify base point: (Select base point.)
Specify scale factor or [Copy/Reference]: <0'-1">R ENTER
Specify reference length <1>: 5' ENTER
Specify new length or [Points] 6' ENTER
```

STRETCH

The **Stretch** command allows you to enlarge or shrink drawing objects. This can be especially helpful when editing the shape and size of a floor plan to meet the changing needs of a client, as seen in Figure 10.10. Drawings made with lines, arcs, traces, polylines (chapter 11), and solids (chapter 14) can all be stretched. To start the command, use one of the following methods:

- Click the **Stretch** button on the **Modify** toolbar.

- Type **S** ENTER at the Command prompt.

- Select **Stretch** from the **Modify** menu.

Objects to be stretched must be selected using one of the window modes. The command sequence is as follows:

```
Click the Stretch button (Or type S ENTER.)
Select objects to stretch by crossing-window or crossing-
    polygon...
Select objects:(Select object to be stretched using automatic
    crossing window.)
Specify opposite corner: (Select the second point for the corner
    so that the object to be stretched is included in the
    window.)
Select objects: ENTER
```

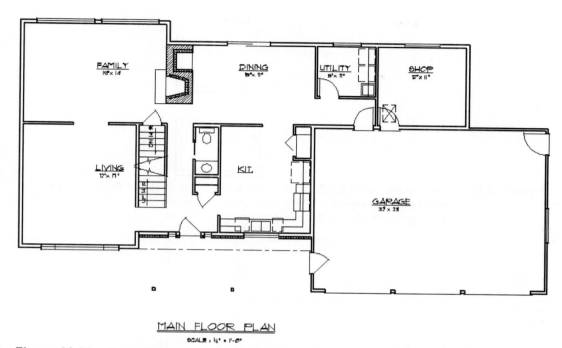

MAIN FLOOR PLAN
SCALE : ¼" = 1'-0"

Figure 10.10 *The Stretch command can be used to elongate an object or group of objects. Compare the garage size with the garage in Figure 9.16b.*

The automatic selection window only needs to cross a portion of the desired object. If the entire object is selected, **Stretch** will function like the **Move** command and move the entire object. AutoCAD now allows multiple selection sets to be created for stretching at the same time. Once the selection set is complete, the prompt will display the following:

```
Specify base point or [displacement] <Displacement>: (Select a
     point to serve as a reference. Be sure OSNAP is
     toggled ON.)
Specify second point or <use first point as displacement>:
     (Select the location that you would like to stretch the base
     point to.)
```

As the cursor is moved, the effect is displayed as the object is stretched between the base point and new point. The process can be seen in Figures 10.11a and 10.11b. **Stretch** can also be used to stretch some objects and move others, as seen in Figures 10.12a and 10.12b.

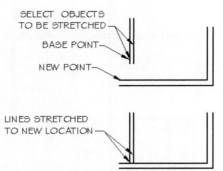

Figure 10.11a *The Window option is used to form the selection set of objects that are to be stretched. Once objects are selected, prompts will be displayed for the base point of the original objects and the new position they are to be stretched to.*

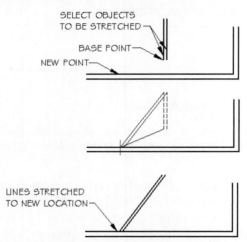

Figure 10.11b *In addition to adding length to line segments, the* **Stretch** *command can be used to alter the angle of the selected lines.*

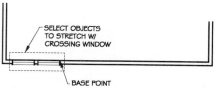

Figure 10.12a   The **Stretch** command can also be used to alter the location of objects.

Figure 10.12b   The window and walls are stretched to alter the location of the window. Lines on the left side of the window are stretched, while the lines on the right side are shortened. This is much more efficient than moving the window, stretching the left lines, and trimming the right lines.

LENGTHEN

The **Lengthen** command is similar to the **Stretch** command. **Stretch** is used to edit the length of groups of objects and closed polygons. **Lengthen** is used to edit the length of objects but not closed polygons. To start the **Lengthen** command, use one of the following methods:

- Click the **Lengthen** icon from the **Modify** menu.

- Type **LEN** ENTER at the Command prompt.

Note: Completing the **Lengthen** command sequence using the command line display offers greater advantages than using the Dynamic Display. With the Dynamic Display, the current length is not displayed. Knowing the length will be an aid for most of the options for completing the command. Use the **Tools** menu to activate the command line display.

The command sequence using the command line display is as follows:

```
Command: LEN ENTER
LENGTHEN
Select an object or [DElta/Percent/Total/DYnamic]: (Select an
    object to edit.)
Current Length: 10'-0 (Length will vary.)
Select an object or [DElta/Percent/Total/DYnamic]<Select object>:
    (Enter an option)
```

Using the Delta Option of Lengthen

Typing **DE** ENTER at the Select object prompt will create the following sequence:

```
Select an object or [DElta/Percent/Total/DYnamic]<Select object>:
    DE ENTER
Enter delta length or [Angle] <0'-0)>: 5' ENTER (Enter the length
    to add.)
Select an object to change or [Undo]: (Select the object again.)
Select an object to change or [Undo]: ENTER
Command:
```

The command sequence can be seen in Figure 10.13. When DElta is entered at the Select prompt, a choice can be made between length and angle. The default is to provide a length. Entering a positive number lengthens the line. Providing a negative response shortens the line. With each option of **Lengthen**, the end of the object that is selected will be the end that is edited.

 Note: The length entered is the length that will be added to the selected line. The value will not become the length for the total line length.

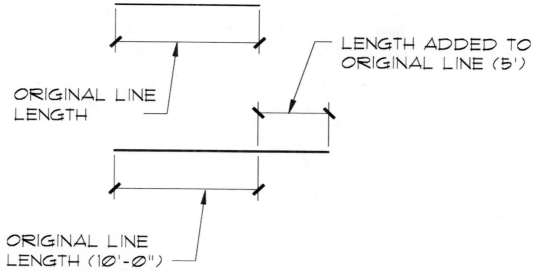

Figure 10.13   *The DElta option of the Lengthen command can be used to lengthen or shorten a line.*

Using the Angle Option of Lengthen

The Angle option can be used to change the included angle of an arc. The effects of this command sequence can be seen in Figure 10.14. The command sequence is as follows:

```
Command: LEN ENTER
LENGTHEN
Select an object or [DElta/Percent/Total/DYnamic]: (Select an arc
    to edit.)
Current Length: 10'-0, included angle: 90 (Length and angle will
    vary.)
Select an object or [DElta/Percent/Total/DYnamic]<Select object>:
    DE ENTER
Enter delta length or [Angle] <5'-0'>: A ENTER
Enter delta angle<0>: 60 ENTER
Select an object to change or [Undo]: (Select arc to be edited.)
Select an object to change or [Undo]: ENTER
Command:
```

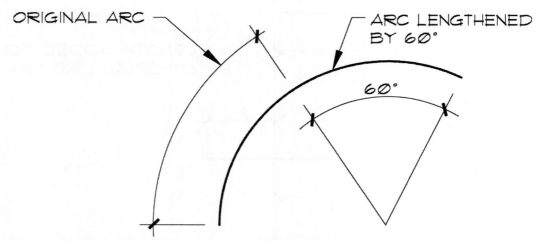

Figure 10.14 *The DElta and Angle options of the **Lengthen** command can be used to lengthen or shorten an angle.*

Using the Percent Option of Lengthen

The Percent option allows the length of a line or arc to be altered by a percentage of the existing length. The existing line length represents 100%. To lengthen the line, provide a value greater than 100%. Providing a percentage of less than 100% will shorten the line. The command sequence to lengthen a 2 line to a 3 line, using the Percent option, is as follows:

```
Command: LEN ENTER
LENGTHEN
Select an object or [DElta/Percent/Total/DYnamic]: (Select an
    object to edit.)
Current Length: 10'-0 (Length will vary.)
Select an object or [DElta/Percent/Total/DYnamic]: P ENTER
Enter percentage length <100>: 150 ENTER
Select an object to change or [Undo]: (Select line to be
    edited.)
Select an object to change or [Undo]: ENTER
Command:
```

The effects of the Percent option can be seen in Figure 10.15. When used with an arc, the angle of the arc will be edited; based on a percentage of the total angle of the specified arc.

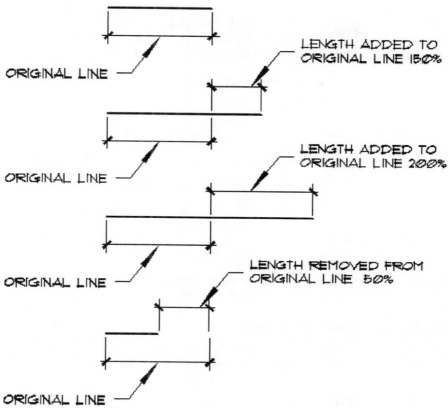

Figure 10.15   *The length of a line or arc can be altered by entering a percentage of the original object length. Percentages less than 100% will shorten the line or arc, and those greater than 100% will lengthen the line or arc.*

Using the Total Option of Lengthen

The Total option allows the length of a line or angle of an arc to be edited, by providing the desired result. The command sequence to alter an existing line to a 35' line is as follows:

```
Command: LEN ENTER
LENGTHEN
Select an object or [DElta/Percent/Total/DYnamic]: (Select an
     object to edit.)
Current Length: 10'-0 (Length will vary.)
Select an object or [DElta/Percent/Total/DYnamic]: T ENTER
Specify total length or [Angle]: 35' ENTER
Select an object to change or [Undo]: ENTER
Command:
```

The result of this sequence can be seen in Figure 10.16. When prompted to select an object, you might prefer to select an object before choosing the Total option. When a line is selected first, the length of the existing line will be displayed.

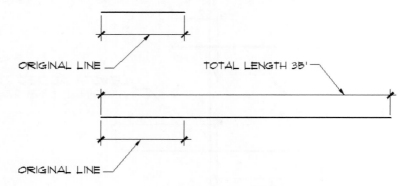

Figure 10.16 *The Total option of **Lengthen** allows a line or angle of an arc to be lengthened, based on the desired total length or angle.*

Using the Dynamic Option of Lengthen

The DYnamic option allows the selected line or angle to be dragged to the selected point to achieve the required length. The results of the DYnamic option can be seen in Figure 10.17.

```
Command: LEN ENTER
LENGTHEN
Select an object or [DElta/Percent/Total/DYnamic]: (Select an
    object to edit.)
Current Length: 10'-0 (Length will vary.)
Select an object or [DElta/Percent/Total/DYnamic]: DY ENTER
Select an object to change or [Undo]: (Select line to be
    edited.)
Specify new end point: (Select new end point for line to be
    edited.)
Select an object to change or [Undo]: ENTER
Command:
```

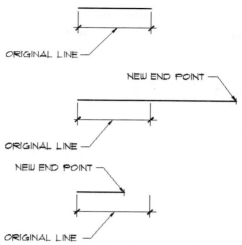

ORIGINAL LINE

NEW END POINT

ORIGINAL LINE

NEW END POINT

ORIGINAL LINE

Figure 10.17 *The DYnamic option of Lengthen allows a line or angle of an arc to be dragged to the desired length by selecting a new ending point.*

MODIFYING OBJECTS BY ALTERING LINES

Up to this point, you've learned to modify objects by making additional copies and by altering the position or size of an object. The commands contained in this portion of the **Modify** menu will allow you to modify a drawing by altering intersections of lines, circles or arcs, or by removing a portion of an object. The commands include **Trim**, **Extend**, **Break**, **Chamfer**, and **Fillet**.

TRIM

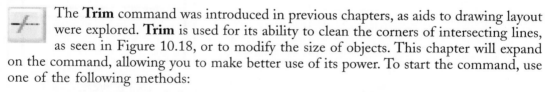 The **Trim** command was introduced in previous chapters, as aids to drawing layout were explored. **Trim** is used for its ability to clean the corners of intersecting lines, as seen in Figure 10.18, or to modify the size of objects. This chapter will expand on the command, allowing you to make better use of its power. To start the command, use one of the following methods:

- Click **Trim** on the **Modify** toolbar.
- Type **TR** ENTER.
- Select **Trim** from the **Modify** menu.

The command allows you to edit lines, arcs, circles, or polylines using a cutting edge. Any portion of the selected object extending past the designated cutting edge will be removed. Lines, arcs, circles, and polylines can be used as cutting edges. Once a cutting edge is defined, the command will continue to prompt for the selection of more cutting edges until ENTER is pressed. Any object that extends past the cutting edge can be trimmed. Notice that an object can be selected to be a cutting edge, and it can also be an object to be trimmed. Completing the **Trim** command from the command line prompt will provide information that is not provided when the command is executed at the Dynamic Display.

The following use of the command assumes use of command line entry. The command sequence to trim an object is as follows:

```
Click the Trim button (Or type TR ENTER.)
TRIM
Current settings: Projection=UCS Edge=Extend
Select cutting edges... (Select the cutting edge using any object
      selection method.)
Select objects or <select all>: 1 found
Select objects: ENTER
Select object to trim or shift-select to extend or
[Fence/Crossing/Project/Edge/eRase/Undo]: (Select object to
      trim.)
Select object to trim or shift-select to extend or
[Fence/Crossing/Project/Edge/eRase/Undo]: ENTER
Command:
```

The Window and Crossing window options work well for selecting multiple objects for cutting edges. Selecting the All option will use all drawing objects as trim edges. To use the All option, you must type **ALL** **ENTER** (not **A**).

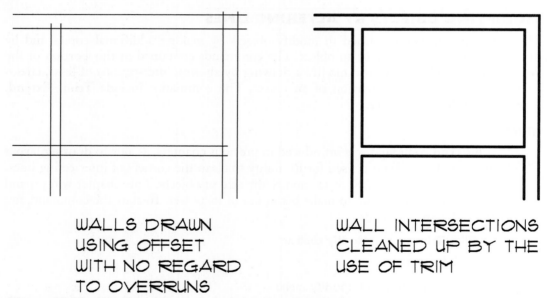

WALLS DRAWN
USING OFFSET
WITH NO REGARD
TO OVERRUNS

WALL INTERSECTIONS
CLEANED UP BY THE
USE OF TRIM

Figure 10.18 The **Trim** command can be used to eliminate line overruns that were created using the Offset command.

Using the Trim Edge Option

The Edge option of **Trim** can be used to trim two objects that do not actually intersect but would intersect if one of the objects were extended. AutoCAD refers to this as an implied extension. Once the edges have been selected, type **E** ENTER when prompted to select another object. This will now display the option

```
Enter an implied edge extension mode [Extend/No extend] <Extend>:
```

When Extend is selected, AutoCAD checks the cutting edge object to verify if it will intersect when extended. If it would touch, the implied intersection is used to trim the second object. This option is shown in Figure 10.19. The command sequence is as follows:

```
Click the Trim button (Or type TR ENTER.)
TRIM
Current settings: Projection=UCS Edge=None
Select cutting edges...(Select the cutting edge using any object
      selection method.)
Select object or <select all>:1 found
Select object: ENTER
Select object to trim or shift-select to extend or [Fence/
      Crossing/Project/Edge/ eRase/Undo]: E ENTER
Enter an implied edge extension mode [Extend/No Extend] <No
      Extend>: E ENTER
Select object to trim or shift-select to extend or [Fence/
      Crossing/Project/Edge/ eRase/Undo]: (Select object to trim.)
Select object to trim or shift-select to extend or [Fence/
      Crossing/Project/Edge/ eRase/Undo]: ENTER
Command:
```

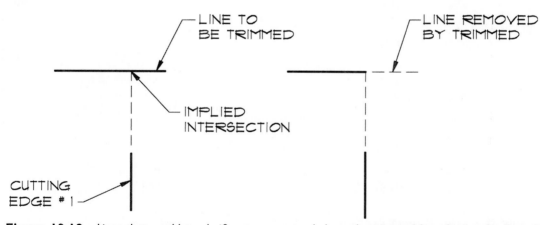

Figure 10.19 *Lines that would touch, if one were extended, can be trimmed by selecting the Extend mode of the **Trim Edge** option.*

Using the Trim Extend Option

The Extend option of **Trim** can be used to extend objects that do not actually intersect but would if one of the objects were extended. AutoCAD refers to this as an implied intersection. To use the Extend option, select the line to be used as the destination. When prompted to select another object, press and hold SHIFT and select the line to be extended with the select button. The command sequence is as follows:

```
Click the Trim button (Or type TR ENTER.)
TRIM
Current settings: Projection=UCS Edge=Extend
Select cutting edges...
Select cutting edges or <select all>: 1 found (Select the
    cutting edge using any object selection method.)
Select objects: ENTER
Select object to trim or shift-select to extend or [Fence/
    Crossing/Project/Edge/ eRase/Undo]: Hold SHIFT and select
    the line to extend)
Select object to trim or shift-select to extend or [Fence/
    Crossing/Project/Edge/ eRase/Undo]: ENTER
Command:
```

Using the Trim Undo Option

Occasionally you might trim an object and not be pleased with the results. The **Trim Undo** option will allow an unsatisfactory **Trim** to be undone, while still in the command sequence. The command sequence is as follows:

```
Click the Trim button (Or type TR ENTER.)
TRIM
Current settings: Projection=UCS Edge=Extend
Select cutting edges... (Select the cutting edge using any object
    selection method.)
1 found
Select objects: ENTER
Select object to trim or shift-select to extend or [Fence
    /Crossing/ Project/Edge/eRase/Undo]: (Select object to be
    trimmed.)
Select object to trim or shift-select to extend or [Fence
    /Crossing/ Project/Edge/eRase/Undo]: U ENTER
Command has been completely undone.
Select object to trim or shift-select to extend or [Fence
    /Crossing/ Project/Edge/eRase/Undo]: (Select object to be
    trimmed.)
Select object to trim or shift-select to extend or [Fence
    /Crossing/ Project/Edge/eRase/Undo]: ENTER
Command:
```

Using the Trim Project Option

The **Trim Project** option is primarily a 3D option. Refer to the **Help** menu for further information. The option can be used, just as the Edge option, as a means of trimming objects that extend past a projection of the selected cutting edge.

Practical Use of Trim

As you discovered earlier, combining the **Line, Offset,** and **Trim** commands will allow almost any object to be drawn easily. Draw two perpendicular lines to represent the corner of an object without regard to the exact length. The exact size of the object can be determined with **Offset** or **Trim**. Open the **FLOOR10** file that was started earlier. With the use of **Trim,** this partial floor plan can be completed so that it more closely resembles the drawing in Figure 9.16b.

EXTEND

 The **Extend** command allows an object to be lengthened to an exact boundary point. To start the command, use one of the following methods:

- Click the **Extend** button from the **Modify** toolbar.

- Type **EX** ENTER at the Command prompt.

- Select **Extend** on the **Modify** menu.

The command will ask you to designate boundary edges and then select objects to be extended to that edge. The boundary edge can be a line, arc, circle, or polyline. Completing the **Extend** command from the command line prompt will provide helpful information that is not provided when the command is executed at the Dynamic Display. The following use of the command assumes use of command line entry. The process can be seen in Figure 10.20. The command sequence is as follows:

```
Click the Extend button (Or type EX ENTER.)
EXTEND
Current settings: Projection=UCS Edge=Extend
Select boundary edges... (Select the edge using any object
    selection method.)
1 found
Select objects: ENTER
Select object to extend or shift-select to trim or [Fence/
    Crossing/ Project/Edge/Undo]: (Select object to be extended
    using any selection method.)
Select object to extend or shift-select to trim or [Fence/
    Crossing/ Project/Edge/Undo]: ENTER
Command:
```

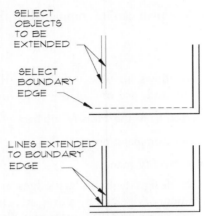

SELECT
OBJECTS
TO BE
EXTENDED

SELECT
BOUNDARY
EDGE

LINES EXTENDED
TO BOUNDARY
EDGE

Figure 10.20 *The **Extend** command can be used to lengthen lines to a selected boundary edge.*

Using the Extend Edge Option

The Extend Edge option is similar to the Edge option presented with the **Trim** command. Edge will allow lines that do not presently touch but have an implied intersection to be extended. In the default setting, the command will automatically extend, using an implied edge. If the Edge setting has been altered, the sequence to extend to an implied edge is as follows:

```
Click the Extend button (Or type EX ENTER.)
EXTEND
Current settings: Projection=UCS Edge=Extend
```

```
Select boundary edges... (Select the edge using any object
     selection method.)
Select objects or <select all> 1 found
Select objects: ENTER
Select object to extend or shift-select to trim or [Fence/
     Crossing/ Project/Edge/Undo]: E ENTER
Enter an implied edge extension mode [Extend/No Extend]
     <Extend>:ENTER
Select object to extend or shift-select to trim or [Fence/
     Crossing/ Project/Edge/Undo]: (Select object to be extended
     using any selection method.)
Select object to extend or shift-select to trim or [Fence/
     Crossing/ Project/Edge/Undo]: ENTER
Command:
```

The result of the command can be seen in Figure 10.21.

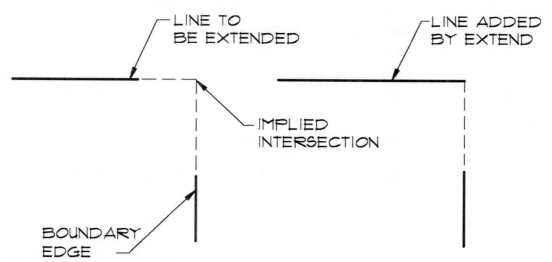

Figure 10.21   *Lines that would touch, if both were extended, can be extended by selecting the Edge option of **Extend**.*

Using the Extend Trim Option

The Trim option of **Extend** allows an object to be trimmed, while using the **Extend** command. The command sequence to use the Trim mode is as follows:

```
Click the Extend button (Or type EX ENTER.)
EXTEND
Current settings: Projection=UCS, Edge=Extend
Select boundary edges... (Select the edge using any object
     selection method.)
Select objects or <select all> 1 found
Select objects: ENTER
Select object to extend or shift-select to trim or [Fence/
     Crossing/ Project/Edge/Undo]: (SHIFT + Select object to be
     trimmed using any selection method.)
```

```
Select object to extend or shift-select to trim or [Fence/
     Crossing/ Project/Edge/Undo]: ENTER
Command:
```

Using the Extend Undo Option

The Extend Undo option allows a sequence that has not produced satisfactory results to be undone, while still in the command sequence.

Using the Extend Project Option

This mode is used for 3D drawings and will not be discussed in this text. Use the **Help** menu for further information on this subject.

BREAK

The **Break** command can be used to remove a portion of a line, circle, arc, or polyline. To start the **Break** command, use one of the following methods:

- Click the **Break** button on the **Modify** toolbar.
- Type **BR** ENTER at the Command prompt.
- Select **Break** from the **Modify** menu.

Break can be completed using two different methods, but their effects are similar.

Select Object, Second Break Point

The default option of **Break** allows you to remove a segment of an object by using two selection points. The first selection point selects the segment to be broken and specifies where the break will begin. The second point specifies where the break will end. The command sequence at the Dynamic Display is as follows:

```
Click the Break button (Or type BR ENTER.)
Select object: (Select an object to break.)
Specify second break point or [First point]: (Select a second
     point.)
```

Once the second point is selected, the portion of the object from the first to the second point will be removed, and the Command prompt will be returned. The sequence can be seen in Figure 10.22. If a circle is being edited, the break occurs in a counterclockwise direction from the first point to the second point, as seen in Figure 10.23. Type **F** ENTER when the second point is requested to allow the procedure to follow the Select Object, Two Break Points method.

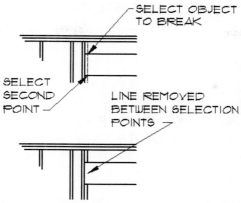

Figure 10.22 *The default option of **Break** allows a selected object to be broken. The point that is picked when the object is selected will become the first break point, and the second point will designate the amount of the segment to be removed.*

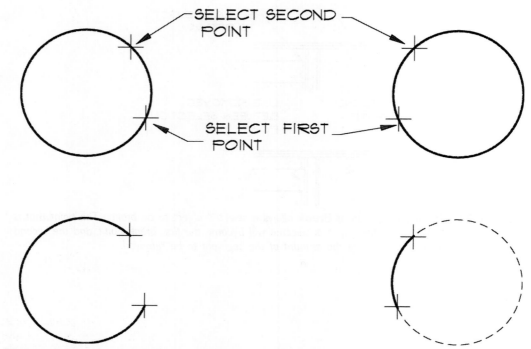

Figure 10.23 *When a break is specified for a circle, the break will occur in a counterclockwise direction.*

Select Object, Two Break Points

This option allows you to remove a segment of an object, by using three selection points. The first request selects the object to be broken. Once the segment has been defined for editing, the next selection point specifies where the break will begin and the final selection point specifies where the break will end. The command sequence is as follows:

```
Click the Break button (Or type BR ENTER.)
Select object: (Select an object to break.)
Specify second break point or [First point]: F ENTER
Specify first break point: (Select a point to start the break.)
Specify second break point: (Select a point to end the break.)
```

The command sequence can be seen in Figure 10.24.

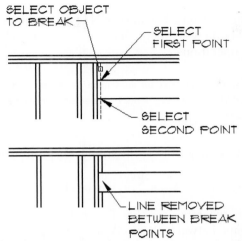

Figure 10.24 *The* **Break** *command's First option can be used to remove a segment from a line. Specified points are used to define which part of the line will be removed.*

BREAK AT POINT

 This option will break a segment, but no apparent change will take place. **Break @** will divide an object into two portions, but no space will be left between the portions. The **@** option is useful for erasing only a portion of an object, by using **Break @** and then using the **Erase** command. To start the command, click the **Break at point** icon on the **Modify** toolbar.

The command sequence is as follows:

```
Command: Click the Break at point button.
Select object: (Select an object to break.)
Specify first break point:
```

JOIN

The **Join** can be used to join line, polyline, arc, elliptical arc, or spline segments that lie on the same plane into one continuous object. To start the command, use one of the following methods:

- Click the **Join** from the **Modify** toolbar.

- Select **Join** from the **Modify** menu.

- Type **JOIN** ENTER at the Command prompt.

The command sequence to join three lines into one segment is:

```
Click Join from the Modify toolbar
Select source object: (Select the first line)
Select lines to join to source: (Select line to merge with
     original)
Select lines to join to source: ENTER
```

No changes will occur immediately, but once ENTER is pressed, the lines will merge to form one line segment. The new segment that has been formed will now function as one object for future editing operations.

> **Note:** Remember that the selected lines must lie on the same imaginary plane for them to be joined. When lines segments are selected where one line is above another, nothing will happen!

Arc

If an arc is selected as the source object, the following prompt will be displayed:

```
Select arcs to join to source or [cLose]: Select one or more
      arcs and press ENTER, or enter L
```

The option will join two separate arcs that lie on the same imaginary circle. The **Arc** option will remove the gaps between them. When joining two or more arcs, the arcs are joined counterclockwise, beginning from the source object. The **Close** option converts the source arc into a circle. The Elliptical arc option functions in a similar manner as the **Arc** option.

EDITING WITH THE CHAMFER COMMAND

 A chamfer is an angled edge formed between two intersecting surfaces, typical of what might be found on the edge of a countertop. The **Chamfer** command trims two intersecting lines and provides a mitered corner. You'll be asked to select two lines and two distances. This will form a new line segment between the two existing line segments. To start the command, use one of the following methods:

- Click the **Chamfer** button on the **Modify** toolbar.
- Type **CHA** ENTER at the Command prompt.
- Select **Chamfer** from the **Modify** menu.

The command sequence to alter the default values is as follows:

```
Click the Chamfer button (Or type CHA ENTER.)
(TRIM mode) Current chamfer Dist1 = 0'-0", Dist2 = 0'-0"
Select first line or [Undo/Polyline/Distance/Angle/Trim/mEthod/
    Multiple]: (Select first line to chamfer)
Select second line or shift-select to apply corner: (Select
    second line to chamfer)
```

The first value will become the default for the second value. When the distances are zero, a square corner will be formed. When the distances are equal, the results will produce a 45° miter. Once the second value or ENTER is entered, the command sequence will continue. Using the default chamfer distance will resemble Figure 10.25.

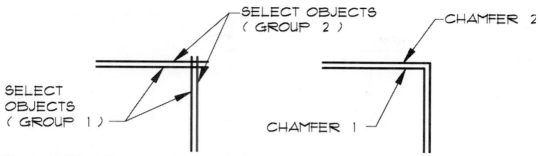

Figure 10.25 *With a value of 0,* **Chamfer** *can be used to trim corners.*

To provide a mitered corner, the sequence is similar. For the following example, a distance of 36" is used—although any numeric value can be used, as long as the value is no longer than that of the line being chamfered. The values that are entered will remain the default values, until new values are entered. If the value exceeds the length, an error message will be displayed. The command sequence is as follows:

```
Click the Chamfer button (Or type CHA ENTER.)
Select first line or [Undo/Polyline/Distance/Angle/Trim/mEthod/
    Multiple]: D ENTER
Specify first chamfer distance <0'-0>: 36 ENTER
Specify second chamfer distance <3'-0>: ENTER
Select first line or [Undo/Polyline/Distance/Angle/Trim/mEthod/
    Multiple]: (Select first line to chamfer)
Select second line or shift-select to apply corner: (Select
    second line to chamfer)
```

As the second line is selected, the chamfer will be performed. The effects of the command sequence can be seen in Figure 10.26. Figure 10.27 shows an example of two different distance values. **Chamfer** can be used to square or chamfer two intersecting lines, as well as to square or chamfer two lines that do not touch. Selecting the Multiple option of **Chamfer** allows multiple chamfers to be performed without having to restart the command. Selecting the Undo option allows the Chamfer to be undone. The Polyline option will be discussed in future chapters.

Figure 10.26 *When mitering a corner, prompts will be displayed for the "first distance," and the "second distance." Once the distances are provided, a new segment will be created between the two points. The distances that are entered will become the default values until new distances are entered.*

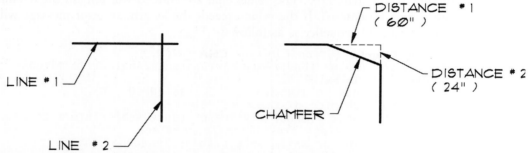

Figure 10.27 *Providing different values for the chamfer distances will affect the angle of the chamfer.*

Selecting the Chamfer Angle

Chamfer also allows the chamfer to be created using one distance and an angle. Start this option by typing **A** ENTER when prompted for the first line. The command sequence to place a 30° angle 4 from a corner is as follows:

```
Click the Chamfer button (Or type CHA ENTER.)
Select first line or [Undo/Polyline/Distance/Angle/Trim/mEthod/
    Multiple]: A ENTER
Specify chamfer length on the first line <0'-0>: 4 ENTER
Enter chamfer angle from the first line <0>: 30 ENTER
Select first line or [Undo/Polyline/Distance/Angle/Trim/mEthod/
    Multiple]: (Select first line to chamfer)
Select second line or shift-select to apply corner: (Select
    second line to chamfer)
```

The process can be seen in Figure 10.28.

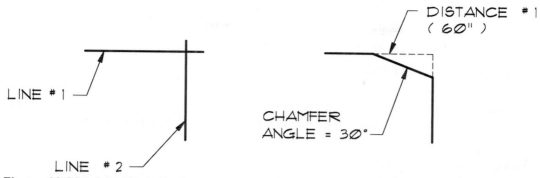

Figure 10.28 *Selecting the Angle option provides a prompt for one length and an angle. AutoCAD converts the angle to a length.*

Selecting the Chamfer Trim Mode

The common method of using the **Chamfer** command is to create the chamfer between two lines and remove a portion of the lines where the chamfer is created. The No Trim option allows for the choice of not removing the lines by the chamfer (by default, the lines are trimmed). The command sequence to form a chamfer and leave the existing lines is as follows:

```
Click the Chamfer button (Or type CHA ENTER.)
Select first line or [Undo/Polyline/Distance/Angle/Trim/mEthod/
    Multiple]: T ENTER
Enter Trim mode option [Trim/No trim] <Trim>: N ENTER
Select first line or [Undo/Polyline/Distance/Angle/Trim/mEthod/
    Multiple]: (Select first line to chamfer)
Select second line or shift-select to apply corner: (Select
    second line to chamfer)
```

Note: If you alter the trim mode of **Chamfer**, the chamfer will be performed but no lines will be removed. The selected Trim mode will remain the default, until it is changed. The **Chamfer Trim** mode will also affect the **Fillet** command. **Fillet** will be discussed in the next section of this chapter. Most users find the **Trim** option (the default) more useful than the No Trim option.

Selecting the Chamfer Method

When a chamfer is placed using the Distance or Angle option, the values are kept as the defaults until you change them. Using the mEthod option allows for a choice of entering two distances or an angle and a distance. With this option, the values selected for one option will not affect the other option, and you can switch between using the Distance or Angle method. The command sequence is as follows:

```
Click the Chamfer button (Or type CHA ENTER.)
Select first line or [Undo/Polyline/Distance/Angle/Trim/mEthod/
    Multiple]: E ENTER
Enter Trim method [Distance/Angle] <Trim>: D ENTER
```

```
Select first line or [Undo/Polyline/Distance/Angle/Trim/mEthod/
    Multiple]: (Select first line to chamfer)
Select second line or shift-select to apply corner: (Select
    second line to chamfer)
```

Selecting the Multiple Chamfer

Using the Multiple option will chamfer the edges of more than one set of objects. The command displays the main prompt and the Select Second Object prompt repeatedly, until you press ENTER to end the command. The command will be repeated, until ENTER is pressed.

EDITING WITH THE FILLET COMMAND

 The **Fillet** command is used to connect two lines to form a corner or allow two lines, arcs, or circles to be fitted together smoothly in rounded corners. To start the command, use one of the following methods:

- Click the **Fillet** button on the **Modify** toolbar.
- Type **F** ENTER at the Command prompt.
- Select **Fillet** from the **Modify** menu.

Once the command is started, set the desired radius using a process similar to setting the chamfer distance. Once the two objects are selected, the radius will be applied and the Command prompt will be returned. Selecting the Multiple option of **Fillet** allows multiple fillets to be performed without having to restart the command. The fillet process can be seen in Figure 10.29. If the Trim option is set to No Trim, the results of the fillet will not be evident. Be sure the Trim option is set to Trim. The command can be used to form a radius between intersecting lines or two lines that do not touch.

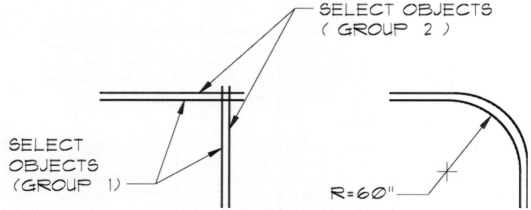

Figure 10.29 *By accepting the default or assigning a new radius value, the Fillet command can be used to round square corners.*

For the command to function with crossing lines as you intend, you need to be careful which portion of the lines you select. Draw two more intersecting lines similar to those in Figure 10.29. Start the **Fillet** command and select the short ends of the lines. Once the command is completed, only the small portion of the lines will remain, as shown in Figure 10.30.

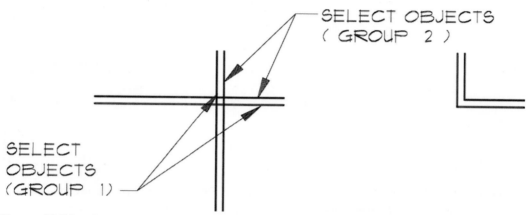

Figure 10.30 *Care must be taken, when selecting objects to be edited by the Fillet command. Selecting the short ends of the lines removes the other portions of the lines.*

Altering the Fillet Radius

The fillet radius can be altered using the Radius option. Once the first object is selected, use the Radius option to alter the radius of the fillet. Typing **R** ENTER at the first prompt allows a radius to be entered for rounded corners. To transform two existing perpendicular lines into a 60" radius corner, the command sequence is as follows:

```
Click the Fillet button (Or type F ENTER.)
Select first object or [Undo/Polyline/Radius/Trim/Multiple]:
   R ENTER
Specify fillet radius <0'-0>: 60 ENTER (Or 5' ENTER)
Select first object or [Undo/Polyline/Radius/Trim/Multiple]:
   (Select first line)
Select second line or shift-select to apply corner: (Select
   second line to chamfer)
```

As the second object is selected, the radius will be formed between the two selected lines, similar to Figure 10.29. The portion of the object that is selected will affect how the fillet will be drawn. Don't speed past the Command prompt. If you've entered a radius value that exceeds the length of the lines, no change will be made to the drawing, and a drawing error prompt will be displayed. A smaller radius value will need to be selected to make the command work.

The Undo option of Fillet can be used to undo a previous action of the command. Similar to the Chamfer command, the Multiple option of Fillet can be used to make unlimited fillets. One of the most common uses of **Fillet** is to form a corner with no radius. Setting the radius to 0 and then proceeding with the command will allow the command to be used to form corners. (See Figure 10.31.)

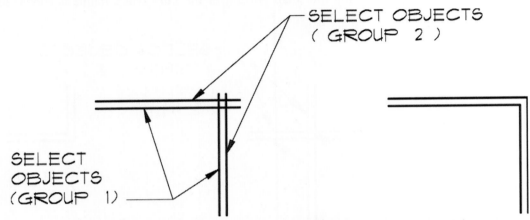

Figure 10.31 *With the value of the Radius set to 0, the **Fillet** command can be used to square corners of crossing lines.*

Practical Uses for Combining Offset, Fillet and Trim

Open the **FLOOR10** drawing that you started in chapter 9. Because you've seen the finished floor plan in Figure 9.16b, it's not too hard to visualize the lines in Figure 9.19 representing the rooms of the finished plan. Using a combination of **Offset** and **Fillet** will make it much easier to visualize. Figure 10.32 shows two examples of how **Fillet** can be used to clean up the floor plan. Notice that if **Fillet** is used at Fillet 3, the lower portion of the line will be removed and needs to be redrawn. This would be a good opportunity to use **Trim** instead of **Fillet**.

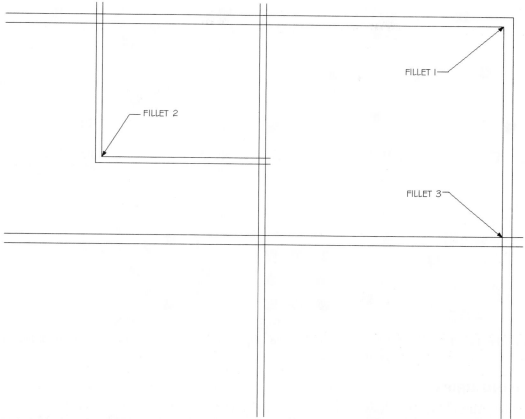

Figure 10.32   *The **Fillet** command was used to clean up the corners at Fillets 1 and 2. If Fillet had been used at Fillet 3, the lines that will be used to form the garage would have been removed.*

SELECTING OBJECTS WITH GRIPS

In chapter 9 you were introduced to several methods of selecting objects for editing. Grips provide an additional method of selecting objects for editing, as well as a method of combining several of the editing procedures introduced in this chapter. Grips are similar to moving a heavy object by its handles. You can move the object without using the handles, but they make life so much easier. Grips are handles placed on objects for easy manipulation, similar to the OSNAP points with **Stretch, Move, Rotate, Scale**, and **Mirror**. The default setting for grips is **ON**. Grips can be toggled **ON/OFF** on the **Selection** tab of the **Options** dialog box. With the **Enable grips** check box selected, grips are displayed when an edit or inquiry command is given; grip boxes are displayed at various object-specific positions, as seen in Figure 10.33. When the cursor is moved over a grip, the cursor snaps to the grip for editing.

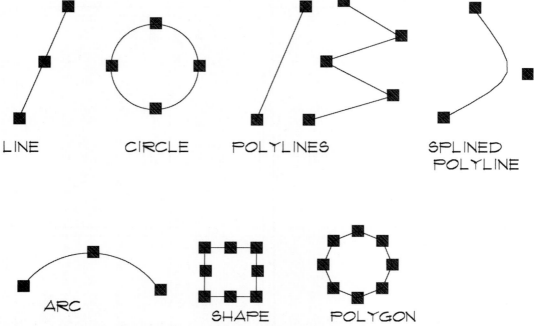

LINE   CIRCLE   POLYLINES   SPLINED POLYLINE

ARC   SHAPE   POLYGON

Figure 10.33 *With the default setting **Enable grips** check box toggled to ON, grips are displayed for each geometric shape.*

USING GRIPS

Enter AutoCAD and draw a line, an arc, and a circle. Move the cursor so that it is over the line and click the select button. Three boxes or grips are now displayed. Move the cursor to the center grip, and click the select button. This grip is now said to be hot. Notice that the Command prompt has also changed to display the following prompt:

```
Specify stretch point or [Base point/Copy/Undo/eXit]:
```

We'll come back to the prompt later. With the center grip hot and the prompt reflecting Stretch, move the cursor slightly. The hot grip in the center of the line functions like the **Move** command. As the cursor moves, so does the line. Clicking the select button will move the line to this new location. Press ESC to cancel the command. Select the line and make one of the end grips hot. Now the grip at the other end remains at a stationary point, and as the hot grip is moved, the selected line is stretched. The effect of **Stretch** on a line can be seen in Figure 10.34. Grips function in a similar manner with arc and circles. Figures 10.35a and 10.35b show the effect of stretching each. If you decide you don't want to edit a selected line, press ESC to deactivate the grips.

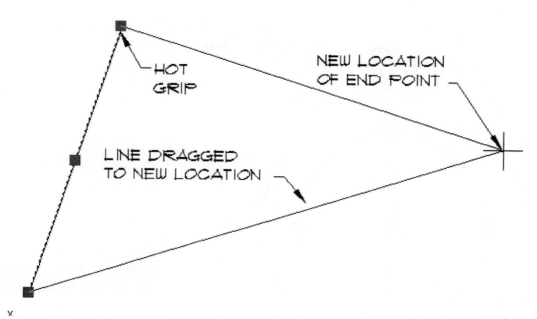

Figure 10.34   *Select a drawing object to display its grips. Moving to one of the grips and clicking the select button will make that grip hot. The object can now be edited by **Stretch**, **Move**, **Rotate**, **Scale**, and **Mirror**.*

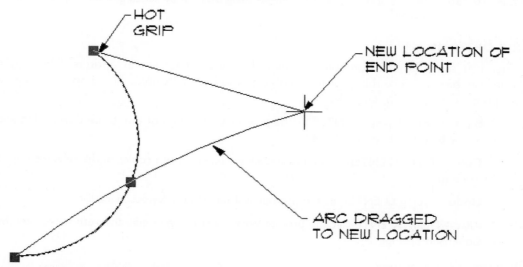

Figure 10.35a   *The **Stretch** command applied to an arc using the hot grip.*

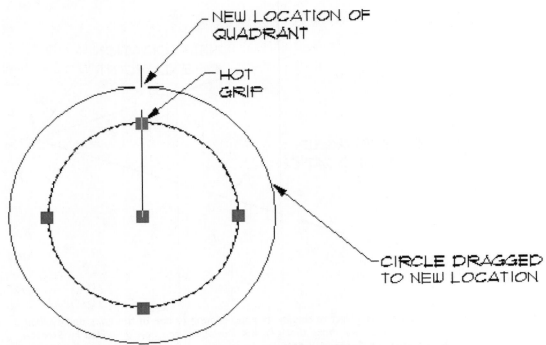

Figure 10.35b *The size of a circle can be altered using* **Stretch** *by selecting a hot grip and a new location.*

GRIP OPTIONS

Earlier you saw that the command displayed in grips has several options. These grip options can be used once the grips for an object have been activated. These options include Base point, Copy, Undo, and eXit.

> **Base Point**—Typing **B** ENTER at the prompt allows the hot box to be disregarded and a new base point to be selected.

> **Copy**—Typing **C** ENTER at the prompt allows one or more copies of the selected object to be made.

> **Undo**—Typing **U** ENTER at the prompt will undo the previous selection.

> **eXit**—Typing **X** ENTER at the prompt terminates the grip edit and returns you to the Command prompt.

Changing Grip Options

Once a grip box has been selected and made hot, a right-click will produce a shortcut menu. The menu allows the **Mirror, Rotate, Scale, Move,** and **Stretch** commands to be used on the selected object grip.

SELECTING OBJECTS FOR MULTIPLE EDIT COMMANDS

Often an object or group of objects might need to be edited with several different editing commands. If you were arranging units in a multi-building apartment complex, a unit may need to be mirrored, moved, and rotated to form a new unit. Up to this point, you would have to select the objects to be mirrored, use the **Mirror** command, select the objects to be moved, use the **Move** command, select the objects to be rotated, and use the **Rotate** command. It works, but you'll never impress the boss as worthy of the big bucks. The **Select** command will allow an object or group of objects to be pre-selected and used for one or more editing functions. The command sequence to select is as follows:

```
Command: SELECT ENTER
Select objects: (Select objects using the automatic selection
      window or any of the other selection methods.)
Specify opposite corner: (Select opposite corner.)
28 found (Number of objects will vary)
Select objects: ENTER
```

The Command prompt is returned but no apparent changes have occurred. The objects that were selected have been grouped as a set and will be affected by any editing command. Start the **Mirror** command, and, when prompted to select objects, type **P** ENTER to automatically select the objects that were just placed in the set. The sequence to mirror, move, and rotate the objects is as follows:

```
Command: Click the Mirror button (Or type MI ENTER.)
Select objects: P ENTER
28 found
Select objects: ENTER
Specify first point of mirror line: (Select first point.)
Specify second point of mirror line: (Select second point.)
Erase source objects? [Yes/No]<N>: Y ENTER
Click the Move button (Or type M ENTER.)
Select objects: P ENTER
28 found
Select objects: ENTER
Specify base point or [Displacement] <Displacement>: (Select base
      point.)
Specify second point of displacement or <use first point as
      displacement>: (Select new location for base point.)
Command: Click the Rotate button (Or RO ENTER.)
Current positive angle in UCS: ANGDIR=counterclockwise ANGBASE=0
Select objects: P ENTER
28 found
Select objects: ENTER
Specify base point: (Select base point.)
Specify rotation angle or [Copy/Reference] <0>: 45 ENTER
```

 Note: If you forget to use **SELECT** ENTER before the editing command, don't worry. The same results can be obtained by entering the commands in the following sequence: complete the **MIRROR**, select your objects, start the **ROTATE** command, and type **P** (for previous) when prompted to select objects.

NAMING OBJECT GROUPS

While **Select** allows a group to be created, the **Group** command allows multiple groups to be created, named, and saved. For instance, an apartment unit could be saved as a group called BASE. It could be mirrored and both units saved as a new group called UNITS (a group that is part of another group is referred to as nested). The UNITS group could also be mirrored and saved as a third group called BLDG1. Start the **Group** command by typing **G** ENTER at the Command prompt. This will produce the **Object Grouping** dialog box shown in Figure 10.36.

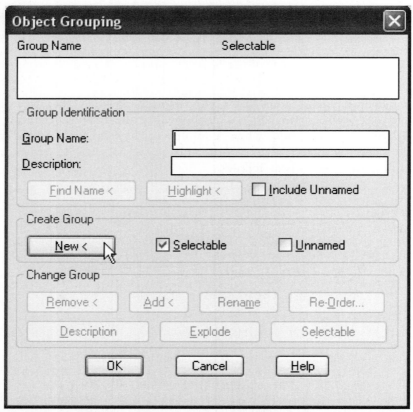

Figure 10.36 *Typing **G** ENTER at the command line displays the **Object Grouping** dialog box. **Group** can be used to select and save drawing objects in groups for editing.*

CREATING AND NAMING A GROUP

In Figure 10.36 you can see that this drawing has no named groups. To create a group, enter the desired name at the keyboard. As you type, the name is entered in the **Group Name** box in the **Group Identification** area. Guidelines for naming groups include:

- Group names can be up to 255 characters long and include numbers, letters, or characters not used by Microsoft Windows or AutoCAD.

- Two separate words can't be used, but compound words can be joined by a dash or underline to create a name such as base_unit.

- Use names that will be easily recognized and that will clearly describe the contents of the group.

Description—A description can be created to help you remember the contents of a specific group. To create a description, click in the **Description** box and type a description.

New—Once the desired name and description have been entered, select the **New** button. This will remove the dialog box from the screen and prompt you to select objects. Use any selection method to select objects to be included in the group. When the selection process is complete, press ENTER to redisplay the dialog box. The group name will be listed in the **Group Name** box. As the group is created, the word **yes** is now displayed beside the group name to indicate that it is selectable. If a group is selectable, selecting one object in the group selects the entire group. Once the group is created, the other areas of the dialog box can be used to alter or use the group.

IDENTIFYING A GROUP

Key elements of the **Group Identification** area can be seen in Figure 10.37. It is used to identify named groups.

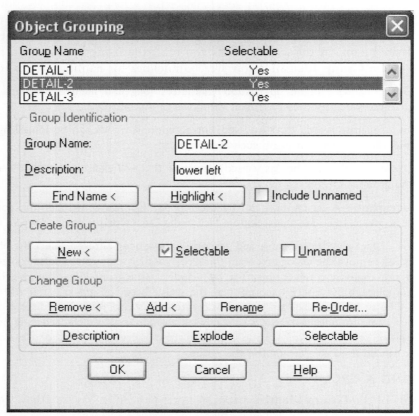

Figure 10.37 *Once groups are identified, selecting one of the groups will activate each portion of the dialog box.*

Find Name

Selecting this button removes the dialog box from the screen and provides the prompt:

```
Pick a member of a group
```

When an object is selected, a **Group Member List** similar to Figure 10.38 is displayed. This box shows all groups that contain the selected object as a member. Clicking the **OK** button returns the **Object Grouping** dialog box. If you select an unnamed group, the **Add** button in the **Change Group** area allows a name to be provided for an unnamed group.

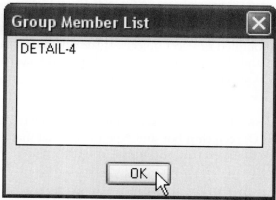

Figure 10.38   *Selecting **Find Name** will display the **Group Member List** box. Select an object on the drawing, and the group that it belongs to will be displayed.*

Highlight

The **Highlight** button is not active as the dialog box is displayed. Selecting one of the named groups activates the button. Clicking the **Highlight** button removes the dialog box from the screen and shows the named group in dashed lines, as well as the **Continue** button. This option is especially helpful on complicated drawings with multiple groups. Click the **Continue** button to restore the **Object Grouping** dialog box.

Include Unnamed

Activate the **Include Unnamed** check box to display unnamed groups in the **Group Name** box. Unnamed groups can be renamed using the **Rename** button in the **Change Group** area.

CREATING A NEW GROUP

The key element of the **Create Group** area can be seen in Figure 10.37. The **New** box was covered previously in the section, "Creating and Naming a Group."

Selectable

Activating the **Selectable** check box determines the status of named groups as selectable. Remember that selecting one object in a named group selects the entire group.

Unnamed

If the **Unnamed** check box is active, AutoCAD will provide names for unnamed groups.

CHANGING A GROUP

Key elements of the **Change Group** area can be seen in Figure 10.37. All elements except **Reorder** are inactive as the dialog box is displayed. Highlighting one of the named groups will activate all of the boxes in the **Change Group** area.

Add

The **Add** button allows drawing objects to be added to an existing group. Once this button is clicked, you will be prompted to select objects. If the selected object is part of an existing group, the group name and description boxes will display the current status of the object. Clicking the **OK** button completes the selection process.

Description

This **Description** button allows the description to be updated and saved. To alter a description, click in the **Description** box and type the desired material to be added. When you have finished, click the **Description** button to update the description and save it for future reference.

Explode

The **Explode** button is used to remove the selected group from the effects of Group select action. The objects that were in the group will not be altered, only the ability to select those objects as a group. Highlight the group to be removed in the **Group Name** list and then click the **Explode** button to explode the group.

Remove

Selecting the **Remove** button allows individual objects to be removed from the grouping. Highlight the group to be edited in the **Group Name** list and click **Remove** to display the Remove objects prompt. The dialog box will be removed and the objects in the group will be highlighted. Objects can be selected for removal using any of the selection methods. When finished removing objects from the set, press ENTER to return to the **Object Grouping** dialog box.

Rename

The **Rename** button can be used to rename an existing group. Highlight the name of an existing group so that the name and description are displayed in the **Group Identification** area. Edit the name as desired and click **Rename** to make the new name the current name.

Reorder

Click the **Reorder** button to produce the **Order Group** dialog box. Notice that the lower portion of the box includes three listings with numbered entries. Objects in a group are numbered in the same order that they were selected as the group was created. Reordering is typically used in CAD/CAM applications of manufacturing and is not often a part of the construction drawings. Options include the following:

Group Name—Displays the name of the group to be reordered.

Description—This selection displays the description of the selected group.

Remove from position (0–x):

For each listing, x will vary depending on the size of the objects in the group. This selection specifies the current numbered position of the object to be reordered.

Replace at position (0–x):

This selection specifies the new position the object will be moved to.

Number of objects (1–x):

This selection lists the range of numbers to be reordered.

CHAPTER 10 EXERCISES

1. Open a template and draw a 38' long × 18 1/2" deep floor truss. Top and bottom chords will be made out of two 3 × 4s with 1" diameter aluminum webs at approximately 45°, equally spaced throughout the truss. Save the drawing as **E-10-1**.

2. Draw four sets of perpendicular crossing lines. On one set of lines use the **Fillet** command to square the corners. On the second set of lines provide a 12" fillet. On the third set of lines use a 12" chamfer. On the fourth set of lines use a 12" and 24" chamfer. Save the drawing as **E-10-2**.

3. Draw a section view of an 8" wide concrete block retaining wall extending 8' above a 4" thick concrete floor slab. Show the wall extending 8" above the finish grade. Thicken the slab to 8" thick at the edge resting on the footing. Use a 16" wide × 12" deep concrete footing, and show a 4" diameter drain on the soil side of the wall. Show a 2 × 6 top plate supporting 2 × 8 floor joists flush with the soil side of the wall. Reinforce the wall with #5 diameter @ 12" o.c. each way, 2" clear of the interior face. Use varied line weights for concrete, grade, sectioned wood, and steel. Adjust the widths to achieve good contrast. Save the drawing as **E-10-3**.

4. Open drawing **E-9-4**. Copy and rotate the pair of doors to provide doors for walls at 45°, 90°, 135°, 180°, and 270° rotations. Save the drawing as **E-10-4**.

5. Open drawing **E-9-5** and make a copy of the column cap. Adjust the base strap so that it is 3" longer and 1" wider. Add one additional bolt. Each end of the side plate must be 2" longer (4" total) and 1 1/2" higher. Save the drawing as **E-10-5**.

6. Open drawing **E-9-7**. Enlarge the foundation plan 8' in each direction. Add required beams and piers to meet the original criteria. Place 4' × 4' diagonal corners in the exterior walls. Save the drawing as **E-10-6**.

7. Open drawing **E-7-1** and make three additional copies. In the original drawing, provide a 6" line from each intersection and perpendicular to the opposite side of the triangle. Rotate the new triangles and place a point of each new triangle tangent to the existing triangle and perpendicular to the 6" lines. Draw a 10" diameter circle with a center point in the center of the original triangle. Trim all elements that extend beyond the circle. When complete, your drawing should resemble the attached drawing. Save the drawing as **E-10-7**.

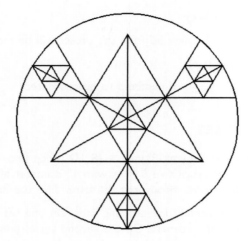

8. Start a new drawing and draw a metal chimney cap that extends a total of 12" above a 4' wide, wood chimney chase. Trim the chase with 3" wide trim and show 6" wide horizontal siding. The top of the chimney cap is 14" wide × 1". The main portion of the cap is 12" × 7". The chimney is 10" wide × 4" high. Use **Mirror**, **Array**, **Offset**, **Trim**, and **Fillet** to complete the drawing. When complete, your drawing should resemble the drawing on the right. Save as **E-10-8**.

9. Start a new drawing and draw a section of a standard one-story foundation similar to the sketch below. Copy and edit the drawing to the proper size for a two-story footing. Save the drawing as **E-10-9**.

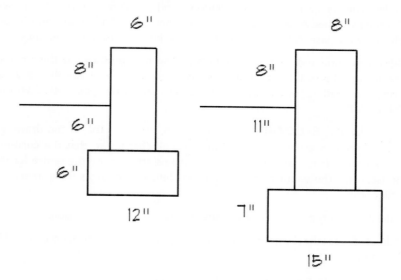

10. Start a new drawing and draw one half of a roof truss with a 5/12 pitch with a 26' span. All chords will be made of 6" wide material; all webs will be made from 4" material. Use a 24" overhang. The webs should be spaced equally at the center of the bottom chord. As a minimum, use **Offset**, **OSNAP**, and **Trim**. Use the **Mirror** command to complete the drawing. Save the drawing as **E-10-10**.

11. Use the floor plan shown in Figure 9.16b and draw the walls and cabinets for this residence. No doors, windows, plumbing fixtures, or appliances need to be drawn at this time. Assume the following:

- All exterior heated walls and toilet walls are 6" thick.

- All exterior unheated walls and interior walls are 4" thick.

- The fireplace is 3'–4" wide. Determine the length by allowing for a 36" wide hallway.

- The stairway, the hall between the stair and bathroom, and the space between the stair and the front wall are 42" wide.

- The bathroom is 36" wide.

  Create separate layers using the guidelines presented in earlier chapters so that the walls, cabinets, fireplace, doors, windows, and plumbing fixtures can be separated from each other. Assign each layer a different color. Add center and hidden lines to the drawing base. Add a dishwasher to the kitchen, using hidden lines.

  Although door and windows symbols should not be drawn at this time, provide the opening in the wall for each door. Provide the outline of all cabinets. Assume 24" wide base and 12" wide upper cabinets. Do not draw appliances. Save this drawing as **EFLOOR**.

12. Open drawing **EFLOOR**. Assume north to be at the top of the drawing so that the front door is presently facing south. Make all changes so that the current wall between the family and living rooms remains straight for the entire length of the house. Alter the width of the wall as needed, as it changes from interior/exterior/toilet wall.

- Add 2'–0" to the family room in the north/south (n/s) direction.

- Add 3'–6" to the living room in the east/west (e/w) direction. Add 18" in the n/s direction.

- Enlarge the bathroom to 42" wide (+6" e/w).

- Remove 24" e/w from the shop.

- Remove the garage doors on the south face of the garage and place them on the east wall.

- Remove the window from the east wall and place (2) 6'–0" wide windows on the south wall.

- In addition to the 2'–8" personnel door on the east wall, enlarge the garage to accommodate (3) 8'–0" wide doors with a 12" minimum space.

- Provide 24" space on each end of the garage between the door and the end of the wall.

  Save the drawing as **E12FLOOR**.

13. Draw five concentric circles, with the smallest having a diameter of 6" and the other sizes increasing in 3" increments. Remove a 1" wide strip going across each of the circles in both the vertical and horizontal directions. The center of the strip to be removed should be at the center of each quadrant. Rotate the remaining portions of the second and fourth circles (starting at the inside) so that the remaining portions of these circles have been revolved 45° while remaining concentric. Save the drawing as **E-10-13**.

14. Use **EFLOOR** as a base. Draw the plan view of a simple tree. Draw the trees in groups of two and three, with each tree in the group having a different size. Create a layer for landscaping and place at least four groups of trees around the house. Create a layer for walkways and design walkways, patios, and a driveway. Save the drawing as **EFLOOR**.

15. Draw concentric shapes of three, four, five, six, seven, and eight sides that are tangent to one circle. Make copies of these objects that are 1/4 size, 1/2 size, full size, 2× normal size, and 90% of the normal size. Save the drawing as **E-10-15**.

16. Open drawing **E-10-4** and make two copies of the door. Scale one door to be 2'–6" wide and the other door to be 2'–8" wide. Leave the window beside the door as it is on each option. Once the sizes are altered, make copies so that double doors are available for each option. A total of four door arrangements should be available when complete. Save the doors as **E-10-16**.

17. Open Drawing **E-9-10** and make two additional copies of the window. Enlarge one window to 5'–0" wide. Reduce the other window to 75% of the original size. Save the drawing as **E-10-17**.

18. Draw a U-shaped kitchen with a 30" × 48" island. Provide a 32" × 21" sink, a 24" × 24" dishwasher, a 34" × 24" refrigerator, and 27" × 27" oven in the kitchen, with a 30" × 22" cook top in the island. Create a second copy of the kitchen that will be flopped and set at a 45° angle to the first kitchen. Save the drawing as **E-10-18**.

19. Using the drawing on the next page as a guide, lay out the outline of the office floor plan. Use 8" wide exterior walls. Copy the plan. Save the drawing as **E-10-19**.

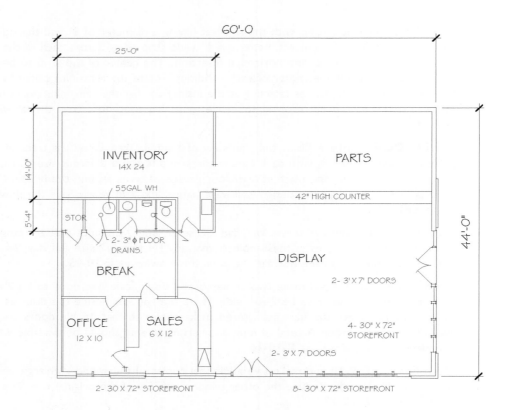

20. Draw a 3" diameter circle, a 6" line, and a polygon with eight sides that will fit in a 5 3/4" diameter circle. Save this portion of the drawing as **E-10-20a**. Make enough copies of each object and then use each option of each grips editing function to edit each object. Save the drawing as **E-10-20b**.

21. Use **E-10-20a** as a template drawing to complete this assignment. Make a second copy of these objects. Use **Select** to create a group consisting of the two circles and two polygons. Use **Mirror**, **Rotate**, and **Scale** to edit the group. Save the drawing as **E-10-21**.

22. Use Drawing **E-10-20a** as a template drawing to complete this assignment. Make a second copy of these objects and then mirror both groups so that you have a total of four circles, lines and polygons. Use **Group** to name and save at least 6 different groups. Save the file as **E-10-22**.

23. Using the drawing below as a guide, draw and array a pattern to represent Spanish claytile roofing. Save the drawing as **E-10-23**.

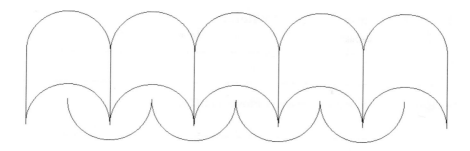

24. Use a reference guide to determine the size of a W8 x 35 steel columns. Draw a plan for a 38' x 76' structure showing three rows of columns. Each of the two exterior rows will be inset 8" from the exterior face of the structure. The interior row of columns will be 18' in from the front (the 76' long surface) of the structure. Columns will be placed at 19' o.c., equally spaced from each exterior wall. Save the drawing as **E-10-24**.

CHAPTER 10 QUIZ

DIRECTIONS

Answer the following questions with short complete statements. Type your answers using a word processor.

1. Place your name, chapter number, and the date at the top of the sheet.

2. Type the question number and provide the answer.

 Warning: Some of the questions have not been covered in the reading material and will require the use of the help menu. You may have to do some exploring to answer the questions.

1. Explain the uses of the **Break** command.

2. Describe two different uses of the **Fillet** command.

3. List the command sequence for **Rotate**.

_____

_____

4. Explain the difference between **Move** and **Copy**.

_____

_____

5. List the command sequence to extend a line to meet another line, and sketch and label the process.

_____

_____

6. Two perpendicular lines cross at a corner. List four methods to edit the lines to form a non-crossing corner.

_____

_____

7. A line has been drawn short of its required destination. How can the error be corrected?

_____

_____

8. What two options are available for offsetting an object?

_____

_____

9. In the **Chamfer** command, what is the relationship between the distances and the selection of lines?

_____

_____

10. You are preparing to move some objects, so you type **P** ENTER at the Select objects prompt. What will be moved?

   _____

   _____

11. Use the **Help** menu to determine what **Setvar** controls a preset **Fillet** value?

   _____

   _____

12. How can an object be rotated in a clockwise direction?

   _____

   _____

13. A door is drawn at a 45° angle. It should be 49°. How can this error be corrected?

   _____

   _____

14. Six objects need to be moved. As you select the objects, an extra object is included in the selection set, and an intended object is not included. Describe the command process and prompts to move these objects.

   _____

   _____

15. List and describe the three **Break** options.

   _____

   _____

16. Describe how each of the editing commands presented in this chapter could be used in drawing a floor plan.

   _____

   _____

17. What will happen if you try to place a 60" radius fillet in two intersecting lines that are 48" long?

_____

_____

18. Explain how **Chamfer** will affect two lines that do not touch if they are selected as the objects of this command.

_____

_____

19. What will be the effect of the Trim option of **Chamfer** on two intersecting lines?

_____

_____

20. What will be the effect of the **Trim Edge** option on two lines that do not intersect?

_____

_____

21. How will the Edge option of **Extend** affect two lines?

_____

_____

22. How will Window and Crossing window affect objects selected for **Stretch**?

_____

_____

23. What are the options that are given if the DElta option of **Lengthen** is selected?

_____

_____

24. A 6" long line is selected to be edited using **Lengthen**. What will be the result of entering 50%? 150%? 200%?

_____

_____

25. How many selection points will be required when you break an arc using **Break@**?

_____

_____

26. Provide an example of how the **Align** command could save time.

_____

_____

27. List the commands used to create E-10-9.

_____

_____

28. What must be done to display grips on a line?

_____

_____

29. What five editing functions can be executed after a grip is hot?

_____

_____

30. What is a hot grip and how do you create one?

_____

_____

31. You're exploring options and select **Gripblock**. What will this option do, and what can you do if you don't want this option?

_____

_____

32. What advantage does **Select** have over grips?

_____

_____

33. Describe the process of selecting and naming a group of objects using **Group**.

_____

_____

34. What is a nested group?

_____

_____

35. What is a selectable group?

_____

_____

36. You've named a group, but you come to realize the name does not describe the drawing well. How do you fix the problem?

_____

_____

37. Save a group of objects without providing a name. How does AutoCAD keep track of this group?

_____

_____

38. You're exploring again and you use the Explode option of **Group**. What effect will this have on your computer? What effect will it have on the current drawing? What effect will it have on the selected group?

_____

_____

39. How do you make the options in the **Change Group** area of the **Object Grouping** dialog box active?

_____

_____

40. Use the **Help** menu and determine what variables affect grips. What are the options and their effects?

_____

_____

Polylines

INTRODUCTION

This chapter will introduce drawing individual line segments that function as one line by the use of polylines. The chapter will explore the **Pline** command and examine methods of editing polylines.

Commands to be examined in this chapter include:

- Pline
- Explode
- Pedit
- Revcloud

One of the best ways to get objects to stand out in a complicated drawing is to vary the line thickness by adjusting the lineweight. Thickness can also be assigned easily with the **Pline** command, which creates polylines. A polyline is a connected sequence of line segments that function as a single object. Figure 11.1 shows examples of polylines used to represent steel reinforcing in concrete details.

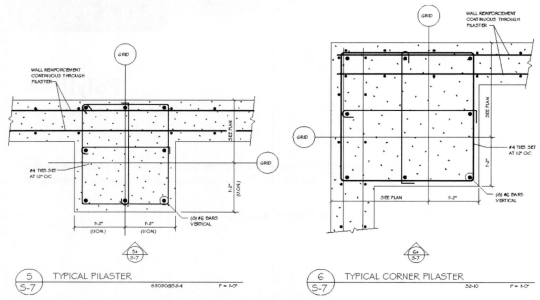

Figure 11.1 *The **Pline** command can be used to distinguish between various materials. The steel is drawn with a polyline and can be easily separated from the edge of the concrete. Not only can polylines be thicker, but they move as one object rather than as individual lines. (Courtesy Van Domelen Looijenga/ McGarrigle/Knauf Consulting Engineers.)*

In addition to being able to display varied widths, a polyline has several other unique features. When the line is drawn using the **Line** command, each segment is a separate object. When multiple segments are created as polylines, each individual line segment functions as one unit. In Figure 11.2 the horizontal line, the arc, and the vertical line forming the L-shaped rebar have been selected for editing. Because these features were drawn using the **Pline** command, they function as one object.

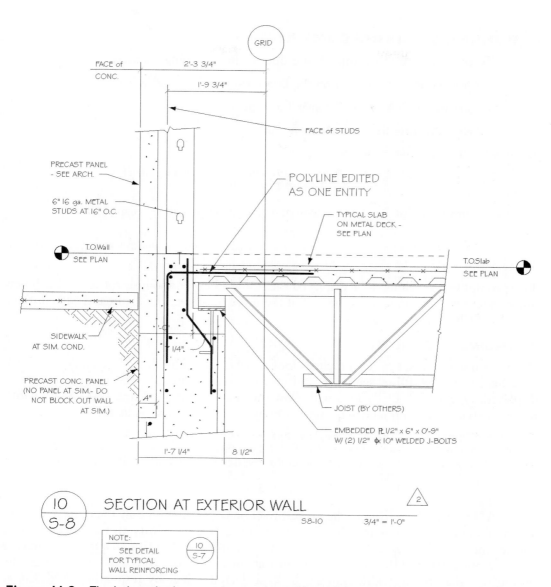

Figure 11.2 *The L-shaped rebar was drawn using polylines. Selecting the rebar at any point will select the horizontal line, the arc, and the vertical line as one object. (Courtesy Van Domelen/Looijenga/McGarrigle/ Knauf Consulting Engineers.)*

EXPLORING THE PLINE COMMAND

To start the **Pline** command, use one of the following methods:

- Click the **Polyline** button on the **Draw** toolbar.
- Type **PL** ENTER at the Command prompt.
- Select **Polyline** from the **Draw** menu.

The command sequence is as follows:

```
Command: Click the Polyline button (Or type PL ENTER)
Specify start point: (Select any point.)
```

Once a beginning point has been selected, a new prompt will be displayed:

```
Current line-width is 0'-0"
Specify next point or [Arc/Halfwidth/Length/Undo/Width]: (Specify
    endpoint line.)
```

This prompt will continue to be displayed each time a new endpoint is specified. Each new line segment will continue from the end of the preceding line. The command will continue until ENTER is pressed.

CHANGING THE WIDTH

With the line width set to "0," polylines look just like any other line. To change the width of a line, start the **Pline** sequence and select a start point. As the option prompt is displayed, type **W** ENTER. This will allow information for a starting and ending width to be entered. The command sequence to draw a 1/8" wide line is as follows:

```
Command: Click the Polyline button (Or type PL ENTER)
Specify start point: (Select any point.)
Specify next point or [Arc/Halfwidth/Length/Undo/Width]: W ENTER
Specify starting width <0'-0">: .125 ENTER
Specify ending width <0'-0 1/8">: ENTER
```

AutoCAD rounds the option to the nearest fraction depending on how the units have been set. If you enter a fraction and the default returns as 0, enter **Units** on the **Format** menu and set the fractional values as desired. For the examples used in this chapter, the units are set to architectural and the denominator for fractions is set to 1/8" (.125"). Notice that the prompt for the ending width shows 1/8" as a default value. By responding ENTER at the prompt, the option prompt will return allowing for selection of the endpoint. With the width selected, the command sequence is as follows:

```
Specify next point [Arc/Close/Halfwidth/Length/Undo/Width]:
    (Specify starting point.)
```

With the starting and ending width the same, a line with uniform thickness will be drawn, as shown in Figure 11.3.

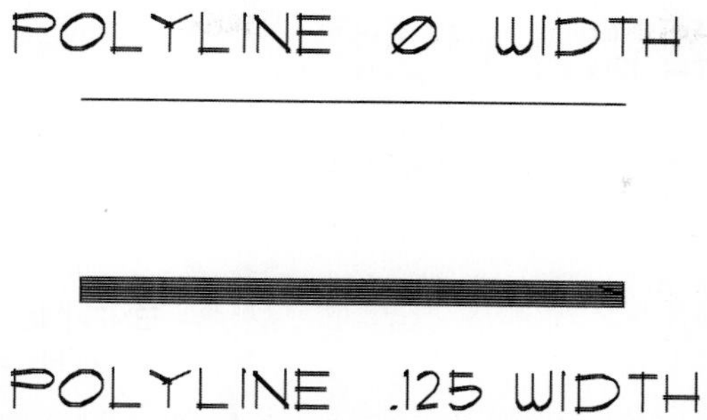

POLYLINE Ø WIDTH

POLYLINE .125 WIDTH

Figure 11.3 *The **Pline** command allows thickness to be assigned to polylines. If drawn with a width value of 0, the polyline will appear as a line segment drawn with the default line width.*

Drawing Tapered Lines

The **Pline** Width option allows the width of a polyline to be changed for each segment or from one end of a segment to the other. A tapered polyline with a starting width of 1/8" and an ending width of 1/4" can be drawn by using the following command sequence:

```
Command: Click the Polyline button (Or type PL ENTER.)
Specify start point: (Select any point.)
Specify next point or [Arc/Halfwidth/Length/Undo/Width]: W ENTER
Specify starting width <0'-0">: .125 ENTER
Specify ending width <0'-01/8">: .25 ENTER
Specify next point [Arc/Halfwidth/Length/Undo/Width]: (Select
    endpoint.)
Specify next point [Arc/Close/Halfwidth/Length/Undo/Width]: ENTER
```

As the endpoint is entered, the line will be drawn and the option prompt will be returned to allow continuing a polyline. Notice that if the command is continued, the default for the next line segment is the ending width of the one just completed. Figure 11.4 shows an example of varied line width.

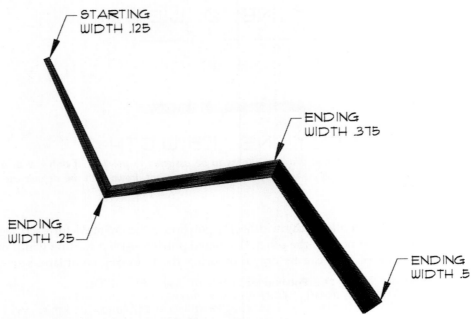

Figure 11.4 *Polylines can drawn with varied widths.*

Choosing a Line Width

When you experiment with **Pline**, the width selection is not terribly important. As you begin to work on more involved projects that will be printed or plotted, the width selection is critical. One of the considerations in selecting a line width is the scale at which the drawing eventually will be printed. A 1/4" polyline will look huge on the screen when you've zoomed in on a feature, but if printed or plotted at 3/8"=1'–0", it will appear as a 0 width line.

It's also important to realize that line width can be controlled in the plotting process. For now, the key is to realize that polyline width can easily be altered.

Halfwidth Option

The Halfwidth option allows the width of a polyline to be set from the center of the polyline to an edge, and is very similar to setting the width. The command sequence is as follows:

```
Command: Click the Polyline button (Or type PL ENTER.)
Specify start point: (Select any point.)
Specify next point or [Arc/Halfwidth/Length/Undo/Width]: H ENTER
Specify starting half-width <0'-01/8">: ENTER
Specify ending width <0'-0 1/8">: ENTER
Specify next point [Arc/Halfwidth/Length/Undo/Width]: (Select
    endpoint.)
Specify next point [Arc/Close/Halfwidth/Length/Undo/Width]: ENTER
```

Length Option

This option allows a polyline to be drawn with a specified length parallel to the last polyline drawn. The new line will extend in the same direction as previous lines. To change the direction in which the line is drawn, enter a negative length. This process can be seen in Figure 11.5. If the last segment was an arc, a line drawn with **L** will be tangent to the arc.

```
Command: Click the Polyline button (Or type PL ENTER.) PLINE
Specify start point: (Select any point.)
Current line-width is 0'-0 1/8"
Specify next point or [Arc/Close/Halfwidth/Length/Undo/Width]:
    L ENTER
Specify length of line: (Enter desired length.) 2 ENTER
Specify next point or [Arc/Close/Halfwidth/Length/Undo/
    Width]: ENTER
```

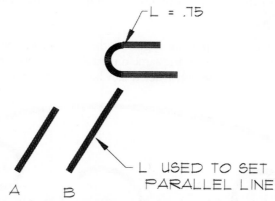

Figure 11.5   *The Length option of **Pline** creates a polyline with a specific length parallel to the most recently created polyline.*

POLYLINE ARCS

When the Arc option is selected, AutoCAD displays the Arc submenu and displays the prompt:

```
Command: Click the Polyline button (Or type PL ENTER.)
PLINE
Specify start point: (Select starting point.)
Current line-width is 0'-0 1/8"
Specify next point or [Arc/Halfwidth/Length/Undo/Width]: A ENTER
Specify endpoint of arc or [Angle/CEnter/Direction/Halfwidth/
    Line/Radius/Second pt/Undo/ Width]: (Select any point.)
Specify endpoint of arc or
[Angle/CEnter/CLose/Direction/Halfwidth/Line/Radius/Second
    pt/Undo/ Width]: ENTER
```

The Arc option will continue to prompt for arc endpoints until ENTER is pressed.

Continuous Polyline Arcs

If an arc is started from a previous polyline segment, the new arc will be tangent to the existing line by default. The arc will continue in the same direction as the most recent polyline segment. An example of how the arc will be constructed can be seen in Figure 11.6.

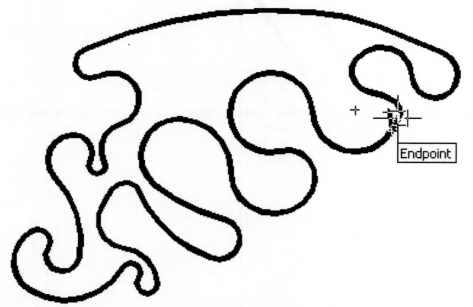

Figure 11.6 *Arcs can be constructed of polylines by using the Arc option of Pline.*

Polyline Arc Angles

This option will allow the start point, center, and angle that the polyline arc is to span (included angle) to be specified. By default, the angle will be drawn counterclockwise.

By entering a negative value, the angle will be drawn clockwise. The command process to draw an arc with a 45° included angle is as follows:

```
Command: Click the Polyline button (Or type PL ENTER.) PLINE
Specify start point: (Select starting point.)
Current line-width is 0'-0 1/8"
Specify next point or [Arc/Halfwidth/Length/Undo/Width]: A ENTER
Specify endpoint of arc or
[Angle/CEnter/Direction/Halfwidth/Line/Radius/Second pt/Undo/
    Width]: A ENTER
Specify included angle: 45 ENTER
Specify endpoint of arc or [CEnter/Radius]: (Select desired
    endpoint.)
Specify endpoint of arc or
[Angle/CEnter/CLose/Direction/Halfwidth/Line/Radius/Second
    pt/Undo/ Width]: ENTER
```

As the endpoint is selected, the center is rotated as needed based on the two endpoints. This process can be seen in Figure 11.7.

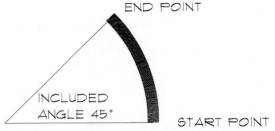

Figure 11.7   *Selecting the **Angle** option of the polyline **Arc** menu allows an arc to be drawn based on the provided angle.*

Arc Center

A polyline arc can be drawn by specifying the endpoint and the center point. When **CE ENTER** is entered at the prompt, the first endpoint and the center point of the arc can be specified. The first steps of the command sequence are identical to other Arc options. The command sequence for drawing a polyline arc based on the center point is as follows:

```
Command: Click the Polyline button (Or type PL ENTER.) PLINE
```

```
Specify start point: (Select starting point.)
Current line-width is 0'-0 1/8"
Specify next point or [Arc/Halfwidth/Length/Undo/Width]: A ENTER
Specify endpoint of arc or
[Angle/CEnter/Direction/Halfwidth/Line/Radius/Second pt/Undo/
    Width]: CE ENTER
Specify center point of arc: (Select desired center point.)
Specify endpoint of arc or [Angle/Length]: A ENTER
Specify included angle: 45 ENTER
Specify endpoint of arc or
[Angle/CEnter/CLose/Direction /Halfwidth/Line/Radius/Second
    pt/Undo/Width]: ENTER
```

Arc Radius

Typing **R** ENTER will allow for specifying the first endpoint and the radius to be selected. The command sequence is as follows:

```
Command: Click the Polyline button (Or type PL ENTER.) PLINE
Specify start point: (Select starting point.)
Current line-width is 0'-0 1/8"
Specify next point or [Arc/Halfwidth/Length/Undo/Width]: A ENTER
Specify endpoint of arc or
[Angle/CEnter/Direction/Halfwidth/Line/Radius/Second pt/Undo/
    Width]: R ENTER
Specify radius of arc: (Enter desired radius.) 3.25 ENTER
Specify endpoint of arc or [Angle]: A ENTER
Specify included angle: (Specify desired angle) 45 ENTER
Specify direction of chord for arc <30>: (Select direction)
Specify endpoint of arc or
[Angle/CEnter/CLose/Direction/Halfwidth/Line/Radius/Second
    pt/Undo/ Width]: ENTER
```

If an angle for the chord is entered, the arc will be drawn, the prompt will be redisplayed for continuing the arc command, and a new arc will be displayed on the screen. Press ENTER to stop the command or enter the letter representing the desired option to continue.

Arc Direction

The default for **Pline** is to draw the arc tangent to the preceding segment. The Direction option will allow you to override this action and specify a starting direction for the arc that is not tangent to the previous line. The process can be seen in Figure 11.8. The sequence is as follows:

```
Command: Click the Polyline button (Or type PL ENTER.) PLINE
Specify start point: (Select starting point.)
Current line-width is 0'-0 1/8"
Specify next point or
[Arc/Halfwidth/Length/Undo/Width]: A ENTER
Specify endpoint of arc or
[Angle/CEnter/Direction/Halfwidth/Line/Radius/Second pt/Undo/
    Width]: D ENTER
```

```
Specify the tangent direction for the start point of arc:
     (Select a point to indicate the desired arc direction.)
Specify endpoint of arc: (Specify desired endpoint.) ENTER
Specify endpoint of arc or
[Angle/Center/Close/Direction/Halfwidth/Line/Radius/Second
     pt/Undo/ Width]: ENTER
```

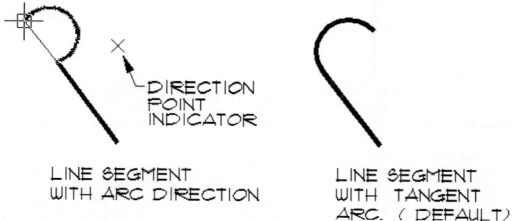

Figure 11.8 *The **Direction** option of the polyline **Arc** menu allows an arc to be drawn that is not tangent to an existing polyline.*

Three-Point Arc

A polyline arc can be drawn by entering a second point between each endpoint. This allows for better placement of the arc. The sequence can be seen in Figure 11.9. The command sequence is as follows:

```
Command: Click the Polyline button (Or type PL ENTER.) PLINE
Specify start point: (Select starting point.)
Current line-width is 0'-0 1/8"
Specify next point or [Arc/Halfwidth/Length/Undo/Width]: A ENTER
Specify endpoint of arc or
[Angle/CEnter/Direction/Halfwidth/Line/Radius/Second pt/Undo/
     Width]: S ENTER
Specify second point on arc: (Specify desired second point for
     arc to pass through.)
Specify endpoint of arc: (Specify desired endpoint of arc.)
Specify endpoint of arc or
[Angle/CEnter/CLose/Direction/Halfwidth/Line/Radius/Second
     pt/Undo/ Width]: ENTER
```

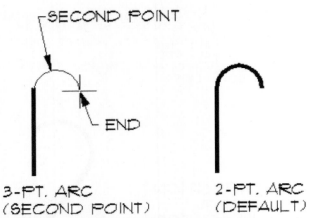

Figure 11.9 *Using the Second point option allows an arc to be drawn based on three points.*

CLOSING A POLYLINE

Polylines are ended by using automatic OSNAP or by typing **C** ENTER. Each option provides a line segment from the present location back to the original starting point. These two options are also available with the **Pline** Arc option. Typing **CL** ENTER at the prompt also allows a polygon constructed of polylines to be closed. Remember that **CL** must be entered when in the arc submenu rather than **C** to distinguish between CLose and CEnter. The results can be seen in Figure 11.10.

 Note: Just because AutoCAD gives you an option doesn't mean you need to use it. Automatically snapping to an endpoint is a much faster method for closing a polygon, but this method will affect how fillets are applied to polylines.

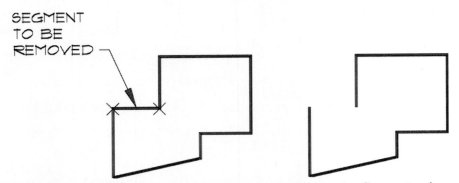

Figure 11.10 *Polygons drawn using polylines can be completed using the Close option. Automatic OSNAP and C ENTER can also be used to close a polygon constructed of polylines.*

UNDOING A POLYLINE

As with other drawing options, **Pline** allows the previous segment or segments to be removed in the reverse order from the way they were drawn. This can be done with the **Undo** option by typing **U** ENTER at the prompt.

```
Specify end point of arc or
Specify next point or [Arc/Halfwidth/Length/Undo/Width]: U ENTER
```

This will remove the last polyline segment that was drawn. Undo can be used repeatedly until just one point is left in the polyline. Separate from the Undo option, the **Undo** command can be used once the drawing command has been terminated. The **Undo** command removes an entire command sequence and will be covered in later chapters.

EDITING A POLYLINE

Polylines can be edited with the commands introduced in previous chapters. Because polylines are segments drawn as a single object, they are edited as a single object. This can facilitate the selection process because only one segment needs to be selected, rather than each of the segments that make up the entire object. Most of the commands explained in chapter 10 will modify a polyline in exactly the same way as another element. **Fillet** and **Chamfer** each have an added feature for a polyline.

ALTERING POLYLINES WITH FILLET

The **Fillet** command can be used to fillet two different polylines in the same way it is used it to fillet two line segments. Once a polyline is selected, the entire object will be highlighted. Prompts are then given to define the specific intersection to receive the fillet. The effects of the **Fillet** command on a polyline can be seen in Figure 11.11. A fillet can also be applied to all edges of the polygon in one command sequence. To edit an entire polyline, select **P** when prompted to select the first object. The **Fillet** command *will not* work on the last intersection created on a polygon unless the **Close** option of **Polyline** is used. See Figure 11.11. Once the radius is set, the command sequence is as follows:

```
Command: F ENTER (Or click the Fillet button.) FILLET
Current settings: Mode = TRIM, Radius = 0'-0 3/8"
Select first object or [Undo/Polyline/Radius/Trim/Multiple]:
    P ENTER
Select 2D polyline: (Select object to receive fillets.)
4 lines filleted (Quantity will vary with each object.)
```

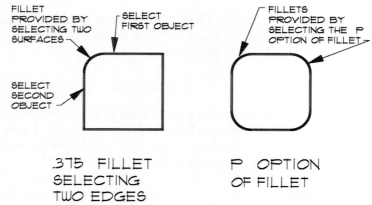

FILLET PROVIDED BY SELECTING TWO SURFACES

SELECT FIRST OBJECT

FILLETS PROVIDED BY SELECTING THE P OPTION OF FILLET

SELECT SECOND OBJECT

.375 FILLET
SELECTING
TWO EDGES

P OPTION
OF FILLET

Figure 11.11   *The **Fillet** command can be used to modify individual corners of polygons constructed of polylines, or the Polyline option of **Fillet** can be used to fillet all edges of a polygon. The last corner created on the polygon will only receive a fillet if it was placed using the **Close** option.*

ALTERING POLYLINES WITH CHAMFER

The **Chamfer** command will edit polylines in the same way that line segments are edited. Once the distances are set, the command will edit a polyline as two line segments. If the Polyline option is selected, all corners of the selected polyline will receive chamfers. The command will only edit the last polygon intersection if the Close option of **Polyline** was used to draw the polygon. The command sequence to edit all edges of a polygon with a radius of .375 is:

```
Command: Click the Chamfer button (Or type CHA ENTER.)
CHAMFER
(TRIM mode) Current chamfer Dist1 = 2'-0", Dist2 = 2'-0"
Select first line or [Undo/Polyline/Distance/Angle/Trim/mEthod/
    Multiple]: D ENTER
Specify first chamfer distance <2'-0">: .375 ENTER
Specify second chamfer distance <0'-0 3/8">:
Select first line or [Undo/Polyline/Distance/Angle/Trim/mEthod/
    Multiple]: P ENTER
Select 2D polyline:
4 lines were chamfered
```

The effect of chamfering a polygon can be seen in Figure 11.12.

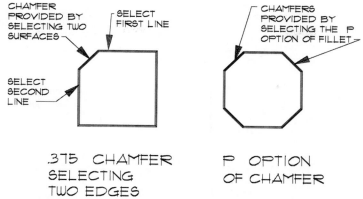

.375 CHAMFER
SELECTING
TWO EDGES

P OPTION
OF CHAMFER

Figure 11.12   *The **Chamfer** command can be used to modify crossing polylines and polygons constructed of polylines, or the Polyline option of **Chamfer** can be used to chamfer all edges of a polygon.*

EXPLODING POLYLINES

One of the benefits of a polyline is also one of its drawbacks. Whether you drawn 2 or 2,000 connected polyline segments, they all function as one object. That's great if you want to move 2,000 lines at once by selecting only one option, but a hindrance if you want to change just one of the segments. A fast way to overcome this obstacle (but not the best way) is to use the **Explode** command. This command explodes groups of drawing objects. To access the command, use one of the following options:

- Click the **Explode** button (the stick of dynamite) on the **Modify** toolbar.

- Type **X** ENTER at the Command prompt.

- Select **Explode** from the **Modify** menu.

The command sequence is as follows:

```
Command: X ENTER (Or click the Explode button.)
Select objects: (Select objects to be edited.)
Select object: ENTER
```

Once ENTER has been pressed, the polyline will be transformed into individual segments. If the polyline had a specified width, the width will revert to 0.

The effects of **Explode** on a polyline can be seen in Figure 11.13. If you can tolerate the loss of line width, proceed. If the line width is important to the drawing, typing **U** ENTER at the Command prompt will restore the polyline to its unedited form. If you want the polyline to function as separate lines but retain their width, **Explode** the polyline and assign width using the **Lineweight** and the **Properties** commands.

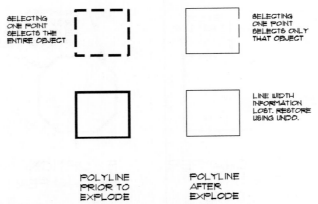

Figure 11.13 *The **Explode** command allows segments of a polyline to be treated as individual objects.*

ALTERING POLYLINES WITH PEDIT

In addition to using the standard editing options, you can edit polylines using the **Pedit** command. To access the Pedit command, use one of the following methods:

- Type **PE** ENTER at the Command prompt.

- Click the **Edit Polyline** button on the **Modify II** toolbar.

Options for the command will vary slightly depending on whether the polyline forms a closed shape or a line. Options for the command when a closed polyline is selected are as follows:

```
Command: Click the Edit Polyline button (Or type PE ENTER.)
Select polyline or [Multiple]: (Select polyline to be edited.)
Enter an option [Open/Join/Width/Edit vertex/Fit/ Spline/
     Decurve/Ltype gen/ Undo]:
```

If an open polyline is selected, the **Open** option is replaced with the **Close** option. Pressing ENTER will restore the command line. You can select objects by any of the selection methods introduced in previous chapters. Notice, when you are prompted to select objects, that the cursor changes to a selection box to allow you to select polylines individually. The **Window** and **Crossing** methods are also excellent methods of selecting polylines.

Occasionally, segments might be selected for **Pedit** that are not polylines. If the **Peditaccept** system variable is set to 1, you can select non-polylines and join them. With a setting of 0, the following prompt will be given:

```
Object selected is not a polyline
Do you want to turn it into one? <Y>:
```

If the default is accepted, the segment will become a polyline and the **Pedit** prompt will be displayed, allowing options of the new polyline to be altered. A circle cannot be changed to a polyline, but it can be drawn using the 360° arc of the **Pline Arc** option. The Donut command can also be used to draw a circle that would have qualities similar to a polyline.

```
The editing options for a polyline include:
Open, Join, Width, Edit vertex, Fit, Spline, Decurve, Ltype gen,
      Undo, and eXit.
```

Open

The Open option will remove the closing segment of a polyline. If a line was drawn back to the starting point without using the Close option, opening the polyline has no visible effect. The results can be seen in Figure 11.14. If **O** is selected for an open polyline, the reverse happens—it will be closed.

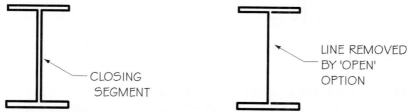

Figure 11.14   *The polygon on the left was completed using the Close option. A polyline segment was removed with the Open option on the right.*

Close

The Close option creates a closing segment of the polyline while in the **Pline** command. Once you've moved on to another command sequence, the polyline must be edited with **Pedit C** (Close) option. If you select **C** on a closed polyline, it will be opened. Figure 11.14 shows the effects of the Close option.

Join

The Join option adds lines, arcs, and other polylines that meet a selected polyline. This option can be useful for joining two or more individual polylines so that they function as one polyline. This option can only be used with open polylines and cannot be used to join segments that do not touch the selected polyline. Objects that cross the polyline will not be joined. The option is started by typing **J** ENTER at the Command prompt once the editing options are displayed.

Width

The Width option will allow the current width of a polyline to be changed. Polylines that consist of segments with varying widths or tapers will be changed so that all segments have the new width. Once **W** ENTER is entered at the editing options prompt, a new prompt will be given requesting the following:

```
Specify new width for all segments.
```

The width can be entered by keyboard or by selecting two points with the select button. Once the width is altered, the **Pedit** prompt will be displayed again, allowing for other editing to take place. Press ENTER to end the PEDIT command. Figure 11.15 shows the effect of the Width option.

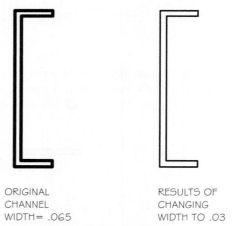

ORIGINAL
CHANNEL
WIDTH= .065

RESULTS OF
CHANGING
WIDTH TO .03

Figure 11.15 *The effect of changing the width of a polyline.*

Edit Vertex

The **Edit vertex** option allows the shape of an existing polyline to be altered. Altering the shape can be done by typing **E** ENTER at the option prompt. An **X** will be placed in the first vertex of the selected polyline, and a new prompt will be displayed with the following options:

```
Enter an option [Open/Join/Width/Edit vertex/Fit/Spline/Decurve/
    Ltype gen/Undo]: E ENTER
Enter a vertex editing option
[Next/Previous/Break/Insert/Move/Regen/Straighten/Tangent Width/
    eXit] <N>:
```

Unless the X is placed in the location you would like to edit, use the **Next** or **Previous** option to move the X to the desired editing location.

Next The Next mode moves the X to the next vertex of the selected polyline. By pressing ENTER, the X marker is moved around the object, as seen in Figure 11.16.

Previous The **Previous** mode moves the X marker to the previous vertex. By pressing ENTER, the marker can be moved around the object in the direction opposite from the one used by Next. The option can be seen in Figure 11.16.

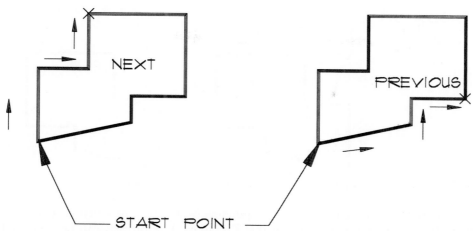

Figure 11.16 *The vertex of a polyline can be selected for editing by moving the X marker using the Next or Previous options.*

Break The Break mode removes a portion of a polyline. The portion to be broken is selected by the use of the **N**, **P** and **G** options. Once the X marker is in the desired position, the prompt reads,

```
Enter an option [Next/Previous/Go/eXit] <N>:
```

As the X marker is moved with **N** or **P**, any point between the original starting location and the present location will be removed. When the X is moved to the end of the segment to be used, type **G** ENTER for **Go**. This will remove the desired segment. If Break is used on a closed polygon, the closing section of the polygon will be removed along with the selected portion. The process can be seen in Figure 11.17. The command sequence to break a polyline is:

```
Enter an option [Open/Join/Width/Edit vertex/Fit/Spline/Decurve/
    Ltype gen/Undo]: E ENTER
Enter a vertex editing option
[Next/Previous/Break/Insert/Move/Regen/Straighten/Tangent Width/
    eXit]<N>: B ENTER
Enter an option [Next/Previous/Go/eXit] <N>: (Continue to select
    ENTER until the marker is at the desired vertex) ENTER
Enter an option [Next/Previous/Go/eXit] <N>: G ENTER
Enter a vertex editing option
[Next/Previous/Break/Insert/Move/Regen/Straighten/Tangent/Width/
    eXit] <N>: X ENTER
Enter an option [Open/Join/Width/Edit vertex/Fit/Spline/Decurve/
    Ltype gen/Undo]: ENTER
```

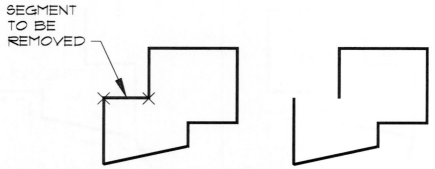

Figure 11.17 *Removing a portion of a polyline using the Break option.*

Insert   The **Insert** mode edits an existing polyline by adding a new vertex. The new vertex is added after the vertex identified by the X marker. The process can be seen in Figure 11.18. The command sequence to insert a vertex is as follows:

```
Enter an option
[Open/Join/Width/Edit vertex/Fit/Spline/Decurve/Ltype gen/ Undo]:
    E ENTER
Enter a vertex editing option
[Next/Previous/Break/Insert/Move/Regen/Straighten/Tangent/Width/
    eXit]<N>: ENTER (Move X to the desired location and press
    ENTER.)
[Next/Previous/Break/Insert/Move/Regen/Straighten/Tangent/Width/
    eXit] <N>: I ENTER
Specify location for new vertex: (Select new location.)
Enter a vertex editing option
[Next/Previous/Break/Insert/Move/Regen/Straighten/Tangent/Width/
    eXit] <P>: X ENTER
Enter an option
[Close/Join/Width/Edit vertex/Fit/Spline/Decurve/Ltype
    gen/Undo]: ENTER
```

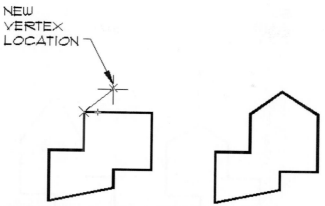

Figure 11.18   *Altering the shape of a polyline by inserting a new vertex.*

Move   The **Move** mode allows an existing polyline vertex to be moved to a new location. The process can be seen in Figure 11.19. The vertex to be moved must be the one currently marked by the X marker. The command sequence to move a vertex is:

```
Enter an option
[Open/Join/Width/Edit vertex/Fit/Spline/Decurve/Ltype/Undo/eXit]:
     E ENTER
Enter a vertex editing option
[Next/Previous/Break/Insert/Move/Regen/Straighten/Tangent/Width/
     eXit] <N>: ENTER (Move X to the desired location and press
     ENTER.)
Enter a vertex editing option
[Next/Previous/Break/Insert/Move/Regen/Straighten/Tangent/Width/
     eXit] <N>: M ENTER
Specify new location for marked vertex: (Select new location.)
Enter a vertex editing option
[Next/Previous/Break/Insert/Move/Regen/Straighten/Tangent/Width/
     eXit] <N>: X ENTER
Enter an option
[Close/Join/Width/Edit vertex/Fit/Spline/Decurve/Ltype
     gen/Undo]: ENTER
```

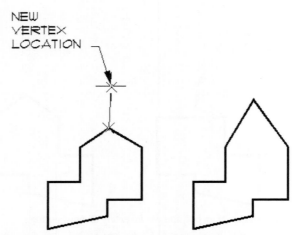

Figure 11.19 *Altering the shape of a polyline by moving an existing vertex to a new location.*

Regen The Regen mode will regenerate the edited polyline.

Straighten The Straighten mode will straighten existing polyline segments that lie between two selected points. Move the marker to the desired location to mark the start of the edit. The X can then be moved to mark the end of the edit. Any arcs or segments between the two marks will be deleted and replaced by a straight segment. The process can be seen in Figure 11.20. The command sequence to straighten segments of a polyline is:

```
Enter an option
[Open/Join/Width/Edit vertex/Fit/Spline/Decurve/Ltype/Undo/eXit]:
    E ENTER
Enter a vertex editing option
[Next/Previous/Break/Insert/Move/Regen/Straighten/Tangent/Width/
    eXit] <N>: ENTER (Move X to the desired location and press
    ENTER.)
Enter a vertex editing option
[Next/Previous/Break/Insert/Move/Regen/Straighten/Tangent/Width/
    eXit] <N>: S ENTER
Enter an option [Next/Previous/Go/eXit] <N>: ENTER (Continue
    pressing ENTER until the X is moved to the desired
    location.)
Enter an option [Next/Previous/Go/eXit] <N>: G ENTER
Enter a vertex editing option
[Next/Previous/Break/Insert/Move/Regen/Straighten/Tangent/Width/
    eXit] <N>: X ENTER
Enter an option
[Close/Join/Width/Edit vertex/Fit/Spline/Decurve/Ltype gen/Undo]:
    ENTER
```

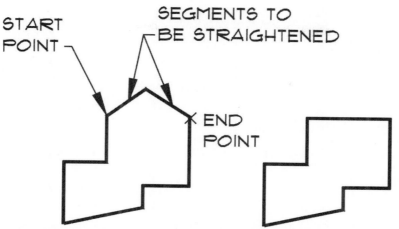

Figure 11.20   *Removing an existing polyline vertex using the Straighten mode of the Edit vertex option.*

Tangent   The **Tangent** mode can be used to attach a tangent direction to the current vertex for later use in curve fitting. The command sequence is as follows:

```
Enter an option
[Open/Join/Width/Edit vertex/Fit/Spline/Decurve/Ltype/Undo/eXit]:
    E ENTER
Enter a vertex editing option
[Next/Previous/Break/Insert/Move/Regen/Straighten/Tangent/Width/
    eXit]<N>: ENTER (Move X to the desired location.)
Enter a vertex editing option
[Next/Previous/Break/Insert/Move/Regen/Straighten/Tangent/Width/
    eXit] <N>: T ENTER
Specify direction of vertex tangent: (Specify a point or enter
    an angle.)
Enter a vertex editing option
[Next/Previous/Break/Insert/Move/Regen/Straighten/Tangent/Width/
    eXit] <N>: X ENTER
Enter an option [Close/Join/Width/Edit vertex/Fit/Spline/Decurve/
    Ltype gen/Undo]: ENTER
```

Either a specific tangent angle can be specified from the keyboard or a point can be selected to mark the direction from the currently marked (X) vertex.

Width   The **Width** mode changes the starting and ending width of the segment following the marked vertex. You've explored the **Width** option of **Pedit** that changes the entire polyline. The **Width** mode of the **Edit vertex** option allows one segment to be edited. The process can be seen in Figure 11.21. The command sequence to edit the width of one segment is:

```
Enter an option
[Open/Join/Width/Edit vertex/Fit/Spline/Decurve/Ltype/Undo/eXit]:
    E ENTER
```

```
Enter a vertex editing option
[Next/Previous/Break/Insert/Move/Regen/Straighten/Tangent/Width/
    eXit]<N>: ENTER (Move X to the desired location.)
Enter a vertex editing option
[Next/Previous/Break/Insert/Move/Regen/Straighten/Tangent/Width/
    eXit] <N>: W ENTER
Specify starting width for next segment <0'-0 1/8">: 0 ENTER
    (Enter desired width.)
Specify ending for next segment width <0'-0">: .375 ENTER (Enter
    desired width.)
Enter a vertex editing option
[Next/Previous/Break/Insert/Move/Regen/Straighten/Tangent/Width/
    eXit] <N>: X ENTER
Enter an option
[Close/Join/Width/Edit vertex/Fit/Spline/Decurve/Ltype gen/Undo]:
```

ORIGINAL SEGMENT
STARTING WIDTH= .25
ENDING WIDTH =.25

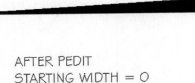

AFTER PEDIT
STARTING WIDTH = 0
ENDING WIDTH = .375

Figure 11.21   *Altering the width of an existing polyline using the Width mode of the Pedit Edit vertex option.*

Exit   The Exit mode exits from Vertex editing and returns you to the **Pedit** prompt.

Fit

The Fit option of **Pedit** allows straight segments to be converted to curved lines. This can be especially helpful on a topography plan where lines are typically drawn from elevation to elevation. The Fit option can also be used to convert zigzag lines to smooth curves, as seen in Figure 11.22.

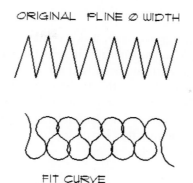

ORIGINAL PLINE Ø WIDTH

FIT CURVE

Figure 11.22   *Changing a polyline using the Fit option.*

Spline

Unlike the Fit curve option that passes a line segment through all of the vertices, a spline curve only passes line segments through the first and last points of the polyline. In between, each curve will be close to each vertex, but they will not pass through the vertex points. The more control points (vertices) that you specify, the smoother the spline curve will be. Figure 11.23 shows a spline curve.

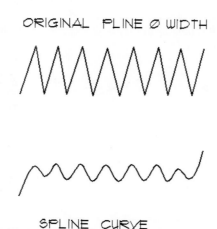

ORIGINAL PLINE Ø WIDTH

SPLINE CURVE

Figure 11.23   *Altering a polyline using the Spline option.*

The **Fit** and **Spline** options are not often used on construction drawings. For further information, remember to use the **Help** menu in AutoCAD. You may also want to research the commands **Splines**, **Splinetype**, **Splinesegs**, **Splinedit**, and **Nurbs** for related types of drawing objects.

Decurve

The Decurve option can be used to remove any vertices that have been inserted by the Fit or Spline option and to straighten segments of a polyline.

Ltype Gen

Figure 11.24 shows a polyline drawn with a centerline linetype. Notice that the pattern extends from vertex to vertex. Each segment is, in effect, a separate centerline. The polyline also can be drawn so that the pattern extends throughout the entire polyline, as seen in Figure 11.25. The Ltype gen option of the **Pedit** command controls how the pattern is displayed. Toggled to the **ON** position, Ltype gen will draw line patterns in a continuous pattern from beginning to end with no consideration for vertices. In the **OFF** position, line patterns will be based on each vertex. Ltype gen does not affect polylines with tapered segments.

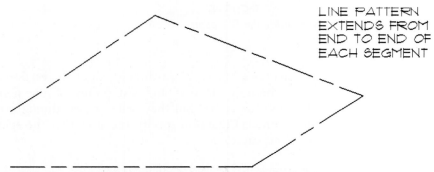

LINE PATTERN
EXTENDS FROM
END TO END OF
EACH SEGMENT

Figure 11.24 *Using the **OFF** setting of the Ltype gen option of **Pedit** will draw a linetype pattern from vertex to vertex.*

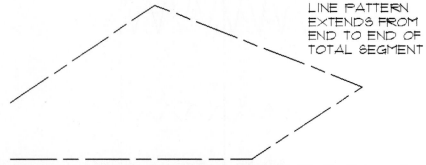

LINE PATTERN
EXTENDS FROM
END TO END OF
TOTAL SEGMENT

Figure 11.25 *Using the **ON** setting of the **Ltype gen** option of **Pedit** will draw a linetype pattern from end to end of the entire line, disregarding the vertex.*

Undo

The Undo option of **Pedit** will undo the most recent **Pedit** operation. Multiple entries of **U** will step back through the drawing, removing previous **Pedit** entries.

Exit

The eXit option is the default for **Pedit**. Pressing ENTER will also end the **Pedit** command and return to the Command prompt.

USING THE REVISION CLOUD COMMAND

Most professionals use a revision cloud to indicate material that has been changed on a drawing since the original printing. Figure 11.26 shows an example of a revision cloud used to highlight changes made to a drawing. AutoCAD uses polylines created using sequential arcs to form the revision cloud. To start the command, use one of the following methods:

- Click the **Revcloud** button from the **Draw** toolbar.
- Select **Revision Cloud** from the **Draw** menu.
- Type **REVCLOUD** ENTER at the command prompt.

The command sequence is as follows:

```
Select the Revision Cloud button.
Specify start point or [Arc length /Object/Style]<object>:
    (Specify the desired arc size or select the desired start
    point.)
```

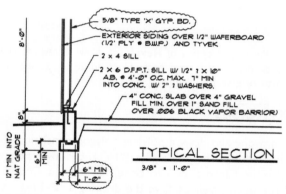

Figure 11.26 *Revision clouds are used by many architectural and engineering offices to highlight changes that have been made to drawings since the original printing.*

By default, an arc length of 1/2" is used to represent the individual arc segments that will form the cloud. The cloud is created by selecting a start point and then moving the crosshairs along the path of the desired cloud. When the ending point is returned to the original starting point, AutoCAD automatically completes the command and displays the revision cloud.

Options when selecting the start point include altering the Arc length, Object or Style. Typing **A** ENTER when prompted to select a start point will allow an alternate arc radius to be selected. Any arc length can be entered for the minimum value, but the maximum arc length cannot exceed three times the minimum arc length. If **O** ENTER is typed when prompted for the start point, AutoCAD allows a closed object such as a circle or rectangle to be converted to a revision cloud. Once the object is converted to a revision cloud, the following prompt is displayed:

```
Reverse direction [Yes/No]:
```

Type **Y** ENTER to alter the direction of the revision cloud or press ENTER to accept the specified revision cloud. Figure 11.27 shows examples of a revision cloud and a reversed cloud. AutoCAD also allows the line style of the revision cloud to be altered by using the normal line weight with a uniform thickness, or by a line that resembles calligraphy. Entering **S** ENTER at the first prompt and accepting **N** for normal or entering **C** for calligraphy determines how the revision cloud will be drawn.

NORMAL
DISPLAY

REVERSE
DISPLAY

Figure 11.27 *Selecting the Object option of **Revcloud** allows a revision cloud to be created by selecting an object such as a circle. Once converted to a revision cloud, an option is provided to reverse the cloud.*

CHAPTER 11 EXERCISES

1. Start a new drawing and draw a polyline polygon consisting of five sides with a 0 width. Set the width of the polyline as 1/16" and draw a polygon with a minimum of four sides. Use the Close option to close the polygon. Change the width to .125" and draw three open polylines. Save the drawing as **E-11-1**.

2. Start a new drawing and draw a polyline with a beginning width of 0.125", a length of 1.5", and an ending width of 0.25". Continue from this segment with a line that is 1.5" long and 0.125" wide. Draw a third segment that will close a polygon with a starting width of 0.25" and ending with a 0 width. Save the drawing as **E-11-2**.

3. Start a new drawing and draw a circle with a 3" diameter and a thickness of 1/16". Use the three-point Arc option to draw a polyline circle with a width of 1/8" and a radius of 1.5". Save the drawing as **E-11-3**.

4. Start a new drawing and use a width of 0 to draw a 3" long line. Draw a 0.125" wide arc with a radius of 1.75" and an angle of 60° on the right end of the original line segment. Use the right end of the straight line segment as the center of a 45° arc that ends at the left end of the arc segment. Save the drawing as **E-11-4**.

5. Start a new drawing and draw a four-sided polygon using **Pline** with a width of 0.125". Copy the polygon so that there are a total of five polygons. Edit one of the polygons so that the line width is 0.25". Edit another polygon so that the width is 0.065". Fillet two corners with a 0.5" radius. Fillet the remaining two corners with a 0.25" radius. Edit a third polygon so that the width of all lines is 0.0 wide. Provide a 0.4" × 0.25" chamfer at each corner. On the fourth polygon, add a vertex at some point so that the polygon has five segments. Save the drawing as **E-11-5**.

6. Open Drawing **E-10-9** and convert the lines that represent the footing to polylines. Set the width as 0.5". Set the line that represents the finish grade as 0.75". Copy the one-story footing and explode the drawing. Save the drawing as **E-11-6**.

7. Open Drawing **E-10-8**. Assume the light source to be in the upper left corner to shade the drawing. Use 0.5" width polylines to create shade on all overhanging materials. On the left side of the horizontal trim, draw a shadow that tapers from 0 up to 1" wide. Save the drawing as **E-11-7**.

8. Start a new drawing and draw a series of seven zigzag polyline 0.0 width lines at approximately 15° from vertical. Make the line segments 10" long. Make two additional copies. Use the Spline option to edit one set of lines and the Fit option to edit the other set. Save the drawing as **E-11-8**.

9. Open a drawing template **E-11-9** and adjust the limits as required to draw a typical wall section that is similar to the drawing on the next page. Use the one level footing to show:

- 2 × 6 sill with a 1/2" × 10" anchor bolt

- 2 × 8 floor joist w/ 2 × 8 rim joist

- 3/4" plywood sub-floor

- 8' high studs with a double top plate and a single base plate

- 2 × 6 ceiling joist

- 2 × 6 rafters @ a 27 1/2° (5/12) pitch

- 2 × 8 fascia and solid eave blocking

Use 0.25" wide polylines to represent any materials that would be cut by the cutting plane, such as the plate, sill, fascia, and blocking. Save the drawing as **E-11-9**. Make a copy of **E-11-9** and draw a revision cloud around any three objects. Save this drawing as **E-11-9REV**.

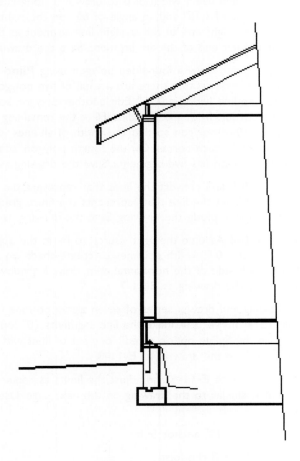

10. Start a new drawing and set the units and limits to draw the grading plan shown below. Layout the grids and approximate the contour lines. Use a 6" wide polyline to represent 5' intervals and a 2" wide polyline to represent 1' intervals. Once all straight line segments have been drawn, edit the plines to provide the most accurate and smooth transitions possible. Save the drawing as **E-11-10.**

C.L.   S.W.   JANICE ANN COURT

| 0+100' | 0+80' | 0+60' | 0+40' | 0+20' | 0+0 |

GRADING PLAN

C.L.   S.W.   14th STREET

CHAPTER 11 QUIZ

Directions

Answer the following questions with short complete statements. Type your answers using a word processor.

1. Place your name, chapter number, and the date at the top of the sheet.

2. Type the question number and provide the answer.

 Warning: Some of the questions have not been covered in the reading material and will require the use of the help menu. You may have to do some exploring to answer the questions.

1. Use the **Help** menu and determine what **Plinegen** controls, and what options are included.

2. Give the command sequence and show all options to draw a polyline with a width of 0.125".

3. List an option for controlling line width other than **Pline**.

_____

_____

4. Major drawings for a residence include the site, floor, and foundation plans, elevations, and sections. Find a set of professional drawings and list common ways
polylines can be used to enhance these drawings.

_____

_____

5. What option would allow a 0.25" wide line to be drawn by entering **.125**?

_____

_____

6. Explain the difference between the Spline and Fit options.

_____

_____

7. Explain the difference between Open and Close **Pedit** options.

_____

_____

8. How does **Explode** affect a polyline?

_____

_____

9. To edit a vertex that is formed between the third and fourth lines drawn, what two
options will be needed?

_____

_____

10. What option will allow an additional vertex to be added to a completed polyline?

_____

_____

11. A polyline has been drawn, but one vertex is 1/2" to the left of where it belongs. What option will fix the problem?

_____

_____

12. What **Pedit** option will remove any vertices that have been inserted by the Fit or Spline option?

_____

_____

13. What **Pedit** option will remove and convert arcs to straight line segments?

_____

_____

14. What option will display an X at the start of a polyline segment, and what three options affect the X?

_____

_____

15. What is the default setting for **Pedit**?

_____

_____

16. How does the Halfwidth option differ from the Width options?

_____

_____

17. A friend needs to draw a 2" long polyline parallel to an existing polyline. How can this be done?

_____

_____

18. How can a circle be drawn using the **Pline** command?

_____

_____

19. How can a polyline arc be drawn in a clockwise direction?

_____

_____

20. How can the direction of an arc chord be controlled?

_____

_____

Supplemental Drawing Commands

INTRODUCTION

This chapter will introduce methods for marking and dividing drawing space, and will explore five new methods for drawing lines.

Commands to be introduced in this chapter include:

- **Customize**
- **Point**
- **Divide**
- **Measure**
- **Ray**
- **Xline (Construction line)**
- **U**
- **Undo**
- **Redo**

ADDING ICONS TO A TOOLBAR

You'll notice as you work through this chapter that several of the buttons noted in commands are not found on the default toolbars. You can find these command buttons using the following steps:

- Select **Customize** from the **Tools** menu.
- Select **Interface** or right-click when the cursor is in any open toolbar.

This will display the **Customize User Interface** dialog box. Use the following steps to add an icon to an existing toolbar:

1. Open the **Customize User Interface** dialog box.
2. Display the **Customize** tab.
3. In the **Command List** display, highlight the name of the icon to be placed. *(Select MEASURE)*

4. Select **Toolbars** in the **All Customization File** to display a list of each available toolbar.

5. Drag the command name you want to add to a location just below the toolbar name in the Customizations area of the tab. *(Drag MEASURE to be placed below DRAW.)*

6. Click the plus sign (+) to the left of the toolbar to display the command you just added.

7. Add RAY and XLINE to the DRAW toolbar.

8. Add any other icons you wish to display.

9. Click **Apply** and then **OK** to make your changes.

Selecting **OK** will add the desired icons to the specified toolbars and close the dialog box. See additional information in the **Help** menu for creating toolbars from scratch.

MARKING AND DIVIDING SPACE

Three commands are available in AutoCAD for marking and dividing space. Although these commands can't be used to draw lines, each can be used to mark space in a drawing and provide a location for placing lines.

PLACING POINTS

Often, a point needs to be marked on a drawing. For example, the loads on a column on the upper level of a structure will need to be carried down through several floors into the foundation. AutoCAD will allow you to mark these load locations on a drawing by drawing a point. Be sure to place the point on a non-plotting layer. To place a point in a specific location, use one of the following methods:

- Click the **Point** button on the **Draw** toolbar.
- Type **PO** ENTER at the Command prompt.
- Select **Point** from the **Draw** menu.

If the command is selected from the **Draw** menu, the command will allow multiple points to be located. Each method will produce a prompt:

```
Command: Click the Point button (Or type PO ENTER.)
Specify a point: (Select a point location or enter coordinates.)
```

The prompt is now waiting for a location to be specified. The location of the point can be specified by the cursor or by keyboard.

Controlling the Point Display

The shape that is used to identify a point can be altered through the **Point Style** dialog box, shown in Figure 12.1. Display the dialog box by selecting **Point Style** from the **Format** menu. The dialog box allows the style and size of the point to be controlled. The active setting is shown in the upper left corner. The default marker is a point. The marker will be displayed on the screen but will not be plotted. Notice the second box from the left in the top row contains no marker. If this option is selected, the point will be marked with an invisible marker. Although the point can't be seen, it will be selected when the **Node** mode of **Object Snap** is active.

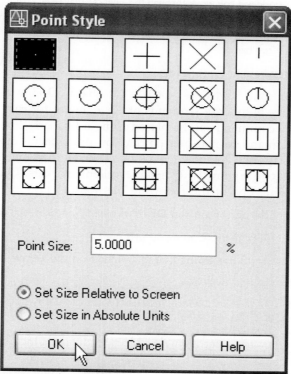

Figure 12.1   *Selecting **Point Style** from the **Format** menu will produce the **Point Style** dialog box. The box can be used to alter the size and shape of the point marker.*

The default point size is listed as 5% of the display screen size. Point size can be altered to meet the needs of the drawing. Just as important as the size is the display method. In the default setting, the size of the point will be displayed relative to the size of the display area. As you zoom in and out of a drawing, points will remain a set percentage of the drawing display size. If you select the display to be in absolute units, points will disappear as you zoom out and become huge as you zoom in. Entering a positive value for the point size will display the point relative to the drawing area. Use the following procedure to display a point that is marked by crosshairs surrounded by a circle and a square, and is displayed at 10% of the screen size:

1. Select **Point Style** from the **Format** menu.

2. Select a marker from the **Point Style** dialog box.

3. Change the display in the **Point Size** edit box to read 10%.

4. Click the **OK** button.

 Note: You'll notice as you work through this chapter that several of the buttons noted by commands are not found on the standard toolbars. Review the previous section of this chapter to place the buttons for each of the commands introduced in this chapter on the **Draw** toolbar.

EXPLORING THE DIVIDE COMMAND

Although this command does not edit an object, **Divide** allows an object to be divided into any number of segments of equal length for easy editing. The markers for the Divide command can be placed along a line, arc, circle, or polyline. Before using the **Divide** command to mark an object, you must decide how each mark will be indicated. The crosshairs, the X, and the crosshairs surrounded by a circle, selected in the **Point Style** dialog box in Figure 12.1, make excellent markers. Once the marker is set, start the command using one of the following methods:

- Click the **Divide** button on the **Draw** toolbar (see previous discussion.)

- Type **DIV** ENTER at the Command prompt.

- Select **Divide** from the **Point** cascading menu on the **Draw** drop-down menu.

The command sequence is as follows:

```
Command: Click the Divide button (Or type DIV ENTER.)
Select object to divide: (Select the objects.)
```

This is your opportunity to select a single object to divide. Once an object is selected, a prompt will be displayed:

```
Enter the number of segments or [Block]: (Enter number.) 14 ENTER
```

Typing a number between 2 and 32,767 and pressing ENTER will cause the object to be divided into the number of segments specified. The object is not physically divided into separate segments—the markers are just placed so that exact points can be selected. The process can be seen in Figure 12.2. The **Divide** command can also be used to insert a group of objects, called a block, at repeated distances. Blocks will be covered in a later chapter.

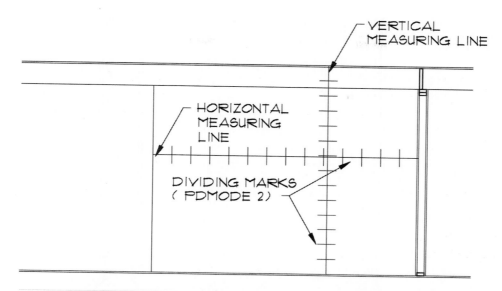

Figure 12.2 *Points are used to divide lines into segments in preparation for a stair layout.*

EXPLORING THE MEASURE COMMAND

 The **Measure** command places markers at specific distances along a line, arc, circle, or polyline at a repetitive spacing. Figure 12.3 shows an example of how **Measure** could be used as a drawing aid. To start the command, use one of the following methods:

- Click the **Measure** button on the draw toolbar.

- Type **ME** ENTER at the Command prompt.

- Select **Measure** from the **Point menu** of the **Draw** menu.

Be sure to change the point marker to something that will be visible when placed on a line. The command sequence to measure an object is as follows:

```
Command: Click the Measure button (Or type ME ENTER.)
Select object to measure: (Select object.)
```

This is your opportunity to select a single object to measure. Once an object is selected, a prompt will be displayed:

```
Specify length of segment or [Block]: (Enter distance.) 16 ENTER
```

A specific distance can be entered by keyboard, or two points can be entered by the cursor. The markers will be placed, starting at the selected end of the line of the object to be measured. This process can be seen in Figure 12.3. To use the select button to enter the distances, move the cursor to the desired location and click the select button.

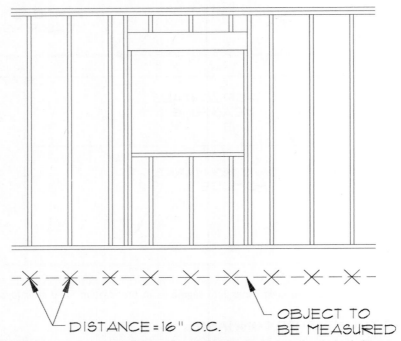

Figure 12.3 *Once a segment has been selected for measurement, the desired distance to be measured can be entered by keyboard or by selecting the distance with the select button.*

DRAWING LINES OF VARIED WIDTH AND LENGTH

Each command introduced in this section can be used as an alternative to the **Line** command. The **Ray** and **Xline** commands allow semi-infinite and infinite lines to be added to a drawing. The **Sketch** command allows freehand sketching to be inserted in a drawing.

CREATING LINES USING THE RAY COMMAND

A line created by the **Ray** command is a line with a defined starting point that extends into infinity in a specified direction. Rays are especially helpful in projecting features from one drawing to another. If you were to use the **Line** command to project from a floor plan to develop an elevation, it would have to have two endpoints. A zoom might be required to go from the first point to the next point. A transparent zoom ('**Z**) could be used, but this is time consuming when compared to using a ray. A ray will provide a line that extends from the selected point indefinitely, no matter how many **Zoom** previous operations are required.

To start the **Ray** command, use one of the following methods:

- Click the **Ray** button from the **Draw** toolbar.
- Type **RAY** ENTER at the Command prompt.
- Select **Ray** from the **Draw** menu.

The command sequence to draw a ray is as follows:

```
Command: Click the Ray button (Or type RAY ENTER.)
Specify start point: (Select desired starting point of ray.)
Specify through point: (Select desired direction of ray.)
Specify through point: ENTER
```

Figure 12.4 shows the layout of an elevation using the **Ray** command. A ray will show when the drawing is plotted. Rays should be kept on a layer separate from the drawing being created so they can be removed from the drawing base by freezing the layer or by using the **Do not Plot** option for the layer.

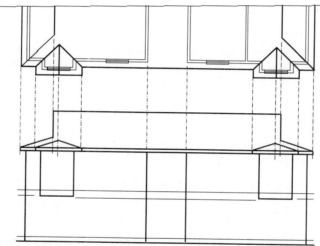

Figure 12.4 *The **Ray** and **Xline** commands are excellent for creating projection lines. Because of their length, **Zoom** and **Stretch** operations can be minimized. The vertical projection lines in this drawing were created using the **Ray** command and then edited using the **Trim** command.*

CREATING LINES USING THE XLINE COMMAND

The **Xline** command creates a line that extends through a point to an infinite distance in two directions. To access the command, use one of the following methods:

- Click the **Construction Line** button on the **Draw** toolbar.
- Type **XL** ENTER at the Command prompt.
- Select **Construction Line** from the **Draw** menu.

Xline offers six options to place the line. The default setting is to select a point for the line to pass through.

Through Point

This option will display a line through a specific point. Once the point is selected, the line will be rotated to the desired angle through a second (the through) point. The command sequence will continue to prompt for additional lines to be drawn through the original point. End the command by clicking ENTER. The command sequence is as follows:

```
Command: Click the Construction Line button (Or type XL ENTER.)
XLINE Specify a point or [Hor/Ver/Ang/Bisect/Offset]: (Select
    desired option.)
Specify through point: (Select a direction for the line to
    extend.)
Specify through point: ENTER
Command:
```

Horizontal and Vertical Options

Choosing the **Hor** option will allow a horizontal line to be drawn through a specified point. The line will remain horizontal, regardless of the ORTHO setting. The **Ver** option will pass a vertical line through the selected point. With ORTHO ON, similar results will be achieved.

Ang Option

The Ang option allows a specific angle relative to horizontal to be specified. The command sequence is as follows:

```
Command: Click the Construction Line button (Or type XL ENTER.)
XLINE Specify a point or [Hor/Ver/Ang/Bisect/Offset]: A ENTER
Enter angle of xline (0) or [Reference]: 30 ENTER
Specify through point: (Select the point for the line to pass
    through.)
Specify through point: ENTER
Command:
```

Reference Option   The **Reference** option of the command draws a line at a specified angle to another line. Once the reference angle is selected, a location for the line to pass through needs to be indicated. The command sequence is as follows:

```
Command: Click the Construction Line button (Or type XL ENTER.)
XLINE Specify a point or [Hor/Ver/Ang/Bisect/Offset]: A ENTER
Specify angle of xline (0) or [Reference]: R ENTER
Select a line object: (Select the line to be used as the
    reference.)
Enter angle of xline <0>: 30 ENTER
Specify through point: (Select the point for the line to pass
    through.)
Specify through point: ENTER
Command:
```

Bisect Option

The **Bisect** option can be used to draw a construction line that bisects an existing angle. The line is created by selecting three points. The first point to be selected is the

Angle vertex point. The second and third points are located on the lines forming the angle to be bisected. Figure 12.5 shows a construction line used to bisect an angle. The command sequence to bisect an angle is as follows:

```
Command: Click the Construction Line button (Or type XL ENTER.)
XLINE Specify a point or [Hor/Ver/Ang/Bisect/Offset]: B ENTER
Specify angle vertex point: enter (Select the point.)
Specify angle start point: (Select the end of one of the lines
        to be bisected.)
Specify angle end point: (Select the end of the other line to be
        bisected.)
Specify angle end point: ENTER
Command:
```

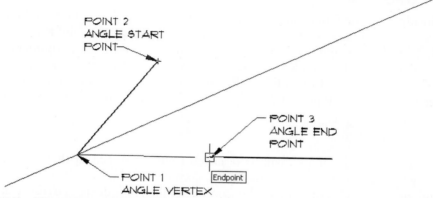

Figure 12.5   *The **Bisect** option of **Xline** can be used to divide an angle in half by selecting three points.*

Offset Option

The **Offset** option draws construction lines that are parallel to the selected line, based on a specified offset distance or point. This option functions like the **Offset** command. You'll be prompted for a line, a distance, and a side to offset. Select an offset distance, and then you will be prompted for a line to be offset and for a through point. The command sequence to offset a line 12" is as follows:

```
Command: Click the Construction Line button (Or type XL ENTER.)
XLINE Specify a point or [Hor/Ver/Ang/Bisect/Offset]: O ENTER
Specify offset distance or [Through] <Through>: 12 ENTER
Select a line object: (Select the line to be offset.)
Specify side to offset: (Select the side of the offset.)
Select a line object: ENTER
Command:
```

CREATING LINES WITH THE SKETCH COMMAND

The **Sketch** command allows the equivalent of freehand drawings to be generated using a computer. Line segments are entered as the cursor is moved, rather than providing a "next" point. The command is useful for entering signatures or other irregular material on a drawing. Because each tiny segment of the sketch is recorded as a line segment, drawing file size can become extremely large in a very short time. The use of **Sketch** should be kept to a minimum to avoid filling up drawing space on a diskette. The size of the drawing is limited only by your storage medium.

Access the **Sketch** command by typing **SKETCH** ENTER at the Command prompt. Best results occur when ORTHO and SNAP are toggled to **OFF**, allowing greatest flexibility in line placement. Because the command is generally not used with construction drawings, it will not be covered in this text. Consult the AutoCAD **Help** menu to further explore this line.

THE U COMMAND

You've already used the **Undo** option to remove a drawing object. The **U** option has been used in the middle of a command routine to remove an object that might not be what you intended. The **U** command also can be used to backtrack through a command sequence once the sequence has been ended. Use the command carefully. Start a new drawing, draw a series of several lines, and press ENTER. Draw a circle, a polygon, and an ellipse in the order listed. Now if you type **U** ENTER at the Command prompt, you will see:

```
Command: U ENTER
ELLIPSE GROUP
```

The last item drawn, the ellipse, will be removed from the screen. If the command line is displayed, you'll notice that the most recent command (ellipse) will be listed and the Command prompt will reappear. If the command line is hidden and the command is entered through the dynamic display, the object will be removed with no prompt given. If the process is repeated, the polygon will be removed.

```
Command: U ENTER
POLYGON GROUP
```

By continuing to enter **U**, you can remove the entire drawing.

```
Command: U ENTER
CIRCLE
Command: U ENTER
LINE
```

UNDO

Depending on how the command is accessed, the **Undo** command has two major functions. When entered by selecting the **Undo** button on the **Standard** toolbar, the command will remove drawing objects one item at a time. When entered by using the command line, the command is a powerful tool for editing objects. The command line option of the **Undo** command allows several command sequences to be undone at one time and permits several operations to be carried out as the objects are undone. The **Undo** command should give you the freedom to know you can try anything, and if the outcome is undesirable, enter **Undo** at the Command prompt to restore the drawing to its original state. For the following command, make sure the command line is displayed. The command sequence is as follows:

```
Command: UNDO ENTER
Current settings: Auto = On, Control = All, Combine = yes
Enter the number of operations to undo or [Auto/Control/BEgin/
    End/Mark/Back]<1>:
```

If a number is typed, such as **2** ENTER, the last two commands will be undone. If you were to press F2 and discover the last twelve commands were a mistake, type **12** ENTER to remove the last twelve command entries.

AUTO

The Auto option issues a prompt to toggle between **ON** and **OFF**. When Auto is **ON**, the default value, any group of commands that is used to insert an item, or group of items, is treated as one item and removed by **U**. This will be helpful later when you work with blocks, wblocks, and macros. If Auto is **OFF**, each command in a group of commands is treated as an individual one.

CONTROL

The Control option can be used to limit the **Undo** command, or to completely disable **Undo**. The command sequence is:

```
Command: UNDO ENTER
Enter the number of operations to undo or [Auto/Control/BEgin/
    End/Mark/Back] <1>: C ENTER
Enter an UNDO control option [All/None/One/Combine] <All>:
```

If All is selected, **Undo** will not be limited and it will erase back to the last Mark. The None option will disable **U** and **Undo** commands so that your machine will function as if these commands have never been placed in AutoCAD. You can reactivate **Undo** and **U** by entering Undo and typing **A** ENTER. Selecting the One option limits **U** and **Undo** to a single operation before returning to the command line. Selecting Combine controls whether multiple, consecutive zoom and pan commands are combined as a single operation for undo and redo operations.

BEGIN AND END

The **Begin** option groups a sequence of drawing operations. The group of drawing commands is defined by the use of the Begin option of **Undo**. The end of the group is marked by the **End** option. Within a specified group, AutoCAD will treat grouped commands as a single operation.

BACK

The **Back** option removes everything in the entire drawing. Not some, not most, we're talking *everything*. Do you get the feeling you need to use this option carefully? AutoCAD will even help you ponder the use of this option by displaying the prompt:

```
This will undo everything. OK? <Y>
```

If ENTER is pressed, this will accept the default **Yes** value, and really clean up your drawing. Typing **N** ENTER will ignore the Back option. If you do use the Back option and then change your mind, use the **Redo** command to restore your drawing. The Back option also can be used with the Mark option so that the entire drawing is not erased.

MARK

This option will limit the distance that **Back** will search through a drawing as it erases work. A mark can be placed in a drawing prior to executing several commands. If the outcome of those commands is not what you expected, **Undo Back** will remove all of the commands until the mark is reached. The command sequence is as follows:

```
Command: UNDO ENTER
Enter the number of operations to undo or [Auto/Control/BEgin/
    Mark/Back] <1>: M ENTER
Command: L enter (Now draw several features in an existing
    drawing.)
Command: UNDO ENTER
Enter the number of operations to undo or [Auto/Control/BEgin/
    Mark/Back] <1>: B ENTER LINE
Mark encountered
Command:
```

NUMBER

The Number option of **Undo** will undo the specified number of drawing operations. The effect is the same as if the **U** command had been used the same number of times, but only a single regeneration is required.

REDO

Occasionally you will remove an object with **U** or **Undo** by mistake. The **Redo** command can be used to restore items that have been deleted. **Redo** will only restore objects, however, when used immediately after an **Undo** or **U**. If **U** or **Undo** is followed by a command such as **Line** and then **Redo**, **Redo** will have no effect on the material that was deleted prior to the other command sequence. To start the command, use one of the following methods:

- Click the **Redo** button on the **Standard** toolbar.
- Type **REDO** ENTER at the Command prompt.

CHAPTER 12 EXERCISES

1. Start a new drawing and set the units and limits to be appropriate for this problem. Draw the outline of a 15' × 28' apartment unit. Make the 15' walls 6" thick, with the long walls 8" wide. Use 4" wide lines for interior walls and lay out a 12' × 15' bedroom, and a 5'–6" × 9' bathroom. No windows or doors are required at this time. Save the drawing as **E-12-1**.

2. Start a new drawing and draw a line 27'–6 1/2" long. Divide the line into five equal segments. Copy the original line and place markers at 16" o.c. Save the drawing as **E-12-2**.

3. A stairway is to be drawn between two floors of a townhouse with a distance of 9'–1 1/2" from finish floor to finish floor. Plan a set of stairs with 7 1/2" maximum rise and 10 1/2" minimum run per step. Determine the required number of stairs and the total run. Draw horizontal lines to represent each floor. The upper floor level will be 13 1/2" thick. Divide the space between floors into the required number of equal risers. Divide the total required run into the required individual steps. Using any editing process presented in the last two chapters, edit the grid that has been laid out to show the shape of the stairs. Draw a diagonal line that passes through the front edge of each step. Offset this line down 12" to determine the required depth of the stairs. Save the drawing as **E-12-3**.

   Required risers   Total run

4. Use Figure 12.3 as a guide and draw an elevation showing an 8' high, 2 × 6 stud wall. Use (2) 2 × 6 top plates and a single base plate. Show an opening for a 6' wide × 4' high window. Use a 4 × 8 header over the window, and set the bottom of the header at 6'–8" above the floor level. Place the studs at 16" o.c. Save the drawing as **E-12-4A**. Make a copy of the elevation and show the wall at 9' high with the studs located at 24" o.c. Set the bottom of the header at 7'–10" above the floor. Save the elevation as **E-12-4B**.

5. Start a new drawing and draw three lines 10'–0" long. Set the point marker to show a marker with a circle with an X in the center. Measure the line into seven equal segments. Divide the second line into 16" long segments using a square around the crosshairs marker. Show the third line with perpendicular lines at 19.2" o.c. Save the drawing as **E-12-5**.

6. Start a new drawing and draw a rectangle. Place a point using the crosshairs surrounded by a circle and a square in the center of the rectangle. Draw lines of infinite length that pass through the point in a horizontal, vertical and at 30° to the horizontal and vertical planes. Trim the lines that extend beyond the rectangle. Save the drawing as **E-12-6**.

7. Start a new drawing and draw a circle with a semi-infinite line extending outward from each quadrant. Draw lines that extend from the center point to the edge of the circle and are placed at 15° intervals for a 360° pattern. Save the drawing as **E-12-7**.

8. Draw two lines that intersect. One of the lines is to be horizontal and the second line is to be 67° above the first line. Draw a line that bisects the intersection of the two lines. Save the drawing as **E-12-8**.

CHAPTER 12 QUIZ

DIRECTIONS

Answer the following questions with short complete statements. Type your answers using a word processor.

1. Place your name, chapter number, and the date at the top of the sheet.

2. Type the question number and provide the answer.

 Warning: Some of the questions have not been covered in the reading material and will require the use of the help menu. You may also have to do some exploring to answer the questions.

Start a new drawing and draw the following objects in the order listed. Draw four pairs of line segments, a three-sided polygon, a circle, a polyline with three segments with a width of 0.125, and an ellipse. Copy the circle, move the ellipse, and enlarge the polygon to twice its size.

1. If the **U** command is used after the polygon is enlarged, what will be affected?

_____

_____

2. If **Undo 2** is used after the polygon was drawn, what will be affected?

_____

_____

3. If a mark is placed after the circle is drawn, what will the effect be if the **Undo Back** command option is used after the circle is copied?

_____

_____

4. A whole drawing sequence is an experiment that you might want to delete. List the command and subcommands that would allow the entire sequence to be removed as one object. Explain when these subcommands are used.

_____

_____

5. If 5 is used with the **Undo** Number option, what will the effect be?

_____

_____

6. In using the **Undo** Number 5 option, one too many objects were removed. How can this be corrected?

_____

_____

7. Explain the difference between **Divide** and **Measure** on a line.

_____

_____

8. What is a ray and how could it be used?

_____

_____

9. What is the effect of Hor on a ray if ORTHO is **OFF**?

_____

_____

10. Describe the difference between the two angle options of **Xline**.

_____

_____

Placing Patterns in Drawing Objects

INTRODUCTION

This chapter will introduce methods of representing various materials using different line or shape patterns referred to as hatch patterns.

Commands to be introduced in this chapter include:

- Hatch

- Region

- Hatchedit

HATCHING METHODS

Hatch patterns are repetitive patterns of lines, dots, or other symbols used to represent a surface or specific material. Materials made of concrete are represented with a pattern consisting of dots and triangles, soil as a series of perpendicular lines, and masonry as diagonal or crossing diagonal lines. The majority of hatch patterns are found in details and sections similar to Figure 13.1, where the pattern is used to distinguish between materials. Hatch patterns can be used on elevations to represent siding and roofing materials as well as shades and shadows, as seen in Figure 13.2. Patterns are also used on plan views to represent various materials and changes in floor or ceiling heights and to distinguish between existing and new construction. No matter the drawing, using the National CAD Standard guidelines, the pattern should be placed on a layer with a minor code of **PATT**, with a name such as **A FLOR PATT** or **A ELEV PATT**. The outline that defines the pattern should be placed on a layer such as **A ELEV PATT OUTL**.

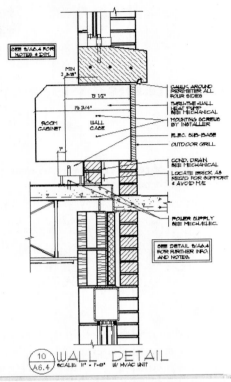

Figure 13.1 *Patterns produced using the **Hatch** command are used to distinguish various materials. (Courtesy G. Williamson Archer, AIA of Archer & Archer, P.A.)*

Figure 13.2 *Hatch patterns have been used to represent the tile roof, wood siding, and the shadows in this exterior elevation. (Courtesy Residential Designs.)*

Patterns are placed in drawings using the **Hatch** command. Hatch patterns are applied using the **Hatch and Gradient** dialog box. To display the dialog box, use one of the following methods:

- Click **Hatch** on the **Draw** toolbar.
- Click **Hatch** on the **2D Draw** dashboard.
- Type **H** ENTER at the Command prompt.
- Select **Hatch** from the **Draw** menu.

USING THE HATCH TAB

The **Hatch and Gradient** dialog box consists of the **Hatch** and **Gradient** tabs. The **Hatch** tab shown in Figure 13.3 is used for placing patterns for common shapes encountered on construction drawings. The display can be expanded to access advanced options. As you prepare to hatch an area using the **Hatch** command, zoom into an area rather than keeping the entire drawing in view. This will eliminate the number of lines that must be examined to determine the boundary set that will define the hatch area.

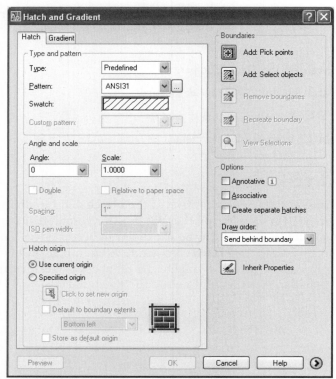

Figure 13.3 *Start the **Hatch** command by clicking **Hatch** on the **Draw** toolbar, or by typing **H** ENTER at the Command prompt. Each method will produce the **Hatch and Gradient** dialog box. Selecting the Expand arrow in the lower-right corner will display additional controls.*

PATTERN TYPE

The **Type** edit box is used to set the pattern type. Options include **Predefined**, **User defined**, and **Custom**. The default value is to use a **Predefined** pattern. Selecting the down arrow displays the other listings. Predefined patterns can be viewed in the **Hatch Pattern Palette** shown in Figure 13.4. For most uses, the predefined patterns will be all that are ever needed. Selecting the **User** defined option creates a pattern based on the current linetype in the drawing. This option allows the angle and spacing of the pattern to be altered. The **Custom** option allows a user-created pattern to be to be added to the drawing base. Methods of creating patterns lie beyond the scope of your beginning cad classes.

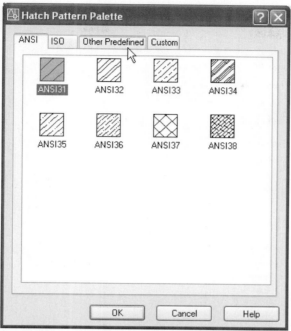

Figure 13.4   *Selecting the **Pattern [...]** button beside the **Pattern** edit box will display the **Hatch Pattern Palette** dialog box.*

SELECTING A PATTERN

The ANSI31 hatch pattern is the default setting shown in the **Pattern** edit box. A sample of the pattern is also shown in the Swatch display. A different pattern can be selected, and it will remain the default until a new pattern is selected. You can select a new pattern using the **Pattern** edit box by three different methods. Selecting the pattern name in the edit box or the edit box arrow will display a listing of the hatch pattern names similar to the list shown in Figure 13.5. The slide bar can be used to scroll through the list of predefined patterns. Clicking the name in the pattern box a second time will close the list without making any changes. Selecting a name from the list will make that pattern the current pattern and close the list. This method of pattern selection works well if you are familiar with the names of patterns.

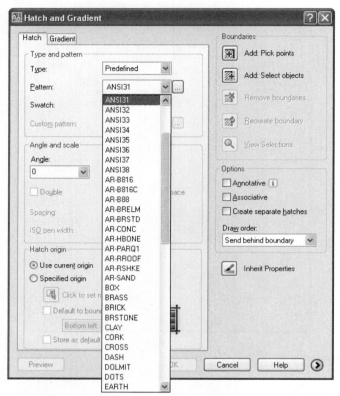

Figure 13.5 *The **Pattern** list can be used to select the pattern to be used for the **Hatch** command.*

If you're a new user, names like ANSI31 or ANSI32 will be meaningless. Select the [...] (ellipsis) button beside the edit box to see a display of each image in the **Hatch Pattern Palette**. The **ANSI** and **Other Predefined** palettes shown in Figure 13.6 provide the patterns associated with construction drawings. The slide bar can be used to move through the listing to select the desired hatch pattern. Once a pattern is selected for use in a drawing, select the **OK** button to return to the **Hatch** tab.

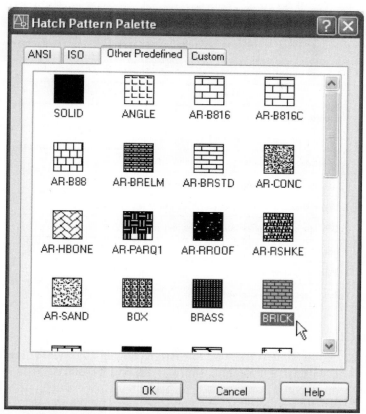

Figure 13.6 *The* **Other Predefined** *tab provides the patterns associated with construction drawings. The slide bar can be used to move through the palette to select the desired hatch pattern.*

A third method of selecting a hatch pattern is to use the **Swatch** display. Clicking the **Swatch** pattern will display the **Hatch Pattern Palette.** If the last pattern used was **Solid**, the **Color** palette will be displayed instead of the Swatch pattern palette.

Several hatch patterns found in the **Hatch Pattern Palette** are specific to architectural drawings. These include the following:

ANSI Tab

Pattern	Common Use
ANSI31	Masonry in plan and section view, thin veneers in section views, shading for elevations
ANSI32	Steel plates in section view
ANSI37	Firebrick in section view

Other Predefined Tab

Pattern	Common Use
SOLID	General usage when a surface needs to be filled solid
AR–B816 &16C	Concrete masonry units in elevation
AR–CONC	Poured concrete in section view
AR–HBONE	Flooring or brick patterns in plan and elevations
AR–PARQI	Flooring or brick patterns in plan and elevations
AR–RROOF	General roofing in elevation (non specific to material)
AR–RSHKE	Cedar shake in elevation (typically not used for roofing)
AR–SAND	EIFS, stucco or plaster in elevations and section
BRICK	Brick in elevation
DOTS	Shading in elevation
EARTH	Soil in section (45-degree rotation)

PATTERN ANGLE

Once the pattern has been selected, use the **Angle** edit box to determine the angle to be used to display the pattern. By default, the pattern will be placed at an angle of 0. The pattern will be reproduced as shown in the **Swatch** display and in the **Hatch Pattern Palette**. Enter a value by keyboard or select a value from the **Angle** list. A value other than zero allows the hatch pattern to be rotated. By entering an angle of 45°, you can use pattern ANSI31 to represent a shadow, as seen in Figure 13.7.

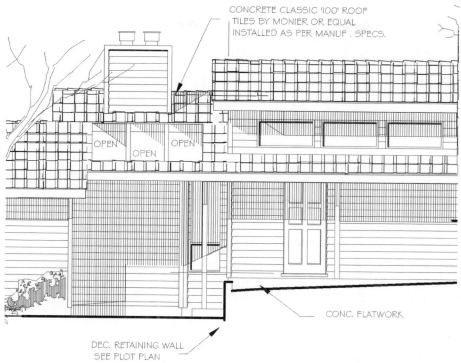

CONCRETE CLASSIC '100' ROOF
TILES BY MONIER OR EQUAL
INSTALLED AS PER MANUF . SPECS.

OPEN   OPEN   OPEN

CONC. FLATWORK

DEC. RETAINING WALL
SEE PLOT PLAN

Figure 13.7 *The hatch pattern ANSI31 is drawn at a 45° angle by default. Providing a rotation angle of 45° will reproduce the pattern in a vertical position. (Courtesy Residential Designs.)*

PATTERN SCALE AND SPACING

The **Scale** option sets the size of the pattern elements with the default of one drawing unit. Adjust the scale value using the same method as used to adjust the angle value, either by keyboard entry or by choosing from the **Scale** list. The scale list contains several common scales ranging from .25 through 2, but any scale can be entered by keyboard. Figure 13.8 shows an example of hatch patterns placed using three different scale values. The hatch patterns for annotative objects can easily be set when working in the 2D drafting annotation workspace. Using the **Annotative** setting of Options will automatically set the scale for hatch patterns. This option will be explored later in this chapter.

The **Relative to paper space** box is active only when you work in a paper space layout. Selecting the **Relative to paper space** box allows patterns to be displayed appropriate to the paper space layout. If a user-defined pattern is being inserted into a drawing, the **Spacing** and **Double** edit boxes will activate. With **Spacing** active, AutoCAD will specify and store the spacing of user-defined patterns. With **Double** active, a second set of lines is located at 90 degrees to the original lines to create a crosshatch.

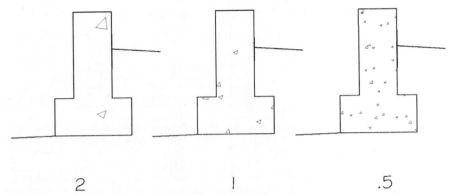

$$2 \qquad\qquad 1 \qquad\qquad .5$$

Figure 13.8 *The effect of altering the scale factor. A scale factor of 2 doubles the pattern size or spacing from a pattern with a factor of 1.*

SELECTING OBJECTS FOR HATCHING

Once the pattern, angle, and scale have been selected, use one of the methods from the right side of the dialog box to choose the area to be hatched. For now, the options are limited to **Add Pick Points** and **Add Select Objects**.

- If the **Add Pick Points** button is selected, only one point will need to be selected to place the pattern.

- If the **Add Select Objects** button is selected, any of the selection methods can be used to define the area.

- Once an object has been selected for hatching, the **Remove boundaries** option is activated. The option removes the boundary definition of any of the objects that were added previously. When **Remove Boundaries** is selected, the dialog box closes temporarily, and the command line displays a prompt to select object for hatching.

AutoCAD does not require the full object to be visible to complete the hatch display. Other methods of selecting the boundary can be selected using the options on the expanded display, which will be discussed later in this chapter.

Using the Pick Points Option

The **Add Pick Points** option allows the object to be hatched to be specified by selecting any point that lies within the object. Selecting the **Add Pick Point** button will remove the dialog box and the drawing will be restored. A prompt is now displayed requesting:

```
Select internal point or [Select objects/remove Boundaries]
```

Select any point that lies within the area to be hatched. As the point is selected, the boundary of the area to be hatched will be displayed as dashed lines. AutoCAD is waiting for you to select additional objects to be hatched.

- Press ENTER or right-click to redisplay the **Gradient and Gradient** dialog box.

- Press ENTER a second time to apply the hatch pattern to the selected object.

Figure 13.9 shows the resulting hatch pattern.

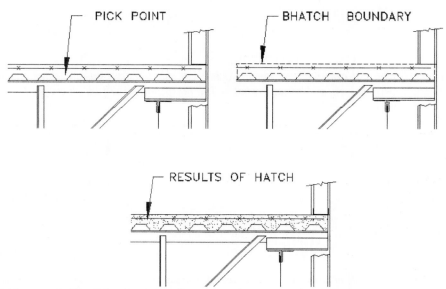

PICK POINT

BHATCH BOUNDARY

RESULTS OF HATCH

Figure 13.9 *Using **Hatch** to add a concrete pattern to a floor slab. Once the area to receive the pattern is selected, the boundary will be highlighted. The AR-SAND pattern at a scale of .5 and an angle of 0 was selected. With this pattern, the angle is unimportant.*

Using the Select Objects Option

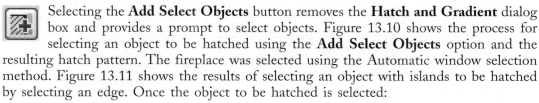 Selecting the **Add Select Objects** button removes the **Hatch and Gradient** dialog box and provides a prompt to select objects. Figure 13.10 shows the process for selecting an object to be hatched using the **Add Select Objects** option and the resulting hatch pattern. The fireplace was selected using the Automatic window selection method. Figure 13.11 shows the results of selecting an object with islands to be hatched by selecting an edge. Once the object to be hatched is selected:

- Press ENTER to restore the **Hatch and Gradient** dialog box.

- Press ENTER a second time to remove the dialog box and apply the pattern.

If the object to be hatched is surrounded by a polyline, the object can be selected by picking the object at any point. The pattern will be the same as the pattern in Figure 13.12. The object to be hatched does not need to form a closed boundary, or it can contain an internal area referred to an island that will not be hatched.

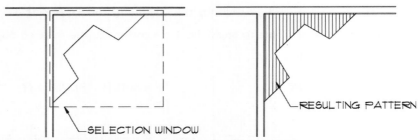

Figure 13.10 With the **Add Select Objects** option, an object can be selected using any selection method, such as Automatic window. The hatch pattern will be placed in the selected objects.

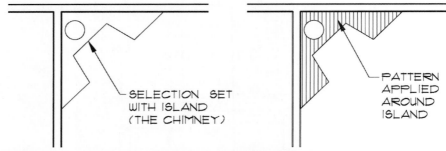

Figure 13.11 The result of using the **Add Select Objects** option on an object with an island. The placement of the pattern can be altered with the **Island Detection Style** option located in the expanded dialog box.

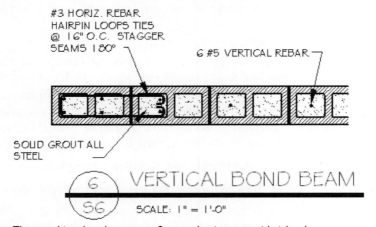

Figure 13.12 The resulting hatch pattern for a selection set with islands.

SELECTING THE PATTERN ORIGIN

The **Hatch origin** area of the **Hatch and Gradient** dialog box offers two methods of determining where the origin of the pattern will be placed. Some hatches, such as brick and concrete block patterns need to be aligned with a point on the hatch boundary.

Use Current Origin

By default, all hatch origins correspond to the current UCS origin. With this option active, the origin is set to 0,0 by default. The bottom of the brick or block may not align with the bottom of the boundary.

Specified Origin

With this option active, a new hatch origin can be specified, allowing the pattern to align with the border. Click this option to make the following options available.

1. **Click to Set New Origin** — Allows the new hatch origin point to be selected directly.

2. **Default to Boundary Extents** — Calculates a new origin based on the rectangular extents of the hatch. Choices include each of the four corners of the extents and its center.

3. **Store as Default Origin** — Stores the value of the new hatch origin in as a system variable.

Figure 13.13 shows an example of brick placed with the current origin and a selected origin. The pattern on the right was placed by selecting **Specified origin,** selecting the **Click to set new origin** button, and then selecting the lower-left corner of the boundary.

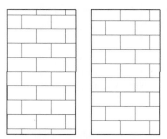

Figure 13.13   *The origin point of the pattern should be considered with patterns such as brick or concrete block. The brick pattern on the left was placed with the **Use current origin** active. The pattern on the right was placed by selecting **Specified origin**, selecting the **Click to set new origin** button, and then selecting the lower-left corner of the boundary.*

USING THE PREVIEW OPTION

If you're unsure of any of the pattern parameters, select the **Preview** button prior to applying the hatch pattern. The **Preview** option provides a display of how the pattern will appear based on the current settings. **Preview** is especially helpful when a large, irregularly shaped boundary is to be hatched. When the **Preview** option is selected, the dialog box is removed and the object with the hatch pattern is displayed as seen in Figure 13.14. The boundary is highlighted by dotted lines to provide a reminder that the pattern has not yet been added to the drawing file. Left-click to restore the **Hatch and Gradient** dialog box. If the pattern is acceptable, right-click, select the **OK** button or press ENTER. If the preview did not display the intended results, any of the pattern options can be altered. In Figure 13.15, a new hatch pattern was selected by returning to the **Boundary Hatch and Fill** dialog box. The **Preview** option was selected. Using **Preview** to test alternatives can go on indefinitely. Because this view was acceptable, ENTER was used to return to the dialog box, and the **OK** button was selected.

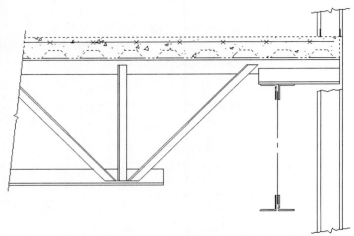

Figure 13.14   *You can display the proposed pattern by selecting the **Preview** button. If the pattern is not what you expected, the properties can be altered before you apply the pattern.*

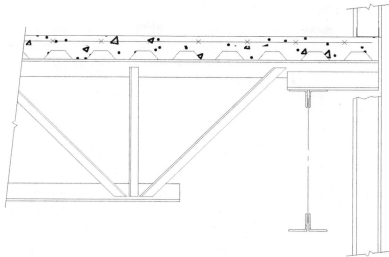

Figure 13.15   *If the **Preview** is acceptable, add the pattern to the drawing base by selecting the **OK** button. Select a new pattern by returning to the **Hatch and Gradient** dialog box.*

Summary of Steps in Placing a Hatch Pattern

The following steps were used to place the hatch pattern in Figure 13.15.

1. Display the **Hatch and Gradient** dialog box using on of the following methods:

   • Click the **Hatch** button on the **Draw** toolbar.

   • Click the **Hatch** button on the **2D Draw** dashboard.

   • Type **H** ENTER at the Command prompt.

   • Select **Hatch** from the **Draw** menu.

2. Choose the **Hatch** tab.

3. Select the [...] button to display the **Hatch Pattern Palette**.

4. Choose the **Other Predefined** tab.

5. Select the **AR-CONC** pattern and select **OK**.

6. Select the **Add Pick Points** button in the **Hatch and Gradient** dialog box.

7. Specify an internal point in the object to be hatched.

8. Select the origin point for patterns that are modular.

9. Select the **Preview** button to display the proposed hatch pattern.

If the pattern is not acceptable, left-click to alter the hatch properties using the **Hatch** tab, and then use **Preview** option again. Right-click to apply the hatch pattern to the drawing if the pattern is acceptable.

REMOVING ISLANDS

Figure 13.16 shows a concrete floor slab with four circular tubes. AutoCAD refers to openings in a hatch pattern as islands. When the boundary is selected, each of the islands will automatically be excluded from the hatch. The hatch pattern was omitted from each of the cavities in the floor on the left side of Figure 13.16. Once the object to be hatched has been defined, selecting the **Remove boundaries** button allows islands to be included and receive the hatch pattern.

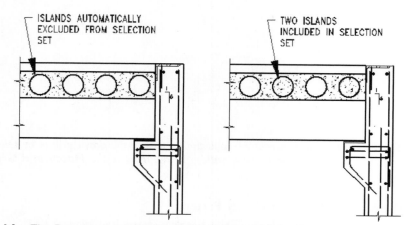

Figure 13.16 *The **Remove boundaries** option can be used to control the placement of the pattern in islands within the boundary.*

VIEW SELECTIONS

With all of the options available for altering the boundary or patterns used in a boundary hatch, you might lose track of what options you've selected. The **View Selections** button will remove the dialog box and display the currently defined boundaries.

INHERIT PROPERTIES

This option is useful when you're hatching an object using an existing pattern but you can't remember the qualities of the existing hatch pattern. Select the object to be hatched, and then select the **Inherit Properties** button. This will close the **Hatch and Gradient** dialog box and return the drawing area. Select the pattern that is to be copied and the dialog box will be restored. Use the preview option to verify the properties.

OPTIONS

The final portion of the **Hatch and Gradient** dialog box to be examined is the **Options** area, which contains toggles for Annotative, **Associative**, **Create Separate Hatches**, and the **Draw Order**. If the 2D Drafting Annotation workspace and the **Annotative** button are active, the process of scaling hatch patterns so that they display and plot at the correct size on the paper is automated. Once objects to be hatched have been selected, click the Annotative button, and then click the OK button.

With the **Associative** button active, as an object containing a hatch pattern is edited, so is the hatch. With the option inactive, if an object with a hatch pattern is altered, the object is altered, but the hatch pattern remains unaffected. Figure 13.17 shows the effect of stretching two steel beams and the effect of an associative pattern.

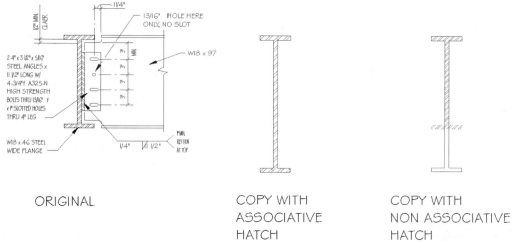

ORIGINAL

COPY WITH
ASSOCIATIVE
HATCH

COPY WITH
NON ASSOCIATIVE
HATCH

Figure 13.17 *With **Associative** active, stretching a steel beam will alter the hatch pattern. With **Associative** inactive, the beam is stretched, but the hatch pattern is unaltered.*

The **Create Separate Hatches** option controls if a single hatch object or multiple hatch objects will be created when several separate closed boundaries are specified.

- With this option **OFF,** if five objects are selected to be hatched, once the pattern is placed in the objects, the pattern in each object will be edited if the pattern in one of the objects is selected for editing.

- With the option **ON,** if five objects are selected to be hatched, once the pattern is placed in the objects, only the pattern in the selected object will be edited.

Draw Order assigns the draw order to a hatch or fill. You can place a hatch or fill behind all other objects, in front of all other objects, behind the hatch boundary, or in front of the hatch boundary.

DEFINING HATCH PATTERNS USING THE EXPANDED TAB

Although the **Hatch** tab can be used to hatch simple objects, expanding the tab provides options to help increase the efficiency of selecting boundary hatch boundaries. The **Hatch** command evaluates all objects displayed on the screen when you select objects to be hatched. If you are working on a complex project, these options can speed the boundary selection process. Select the **Expand** button in the lower-right corner of the dialog box to expand the **Hatch and Gradient** dialog box. Once expanded, selecting the **Collapse** button returns the dialog box to the base size. Key elements of the expanded box include controls for island detection, object type, boundary set, and island detection method. The expanded **Hatch and Gradient** dialog box can be seen in Figure 13.18.

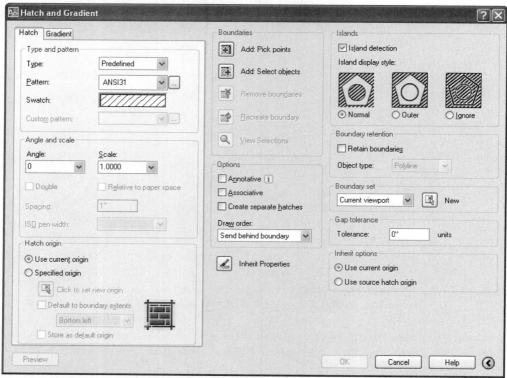

Figure 13.18 *The expanded **Hatch and Gradient** dialog box displays options to help place hatch patterns in complex shapes. Key elements of the tab include controls for island detection, object type, boundary set, and island detection method. Selecting the **Expand** button in the lower-right corner displays all of the advanced options.*

ISLANDS

The **Island** area provides controls for how hatch patterns will be displayed in objects with internal boundaries. Areas include **Island detection** and the **Island display style.**

Island Detection

The **Island Detection** box toggles the detection mode **ON/OFF.** The detection mode controls whether internal closed boundaries, called islands, are detected. If no internal boundaries exist, specifying an island detection style has no effect.

Island Display Style

Three methods are available for hatching islands, including **Normal, Outer,** and **Ignore.** The **Normal** setting is the default and the image tile reflects how the pattern will be displayed. You can change the setting by selecting one of the other radio buttons.

Normal Example A in Figure 13.19 shows how hatching sets made up of multiple objects will be hatched when the default of **Normal** style is selected and each of the objects is selected as a boundary. **Normal** hatching style works inward, starting at the area boundary, and proceeds until another boundary is found. The pattern is turned off until another boundary is discovered. Because a precise set of boundaries can be defined, it is often best to use the **Normal** style.

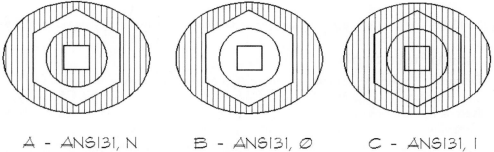

Figure 13.19 *The **Island Detection Style** area controls how the pattern will be displayed around multiple object sets. The **Normal** option works inward until another boundary is found. The pattern is not displayed until another boundary is found. The **Outer** option will display only the pattern in the outer boundary area. The **Ignore** option ignores all boundaries and hatches all areas within the outer boundary.*

Outer Example B in Figure 13.19 shows how hatching sets made up of multiple objects will be hatched when the **Outer** option is selected and each of the objects is selected as a boundary. The **Outer** hatching style works inward, starting at the area boundary and proceeds until another boundary is found. The pattern is turned off at the second boundary.

Ignore Example C in Figure 13.19 shows how multiple objects will be hatched when the **Ignore** option is selected. The **Ignore** hatching style works inward, starting at the area boundary, and passes through all objects within that boundary.

 Note: With the **Normal** setting, if a hatch operation encounters another shape or text, the pattern will be placed around the text. Although text has not yet been added to the drawing, it will be important to remember that text inside a hatch boundary will still be legible (see Figure 13.20).

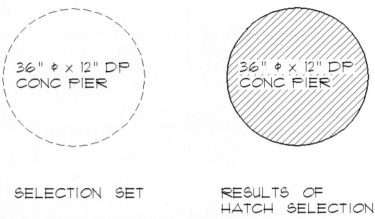

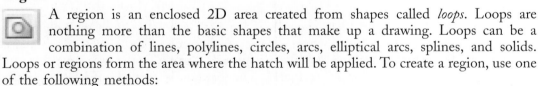

Figure 13.20 *Text will be treated like an island and automatically excluded from the area to be hatched.*

BOUNDARY RETENTION

The **Boundary retention** area controls boundary options for applying the hatch pattern. The options for the boundary include the **Retain Boundaries** check box and the **Object Type** box. The **Retain Boundaries** option specifies whether or not a temporary boundary will be added to the drawing. Boundaries for hatch patterns can be specified by **Object Type**. These options are not active unless the **Retain Boundaries** option is active. **Object Type** controls the type of new boundary to be created with options including **Region** and **Polyline**. You've explored polylines in a previous chapter. A closed shape comprised of a polyline can be used to form a boundary for hatch patterns.

Region

A region is an enclosed 2D area created from shapes called *loops*. Loops are nothing more than the basic shapes that make up a drawing. Loops can be a combination of lines, polylines, circles, arcs, elliptical arcs, splines, and solids. Loops or regions form the area where the hatch will be applied. To create a region, use one of the following methods:

- Click the **Region** button on the **Draw** toolbar.
- Click the **Region** button on the **2D Draw** dashboard.
- Type **REG** ENTER at the Command prompt.
- Select **Region** from the **Draw** menu.

Regions are created using the following command sequence:

```
Click the Region button (Or type REG ENTER.)
Select objects:(Select objects to form the region using any
    selection method and press ENTER.) W ENTER
4 found
Select objects: ENTER
1 loop extracted
1 Region created
Command:
```

The selected lines will now function as a polyline, forming a boundary for future hatch patterns.

BOUNDARY SET

This portion of the dialog box defines how the selection set for the hatch pattern will be selected when the **Add Pick Points** option is selected. By default, when the **Add Pick Point** option is used to define a boundary, the **Hatch** command analyzes all objects in the current viewport extents. By fine-tuning the selection set, the command can function more quickly because fewer objects are examined to define the selection set. Selecting **New** will remove the dialog box from the screen and allow a new boundary to be selected. The existing boundary set will be discarded and you will be allowed to construct a new boundary set to define the area to be hatched.

Once the new boundary set is selected, the dialog box is returned and the default setting is now an **Existing** set. An unlimited number of boundary sets can be selected. The last set selected will remain the default until a new boundary set is selected. When **Current viewport** is selected, the selection set will be created from everything that is visible in the current viewport. Selecting this option when there is a current boundary set will discard the current selection and place the pattern in everything in the current viewport. The normal setting is to place the pattern based on the **Existing boundary** set.

APPLYING FILL PATTERNS WITH GRADIENT

Earlier you learned that AutoCAD allows a solid fill pattern to be selected for the specified pattern to be assigned to an object. As an alternative to a solid pattern, the **Gradient** tab, shown in Figure 13.21, allows varied solid patterns to be created that greatly increase the presentation aspect of a drawing. Key areas of the tab include color selection, shade selection, gradient icons, and angle control.

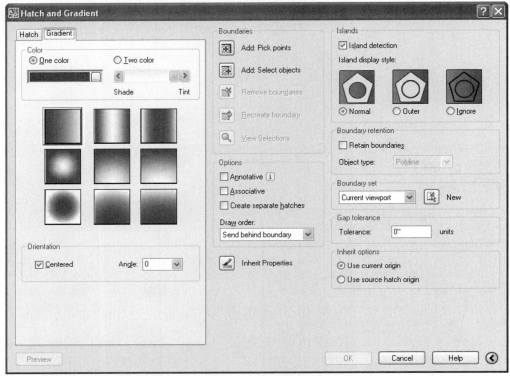

Figure 13.21 *The **Gradient** tab of the **Hatch and Gradient** dialog box can be used to edit solid fill patterns.*

CONTROLLING GRADIENT COLORS

Controls are available to control the number of colors to be used in a hatch pattern as well as how shading will be applied to the pattern.

Color

Buttons are provided to fill patterns with one- and two-color gradient patterns. Selecting the **One color** box provides a fill pattern that uses a smooth transition between darker shades and lighter tints of one color. With this button active, AutoCAD displays a color swatch with the **Browse** button. Selecting the **Browse** button displays the **Select Color** dialog box. Gradient colors can be selected by using an AutoCAD Index color, true color, or color book color. The default color displayed is the current color in the drawing. A **Shade and Tint** slider is displayed below the color swatch and can be used to control the pattern shade and tint. Sliding the control box to the left alters the shade of the color by mixing the selected color mixed with black. Sliding the control box to the right alters the tint of the color by mixing the selected color mixed with white. Once the color has been selected, the fill gradient can be altered by selecting one of the nine gradient patterns.

If the **Two Color** box is selected, a fill pattern that uses a smooth transition between two colors will be provided. With this button active, AutoCAD displays a color swatch with a **Browse** button for each color to be used. Selecting the **Browse** button displays the **Select Color** dialog box allowing colors to be selected as they were when the **One color** box was active. Once the two colors have been selected, HATCH displays a color swatch with a **Browse** button for color 1 and color 2.

Orientation

No matter how many colors are selected, the pattern can be altered using the **Centered** and **Angle** controls. With the **Center** box active, the gradient configuration is symmetrical. With this option inactive, the gradient fill is shifted up and to the left, creating the illusion of a light source to the left of the object. The **Angle** control box can be used to specify the angle of the gradient fill relative to the current UCS. This option is independent of the angle specified for hatch patterns.

EDITING HATCH PATTERNS

 Features of boundary hatch patterns can be edited using the **Hatch edit** command. To start the command, use one of the following methods:

- Select a hatch pattern to edit and then right-click in the drawing area.
- Click **Edit Hatch** on the **Modify II** toolbar.
- Type **HE** ENTER at the Command prompt.

Each method will produce the **Hatch edit** dialog box show in Figure 13.22. The exact display will vary depending on which pattern is selected. The box is very similar to the **Hatch and Gradient** dialog box. Each of the features of the pattern can be altered, and the options function as they do in the **Hatch and Gradient** box.

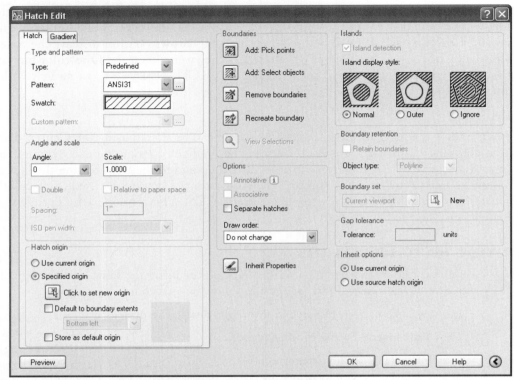

Figure 13.22 *The **Hatch Edit** dialog box can be used to edit the pattern once it has been added to the drawing base. Display the dialog box by selecting a hatch pattern to edit and then right-clicking.*

CHAPTER 13 EXERCISES

1. Start a new drawing and use the following coordinates to draw the required object.

Start: 1,1 B.	= 3,0	C. = –5,.25	D. = –.75,1	E. = –1,1
F.	= .75,1	G. = .5,1	H. = 0,1	J. = .25,.25
K.	= –2.25,0	L. = .25,–.25	M. = 0,–1	N. = .75,–1
P.	= –1,–1	Q. = 0,–1	R. = 1,–.75	S. = C

   Make a copy of the completed shape. Use the **SOLID** pattern to fill one of the objects. Turn **Fill** to **OFF**. Use the **ANSI31** hatch pattern and adjust the scale factor to a suitable scale to hatch the object so that it appears solid. Save the drawing as **E-13-1**.

2. Start a new drawing and use the drawing below as a guide to draw a 15" diameter circle. Draw a 12" diameter circle with a center point 10" above the first center point. Array the 12" circle so that there are a total of six circles. Copy the pattern so that you have a total of three circular patterns. Create layers to separate the

object from the hatching. Use the **Hatch** command to form three completely different hatching styles. Use different hatch patterns for each area to be hatched. Save the drawing as **E-13-2**.

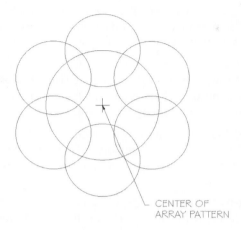

CENTER OF
ARRAY PATTERN

3. Start a new drawing and draw a red square, surrounded by a blue circle, surrounded by a green octagon, surrounded by a cyan triangle. Copy this pattern so that there are a total of three copies. Create a separate layer for hatching. Use **ANSI34** and create a double hatching pattern. Provide examples of **Normal**, **Outer**, and **Ignore** styles of hatching. Save the drawing as **E-13-3**.

4. Start a new drawing and draw a one-story footing. Use a 6" wide stem wall extending from the natural grade 8". Support the wall on a 12" × 6" foundation. Show a 2 × 4 key, a 2 × 6 plate with a 1/2" diameter × 10" anchor bolt. Place wood, steel, and concrete on separate layers with different colors. Use varied line widths to distinguish each material. Use **EARTH** to represent the soil and **AR-CONC** to represent the footing and stem wall. Copy the drawing and show a 2 × 10 floor joist with 3/4" plywood floor sheathing. Show the bottom of a 2 × 6 stud wall with single-wall

construction. On the second copy, show a post and beam floor with 2" decking and a 6" deep beam. Save the drawing as **E-13-4**.

5. Draw the plan view of a 5' wide × 2'-8" deep masonry fireplace with a 36" wide opening centered in the masonry. Provide a 4" wide firebrick lining around the fireplace opening. Use hatch patterns to distinguish between the firebrick of the firebox and normal brick of the chimney. Save the drawing as **E-13-5**.

6. Open drawing **7-4** and create a layer for oak (green), maple (red), and Dutch elm (yellow). Change the trees from their existing layer to these new layers. Tree E is an oak, A is a maple, and the other three are Dutch elm. Hatch each type of tree with a different pattern. Save the drawing as **E-13-6**.

7. Open drawing **E-7-5** and create a layer titled **PATT** and a second layer titled **PATT OTLN**. Assign each layer a separate color. Hatch the glass area with DOTS, rotated at 45°. Adjust the scale so that the glass appears gray rather than black or white. Save the drawing as **E-13-7**.

8. Open drawing **PB-FND** (E-9-7). Create layers for **FND WALLS**, **FND FOOT** (dashedx2), **FND PIER** (hidden), **FND CONC**, **FND ANNO**, and **FND BEAM** (center). Assign each layer a different color. Assume the wall to be made from concrete masonry units and hatch the wall with an appropriate pattern. Offset the wall lines to form a 16" wide footing centered on the wall. Trim the footings as required. Use the **Properties** command to change the footing lines to the appropriate linetype and layer. Save the drawing as **E-13-8**.

9. Open drawing **E-11-9** and complete the drawing by providing suitable hatch material for the soil and concrete. Place hatch patterns on a new layer with an appropriate title. Provide a #4 Ø rebar 3" down from the top of the stem wall, and 3" up from the bottom of the footing. Provide a #4 Ø vertical piece of rebar with a 4" toe that extends from the footing into the stem wall. Save the drawing as **E-13-9**.

10. A residence is 24' x 36' wide. The ridge is 36' long and the roof is a 6/12 pitch with 24" overhangs, with a 2 x 6 fascia. The roof has a 12" overhang at the gable end walls. The finish floor is 18" above the finish grade and the plate height for the walls is 8'-0". You are to design an elevation of the 36' front side, which must include the following:

- 3' x 6'-8" front door (show some type of appropriate decorative pattern)

- 8' x 5' sliding/picture window with decorative grids

- 18" x 60" shutters each side of window

- 12" wide x 6" deep stucco columns each side of window shutters with 6" horizontal siding above and below the window and shutters

- A 24" wide x full-height area of brick on each side and above the front door (remember, brick should be approximately 8" long)

Complete the drawing by representing the following materials:

- Hatch the roof with **AR-ROOF**.

- The area (36") between the doorway brick and the stucco columns is to be **AR-RSHKE** (rows should be approximately 10" wide).

- All other areas are to have vertical siding @ 8" O.C.

- Show a shadow that extends 6" below the top of the door and window and show the shade that would result from the sun being above and to the left of this gorgeous structure. (Hey! Be glad it's not in your neighborhood!)

Create a separate layer for each material and save this work of art as **E-13-10**.

CHAPTER 13 QUIZ

DIRECTIONS

Answer the following questions with short complete statements. Type your answers using a word processor.

1. Place your name, chapter number, and the date at the top of the sheet.

2. Type the question number and provide the answer.

 Warning: Some of the questions have not been covered in the reading material and will require the use of the **Help** menu. You may also have to do some exploring to answer the questions.

1. A circle surrounds a square, which encloses a hexagon. What style is used to hatch the circle and hexagon?

_____

_____

2. In the object described in question 1, what effect will **Ignore** have on each shape?

_____

_____

3. What option should be used to fill an object with a hatch pattern?

_____

_____

4. List the dialog boxes that are associated with **Hatch**.

_____

_____

5. How can the **Hatch Pattern Palette** dialog box be accessed?

_____

_____

6. How can the scale factor of the **Hatch** pattern be controlled by the plotted scale factor?

_____

_____

7. What process is used to keep the hatch pattern from crossing text?

_____

_____

8. Explain the differences in using the **Pick Points** option and the **Select Objects** option.

_____

_____

9. What is direct hatching?

_____

_____

10. Describe the process to create a hatch pattern of vertical lines at 1/16" spacing.

_____

_____

11. What is the effect of using the ANSI32 pattern?

_____

_____

12. Use the **Help** menu to determine what affect the **Explode** command will have on a hatch pattern.

_____

_____

13. What does associative have to do with **Hatch**?

_____

_____

14. You've defined a hatch pattern, but you're not sure how it will appear. List two methods to view the pattern.

_____

_____

15. What is an island?

_____

_____

Inquiry Commands

INTRODUCTION

As the drawings you're working with get more complex, you might find it hard to keep track of some of the drawing parameters that have been established. This chapter will introduce methods for

- Determining information about drawing objects.
- Determining information about drawing files.
- Using the AutoCAD **QuickCalc** calculator.

The **Inquiry** toolbar and the **Inquiry** menu on the **Tools** menu contain commands used to obtain information about drawing objects or an entire drawing. Commands to be introduced in this chapter include:

- **Distance**
- **Area**
- **Massprop**
- **Region**
- **List**
- **Dblist (Data Base List)**
- **ID Point**
- **Time**
- **Status**
- **Purge**
- **QuickCalc**

DETERMINING DISTANCE

The **Distance** command measures the distance and angle between two selected points. This command can be especially helpful when used with OSNAP locations to measure exact locations. **Distance** can be useful in determining the size of a room during the design stage or to check the angle of an object from a baseline. To access the command, use one of the following methods:

- Click the **Distance** button on the **Inquiry** toolbar.
- Type **DI** ENTER at the Command prompt.
- Select **Distance** from **Inquiry** on the **Tools** menu.

The command sequence is as follows:

```
Command: Click the Distance button (Or type DI ENTER.)
Specify first point: (Select a point.)
Specify second point: (Select second point.)
```

The resulting display for the points in Figure 14.1 would be

```
Distance=14'-0" Angle in XY plane=225.0 Angle from XY plane=0
Delta X=-9'-101/2"Delta Y=-9'-101/2"Delta Z=0'-0"
```

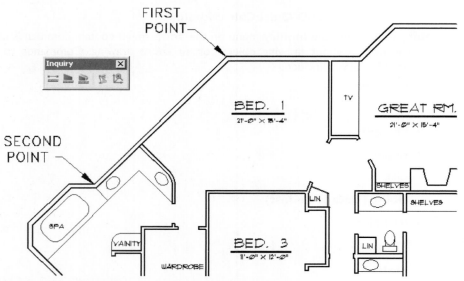

Figure 14.1 *The **Distance** command measures the distance and angle between two selected points.*

DETERMINING AREA

The **Area** command can be used to determine the area and perimeter of an object. If OSNAP is used, the **Area** command can be used to determine exact square-footage sizes of structures and, at the same time, determine wall lengths. A running total can be kept and areas can be added to or deleted from the total. To access the command, use one of the following methods:

- Click the **Area** button on the **Inquiry** toolbar.

- Type **AA** ENTER at the Command prompt.

- Select **Area** from **Inquiry** on the **Tools** menu.

As the command is entered, four options are given:

```
Command: Click the Area button (Or type AA ENTER.)
Specify first corner point or [Object/Add/Subtract]: (Specify a
    point or enter an option.)
```

FIRST CORNER POINT

When you accept the **Area** "first corner point" option, you will be asked to supply a series of next corner points to specify the area to be defined. Points can be easily selected with the cursor with OSNAP in the ON setting to increase the accuracy in computing the area. Figure 14.2 shows an example of computing the area of a retaining wall using the point option. The command sequence is as follows:

```
Command: Click the Area button (Or type AA ENTER.)
Specify first corner point or [Object/Add/Subtract]: (Specify
    first point.)
Specify next corner point or press ENTER for total:
```

Continue to select points as required. In the example in Figure 14.2, press ENTER at point 12. This will produce a display listing the area (in square inches and square feet when units are set for architectural units) and the perimeter. The same results will occur if Point 13 (Point 1) is entered as the ending point. If you stop at point 12, AutoCAD will automatically close the polygon and count the line that would be formed between points 12 and 13 as the perimeter is determined.

OBJECT

The Object option will compute the area of a circle, ellipse, spline, or an object composed of polylines. Selecting a circle will cause the area and the circumference to be displayed. When a closed polyline is selected, the area and perimeter will be displayed. If an open polyline is selected, the area is computed based on an imaginary line extending from each end of the polyline. The command sequence is as follows:

```
Command: Click the Area button (Or type AA ENTER.)
Specify first corner point or [Object/Add/Subtract]: O ENTER
Select object: (Select object.)
Area=27.63 square in. (0.1918 square ft.) Length 1'-15/8"
```

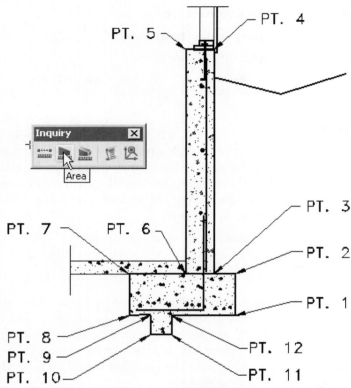

Figure 14.2 *The **Area** command can be used to compute the area of a specified shape. The area and the perimeter are given at the Command prompt.*

Once the object is selected, a display of the area and circumference will be displayed. If an open polyline is selected, AutoCAD will calculate the area as if a line were drawn between the endpoints. (See Figure 14.3.) The line that is projected by AutoCAD is not included in the calculation of the area or perimeter. The command will list the area confined by the polyline and the length of the line.

 Note: As you work with the **Area** command, the area and perimeter will be displayed at the Dynamic Display. If you move the cursor before you've written down the solution, the information can be redisplayed by clicking **F2**. If several different areas need to be determined, activating the Command Prompt will display the entire command sequence and remove the need to use flip screen.

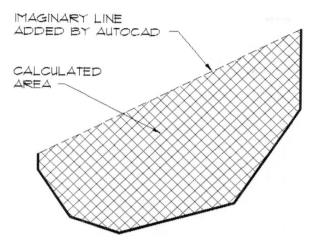

IMAGINARY LINE
ADDED BY AUTOCAD

CALCULATED
AREA

Figure 14.3 *The **Object** option of **Area** can be used to obtain information about an open or closed polyline. AutoCAD will provide the line required to close the polygon and calculate the area without displaying the line.*

ADD AND SUBTRACT

Using the **Add** option allows each area measured in this mode to be displayed and added to a running total. The running total will be displayed after each addition. To begin a running total, activate the command and select the Add option before selecting the objects to be added. Using the Subtract option provides the opposite effect. Each area measured in the Subtract mode will be displayed and then subtracted from the running total. With both the Add and Subtract options, the prompt will be altered to reflect the active mode. The selected option will stay in effect until the command is canceled. Figure 14.4 is an example of computing the usable area of a parcel of land with the Add and Subtract options. The area of the lot was determined using continuous OSNAP to select each corner of the lot, and then subtracting the area of the residence and a garage, which are both outlined by a polyline. The command sequence is as follows:

```
Command: Click the Area button (Or type AA ENTER.)
AREA
Specify first corner point or [Object/Add/Subtract]: A ENTER
Specify first corner point or [Object/Subtract]: (Select
    point 1.)
Specify first corner point or press ENTER for total (ADD mode):
    (Select point 2.)
Specify first corner point or press ENTER for total (ADD mode):
    (Select point 3.)
Specify first corner point or press ENTER for total (ADD mode):
    (Select point 4.) ENTER
Area=864000.0 square in. (6000.000 square ft.), Perimeter=320'-0"
Total area=864000.0 square in. (6000.000 square ft.)
Specify first corner point or [Object/Subtract]: S ENTER
Specify first corner point or [Object/Add]: (Select point 1.)
Specify first corner point or press ENTER for total (SUBTRACT
    mode): (Select point 2.)
```

```
Specify first corner point or press ENTER for total (SUBTRACT
    mode): (Select point 3.)
Specify first corner point or press ENTER for total (SUBTRACT
    mode): (Select point 4.) ENTER
Area=189154.2 square in. (1313.571 square ft.), Perimeter=118'-
    31/2"
Total area=674845.8 square in. (4686.429 square ft.)
Specify first corner point or [Object/add]: (Select point 1.)
Specify first corner point or press ENTER for total (SUBTRACT
    mode): (Select point 2.)
Specify first corner point or press ENTER for total (SUBTRACT
    mode): (Select point 3.)
Specify first corner point or press ENTER for total (SUBTRACT
    mode): (Select point 4.) ENTER
Area=36365.9 square in. (252.541 square ft.), Perimeter=65'-7"
Total area=638479.9 square in. (4433.888 square ft.)
Specify first corner point or [Object/Add]: ENTER
Command:
```

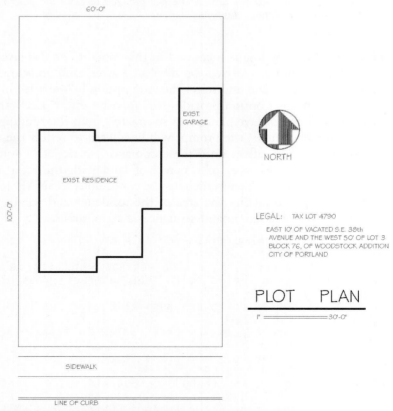

Figure 14.4 *The **Add** and **Subtract** options can be used with the **Area** command to add or subtract areas from a running total.*

DETERMINING THE AREA OF HATCHED AREAS

In addition to using the **Area** command, the area, perimeter, radius and diameter can be determined by using the **Properties** command. The exact information provided in the **Properties** display will depend on the shape of the selected object. To use the **Properties** command to determine the area of an object, click the **Properties** button on the **Standard** toolbar to display the Properties palette. The desired information will be displayed on the **Geometry** tab of the palette once an object is selected.

MASS PROPERTIES

 The **Mass Properties (Massprop)** command can be used to calculate the properties of 2D and 3D objects. Although primarily a 3D command, **Mass Properties** can be used to determine the area, perimeter and several other features of 2D shapes. To start the command, use one of the following methods:

- Click the **Region/Mass Properties** button on the **Inquiry** toolbar.

- Select **Region/Mass Properties** from **Inquiry** on the **Tools** menu.

- Type **MASSPROP** ENTER at the Command prompt.

To use the **Mass Properties** command with a 2D object, the object first must be converted to a region. If an object that is not a region is selected, the following prompt is displayed at the command line display:

```
Click the Mass Properties button.
Select Objects: (Select desired objects using any selection
    method.)
1 found
No solids or regions selected.
Command:
```

To change an object to a region, select the **Region /Mass Properties** button on the **Draw** toolbar or type **REG** ENTER at the command line and select the objects to be converted. **Mass Properties** can now be used to display the properties of the selected region. The command sequence at the command line is:

```
Command: Click the Mass Properties button (Or type REG ENTER.)
Select Objects: (Select desired objects using any selection
    method.)
104 found
Select objects: ENTER
1 loop extracted
1 Region created
Command:
```

Now reselect the original object that was just converted to a region. Once objects are selected, AutoCAD will display the mass properties in the text window. The mass properties for the shape in Figure 14.5 are seen below.

Regions Area:	277.1929 sq in
Perimeter:	91.8271 in
Bounding box:	X: 23.5672 — 46.5362 in
	Y: -3.1626 — 18.0234 in
Centroid:	X: 39.2238 in
	Y: 9.3193 in
Moment of inertia:	X: 33838.1482 sq in sq in
	Y: 433392.8323 sq in sq in
Product of inertia:	XY: 97969.2639 sq in sq in
Radii of gyration:	X: 11.0487 in
	Y: 39.5412 in Principal moments (sq in sq in) and X-Y directions about centroid:
	I: 4704.5411 along [0.5527 -0.8334]
	J: 11989.3890 along [0.8334 0.5527]

```
Write analysis to a file? [Yes/No] <N>: (Press ENTER to
      continue.)
```

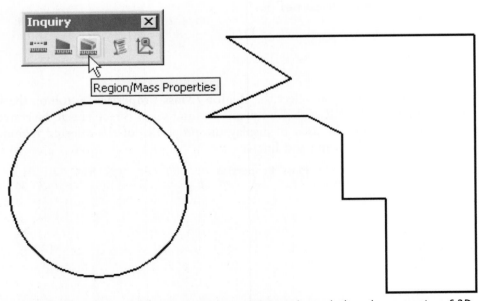

Figure 14.5 *The **Mass Properties** command can be used to calculate the properties of 2D objects, once they have been converted to regions. Once a region is selected, AutoCAD will display the mass properties in the text window.*

Although **Mass Properties** gives more information than may be needed for most 2D situations, the command can be used to get the area and perimeter of an object. Although much of the information provided by this command is beyond the scope of this text, other information includes the following:

Bounding box—The two coordinates that define the bounding box are the diagonally opposite corners that enclose the rectangle.

Centroid—The center of the region.

Moment of inertia—A value used in engineering formulas to compute the distributed loads.

Product of inertia—A property used to determine forces causing motion in an object.

Radii of gyration—A value used to determine the moments of inertia of a solid.

Principal moments—The properties derived from the products of inertia.

LISTING INFORMATION ABOUT AN OBJECT

The **List** command displays a listing of information about an object in the drawing. To access the command, use one of the following methods:

- Click the **List** button on the **Inquiry** toolbar.
- Type **LI** ENTER at the Command prompt.
- Select **List** from **Inquiry** on the **Tools** menu.

The command sequence is as follows:

```
Command: Click the List button (Or type LI ENTER.)
Select objects: (Select an object.)
```

One or several objects can be selected for listing, although the information might not fit on the screen all at once. To view additional pages, press ENTER to resume the output display. Press F2 to terminate the text box and return the Command prompt. Selecting the **Close** button in the title bar will also close the window.

Figure 14.6 shows a partial floor plan with a line selected for **List** and the resulting list for this line. The information that is displayed depends on the type of object that is selected—but the object's type, position relative to the current UCS, the layer location, and either model or paper space is always listed. The color and linetype are listed if either is not set by layer. Four values are always given to describe a line. The X and Y values specify the horizontal and vertical distances of the "first point" and the "next point," from the 0,0 point when the limits were established. The line in Figure 14.6 is 32'–1 1/2" to the right of the X axis. Notice that distances of the "from point" and the "to point" of the Y axis are 34'–8 1/4" and 12'–9 3/8" respectively. The difference between these two values is 21'–10 7/8", which is the specified length.

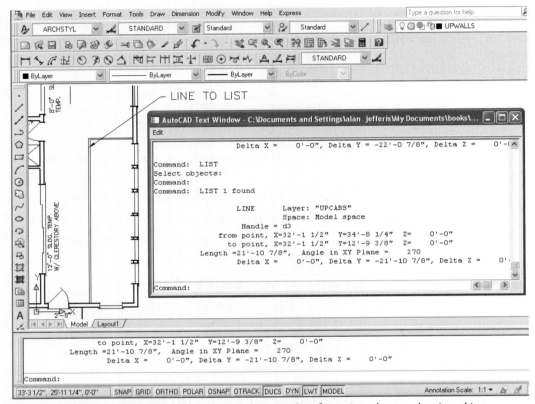

Figure 14.6 *The **List** command can be used to supply information about a drawing object.*

When the List command is used to gather information on a circle, information is based on the center point and radius, the area, and the circumference. Listing a circle is an easy way to determine the area of a pier for doing concrete estimates. An example of a listing for the circle in Figure 14.7 is as follows:

CIRCLE	Layer: FND FOOT
	Space: Model space
Handle	2D
Center point	X=119'-8" Y=103'-6" Z=0'-0"
radius	0'-9"
circumference	4'-89/16"
area	254.47 sq in (1.7671 sq ft)

The **List** command can also be used to provide information about text in the drawing base. (Text will be introduced in the next chapter.)

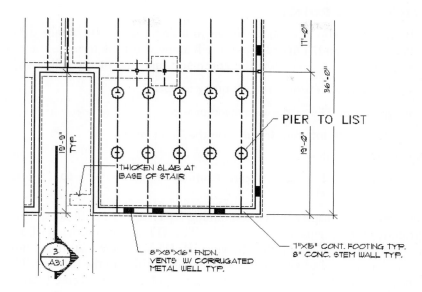

Figure 14.7 *The **List** command provides information based on the center point, the radius, and the diameter. (Courtesy Scott R. Beck, Architect.)*

OBTAINING INFORMATION ABOUT A DRAWING

The **List** command is used to obtain information about selected objects. The **Dblist** command will produce information about the entire drawing. This command can be useful in comparing drawing sequences or information contained in two similar drawings. Information is presented in the same order as used for **List**: one page at a time. The command can be accessed only from the command line. The command sequence is as follows:

```
Command: DBLIST ENTER
```

If the command line prompt is displayed, pressing ENTER will advance through a listing of information for the entire drawing. Pressing ESC will terminate the scrolling of the listing and pressing **F2** will remove the text window and restore the drawing screen. With the Dynamic Display active, entering the command will appear to do nothing. Select **F2** to display the AutoCAD Text window and the results of the command. You can print listings from both the **Dblist** and **List** commands if your terminal is connected to a printer by pressing the PRINT SCREEN key.

IDENTIFYING A POINT

 Information about a specific point in a drawing can be obtained by using the **ID Point** command. This command will allow you to determine the exact location of a specific point. To access the command, use one of the following methods:

- Click **Locate Point** on the **Inquiry** toolbar.
- Type **ID** ENTER at the Command prompt.
- Select **ID Point** from **Inquiry** on the **Tools** menu.

The sequence is as follows:

```
Command: Click the Locate Point button (Or type ID ENTER.)
Specify point: (Select a point.)
```

Entering a point will display the corresponding X, Y, and Z coordinates. This can be helpful in determining the dimension of an object. The **ID Point** command also can be used to identify a specific location in the drawing. If **ID** is entered at the Command prompt, coordinates can be supplied and that point will be identified in the drawing. This can be useful with relative coordinates entry methods to start an object at a specific location.

The **ID Point** listing for the location of intersection of the line in Figure 14.6 is as follows:

```
Command: Click the Locate Point button (Or type ID ENTER.)
Specify point: (Select desired point. For this example the
    handrail in Figure 14.6 was used.)
X=32'-1 1/2"xY=34'-8 1/4"xZ=0'-0"
```

TRACKING DRAWING TIME

The current time is displayed at the bottom of the drawing screen on the task bar. The **Time** command uses the date and time maintained by the computer to keep track of the time related to the drawing session. To access the command, use one of the following methods:

- Click **Time** from **Inquiry** on the **Tools** menu.
- Type **TIME** ENTER at the Command prompt.

Before the command can function properly, the date and time must be set. The **Date/Time** setting on the **Control Panel**, accessed from **My Computer**, can set an accurate date or time. If you reenter a drawing, AutoCAD can now provide the current time, drawing creation time, last update time, and time elapsed while editing the drawing. Each might seem unimportant in an educational setting, but the information is highly useful in keeping track of hours spent on a project for billing purposes and minor bookkeeping functions, such as determining profit margins.

TIME DISPLAY VALUES

Access the time display by typing **TIME** ENTER at the Command prompt. The display will resemble:

```
Current time: 16 Aug, 2007@ 11:15:25.510 AM
Times for this display:
Created: Tuesday, Jul 22, 2005 at 3:3106.970 AM
Last updated: Tuesday, Aug 16, 2006 at 2:2506.970 AM
Total editing time: 0 days 3:27:38.175
Elapsed timer (on): 0 days 1:59:51.510
Next automatic save in: 0 days 00:12:20.70
Enter option [Display/On/Off/Reset]:
Command:
```

Times are listed as day, month, year, hour, minute, seconds, and milliseconds.

- The Current time is the current date and time.

- Created time is marked when the drawing was initially started.

- The Last updated time lists the last editing session when either the **End** or **Save** command was used with the default file name.

- The Total editing time is the amount of time used editing the current drawing, since it was created until Save is used.

If **Quit** is used to end a drawing session, the time spent in that session is not added to the accumulated time.

Time spent printing and plotting is excluded from the Editing time.

This time is updated continuously and cannot be reset or stopped. This timer is most helpful to supervisors in tracking efficiency and logging total hours for billing.

The Elapsed time can be reset and turned **ON** or **OFF**. This timer is best suited for individual use each day to keep track of your time if you will be working on several different jobs. The Next automatic save time indicates when the next save is scheduled to occur (see chapter 3).

PROMPT OPTIONS

After the time display, four options are listed on the prompt line.

- The **Display** option will redisplay the **Time** display and update the current values.

- The **On** option will activate the **User Elapsed** timer. The default value is set to **On** when a drawing is started.

- The **Off** option will stop the **User Elapsed** timer.

- The **Reset** option clears the **User Elapsed** timer to zero.

Pressing ENTER or F2 will restore the command line.

STATUS

The **Status** command will list the current value settings of a drawing. To access the command, use one of the following methods:

- Select **Status** from **Inquiry** on the **Tools** menu.
- Type **STATUS** ENTER at the Command prompt.

The status values of the drawing shown in Figure 14.7 are as follows:

931 objects in FOUND Model Space Limits are:	X: 0'-0"	Y: 0'-0" (off)
	X: 275'-0"	Y: 250'-0"
Model space uses	X: 45'-31/8"	Y: 48'-71/2"
	X:194'-113/8"	Y:152'-73/8"
Display shows	X: 89'-5"	Y: 83'-63/4"
	X: 145'-05/8"	Y:123'-93/8"
Insertion base is	X: 0'-0"	Y: 0'-0" Z: 0'-0"
Snap resolution is	X: 0'-1"	Y: 0'-1"
Grid spacing is	X: 2'-0"	Y: 2'-0"
Current space:	Model space	
Current layer:	FNDFOOT	
Current color:	BYLAYER—1 (red)	
Current linetype:	BYLAYER—DASHED	
Current lineweight:	BYLAYER	
Current plot style:	ByLayer	
Current elevation:	0'-0" thickness:	0'-0"

Fill on Grid off Ortho on Qtext off Snap off Tablet off Object snap modes: Center, Endpoint, Intersection, Node, Extension Free DWG disk (C:) space: 3919.7 Mbytes Free temp disk (C:) space: 3919.7 Mbytes Free physical memory: 166 Mbytes (out of 510.8M) Free swap file space: 962.6 Mbytes (out of 1248 M). Command:

Several elements in the list, such as color, linetype, Snap, Grid, and Ortho, should be very familiar by now. Other terms might seem a little foreign. Remember that much of the information provided by the **Status** command can be obtained using the **Properties** command.

PURGE

This command does not provide information about a drawing, but it can be a useful tool within a drawing file to eliminate unused items and minimize storage space used. Objects that have been named in a drawing session, such as Blocks, Dimstyles, Layer, Linetypes, Shapes, and Text styles, are examples of items that can be purged from a file. **Purge** can be used if a template drawing is used as a base for a new drawing. The base drawing might contain the LAYER listings for a three-story structure with prefixes such as UPPER, MAIN, and LOWER. If only a two-level structure is to be drawn, all layers with the MAIN prefix could be purged.

Items to be purged can be eliminated at any time after the entering drawing file is opened. To start the command, use one of the following methods:

- Type **PU** ENTER at the Command prompt.

- Select **Purge** from **Drawing Utilities** on the **File** menu.

Each method will produce a **Purge** dialog box similar to Figure 14.8.

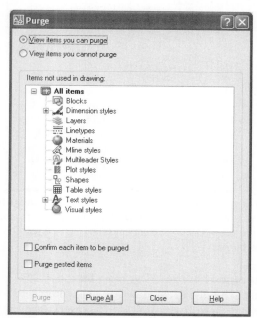

Figure 14.8   *The **Purge** dialog box can be used to delete unused objects and features from the drawings. Display the dialog box by typing **PU** ENTER at the Command prompt or by selecting **Purge** from **Drawing Utilities** of the **File** menu.*

Four choices can be made regarding information to be deleted from the drawing base including:

View items you can purge—In the default setting, this radio button and the **Items not used in drawing** display window are displayed. The window lists categories of named items that have been defined in the drawing but are not presently being used. Double-clicking any of the listed features will display a listing of individual named items for that category that have not been used. For instance, if you created a layer but placed nothing on that layer, the name of the layer would be listed as an item that is not in use. Notice in Figure 14.9 that ten different layers are listed. The **Purge** command will remove these and other unused items from the drawing. Execute the command by selecting the **Purge** button or the **Purge All** button.

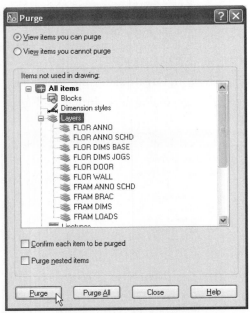

Figure 14.9 *With **View items you can purge** as the default setting, the **Items not used in drawing** list is displayed. The window lists categories of named items that have been defined in the drawing but are not presently being used. Ten layers are listed as layers that can be purged.*

View items you cannot purge—This radio button is **OFF** in the default setting. Notice in Figure 14.10 that the display window has been altered to list features that have been used in the drawing. These layers are listed as layers that cannot be removed from the drawing. If you attempt to remove a listed feature, a warning is displayed in the dialog box: "This Layer cannot be purged if it contains objects."

Confirm each item to be purged—Once an option has been entered, you can decide if you want verification of each object that is to be purged. With this check box active, a **Confirm Purge** warning box will be displayed allowing a **YES** or **NO** answer to be given for each item to be purged.

Purge nested items—A nested object is an object that is part of another object. In later chapters, you'll learn how to save components in groups called blocks. A sink, stove, and dishwasher are examples of items that could be saved as a block. These blocks could then be saved in bigger blocks that represent typical kitchen arrangements. The kitchen sink is said to be nested in the larger kitchen block. This option, if active, will purge drawing items even if they are contained within or referenced by other unused objects.

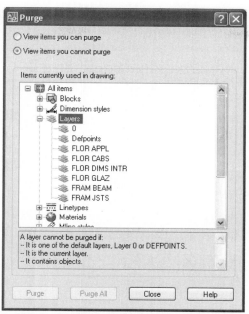

Figure 14.10 *If the **View items you cannot purge** radio button is ON, the window lists features that have been used in the drawing. The listed layers are layers that cannot be removed from the drawing.*

Once the desired settings have been entered, you can remove objects by selecting the **Purge** button or the **Purge All** button.

- Clicking the **Purge** button will remove multiple objects, one at a time. A warning will be displayed before each item is removed from the drawing base.

- Clicking the **Purge All** button still allows you to rethink the process. Selecting the **Yes** button will purge the item and continue prompting for items to purge. Selecting the **Yes to All** button will remove all items not used in the drawing.

If you decide that you've made a mistake, exit the command and select the **Undo** button to restore the purge.

USING QUICKCALC

The calculator of AutoCAD can be used just like any handheld calculator. To access the calculator, use one of the following methods:

- Click the **QuickCalc** icon on the **Standard** toolbar.

- Type **QC** ENTER at the Command prompt.

- Select **QuickCalc** from the **Tools** menu.

The basic portion of the calculator can be seen in Figure 14.11. In addition to the number pad that performs like any standard calculator, the palette contains areas for scientific calculations, unit conversions and variables, as well as tools to complete geometric calculations.

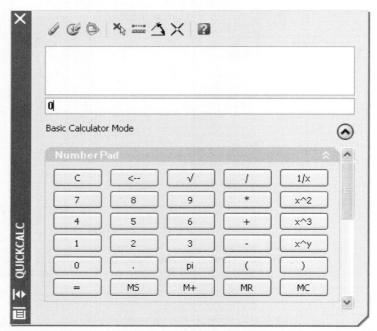

Figure 14.11 *The calculator is displayed by clicking the **QuickCalc** button in the **Standard** toolbar.*

USING THE BASIC CALCULATOR

The basic calculator is divided into the tool bar, history area, and the keypad. The toolbar is used to perform quick calculations of common functions. Buttons found on the toolbar include:

- **Clear**—Removes unwanted information from the Input box. If you make a mistake entering information, the button provides a chance to reenter the correct information. Information can also be removed from the edit box using the backspace key.

- **Clear History**—Clears information from the history area. As problems are completed, the problem and solution are stored in the **History** area.

- **Paste Value to Command Line**—Pastes the value that is in the Input box to the command line. This button is replaced by the **Apply** button if **QuickCalc** is used transparently during a command.

- **Get Coordinates**—Calculates the coordinates of a point location that has been selected in the drawing. Selecting this button removes the palette allowing a point to be identified. Once selected, the palette is redisplayed and the coordinates are displayed on the input display line.

- **Distance Between Two Points**—Calculates the distance between two points that have been selected in the drawing. Selecting this button removes the palette allowing two points to be identified. Once selected, the palette is redisplayed and the distance between the two points is displayed on the input display line.

- **Angle of Line Defined by Two Points**—Calculates the angle of two points that have been selected in the drawing.

- **Intersection of Two Lines Defined by Four Points**—Calculates the intersection of four points that have been selected in the drawing.

- **Help**—Displays Help for **QuickCalc**.

The number pad functions as any handheld calculator. Symbols on the number pad include:

Control	Description
C (Clear)	Clears any entry in the Input box and resets its value to 0.
<– (Backspace)	Moves the cursor one space to the left in the Input box, removing one decimal place or character from the display.
$\sqrt{}$ (Square Root)	Obtains the square root of a value.
1/X (Inverse)	Inverts any number or expression entered in the Input box.
x2 (X to the Power of 2)	Squares a value.
x3 (X to the Power of 3)	Raises any number or expression entered in the Input box to the power of 3.
xy (X to the Power of Y)	Raises a number or expression entered in the Input box to a specified power.
pi	Enters pi to 14 decimal places in the Input box.
((Open Parenthesis)) (Close Parenthesis)	When combined in pairs, groups a portion of the expression. Items contained in a parenthetical grouping are evaluated before the remainder of the expression.
= (Equal)	Evaluates the expression currently entered in the Input box.
MS (Store in Memory)	Stores the current value in the QuickCalc memory.
M+ (Add to Value Stored in Memory)	Adds the current value to the value stored in the QuickCalc memory.
MR (Restore Memory Value)	If a value is currently stored in the QuickCalc memory, the value is restored to the Input box.
MC (Clear Memory)	Clears the value currently stored in the QuickCalc memory.

USING THE SCIENTIFIC CALCULATOR

The scientific portion of the calculator allows advanced equations to be completed. This includes trigonometric, logarithmic, exponential, and other expressions commonly associated with scientific and engineering applications. The scientific portion of the calculator can be seen in Figure 14.12. This portion of the calculator is exposed by selecting the expand arrow on the right end of the title bar. Selecting the collapse arrow will remove this portion of the display from the calculator.

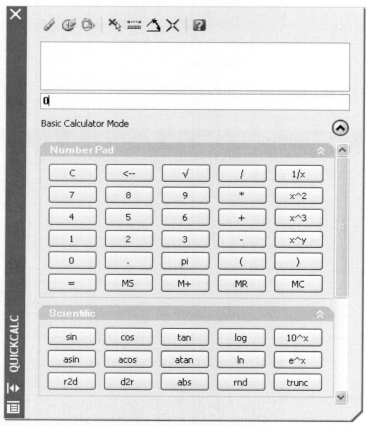

Figure 14.12 *The scientific portion of the calculator allows equations including trigonometric, logarithmic, exponential, and other expressions commonly associated with scientific and engineering applications to be completed.*

USING THE UNIT CONVERSION AREA

Once the area of an object has been determined using the **Area** command, the **Unit Conversion** display can be used to convert an area from one unit to another. The unit conversion area accepts only decimal values without units.

Units Type—This area will display a drop-down menu that includes the option of length, area, volume, and angular values.

Convert From—This area of the conversion display lists the units of measurement from which to convert. Options include acres, square centimeters, square feet, square inches, square kilometers, square meters, square millimeters, and square yards. A drawing object can be selected, the area determined, and then a base unit on measurement will be selected from the **Convert From** menu.

Convert To—Once a **Convert From** value has been selected, this portion of the conversion display lists the units of measurement the existing value will be converted to. Unit options are the same values that are available in the **Convert From** menu.

Value to Convert—Provides a box to enter a value to convert. Values can also be entered in the input box above the basic calculator.

Converted Value—Converts the units entered in the **Value to Convert** and displays the converted value in the selected units.

Calculator Icon—Returns the converted value to the Input box.

Open a drawing and draw a circle. Now determine the area using the **Area** command. For this example, suppose the circle has an area of 45.25 inches. To find the area in millimeters would require the following steps:

1. Select the type of information to be determined using the **Units Type** menu. *(Select AREA)*

2. Enter this value in the **Value to Convert** box. *(Enter 45.25)*

3. Select the desired units for the **Convert From** menu. *(Select **square inches**)*

4. Select the desired units for the **Convert To** menu. *(Select **square millimeters**)*

5. The **Converted Value** is displayed as 29193.49.

In addition to determining area, menu options include Length, Volume, and Angular. Options for each area include:

Length	Volume	Angular
Feet	Cubic centimeters	Circles
Inches	Cubic feet	Degrees
Kilometers	Cubic inches	Grads
Meters	Cubic kilometers	Quadrants
Miles	Cubic meters	Radians
Millimeters	Cubic millimeters	
Survey Feet	Cubic yards	

USING THE VARIABLES AREA

This area of the calculator provides access to predefined constants and functions. Key areas can be seen in Figure 14.13 and include the variable tree and the four buttons on the title bar. The **Variables Tree** is used to store predefined shortcut functions and user-defined variables. Shortcut functions are common expressions that combine a function with an object snap. The following table describes the predefined shortcut functions in the list. Additional constants and functions can also be stored in this area. The sample shortcuts include the following:

Function	Shortcut for	Calculation to be performed
DEE	Dist(end,end)	Distance between two end points
ILLE	ill(end,end,end,end)	The intersection of two lines defined by four endpoints
MEE	(End+end)/2	Midpoint between two endpoints
NEE	Nor (end,end)	Unit vector in the XY plane and normal to two endpoints
RAD	Rad	Gives the radius of a selected circle
VEE	Vec (end,end)	Vector from two endpoints
VEE1	Vec1 (end,end)	Unit vector from two endpoints

Above the **variable tree** are these four buttons:

New Variable Button—Opens the **Variable Definition** dialog box allowing creation of new calculator variables using the shortcut menus.

Edit Variable Button—Opens the **Variable Definition** dialog box allowing changes to be made to the selected variable.

Delete Variable Button—Selecting this button deletes the selected variable.

Calculator Button—Selecting this button returns the selected variable to the **Input box**.

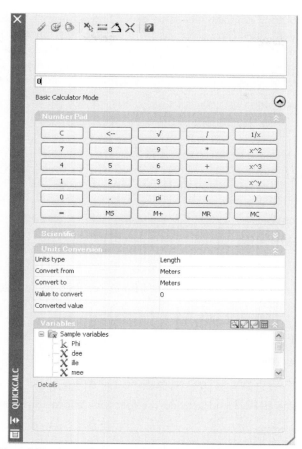

Figure 14.13   *The **Variables** area of the calculator provides access to predefined constants and functions.*

CHAPTER 14 EXERCISES

1. Open drawing **E-13-9** and determine the area of one side of the cross-sectional view of the manhole cover. Assume all unmarked radii to be 0.25". Maintain a thickness of 7/8" at the cover support. Compute the following:

   Surface area of one side the cover _____

   Sectional area _____

   Perimeter of section _____

   Manhole surface area _____

   Circumference _____

2. A steel fabricating company needs to order enough primer to cover (5,000) 21 1/2"
   × 8 1/2" steel plates. Each plate has (4) 3/4" diameter holes with the center points
   located 2" down, 2" from plate edge, and 3" o.c. The plates will be made of 1/2"
   steel. Draw the top, front, and side views of the plate, and save the drawing as
   E-14-2. Use the drawing to determine the following information:

   Surface area 1 side _____

   Surface area 2 of sides _____

   Perimeter _____

   Area of edges _____

   Area of each hole _____

   Total area of holes _____

   Total area of 1 plate _____

   Total area of 5,000 plates _____

3. Open drawing **E-14-2**. Copy the listed times required to complete the project.

   Created _____

   Last updated _____

   Total editing time _____

   Terminate the drawing session using **End** or **Save**.

4. Open drawing **E-12-3** and provide the following information:

   Drawing name _____

   Total objects _____

   Limits _____ _____

   Limits _____ _____

   Current layer _____

   Current linetype _____

   Free disk space _____

   Status of: Grid _____ Snap _____
   Ortho_____

   Qtext _____ Tablet _____

   Terminate the drawing with **Save**.

5. Open drawing **E-7-5** and determine the surface area of the window.

AREA _____ PERIMETER _____

What is the distance between the endpoints of the inner side of the interior arch?

_____

List the values that describe the outer arch at the top of the window.

List the total editing time _____

Current drawing session time _____

Terminate the drawing using **Save** or **End**.

CHAPTER 14 QUIZ

DIRECTIONS

Answer the following questions with short complete statements. Type your answers using a word processor.

1. Place your name, chapter number, and the date at the top of the sheet.

2. Type the question number and provide the answer.

Warning: Some of the questions have not been covered in the reading material and will require the use of the help menu and for you to do some exploring to answer the questions.

1. Give four types of information that will be given if a circle is selected for a **List**.

_____

_____

2. Explain the difference between **List** and **Dblist**.

_____

_____

3. Write the name of the command and describe two different uses for identifying a point.

_____

_____

4. What are the two methods of selecting objects to determine their area?

_____

_____

5. What command is best suited to determine the length of a line between two points?

_____

_____

6. What is the process to set the internal clock of your computer?

_____

_____

7. What value of the status listing keeps track of available drawing space?

_____

_____

8. What is the meaning of DELTA X and DELTA Y?

_____

_____

9. You've opened a drawing, made a few changes, and then discarded the drawing. Will the time spent in this drawing be added to the accumulated time?

_____

_____

10. Explain each of the categories used to measure time.

_____

_____

11. Describe how the **Area** command can be used on an open polyline segment.

_____

_____

12. You need to determine the area of a rectangle. Explain the difference in selecting three and four points.

_____

_____

13. List the symbols used by calculator for add, subtract, multiply, divide, and to indicate a set.

_____

_____

14. Determine the following values for your workstation:
Free drawing space
Free temp disk space
Free physical memory

_____

_____

15. Determine how many millimeters are in 500 inches and how many feet are in 3000 mm.

_____

_____

16. What is the circumference of a 4" diameter circle to the nearest 1/16"?

_____

_____

17. What function of the calculator should be used to determine the radius of an existing circle?

_____

_____

18. Sketch the key symbol on the calculator to determine the following functions:
The cosine of an angle
The square root of a positive number

_____

_____

19. What will the ILLE calculator function do?

_____

_____

20. Research and describe the calculator function to rotate a point about an axis.

_____

_____

Placing Text on a Drawing

INTRODUCTION

This chapter will introduce you to general considerations for placing text on construction drawings, including methods for

- Placing small amounts of text using the **Text** command.
- Placing large amounts of text using the **Mtext** command.
- Creating varied styles and fonts of text.
- Editing text on a drawing.
- Finding and editing text.
- Referencing text to objects in drawings.
- Placing text in template drawings.
- Creating and editing tables.

New commands to be examined in this chapter include:

- **Text**
- **Style**
- **Ddedit**
- **Mirrtext**
- **Qtext**
- **Mtext**
- **Spell**
- **Mleader**
- **Mleaderalign**
- **Mleadercollect**
- **Mleaderedit**
- **Mleaderstyle**

- **Table**
- **Field**
- **Tableedit**
- **Tablestyle**
- **Matchcell**
- **Updatefield**

GENERAL TEXT CONSIDERATIONS

Before text is placed on a drawing, several important factors should be considered. Each time a drawing is started, consider who will use the drawing, the information the text is to define, and the height of the text. Once these factors have been considered, the text values can often be placed in a template drawing and saved for future use, similar to the stock notes seen in Figure 15.1. Architectural and engineering offices often use several different styles of lettering. Many engineering offices that do civil, municipal, or government projects tend to use a simple block lettering shape. Architectural and engineering offices dealing with construction projects use a style of lettering that features thin vertical strokes, with thicker horizontal strokes to give a more artistic flair to the drawing. Other common variations are compressed and elongated letter shapes. In addition to the basic shape of the letters, some offices use a forward or backward slant to make their office lettering more distinctive. You are unlikely to find all of these variations on one professional drawing. Figure 15.2 shows some of the variations that can be found on drawings.

STEEL JOIST
1. STEEL JOISTS SHALL BE DESIGNED, FABRICATED, AND ERECTED PER SJI STANDARD SPECIFICATION.
2. CAMBER ALL JOISTS WITH STANDARD CAMBERS PER ABOVE STANDARD SPECIFICATION UNLESS NOTED ON THE DRAWINGS.
3. PROVIDE BRIDGING AS CALLED FOR BY ABOVE SPECIFICATIONS. DO NOT USE SLOTTED HOLES IN BOLTED CONNECTIONS UNLESS APPROVED BY ENGINEER.
4. PROVIDE SHOP DRAWINGS AND CALCULATIONS THAT INDICATE JOIST SIZES, LOADING, MEMBER SIZES, PANEL POINT LOCATIONS, CAMBERS, BRIDGING, AND ANY OTHER INFORMATION THAT MAY BE PERTINENT TO THE JOB. SHOP DRAWINGS SHALL BE STAMPED BY AN ENGINEER (THE PRODUCT ENGINEER) REGISTERED IN THE STATE OF OREGON.
5. PROVIDE INSPECTION REPORT PREPARED BY THE PRODUCT ENGINEER CERTIFYING THAT THE JOISTS HAVE BEEN FABRICATED ACCORDING TO THE DESIGN ASSUMPTIONS AND REQUIREMENTS. REPORT IS TO ACCOMPANY THE JOISTS ON DELIVERY TO THE PROJECT.
6. JOIST ERECTOR SHALL EXERCISE EXTREME CARE DURING ERECTION OF JOISTS TO PREVENT THE JOISTS FROM BUCKLING LATERALLY. USE SPREADER BARS FOR LIFTING JOISTS AND PROVIDE LATERAL BRACING AS NECESSARY. REMOVE ANY DAMAGED JOISTS FROM THE JOB SITE. DO NOT ATTEMPT TO REINFORCE DAMAGED JOISTS.

Figure 15.1 *A partial listing of standard construction notes.*

ARCHITECTURAL OFFICES USE MANY
STYLES OF LETTERING.
SOME TEXT IS ELONGATED.
SOME OFFICES USE A COMPRESSED STYLE.
NO MATTER THE STYLE THAT IS USED,
IT MUST BE EASY TO READ.

Figure 15.2 *Common types of lettering found on construction drawings.*

TEXT SIMILARITIES

With all of the variation in styling, two major areas are uniform throughout architectural and engineering offices. Text height and capital letters are as close to an industry standard as architects and engineers get. Text placed in the drawing area is approximately 1/8" (3 mm) high. Titles are usually between 1/4" and 1" (6 and 25 mm) in height, depending on office practice. The other common feature is that text is always comprised of capital letters, although there are exceptions based on office practice. The letter "d" is always printed in lowercase when representing the pennyweight of a nail. A typical use would resemble:

USE (3) -16d @ EA. RAFT. / PL CONNECTION

Occasionally manufacturers use lowercase letters to represent the size of special nails or fasteners specific to their products. With CAPSLOCK activated, all lettering will be capitalized unless SHIFT is pressed. Pressing SHIFT will make any letters typed while the lowercase letters key is depressed. Numbers are generally the same height as the text they are placed with. When a number is used to represent a quantity, it is generally separated from the balance of the note by a dash. This can be seen in the nailing specification above. Another common method of representing the quantity is to place the number in parentheses to clarify the quantity followed by the size. This would resemble:

USE (3) -3/4" DIA. M.B. @ 3" O.C.

Fractions are generally placed side-by-side (3/4") rather than one above the other so that larger text size can be used. When a distance such as one foot, two and one half inches is to be specified, it can be done as 1'–2 1/2". The ' (foot) and "(inch) symbols are always used, with the numbers separated with a dash. If metric sizes are to be represented on drawings, the use of a comma as a number separator is somewhat different than with English units.

Metric sizes of four digits are often written with no comma. Three thousand one hundred and fifty millimeters is written as 3150 mm. When a number five digits or larger must be specified, a space is placed where the comma would normally be placed. Fifty-six thousand, three hundred and forty-five millimeters is written as 56 345 mm. Other common methods of placing numbers will be presented as dimension methods are introduced in later chapters.

TEXT PLACEMENT

Text placement refers to the location of text relative to the drawing and within the drawing file. The primary placement for text according to the NCS guidelines is to:

- Keep the text orientation parallel to the bottom of the drawing sheet.

- A second acceptable option is to place text parallel to the right edge of the page.

- When structural material shown in plan view is described, the text can be placed parallel to the member being described.

Figure 15.3 shows methods of placing text in plan views. In plan views such as Figure 15.4, text is placed within the drawing but arranged so that it does not interfere with any part of the drawing. Text is generally placed within 2" of the object being described. A leader line is used to connect the text to the drawing. The leader line can be either a straight line or an arc, depending on office practice. Leader lines will be introduced later in this chapter as the Multileader command is explored.

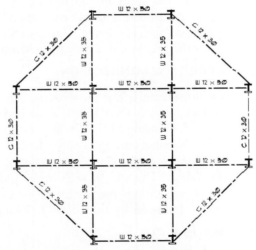

Figure 15.3   *Text should be placed so that it can be read when viewing the drawing from the bottom or right side.*

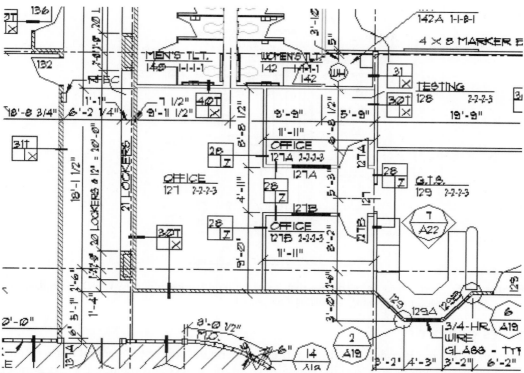

Figure 15.4   *Text can be placed within the drawing, but it should not interfere with drawing clarity. (Courtesy Tom Kuhns, Michael Kuhns, Architects, P.C.)*

On details, text can be placed within the detail if large open spaces are part of the drawing. It is preferable to keep text out of the drawing. Text should be aligned to enhance clarity and can be either aligned left or right. Figure 15.5 shows an example of a detail using good text placement. The final consideration for text placement is to decide on layer placement. Text should be placed on a layer with a major group code of ANNO. Common minor group codes include KEYN (key notes), LEGN (legends and schedules), NOTE, NPLT (non-plotted text), REDL (redline), REVS (revisions), SYMB (symbols), TEXT, and TTLB (border and title block).

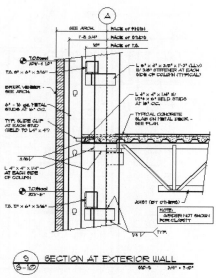

Figure 15.5 *Text should be aligned so that it can be read quickly. (Courtesy Van Domelen/Looijenga/ McGarrigle/Knauf, Consulting Engineers.)*

TYPES OF TEXT

Text on a drawing is considered to be either a general or local note. General notes can refer to an entire project or to a specific drawing within a project and include the categories of:

- **General notes**—notes located in the G series (General drawing sheets) of the drawing set and contain notes that apply to the entire project.

- **General discipline notes**—notes located in the 0 sheet for a specific design discipline. General discipline notes for the structural drawings would be found on sheet S-001 drawing set and contain notes that apply only to the structural drawings.

- **General Sheet notes**—Notes that apply only to the specific sheet where they are located.

Local notes refer to specific areas of a project such as the notes in Figure 15.4. No matter the type of text, two commands are available for placing text including:

- Small amounts of local notes can be placed using the **Text** command. The notes shown in Figure 15.1 are general notes that specify the materials of a framing plan.

- The **Mtext** (multiline) command should be used to place large amounts of text. Because the command provides the convenience of using a word processor and offers more options for placement and editing, it will typically be your first choice for placing large bodies of text.

Practical uses for including each in template drawings will be discussed later in this chapter.

LETTERING HEIGHT AND SCALE FACTOR

In previous chapters, the scale factor required to make line segments the desired size when plotted was considered. Similar factors must be applied to text height to produce the desired 1/8" high text when plotting is finished. Two methods are available for setting the text height, including:

- **The slow way**—Determine the text height based on the scale factor. This method is appropriate when working in the **Classic AutoCAD** workspace.

- **The easy way**—Work with the **2D Drafting Annotation** workspace and use the **Annotative features** of the workspace. This method will automatically adjust the text height to the workspace and will be discussed later in this chapter as styles are explored.

Determining Text Height by Scale Factor

To determine the required text height for a drawing, multiply the desired height (1/8") by the scale factor. The text scale factor is the reciprocal of the drawing scale. For drawings at 1/4"=1'–0", this would be 48. By multiplying the desired height of 1/8" (.125) × 48 (the scale factor), you see that the text should be 6" tall. Quarter-inch-high lettering should be 12" tall. You'll notice that the text scale factor is always twice the line scale factor. Working at a scale of 1"=20', 1/8" high lettering would be required to be drawn as 30" high text (.125" × 240).

Other common text scale heights include the following:

ARCHITECTURAL VALUES		ENGINEERING VALUES	
Drawing scale	Text Scale	Drawing scale	Text Scale
1"=1'–0"	12	1"=1'–0"	12
3/4"=1'–0"	16	1"=10'	120
1/2"=1'–0"	24	1"=100'	1200
3/8"=1'–0"	32	1"=20'	240
1/4"=1'–0"	48	1"=200'	2400
1/16"=1'–0"	64	1"=30'	360
1/8"=1'–0"	96	1"=40'	480
3/32"=1'–0"	128	1"=50'	600
1/16"=1'–0"	192	1"=60'	720

USING THE TEXT COMMAND

 The **Text** command can be used for placing local notes on a drawing. The **Text** command creates text that can be seen on the screen as it is entered at the keyboard. Multiple lines of text can be placed even though the command name of

Single Line Text on the **Draw** menu implies otherwise. The basic process of placing text will be introduced using the **Text** command. Text heights will be determined assuming the use of model space and non-annotative features. Once basic text skills have been mastered, the full power of **Mtext** will be explored. To access the **Text** command, use one of the following methods:

- Select the **Single Line Text** button on the **Text** toolbar.
- Select the **Single Line Text** button on the **Text** dashboard.
- Type **TEXT** ENTER at the Command prompt.
- Select **Single Line Text** from **Text** of the **Draw** menu.

The command sequence is:

```
Select the Single Line Text button.
Specify start point of text or [Justify/Style]: (Select text
    starting point.)
Specify height: <0'-0 3/16">: 6 ENTER
Specify rotation angle of text <0>: ENTER
```

As the command is started, the first display is the text style and the text height. If the command line is displayed, you'll be notified that the current style is referred to as Standard. Methods for creating different styles will be explained after you've explored the **Text** command.

START POINT

The name says it all. This is your opportunity to dictate where the start of the text will be placed. Indicating a point with the cursor will mark the lower-left corner of the proposed text. This type of text placement is referred to as *left justified*. As the prompt for text is displayed, the crosshairs will display the starting point at the lower-left corner of the text. The relation of the text to the starting point can be seen in Figure 15.6. If you don't like the location, the starting point can be altered after the other two prompts are addressed. Two other options are available instead of placing a start point. Each will be introduced after the basic sequence has been explored. Once the start point has been selected, a prompt for the height will be displayed.

Figure 15.6 *The default placement with the **Text** command is to place text on the right side of the start point. This is called left justified.*

HEIGHT

The second option specifies the text height, with the default of 0'-0 3/16". If you'll be plotting this drawing at 1/4"=1'-0", type **6** ENTER for the height. If you plan to use annotative features, enter a value of 1/8". You can also adjust text height by dragging the pointing device until the distance between the cursor and the insertion point indicates the desired text height. If you have not adjusted the initial drawing limits, any text you create is going to appear to be of gigantic proportions. If the height is set to zero, the text height will be controlled by the **Style** height value. The **Style** command will be introduced later in this chapter.

ROTATION ANGLE

The third option to be adjusted prior to entering text is the rotation angle of the text. The rotation angle can be visualized by thinking of a line extending from the start point at the specified angle, with the text above this line. The rotation angle can be entered at the Command prompt by typing the desired rotation angle. You can accept the default of 0 rotation by pressing ENTER, or you can type a numeric value followed by ENTER. X and Y coordinates can also be entered using the cursor. To use the cursor, drag the pointing device until the angle between the cursor and the insertion point represents the desired text rotation angle. Figure 15.7 shows an example of text rotated at various angles. Although text is read looking from the bottom or right side of the drawing, this option will allow text to be aligned with specific objects of a drawing, as seen in Figure 15.8.

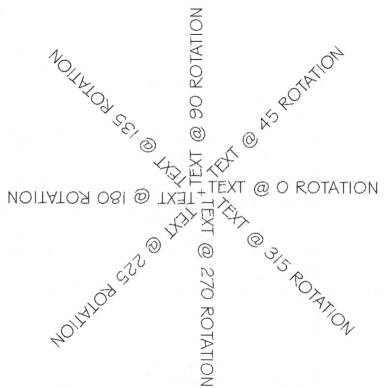

Figure 15.7 *Lines of text can be rotated so that they can be read from any angle.*

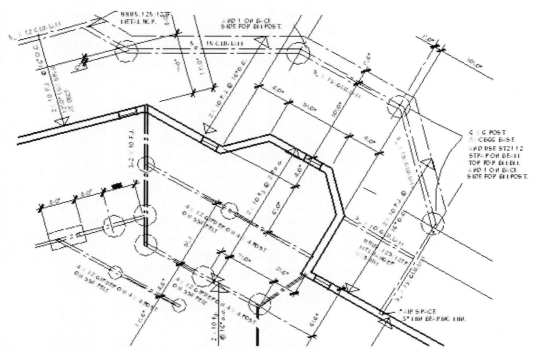

Figure 15.8 *Text should be rotated to be parallel to the structural member when a structure is an irregular shape. (Courtesy Residential Designs.)*

PLACING TEXT

Once the rotation angle is selected, selecting the ENTER key, activates the text cursor at the specified start point. The desired text can now be entered by keyboard. Be sure to have CAPSLOCK activated. Notice that as you type, the text is displayed on the screen. Depending on the zoom setting, you might be unable to read it. To increase the text size, cancel the command and **Zoom** into the area where you will be adding the text. To add additional lines of text, press ENTER once, and the new start position will be adjusted to be directly below the first line of text. The prompts for start point, height, and rotation angle will be skipped and the values for the first line of text will be reused. The command sequence displayed at the command line is:

```
Select the Single Line Text button (Or type TEXT ENTER.)
Current text style: "Standard' Text height: 0'-0 3/16"
Specify start point of text or [Justify/Style]: (Select text
      starting point.)
Specify height: <0'-0 3/16">: 6 ENTER
Specify rotation angle of text <0>: ENTER
(Enter desired text) FINISH FLOOR ENTER
(Enter desired text) ELEV. 100.00' ENTER
```

Figure 15.9 shows the result of this command sequence. Once the desired text has been entered, press ENTER twice to exit the **Text** command. Pressing ESC will destroy any text entered during the active command sequence.

 Note: Compare the command sequence given earlier at the *Dynamic Display* and the command just completed at the command line. Notice that information for the current height and style is given at the command line display. When placing text, it may be wise to work with both displays. For the balance of the chapter, the command line display will be given.

START POINT

Figure 15.9   *The results of the **Text** command using a specified start point, a height of 6", and a rotation angle of 0.*

JUSTIFYING TEXT

As the prompt for the start point was displayed, the options of **Justify** and **Style** were available. To justify text is to determine where and how the text will be located within a defined space. The default for placing text is for left justified text, which places the left edge of each new line of text directly in line with the line of text above. The lower-left starting point is defined, and text will be placed to the right of the starting point. This method is used on construction drawings, such as sections, and details where large amounts of written information can be neatly aligned. Figure 15.5 shows an example of left justified text.

Fourteen other options are available for aligning text. Each option can be accessed by the following sequence:

```
Select the Single Line Text button (Or type TEXT ENTER.)
Current text style: "Standard' Text height: 0'-0 3/16"
Specify start point of text or [Justify/Style]: J ENTER
Align/ Fit/ Center/ Middle/ Right/ TL/ TC/ TR/ ML/ MC/ MR/ BL/
    BC/ BR:R ENTER
```

Once you feel comfortable with each option, the option can be accessed without displaying the Justify menu by typing the letter of the option at the Command prompt. The process to align text from the right can be started by the following sequence:

```
Select the Single Line Text button (Or type TEXT ENTER.)
Current text style: "Standard' Text height: 0'-0 3/16"
Specify start point of text or [Justify/Style]: R ENTER
Specify right endpoint of text baseline:
```

Even though **Right** is not listed as one of the current options, **R** will be accepted and will change the prompt to the right endpoint prompt for the text baseline. This prompt will then be followed by the height and rotation prompts. As text is entered using the **Right** option, it will be placed just as it was when **Left** justified was used. When ENTER is pressed to place the text, it will be added to the drawing as right justified text.

Align

The **Align** option allows two useful options to be accessed. The starting and ending point can be selected so that text can be aligned with an inclined object similar to the way the rotation angle option is used. This method of selection is superior to using **Rotation** when the angle is not known. The second benefit of **Align** is the ability to select the start and endpoints of the line of text. The drawback to this option is that text height is assigned based on the amount of text to be placed between the start and endpoints. Figure 15.10 shows an example of aligned text. The command sequence is as follows:

```
Select the Single Line Text button (Or type TEXT ENTER.)
Current text style: "Standard' Text height: 0'-0 3/16"
Specify start point of text or [Justify/Style]: A ENTER
Specify first endpoint of text baseline: (Select starting point.)
Specify second endpoint of text baseline: (Select starting
    point.)
(Enter desired text.)
```

Figure 15.10 *Text can be aligned by specifying two points. The desired text will now be placed between these points. If more than one line of text is required, the height of each line may vary.*

Fit

The **Fit** option is similar to the **Align** option. Using **Fit**, the text height will remain constant throughout several different lines of text, but the width will be varied. This option uses a command sequence similar to Align, but it allows for a height to be specified prior to placing the text. Figure 15.11 shows an example of text placed using the **Fit** option.

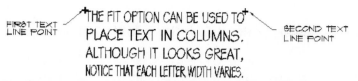

Figure 15.11 *The **Fit** option will place text between two specified points, but the height will remain constant if more than two lines of text are required.*

Center

The **Center** text option will place a line of text centered on a specified point, as seen in Figure 15.12. To place several lines of text using the **Center** option, press ENTER to move the cursor to start the second line below the first line. Type the second line of type and press ENTER when the line is complete. This process can be continued indefinitely.

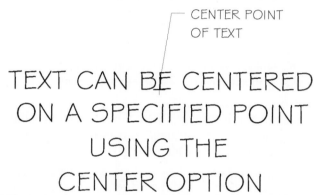

Figure 15.12 *The Center option allows text to be spaced equally around a specified center point.*

Middle

This option is similar to the effects of Center, but text is centered vertically and horizontally on the start point, as seen in Figure 15.13. The command sequence is similar to the sequence used for the **Center** option, once the Middle point is selected.

Figure 15.13 *The Middle justified option will center text along an imaginary line.*

Right

This option is the reverse of the default left justified option. Figure 15.14 shows an example of right justified text.

Figure 15.14 *The Right justified option will align the right ends of lines of text.*

Letter Options

The remaining options can be accessed by typing the designated letters at the **Justify** prompt:

TL—top left	TC—top center	TR—top right
ML—middle left	MC—middle center	MR—middle right
BL—bottom left	BC—bottom center	BR—bottom right

The effects of each can be seen in Figure 15.15. The command sequence is similar to the sequence used for left justified text, although the start point is referred to by the new location. For instance, the command sequence for BR is as follows:

```
Select the Single Line Text button (Or type TEXT ENTER.)
Current text style: "Standard' Text height: 0'-0 3/16"
Specify start point of text or [Justify/Style]: BR ENTER
Specify bottom right point of text: (Select desired start point.)
Specify height <0'-0 3/16">: (Enter desired text height.)
Specify rotation angle of text <0>: (Enter desired angle.)
(Enter desired text.)
```

The result of this sequence can be seen in Figure 15.16.

Figure 15.15 *The start points for each of the nine letter options for placing text.*

BOTTOM RIGHT JUSTIFICATION

Figure 15.16 ***BR*** *(bottom right) justification is often used when notes must be placed on the left side of an object.*

STYLE OPTION

The **Style** command is used to determine the appearance of the text characters. The command is selected by typing **S** ENTER at the first text prompt. The sequence and options include the following:

```
Select the Single Line Text button (Or TEXT ENTER.)
Current text style: "Standard' Text height: 0'-0 3/16"
Specify start point of text or [Justify/Style]: S ENTER
Enter style name or (?) <STANDARD>:
```

Standard will remain the default style until the **Style** command is used to create other styles.

WORKING WITH STYLES

The term *style* in CAD drafting describes the characteristics of a group of text such as the font, height, width, and the angle used to display the text. One style can be used for all drawing text, a second style used for the drawing titles, and a third style used for all information in the title block. The **Style** command can be used to add different fonts to your drawing as well as change the properties of the letter shapes. As styles are created, they should be stored in drawing templates for future use with all similar projects. Assigning a style name, selecting a font, and defining style properties are the first three areas to be considered when creating a new text style.

NAMING A STYLE

To access the **Style** command, use one of the following methods:

- Click the **Text Style** button on the **Text** toolbar.
- Click the **Text Style** button on the **Text** dashboard.
- Select **Text Style** from the **Format** menu.
- Type **ST** ENTER at the Command prompt.

Each option will produce the **Text Style** dialog box, shown in Figure 15.17, which can be used to name a style, assign a font, control the style options, and preview the results.

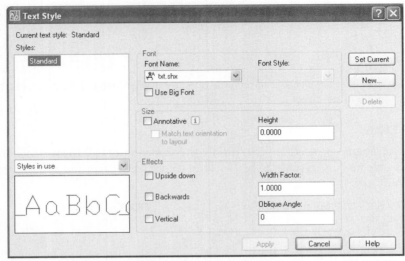

Figure 15.17 *Selecting the **Text Style** button on the **Text** toolbar, typing **ST** ENTER at the Command prompt, or selecting **Text Style** from the **Format** menu produces the **Text Style** dialog box that can be used to name and control text styles.*

New styles can be created in the **Text Style** dialog box by selecting the **New** button. This will produce the **New Text Style** dialog box shown in Figure 15.18. The new style is named **style1,** by default. A new name can be assigned by using the same procedure used to rename a drawing or create a layer name. Type the desired name, in this case TITLES, in the edit box and then accept the name by clicking the **OK** button. The **New Text Style** dialog box will close and TITLES will now be displayed as the current name in the **Style Name** edit box. Selecting the **Cancel** button will remove the **Rename Text Style** dialog box and allow a different aspect of the **Style** command to be addressed or allow the command to be terminated.

Figure 15.18 *Selecting the **New** button produces the **New Text Style** dialog box, allowing a new text style to be named.*

Choosing the Style Name

The name of the text style to be created should reflect the use of the text. The name can contain up to 255 letters, numbers, spaces, and special characters. (Names longer than 31 characters will spill out of the edit box.) The same guidelines that were used to name linetypes and layers should be used to name text styles. Some offices use the name of the text font for the style name. Names such as TITLE BLOCK, GENNOTES (general notes), FNDNOTES (foundation notes), or SITENOTES are examples of names that could be used to describe different styles of text. Names that include the text font and usage are helpful. Most of the lettering in this text was created using the font STYLUS BT, which is available in AutoCAD. Other art such as Figure 15.1 was created with a style called ARCHSTYLE, which is created by a third-party vendor for use with AutoCAD.

LETTERING FONTS

The word *font* is written in an italic font. The term font is used to describe a particular shape of text that includes all lowercase and uppercase text and numerals. The shape of the letters is one of the defining differences between architectural and mechanical lettering. Construction drawings usually contain text fonts that are more artistic than the gothic block lettering used on mechanical drawings. Most offices use different fonts throughout a drawing to distinguish between titles and text or to highlight specific information.

AutoCAD allows all of the fonts loaded into your computer to be accessed by the **Style** command. As a font is selected for use, a sample pattern is displayed in the preview window of the **Text Style** dialog box. Many engineering offices use fonts, such as ROMANS, ROMANC, ROMANT, or ITALICC, to letter drawings. In addition to these fonts, some architectural offices use fonts that more closely resemble freehand lettering, such as Stylus BT, City Blueprint and Country Blueprint, for notes on a drawing and Romantic (normal), Romantic (Bold), or Sansserif for titles. Examples of these fonts can be seen in Figure 15.19. It is important to select fonts that are easy to read.

THE STYLE 'STANDARD' IS COMPOSED
OF THE TXT FONT.

THE STYLE 'NOTES 'IS COMPOSED
OF THE ROMANS FONT.

THE 'TITLERESPONSE' STYLE IS COMPOSED
OF THE ROMANC FONT.

**THE TBLOCK STYLE IS COMPOSED
OF THE FONT ROMANT.**

*THE STYLE 'GENERALNOTES' IS
COMPOSED OF THE FONT ITALICC.*

**THE 'TITLEBLOCK' STYLE IS COMPOSED
OF THE FONT ROMANTIC.**

**THE STYLE 'ENGINEERSTAMP' IS
COMPOSED OF THE FONT SANS SERIEF.**

THIS TEXT IS A SAMPLE OF THE
'STYLUS BT' FONT.

*THE STYLE 'FNDNOTES IS COMPOSED
OF THE FONT COUNTRY BLUPRINT.*

THE 'TEXT' STYLE IS COMPOSED OF THE
FONT CITY BLUEPRINT.

THIS TEXT IS A SAMPLE OF TEXT BY A
THIRD PARTY DEVELOPER. THE FONT
IS 'ARCHSTYL.'

Figure 15.19 *Common fonts used in architectural and engineering offices.*

Lettering Programs

One of the benefits of AutoCAD is that many third-party suppliers have developed material to supplement the standard program. Many offices add text libraries to the basic program. You'll notice that much of the text used throughout the illustrations in this text is not found in the AutoCAD font files. This lettering is produced using material created by third-party vendors. Consult advertisements in local trade magazines for other available text libraries. Once loaded into AutoCAD, the **Text Style** dialog box can be used to select the font to be used with each lettering style. The box will be discussed in detail shortly, but it's important to remember that in addition to the fonts contained in AutoCAD, TrueType fonts contained in additional programs in your computer can be used in AutoCAD drawings. Methods of accessing these files will be discussed as ways of creating text styles are introduced.

Choosing a Font

The second aspect of defining a text style is choosing a font. Once the style name has been entered, selecting the down arrow beside the **Font Name** edit box will produce a list of all of the text fonts available to you. A sample of the list can be seen in Figure 15.20. The display will depend on the status of the **Use Big Font** check box.

- With the box inactive, all fonts in all programs on your computer will be displayed.

- The AutoCAD fonts are preceded by the Divider icon and followed by .SHX.

- The TrueType icon precedes other fonts available to AutoCAD.

- With the box active, only the fonts contained in AutoCAD will be displayed in the list.

Scroll through the list and select **STYLUS BT** font or an architectural font of your choice. As the name of the font is selected in the list, the name will be displayed in the edit box, the list will be closed, and a sample of the font will be displayed in the **Preview** window.

 Note: It's important to remember that even though you can add third-party fonts to your drawings, or fonts contained in other common Windows programs, these fonts may cause problems if you send your drawings to firms that do not have these fonts. If a drawing is opened on a machine that does not contain the font, AutoCAD will substitute the TXT font, and some special symbols associated with the original font will be lost.

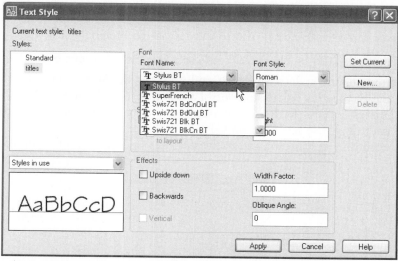

Figure 15.20   *Selecting the down arrow beside the **Font Name** edit box will produce a listing of the AutoCAD and the True Type font files. Scroll through the desired list and select the name of the font to be used with the style being created.*

DEFINING STYLE PROPERTIES

Once the style has been named and the font has been assigned, seven options are available to further define the style being created. These options included **Font Style, Size, Upside-down, Backwards, Width Factor, Oblique Angle**, and **Vertical**.

Font Style

So far you've named a style (TITLE) and assigned the STYLUS BT font to the text style. The **Font Style** edit box is activated once a font is named. This edit box will allow characters to be formatted with bold, italic, oblique or regular text. Options will vary depending on the selected font. With a font of **STYLUS BT, Roman** is the only option available.

Height

If a height is entered at this prompt, all text created using this style of text will be the same height. If all TITLES throughout the drawing are to be the same size, the height variable of the **Style** command can be set. When height is set with the **Style** command, the variable will not have to be set each time the font is used. The height of text is best controlled with the Height option of the command used to create the actual text. Entering a height now will delete the Height prompt in the **Text** and **Mtext** command sequences when this style is used. Responding with a **0** will allow the height to be determined each time this style is used.

Annotative   A second method of controlling text is to make the text an annotative object. Use the following steps to assign annotative features to text:

- Click the **Annotative** button in the **Size** box.
- Enter the value to be assigned to the text in the **Height** edit box.
- If this style is to be the current style, click the **Set Current** button.

Once all features of the style have be set and accepted, an additional setting needs to be adjusted on the **Properties** palette to control annotative text. Figure 15.21 shows a portion of the **Properties** palette that can be used to control text. Open the palette, select Annotative, and toggle the setting from OFF to ON. See chapters 23 and 25 for additional information regarding the use of annotative features in layouts and while plotting.

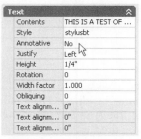

Figure 15.21   *The **Text** area of the **Properties** palette can be used to toggle the annotative properties of text ON or OFF.*

Upside-Down

The default for this option will place text in the normal position. Activating this box will place text upside-down. Although this option is not typically used in construction drawings, an example of upside-down text can be seen in Figure 15.22.

NORMAL ORIENTATION
�078IDE-DOWИ ORIENTATIONᴎ

Figure 15.22 *Text can be printed in both normal and upside-down orientation.*

Backwards

The default value for this option, inactive, will produce text that is read from left to right. Activating this box will produce text that is read from right to left or backwards. Figure 15.23 shows an example of text printed backwards.

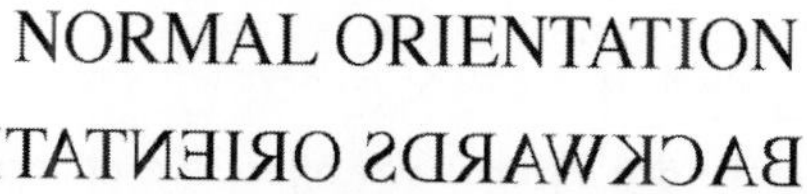

NORMAL ORIENTATION

ИОITATИƎIЯO 2ᗡЯAWᑋƆAᗺ

Figure 15.23 *Text can be printed in both the normal and backward orientation.*

Vertical

This option is not active if a TrueType font has been selected. The option will display .SHX fonts in the vertical position, as shown in Figure 15.24.

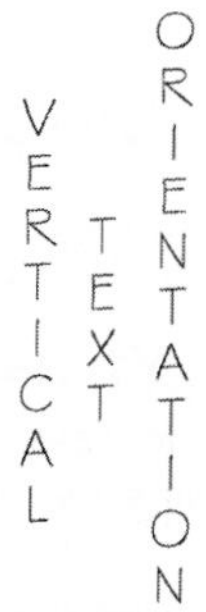

V E R T I C A L   T E X T   O R I E N T A T I O N

Figure 15.24 *Placing text in a vertical position might be required on some construction drawings.*

Width Factor

Font characters are displayed and plotted based on a ratio of their height and width. No matter what the actual height is, the width is assigned a value of 1. The default value for the width factor is also 1, making the width equal to the height of the letter. If a value of .5 is entered, text will be half as wide as it is tall. A value of 2 will produce an elongated style of lettering. Figure 15.25 shows a comparison of various width factors.

WIDTH FACTOR = .5

NORMAL WIDTH FACTOR = 1.00

WIDTH FACTOR = 2.0

Figure 15.25 *Adjusting the **Width Factor** will affect the appearance of text.*

Oblique Angle

This option allows text to be tilted from true vertical. The default angle of **0** displays text in the normal position with the vertical leg of each letter in a vertical position. Entering a positive value will slant text to the right. Entering a negative value will produce a back slant. Figure 15.26 shows examples of altering the obliquing angle.

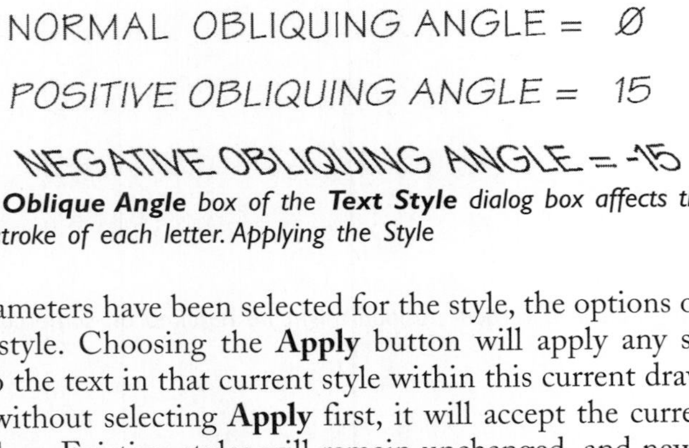

NORMAL OBLIQUING ANGLE = Ø

POSITIVE OBLIQUING ANGLE = 15

NEGATIVE OBLIQUING ANGLE = -15

Figure 15.26 *The **Oblique Angle** box of the **Text Style** dialog box affects the angle of the vertical stroke of each letter. Applying the Style*

Once all of the parameters have been selected for the style, the options of **Apply** and **Close** remain to set the style. Choosing the **Apply** button will apply any style changes made in the dialog box to the text in that current style within this current drawing. If you choose the **Close** button without selecting **Apply** first, it will accept the current style values and remove the dialog box. Existing styles will remain unchanged, and new text will be added using the current settings.

SPECIAL CHARACTERS

In addition to the symbols on a keyboard, it is often necessary to print special symbols, such as a degree or diameter symbol. Special symbols are added using the Text command by typing a code letter preceded by two % (percent) symbols. Common symbols of AutoCAD include the following:

```
%%C=diameter symbol
%%D=degree symbol
%%O=overscore text
```

```
%%P=plus/minus symbol
%%U=underscore text
```

The symbols for overscore and underscore text are toggles. Type **%%U** preceding the text to be underscored to activate underscoring. The symbol must also be typed at the end of the phrase to turn off underscoring. An example of the sequence is:

%%UPROVIDE AN ALTERNATE BID FOR ALL ROOFING MATERIALS%%U

As the text line is displayed at the command line, the %%U will be displayed in addition to the desired text. When ENTER is pressed, the %%U will be removed from the screen, and the text will be underscored. Although a single percent symbol can be created by using the percent key (SHIFT+5), when the percent symbol must proceed another special character, %%% must be used. To write 50°+/−5% you must enter the following sequence:

50%%D%%P5%%%

Note: Remember that these symbols apply to the fonts in AutoCAD and do not apply to text created by third-party vendors. It's also good to remember there is an easier method of applying special text: Use the *Mtext* command.

EDITING EXISTING TEXT

Text can be edited by using the backspace key and entering the corrections before the text is added to the drawing base. Text can also be edited once it is part of the drawing base using the **Ddedit** command. To start the command, use one of the following methods:

- Click the **Edit Text** button on the **Text** toolbar. *(The button will need to be added to the dashboard if you're working in the 2D Drafting and Annotation workspace.)*
- Type **ED** ENTER at the Command prompt.
- Select **Edit** from the **Text** submenu from **Object** of the **Modify** menu.

Each entry method will produce the following prompt:

```
Select the Edit Text button (Or type ED ENTER.)
Select an annotation object or [Undo]:
```

Select the text to edit by moving the cursor to the line of text and selecting it with the select button. The error does not have to be selected, only the line of text containing the error. As the text is selected it will become highlighted. Editing can be performed by moving the arrow to the desired location and using the select button. The steps to edit text can be seen in the following example:

aLL FRAMINGG LUMBER IS TOBE DFL # 2 OR BETTER UNLESSNOTED.

The "a" of "aLL" should be in capital letters, "FRAMINGG" is misspelled, and a space should be added in "TOBE" and "UNLESSNOTED." To fix the "a" move the cursor to the right side of the "a" and click the mouse select button. Use backspace to eliminate the "a." Without moving the cursor, the error can be fixed by typing an **A**.

Use the arrow key or the mouse to move the cursor to the right side of "FRAMINGG" and click the select button. Press BACKSPACE to eliminate the last "G." Now if the cursor is placed between the "O" and "B" of "TOBE," you can insert a space by pressing SPACEBAR. "UNLESSNOTED" can be corrected in the same manner. The corrected text will now read as it was intended:

ALL FRAMING LUMBER IS TO BE DFL # 2 OR BETTER UNLESS NOTED.

If the edit is not what you desired, the **Undo** option can be used to restore the line of text to its previous status. Once you are satisfied with the text, click ENTER to add the text to the drawing base. The **Ddedit** command is still active. If no other editing needs to be done, press ENTER a second time to exit the command.

QUICK TEXT

Quick text is not a method of creating text, but a useful way to display it. As you're editing drawings containing large amounts of text, you might find the text distracting. The **Qtext** command will display lines of text as rectangles. The height of the rectangle is the same as the text height, but the length of the rectangle is typically longer than the line of text. The command can be accessed using one of the following methods:

- Type **QTEXT** ENTER at the command prompt.
- Select **Show text boundary frame only** in the Display Performance area of the **Display tab** of the **Options** menu.

Once the command is started, options include On and Off. With **Qtext** set to **ON**, text will be displayed as a rectangle. When **Qtext** is set to **OFF**, text that was represented by the text boundary frame will retain its current setting until a **Regen** is performed. Figures 15.27a and 15.27b show an example of **Qtext** in both the **ON** and **OFF** settings.

 Note: If *Qtext* is *ON* during a plot, the rectangles, not the text, will be plotted.

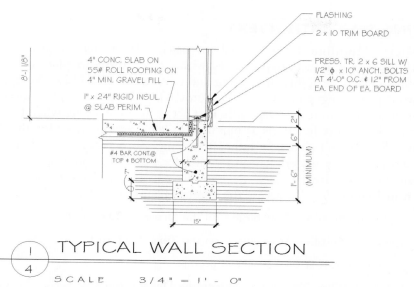

Figure 15.27a *An example of normal text. Figure 15.26b shows the same drawing with quick text. (Courtesy Piercy & Barclay Designers, Inc., A.I.B.D.)*

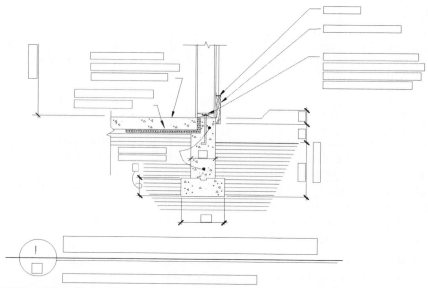

Figure 15.27b *The effect of **Qtext** on the drawing in Figure 15.27a. (Courtesy Piercy & Barclay Designers, Inc., A.I.B.D.)*

CREATING TEXT WITH MTEXT

A   **Mtext** (**Multiline Text**) is ideal for placing large areas of text on a drawing. Multiple lines of text can be created with the **Multiline In-place** text editor, which offers greater flexibility in creating and editing the text. Using **Mtext** is similar to using a word processing program within AutoCAD. **Mtext** also offers advantages when editing the text. Lines of text placed with **Text** act as individual lines when they are moved. Text placed with **Mtext** functions as one object. A paragraph created with Mtext can be moved by selecting one character rather than the entire paragraph. Multiline text also allows text to be easily underlined and varied fonts, colors, and heights to be assigned throughout the text. **Mtext** is ideal for entering and editing large bodies of text such as those seen in Figure 15.1. To access the command, use one of the following methods:

- Click the **Multiline Text** button on the **Draw** or **Text** toolbars.
- Click the **Multiline Text** button on the **Text** dashboard.
- Type **T** ENTER at the Command prompt.
- Select **Multiline Text** from **Text** on the **Draw** menu.

Each will produce the text prompt (the crosshairs with ABC) and the following prompt:

```
Command: T ENTER
Specify first corner:
```

If the command line is active, the display will list the style and text height. AutoCAD is waiting for you to select where you would like to place the text. Once you select the first corner, you will be prompted:

```
Specify opposite corner or [Height/Justify/Line spacing/
    Rotation/Style/Width/Columns]:
```

Notice in Figure 15.28, as the box is being dragged to indicate the second corner, an arrow is shown indicating the direction that excess text will flow out of the box. Methods of altering the direction will be introduced shortly. By prompting for a second point, AutoCAD is asking you to provide a window to place the text in. The width of the window will control the width of the paragraph. Extra text will spill out of the top or bottom of the window depending on the justification, so that the depth of the window is not critical. Another key element of the text window is its zoom feature.

- If the current display of the drawing is appropriate to the selected text size, the text editor will display the text at its specified size.
- If zoom has been minimized so that it is too small to read, the text editor will display the text in an enlarged view.
- If zoom has been maximized so that the text is too large, the text editor will display the text in a reduced view.

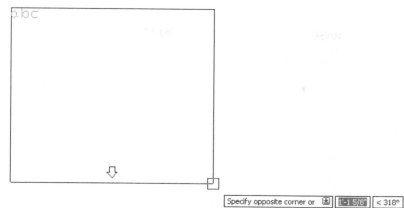

Figure 15.28   *Multiline text is started by selecting a box to define the width that the text will occupy. If the text does not fit into the box, it will overflow the bottom of the box in the current setting (the down arrow). The direction of overflow can be adjusted.*

Selecting the second corner will display a rectangle, a text ruler, and the **Text Formatting** toolbar. Each can be seen in Figure 15.29. The rectangle defines the location and size of the text. The **Text Formatting** toolbar provides quick access to text editing tools that can be used to create or edit text. The ruler can be used to set and modify paragraph tabs and indents. Text can now be entered in the edit window using procedures similar to those used with other text editors. AutoCAD allows a third-party text editor to be used if you select the editor from the **File** tab in the **Options** dialog box. See the AutoCAD User's Guide in the **Help** menu to select a third-party editor. The balance of this section will introduce the many capabilities of the **Text Formatting** toolbar.

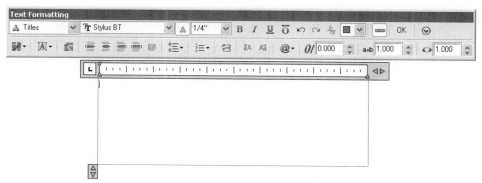

Figure 15.29   *The **Text Formatting** toolbar is used to control text placed in a drawing using **Mtext**. The toolbar can be used to assign text, various colors, heights, underlining, and other alterations.*

As text is entered at the keyboard, it will be displayed in the edit box. Text will wrap to the next line, as the specified width is reached, just as with any word processing program. When the desired text has been entered, select the **OK** button on the right end of the **Text Formatting** toolbar or move the cursor outside the text box and single-click to end the command. Either option will save any changes that have been made and return to the drawing. Pressing ESC will close the **In-place Text Editor** without saving the text.

CONTROLLING TEXT WITH THE TEXT FORMATTING TOOLBAR

Each of the qualities introduced with text are available in **Mtext**. The qualities of **Mtext** are just much easier to use and alter. Qualities controlled on the **Text Formatting** toolbar include style, font, height, bold, italic, underline, undo, stack, and color. You will also find the ruler, and the **Display** menu options button on the toolbar.

Style

The **Style Edit box** allows an existing text style to be applied to a new or selected text body. Character formats for font, height, bold, and italic can be overridden by applying a new style to existing multiline objects. Information can be typed in one style, and altered before the text is placed in the drawing base. To alter the current text style, click the arrow next to the **Style** control on the toolbar and then select a style. Figure 15.30 shows an example of the **Style** list.

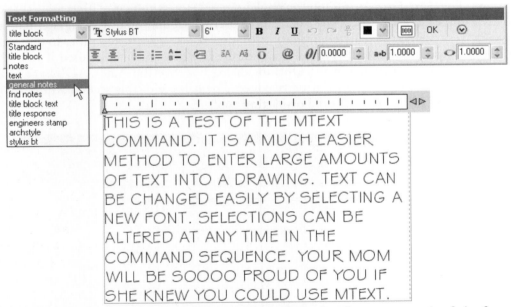

Figure 15.30 *Selecting the down arrow for **Style** displays a list of current text styles. Styles for multiline text are created using the same procedures as used to create a style for text.*

Font

Clicking the down arrow beside the **Font** edit box displays a list similar to the font list displayed with the **Style** command. The list allows a font for new text to be selected or the font of selected text to be changed. The current font for multiline text is the same as the current font for the **Style** command. With **Mtext**, the style can be altered at any time throughout the use of the command, including mid-sentence. Figure 15.31 shows an example of text that was altered to provide different fonts within the same command sequence.

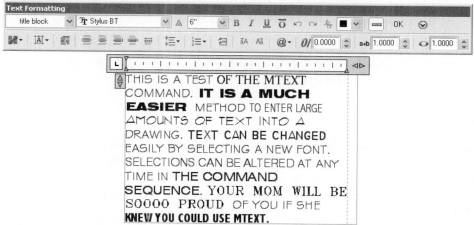

Figure 15.31   *The text font of one word or a string of words can be altered at any point of the drawing's life.*

Annotative

This button serves as a toggle for annotative features for the current Mtext object.

Height

This option controls the height of the text just as with the **Text** command. The default value is the same as the current style. The value for text height can be changed at any point of the command, allowing words to be printed at varied heights. Selecting the **Height** edit arrow will display the heights of all current styles. To provide a new height, select the **Height** edit box and click the cursor to highlight the current height. Remove the current height with the backspace arrow and then enter the desired height value. Text height can be altered before the text is added to the drawing by highlighting the text to be altered and then providing the required height in the edit box. Figure 15.32 shows an example of how text height can be altered.

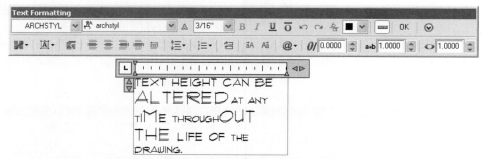

Figure 15.32 *Multiline text height can be altered at any time throughout the life of the drawing. Selecting the **Height** edit arrow will display a listing of current heights. New heights can be entered by replacing the current value with the desired value.*

Bold

The **Bold** option is active only for TrueType fonts. Selecting the **Bold** button toggles between bold and normal text. Bold text can be created by selecting the **Bold** button before text is entered. Existing normal text can be changed to bold by highlighting the desired text to be changed and then clicking the **Bold** button. Figure 15.33 shows an example of text that was altered to provide bold text.

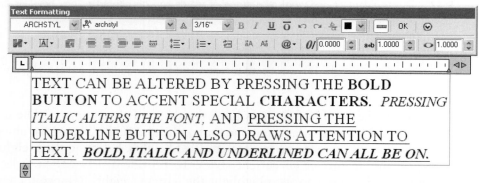

Figure 15.33 *Text can be changed by selecting the **Bold**, **Italic**, or **Underline** button. One or more of the features can be used in combination to affect the font.*

Italic

Selecting the **Italic** button toggles italic formatting for text. As with other options, **Italic** can be applied to text being created as well as used to modify existing text. The option is available only for TrueType fonts.

Underline

Selecting this button allows the underline function to be toggled **ON** and **OFF**. This option of **Mtext** is so much easier than %% something that is required with the **Text** command.

Overline

This button will place a line over selected text. Highlighting the desired text and then selecting the **Overline** button will place a line over the specified text.

Undo and Redo

The **Undo** option can be used to undo the last sequence of the text editor. Changes can be made to the text or to one of the formatting options. The **Redo** option reverses actions of **Undo** to the last sequence of the text or to one of the formatting options.

Stack/Unstack Text

The **Stack** option can be used to place a portion of text over other text. This is useful in writing stacked numbers such as 1/4 rather than ¼ or in writing equations. When a fraction is entered into the text editor, the **AutoStack Properties** dialog box shown in Figure 15.34 is automatically displayed. The box can be used to set the properties of fractions.

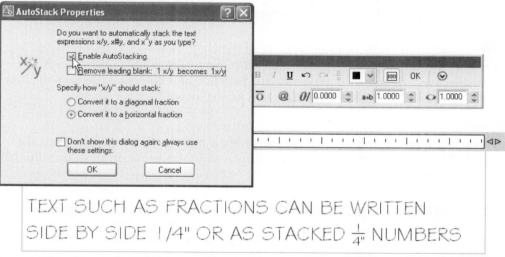

Figure 15.34 *Text such as fractions can be written side-by-side or stacked depending on office preference.*

Color

Selecting the **Color** button allows the color for new text to be assigned or for existing text to be altered. By default, multiline text will be assigned the color of the current layer. To select a new text color, select either the color box or the edit arrow beside the **Color** edit box, and select the desired color. Selecting the **Select Color** option will produce the **Select Color** dialog box and allow for a full range of colors to be selected.

Adding Background Color for Mtext Text can have colored backgrounds to improve legibility when placed over a drawing object. Select the **Options** button or right-click to display the shortcut menu. Select **Background Mask** from the shortcut menu to open the **Background Mask** dialog box as shown in Figure 15.35. This box will allow a border offset factor to be specified. This is the distance the background mask extends beyond the edge of the text. A specific background fill color can be selected, or the same color as the display background can be selected.

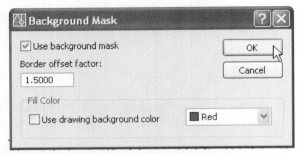

Figure 15.35 *The **Background Mask** dialog box can be used to highlight text. The dialog box can be used to specify the background color and the size of a box that will surround text.*

Ruler

The ruler button toggles the ruler above the text editor **ON/OFF**. With the ruler activated, left and right arrows are displayed. Selecting and holding an arrow will allow the text window to be stretched in the direction of the arrow.

Options

Selecting the **Options** button displays the shortcut menu of formatting options shown in Figure 15.36. A similar shortcut menu is displayed by right-clicking while in the **Mtext** command.

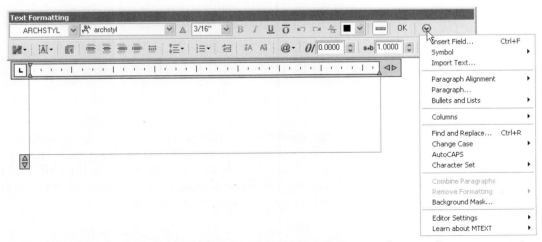

Figure 15.36 *Selecting the **Options** button displays a shortcut menu for controlling text properties.*

CONTROLLING TEXT WITH THE LOWER LEVEL OF THE TOOLBAR

The lower level of the toolbar contains controls for columns, justification, paragraphs, line spacing, numbering, insert fields, uppercase and lowercase text, symbols, oblique angle, tracking, and width factor.

Columns

Clicking the **Columns** button displays a menu of options to add and control columns in a body of mtext. The default setting is **No Columns**. Other options include controls for dynamic and static columns, column breaks, and a selection for additional column settings. Each can be seen in Figure 15.37.

- **Dynamic Columns**—Selecting this option activates the dynamic column feature. Dynamic columns adjust the text flow, causing columns to be added or deleted. Options of the setting include **Auto Height** and **Manual Height**.

- **Static Columns**—Selecting this option activates the static column mode, allowing the total width and height and number columns of the mtext to be set. Options range from 2 through 6. Selecting the **More** option displays the dialog box shown in Figure 15.37.

- **Insert Column Break**—If the **No Columns** option is deactivated, this option will insert a manual column.

- **Column Settings**—Selecting this option will display the **Column Settings** dialog box shown in Figure 15.38. The display contains radio buttons and edit boxes to control the settings required for each of the other options.

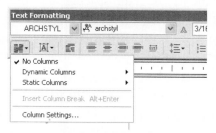

Figure 15.37   *Selecting the **Columns** button provides a list of controls for adding columns to a group of notes. Selecting the **Column Settings** options displays the additional column controls shown in Figure 15.38.*

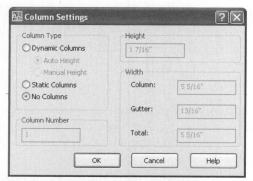

Figure 15.38 *Selecting the **Column Settings** button provides additional column controls.*

Justification

Justification with **Mtext** is similar to that of the **Text** command. As with the Text command, selecting a **Justification** option for Mtext allows the alignment and flow of text to be altered. The default setting places the cursor in the top-left corner of the text box. Options include Left, Center, Right, Top, Middle, Bottom, and combinations of these six options.

Paragraphs

Clicking the **Paragraph** button displays the **Paragraph** dialog box similar to the display in Figure 15.39. These controls can be used to set the size of paragraph indentations and to control the locations of the first line of text. Options include tab stops, indent controls, and controls for paragraph alignment, paragraph spacing, and for line spacing with a paragraph. Components of the dialog box include:

- **Specify Decimal Style**—This option sets the decimal style based on current user locale. Options include setting the decimal style as a period, comma, or space.

- **Left Indent**—These values set the indent value for the first line of text and for hanging lines of text. These values can be applied to selected or current paragraphs.

- **Right Indent**—This value sets the indent value for the entire selected or current paragraph.

- **Paragraph Alignment**—Sets the alignment properties for the current or selected paragraphs. Options include Left, Center, Right, Justified, and Distributed. Buttons for each of these options are also located on the lower line of the tool bar.

- **Paragraph Spacing**—Specifies the spacing before or after the current or selected paragraphs. The distance between two paragraphs is determined by the total of the **After paragraph spacing** value of the upper paragraph and the **Before paragraph spacing** value of the lower paragraph.

- **Paragraph Line Spacing**—Sets the spacing between individual lines in the current or selected paragraphs. Clicking the **Line spacing** button on the **Text Formatting** toolbar will also control line spacing. Selecting the **More** option will display the Paragraph dialog box shown in Figure 15.39.

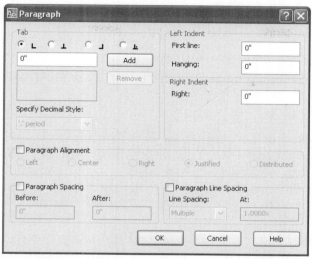

Figure 15.39   *Clicking the **Paragraph** button displays the **Paragraph** dialog box allowing paragraph indentations and the locations of the first line of text to be controlled.*

Controlling Lists

The **Mtext** allows for the creation of numbered, bulleted lists, and lists defined by letters. Clicking the **Numbering** button allows a list to be created with each major point preceded by a letter and a period, a number and a period, or a bullet. Additional options include:

- **Off**—With this option active, bullets, letters, and numbers will be removed from the selected text, but the indentation associated with these markers will remain.

- **Restart**—With this option active, a new letter or number sequence will be started.

- **Continue**—This option adds selected paragraphs to the last list above and continues the sequence.

- **Allow Auto-list**—This option applies list formatting as text is entered at the Mtext prompt. With this option on, the following characters can't be used as bullets, but can be used as punctuation after letters and numbers:

period (.),	comma (,),	close parenthesis ()),
close angle bracket (>),	close square bracket (]),	close curly bracket (}).

- **Use Tab Delimiter Only**—Limits the **Allow Auto-list**, **Allow Bullets**, and **Lists** options. List formatting is applied to text only when the space after the letter, number, or bullet character was AB, not SPACEBAR. This option is selected by default.

- **Allow Bullets and Lists**—When this option is active, list formatting is applied to all plain text in the multiline text object that looks like a list. Text that meets the following four criteria is considered to be a list:

- The line begins with one or more letters, numbers or a symbol, followed by
  - Punctuation after a letter or number
  - A space created by pressing TAB
  - Some text before the line is ended by ENTER or SHIFT+ENTER.

Insert Field

Fields are placeholders in Text objects for data that might change during the life of a drawing. This is useful because when the object changes the data field will be updated to display the latest data. Field data can be linked to a variety of information, including callouts, labels, and sheet properties such as date, project number, and sheet number. As an example, a field in a room could show the square footage; as the room changes, the area is updated.

Selecting the **Insert Field** button displays the **Field** dialog box shown in Figure 15.40, where you can select a field to insert in the text. When the dialog box closes, the current value of the field is displayed in the text. Key portions of the dialog box include the **Field Category**, **Field Names**, **Field Expression**, **Author**, and **Format**. Field categories include Date & Time, Document, linked, objects, others, plot, and set sheets.

Use the following steps to create a field:

1. Click the **Multiline Text** icon on the **Draw** toolbar. Specify the text window by selecting the first corner and the opposite corner.

2. Select **Insert Field** button from the toolbar.

3. Select the field name (in this example, **Object**), and click the **Select object** button as shown in Figure 15.40.

4. This will return you to the drawing session, prompting you to select object(s). Select the room diameter; this will work only if the room diameter is one entity (a Pline).

5. By default the Property will be Area, and you need to select Format type of Architectural and select **OK**. The room square footage will appear with a background color. The background color does not plot; it is a reminder that this text is a field.

As the room changes in size, you can update the field by selecting **Regen** from the **View** menu. By default, field values automatically update when you open, save, plot, eTransmit, or regenerate a drawing; this ensures that you always have the most current data.

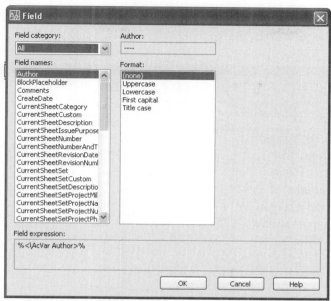

Figure 15.40   *Clicking the **Insert Field** button displays the **Field** dialog box. The dialog box can be used to select a field to be inserted in multiline text.*

Upper and Lowercase

These buttons can be used to alter the case of text. Use the following steps to change lowercase text to all caps:

- Highlight the desired lowercase text to be altered.
- Click the **Uppercase** button and the selected text will be changed to be all uppercase.

Select the **Lowercase** button to change text from all caps to lowercase letters.

Symbols

Selecting the **Symbol** button displays the list of symbols shown in Figure 15.41. The degree, plus/minus, angle, centerline, and diameter are often used on construction documents. Each can be inserted directly into multiline text, by selecting the **Symbol** button and then selecting the desired symbol. If this seems too easy, the symbols can also be inserted using the special characters that were discussed as the **Text** command was introduced. Selecting the **Other** option will display the **Character Map** dialog box shown in Figure 15.42 and allow one of the symbols from the **Character Map** to be inserted. Symbols will vary, depending on the current font. Use the following sequence to insert a symbol into a drawing:

1. Select the desired symbol and then click the **Select** button. This will display the symbol in the **Characters to Copy** edit box.
2. Select the **Copy** button and then minimize the dialog box.

3. Right-click to display the shortcut menu.

4. Select **Paste** to place the symbol in the drawing text.

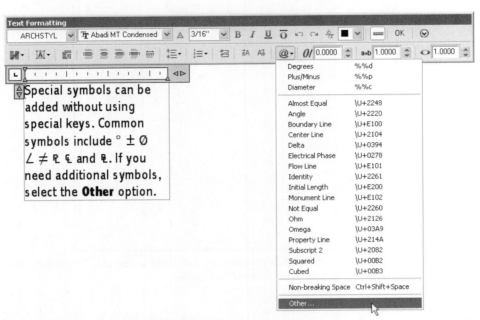

Figure 15.41   *Using **Mtext**, special characters such as the degree, plus/minus, diameter, angle, not equal, property line, centerline, and boundary line can be added without using any %% special characters.*

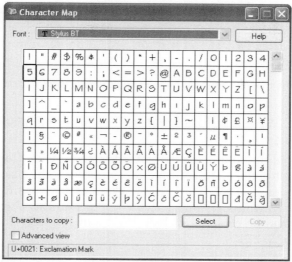

Figure 15.42 *Selecting the **Other** option from the **Symbols** list displays the character map. Options will vary depending on the current font.*

Oblique Angle

The **Oblique Angle** option determines the forward or backward slant of the text. The angle represents the offset from 90 degrees of the vertical leg of the text. This option does not alter the angle of a line of text. Entering a value between -85 and 85 makes the vertical leg of the text oblique. A positive obliquing angle slants text to the right. A negative obliquing angle slants text to the left. The value can be entered by entering a value in the edit box or by scrolling the listed values with the edit arrows.

Tracking

The Tracking option decreases or increases the space between the selected characters. The 1.0 setting is normal spacing. Set to more than 1.0 to increase spacing, and set to less than 1.0 to decrease spacing. The value can be entered by entering a value in the edit box or by scrolling the listed values with the edit arrows.

Width Factor

This option can be used to increase or decrease the size of selected characters. The 1.0 setting represents the normal width of the letter in this font. Entering a value of 2 will double the width. A width factor of 0.5 will decrease the width by half.

EXPLORING THE OPTIONS MENU

The **Multiline Text Editor Options** menu shown in Figure 15.36 and the Mtext shortcut menu provide many of the toolbar text controls and additional options for controlling new text and revising existing text.

- The **Options** menu is available by selecting the **Options** button on the right end of the toolbar.
- The shortcut menu is available by right-clicking in the **Multiline Text Editor** as text is being added.

In addition to basic editing options such as **Undo, Redo, Cut, Copy**, and **Paste** on the shortcut menu, several options are available for controlling the placement of text. These options are similar to the controls of Microsoft Word or other word processing programs. Several options on the menus are similar to the options on the **Text Formatting** toolbar. Key options that should be explored include **Find and Replace**, and **Import text**.

Find and Replace

Selecting the **Find and Replace** option displays the **Replace** dialog box shown in Figure 15.43. The dialog box can be used for searches for specified text strings and replaces them with new text. This can be especially helpful in finding a key word in a long list of specifications. A similar dialog box is displayed by clicking the **Find and Replace** button on the **Text** toolbar. That dialog box will be explored later in this chapter. Components of the dialog box include the following:

Find what—Defines the word or text string to search for.

Replace with—The **Replace with** option can be used in conjunction with **Find what** to alter a word or word string. Once a word has been found, enter a word in the **Replace with** box and click the **Replace** button.

Find—Starts a search for the text string in **Find what**. To continue the search, click **Find** again.

Replace—Replaces the highlighted text with the text that is in the **Replace with** edit box.

Replace All—Finds all instances of the text specified in **Find what** and replaces it with the text in **Replace with**.

Match whole word only—With this box inactive, any use of the selected word will be identified with the **Find** option. With "truss" indicated in the **Find what** box, "Truss–joist" would be identified in the document. Selecting this option will identify the word only if it is a single word. When the word is part of another text string, it will be ignored.

Match case—With this box checked, only the exact word with the same case will be identified. With the box inactive, placing "Truss" in the **Find what** box would identify truss in the body of text. With the **Match case** box active, "truss" would not be identified. Only uses of the word or string that are identical will be located.

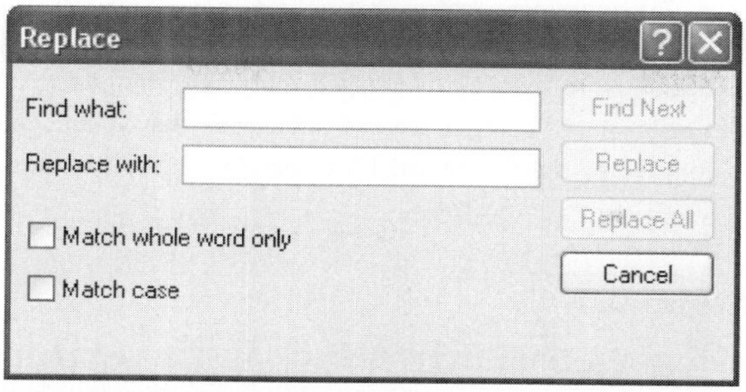

Figure 15.43 *Selecting the **Find/Replace** option provides options for editing multiline text.*

To find a word, select the **Find what** box by placing the cursor in the box and then type the desired word. Figure 15.44 shows an example of using **Find** to locate the word JOISTS. Clicking the **Find Next** button starts the search for the text in the **Find what** box. Press ENTER to find the next occurrence of the selected word.

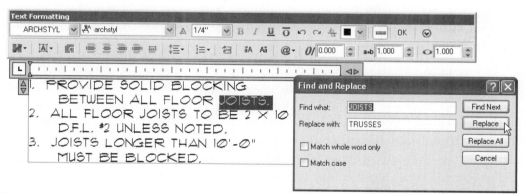

Figure 15.44 *Selecting the **Find what** box allows a word or word string to be specified for a word search. The word JOISTS has been selected for the search. Clicking the **Find Next** button will highlight the first use of the specified word. Pressing **REPLACE** will change the words and move to the next occurrence of the selected word or string.*

Import Text

The **Import Text** option allows ASCII (American Standard Code for Information Interchange) and RTF (Rich Text Format) text files to be inserted in a drawing file. This might seem unimportant as you're learning AutoCAD now, but it means you can create schedules or large bodies of text up to 32K in size, using your second-favorite software, and then import the file into AutoCAD once you feel more comfortable with multiline text. Imported text retains its original character formatting and style properties, but it can be edited as if it had been created within **Mtext**.

Note: Files larger than 32K can be inserted into AutoCAD if they are not imported directly into AutoCAD. To insert a file larger than 32K, minimize AutoCAD, and then open the file to be inserted using the word processor that was used to create the file. Now save the text file to the Clipboard. Maximize AutoCAD and paste the file into the drawing.

EDITING MULTILINE TEXT

 Multiline text can be edited using the **Copy, Erase, Grips, Mirror, Move,** and **Rotate** commands. Multiline text can also be edited using either the **Ddedit** or **Properties** commands or by using the **Text Style, Scale, Justify,** and the **Convert distance between text** buttons on the **Text** toolbar.

USING DDEDIT

Use the **Ddedit** command with multiline text in the same way it was used when editing text created with the **Text** command. Open the command using one of the methods introduced as text was edited. Double-clicking the body of text to be edited will also open the command. When the paragraph to be edited is selected, the text will be redisplayed in the **In-Place text Editor**. Each of the options that were used to create the text can be used to alter the selected text. To edit text, move the cursor to highlight the text to be changed. With the text highlighted, any one of the features set using the text editor can be altered. When the changes are made, click the **OK** button to return to the drawing screen. As the screen is restored, the prompt for selecting additional annotation is still displayed. The program is waiting for you to select additional text to be edited. Press ENTER a second time to restore the Command prompt.

In addition to the **In-Place Editor** for editing text, standard Windows control keys can be used to alter the text in the editor. The control keys include the following:

CTRL+A	Selects all text in the **Multiline Text Editor**.
CTRL+B	Applies or removes bold format for selected text.
CTRL+C	Copies selected text to the Clipboard.
CTRL+I	Applies or removes italic format for selected text.
CTRL+SHIFT+L	Converts selected text to lowercase.
CTRL+U	Applies or removes underline format for selected text.
CTRL+SHIFT+U	Converts selected text to uppercase.
CTRL+V	Pastes Clipboard contents to cursor location.
CTRL+X	Cuts selected text to the Clipboard.
CTRL+SPACEBAR	Removes character formatting in selected text.
ENTER	Ends the current paragraph and starts a new line.

EDITING WITH PROPERTIES

 The **Properties** palette can be used to alter the properties of multiline text. The command has been introduced earlier, but its power to alter text should not be forgotten. The properties of multiline text can be displayed using the following steps:

1. Select the body of text to be edited.

2. Select the **Properties** icon from the **Standard** toolbar or right-click in the drawing area to display the shortcut menu.

You can also start the command by typing **MO** ENTER at the Command prompt. Each method will display the **Properties** palette similar to Figure 15.45. When the properties are listed by category, general properties affecting the text and specific properties are listed for multiline text.

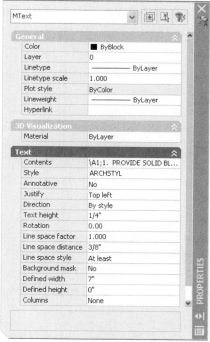

Figure 15.45 *The **Properties** palette can be used to edit multiline text. Select the **Properties** button on the **Standard** toolbars select the text to be edited, right-click, and select **Properties** or type **MO** ENTER at the Command prompt to produce the **Properties** palette.*

General Qualities

Each of the general properties are used to display or set the specified property. For instance, if you select a block of multiline text and then select **Color**, the color property will be highlighted, and the edit arrow will be displayed. Selecting the edit arrow will display the color palette, allowing the color of the text to be altered. Each of the other general properties of the text can be altered using the same procedure.

Specific Text Properties

Each of the multiline text features that are controlled in the **Multiline Text Editor** dialog box can be controlled using the **Properties** palette. To alter the current **Style**, select the text to be altered. With the text highlighted, selecting **Style** will display a listing of the current styles. Selecting the name of the new style will automatically update the selected text to the indicated style. Each of the other properties can be altered using the same procedure. Other text specific properties include the following:

Contents—The edit box specifies the selected text string. Only a portion of the text will be displayed for long text strings.

Style—Specifies the style name of the selected text body and allows a new style to be specified. Selecting the **Text Style** button on the **Text** toolbar can also alter the style of existing text.

Annotative—Assigns or removes annotative properties to the selected text.

Justify—Specifies the attachment point of the text body using the nine common justification options for placement.

Height—Specifies and alters the height of the selected text. Text can be enlarged or reduced by altering the current value. Selecting the **Scale Text** button on the **Text** toolbar can also alter the scale of existing text.

Rotation—This option specifies the rotation angle of the text body. Enter the desired rotation angle by keyboard and press ENTER. The text will be rotated around the current justification point.

Width factor—Allows the width of multiline text to be edited. With the text highlighted, a scale factor can be entered to lengthen or shorten the text. **Obliquing**—Alters the forward and backward slant of text. A positive angle value will slant text to the right, and a negative angle value will slant text to the left.

Text alignment x, y, and z—Allows the alignment in the specified direction to be altered.

Note: Keep in mind that the **Properties** palette can be displayed beside the drawing area and used to alter more than just multiline text. With the palette open, the Properties listing will be altered as each new object is selected. The name of the listed object will be displayed in the object edit box located at the top of the dialog box.

USING THE SPELL CHECKER

The **Spell** command can be used to correct spelling errors in text, multiline text, and text placed with the Leader and Attdef commands. Access the command by using one of the following methods:

- Click the **Spell Check** button on the **Text** toolbar.

- Click the **Spell Check** button on the **Text** dashboard.
- Type **SP** ENTER at the Command prompt.
- Type **'SP** ENTER at the Command prompt.
- Select **Spelling** from the **Tools** menu.

Each option will open the **Check Spelling** dialog box shown in Figure 15.46. The **Where to Check** edit box allows the limits of the spell check to be specified. Options include **Entire drawings**, **Current Space/layout** and **Selected objects**. If no spelling errors are contained in the selected text, AutoCAD will display a message verifying that the spell check is complete. Clicking **OK** or pressing ENTER will return you to the drawing.

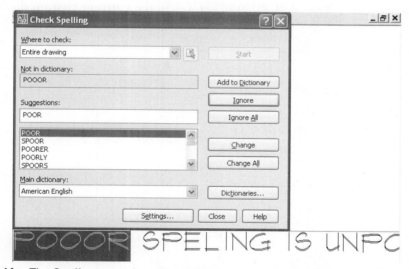

Figure 15.46 *The **Spell** command can be used to check text and multiline text. The current word "pooor," four suggestions, and the current context of the word are displayed.*

Once the dialog box is displayed, the **Where to Look** option will dictate how the command is executed.

- If the search includes the **Entire Drawing** or **Current Space/layout**, click the **Start** button to activate the command.
- If the search is limited to **Selected objects**, click the **Select Text Objects** button.

If errors are found in the selection set, the line of text will be displayed in the **Not in Dictionary** display. A possible listing of alternatives can be found in the **Suggestions** box. If the current word is correct, select the **Add to Dictionary** or the **Ignore** button, and the spell checker will move to the next word identified by the checker. If the **Add to Dictionary** option is selected, the highlighted word will be added to the dictionary for future reference. If you would like to use one of the suggested words, move the cursor to highlight the word and then select **Change** or **Change All**. **Change** will only alter the selected word. **Change All** will alter each occurrence of the specified word within the search parameters. The editor will proceed through your text until AutoCAD displays the "spell check complete" message.

CHANGING THE DICTIONARY

Selecting the **Dictionaries** button will produce the dialog box shown in Figure 15.47. The **Main dictionary** box will allow you to change the language that is stored in the dictionary. The **Custom dictionary** can be used to add words, acronyms, client names or vendor names specific to your area of expertise to the dictionary. When creating a custom dictionary, use the same guidelines that were presented in chapter 2 for naming drawing files. An extension of .CUS must be added after the name. To enter a custom name for an architectural dictionary, type ARCHWORD.CUS in the **Current custom dictionary** box.

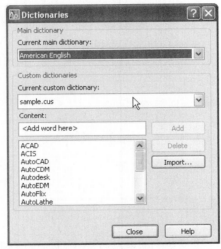

Figure 15.47 *A separate dictionary can be created to contain words specific to your field.*

To add words to a specific dictionary, select the **Custom dictionary words** box. Names of up to 63 characters can be entered in the box. Multiple names can be entered at one time when separated by a comma. Selecting the **Add** button will add the names to the existing list of names contained in the dictionary.

SETTINGS

Clicking the **Settings** button will display the **Check Spelling Settings** dialog box similar to the example in Figure 14.48. This box can be used to specify which text will be examined by spell check. By default, text included as part of dimensions and block attributes will be examined. When a drawing contains referenced drawings (see chapter 22) the **External References** option can be used to extend the spell check to these drawings. In the **Options area,** limits can be placed on spell check related to specific types of words to be examined.

Figure 15.48   *The **Check Spelling Settings** dialog box can be used to specify which text will be examined by spell check.*

FINDING AND REPLACING TEXT

The **Find** command is another tool that can be used to edit text. It functions in the same manner as **Find and Replace** on the **Options** menu of the **Text Formatting** editor. To start the command, use one of the following methods:

- Right-click and choose **Find** from the shortcut menu.
- Select the **Find and Replace** button on the **Text** toolbar.
- Select the **Find and Replace** button on the **Text** dashboard.
- Select **Find** from the **Edit** menu.
- Type **FIND** ENTER at the Command prompt.

Each choice will produce the **Find and Replace** dialog box similar to Figure 15.49. The **Find and Replace** dialog box can be used to select, find, replace and zoom to text created by the **Text** or **Mtext** command. The dialog box can also be used to edit dimension text, block attribute text, hyperlink descriptions, and hyperlinks. Each of these four types of text will be introduced in later chapters. The command can be used in a method similar to the **Find/Replace** tab of the **Multiline Text Editor**. To find text using the dialog box, enter the text or text string that you would like to find in the **Find text string** edit box. The search field can be the entire drawing or limited to the current text selection. Once the text string and the search field are defined, select the **Find** button. If the text is found, it will be displayed in the **Search results** window. Choosing the **Zoom to** button will display the selected text on the screen.

1. PROVIDE SOLID BLOCKING
   BETWEEN ALL FLOOR TRUSSES.
2. ALL FLOOR JOISTS TO BE 2 × 10
   D.F.L. #2 UNLESS NOTED.
3. JOISTS LONGER THAN 10'-0"
   MUST BE BLOCKED.

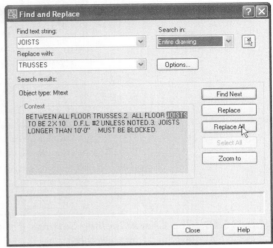

Figure 15.49 *The **Find and Replace** dialog box can be used to select, find, replace, and zoom to text created by the **Text** and **Mtext** commands.*

In addition to locating specific text in a drawing, the **Replace with** edit box can be used to replace a word or phrase. After entering the text to be found, enter a phrase to be used as the replacement text in the **Replace with** edit box. Choosing the **Replace** button will replace the specified text with the replacement text. Choosing the **Replace All** button will replace each usage of the selected text or phrase.

REFERENCING TEXT TO OBJECTS WITH MULTILEADER LINES

The **Multileader** command can be used to place a leader line to connect notes or dimensions to a specific portion of the drawing. Figure 15.5, 27a, and 27b show examples of the use of leader lines. Leaders are used to connect a note to a specific portion of a drawing and are generally placed at any angle except horizontal or vertical. Some architectural offices use horizontal and vertical leaders with a custom arrow to duplicate manual practice. A horizontal leg is placed at the end of the leader line to tie the text to the leader. The horizontal leg should extend from the left side of the first line of text or from the right side of the last line of text. Figure 15.50 shows common methods of placing leader lines.

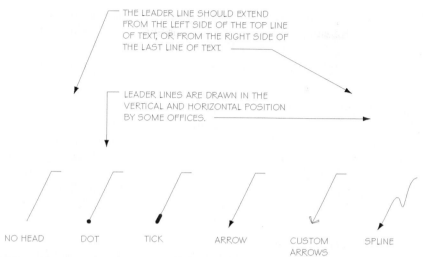

THE LEADER LINE SHOULD EXTEND FROM THE LEFT SIDE OF THE TOP LINE OF TEXT, OR FROM THE RIGHT SIDE OF THE LAST LINE OF TEXT.

LEADER LINES ARE DRAWN IN THE VERTICAL AND HORIZONTAL POSITION BY SOME OFFICES.

NO HEAD   DOT   TICK   ARROW   CUSTOM ARROWS   SPLINE

Figure 15.50 *Common methods of placing and terminating leader lines.*

USING THE MLEADER COMMAND

The **Mleader** command should be used when local notes need to be referenced to the drawing, or when you want to control the width of the space where the notes will be placed. Experienced AutoCAD users will be familiar with the Qleader command, which is still available. The Mleader provides so many other options that the use of the **Qleader** command is no longer recommended. To start the **Mleader** command, use one of the following methods:

- Click the **Mleader** button on the **Multileader** toolbar.
- Click the **Mleader** button on the **Multileader** dashboard.
- Type **MLD** ENTER at the Command prompt.
- Select **Multileader** from the **Dimension** menu.

Each method will produce the prompt:

```
Specify leader arrowhead location or [leader Landing first/
    Content]:
```

AutoCAD is waiting for you to make a decision. Options include:

- Placing the arrowhead first.
- Placing the leader landing (the horizontal tail) first.
- Placing the text first.

Your choice will depend on how you think, and what material has already been supplied. If you're an experienced AutoCAD user, placing the arrowhead first is most likely what you're familiar with. This method allows you to place the arrow on the material to be described and then extended to the note. Selecting the **leader landing** option allows the leader to be started at a note, and extended to the object. The **Text first** option allows a note to be created using the **Mtext Text Formatting** tools, and then a leader line extended to the object. Each option will be explored once creating a simple leader line has been created.

Connecting a note to the Object

In its simplest form, a leader line is used to connect an object to a note that has been placed using the Text or Mtext command. In Figure 15.51 most of the notes were placed with the text command and then referenced to the object using a leader line. Notes with special symbols were placed using the Mtext command. The welding symbol in the left corner was placed using the Mleader command. The lines below (the fillet weld symbol) and at the end of the leader line were added to the leader line using the Line command. The command sequence to place a Mleader (arrow head first) to connect a note to an object is:

```
Click the Mleader button.
Specify leader landing location or [leader arrowHead first/
       Content first/ Options] <leader arrowHead first>:
```

The default option will depend on the last option used. To place the arrow first option, use the following steps to complete the command sequence:

```
Specify leader landing location or [leader arrowHead first/
       Content first/ Options] <leader arrowHead first>:H ENTER
Specify leader arrowhead location or [leader Landing first/
       Content first/Options <leader Landing first>: Select the
       desired arrow head location
Specify leader landing location: Select the desired location
```

As the landing location is selected, the leader tail is added automatically. And the text formatting tools are displayed. Since the leader was used to connect to existing text, select **OK** to close the toolbar. If the command is repeated, the **arrowHead first** option will remain the default.

 Note: Although most new leaders should be placed with the Mleader command, older drawing versions will contain leaders placed with Qleader (quick leader) command. This command is still available in AutoCAD 08, but because of the annotative features available through the current release, Mleader should be used in place of Qleader.

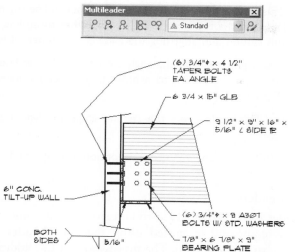

Figure 15.51   *In its simplist form, a leader is used to connect a note to an object.*

Exploring Options

Selecting the **Options** setting at the command prompt provides a list of leader controls. Options include:

Leader type—This option controls the type of leader to be used and includes straight, spline, and none.

Leader landing—This option controls the length of the leader landing. Two prompts will be displayed if this option is selected.

- The first prompt is to decide if a landing will be provided, and provides a [Yes/No] option.

- If **Yes** is provided at the first prompt, a second prompts is provided to specify the length of the leader landing. Most professionals use a landing of between 1/8" and 1/4" long.

Content type—This option specifies the type of content that will be used with the leader. Options include:

- **Block**—This option will specify a block that is stored within the drawings to associate with the leader line to be created. This is an excellent option to use with keyed notes. Blocks available with the command will be explored as **Mleader Styles** are explored.

- **Mtext**—This option will display the **Mtext Text Formatting** tools for placing text at the leader landing. This option should be used when a note has not been previously displayed.

- **None**—This option will place nothing at the end of the leader landing. This option should be used when the leader is placed to reference existing notes.

Maxpoints—This option allows the maximum number of line segments for the leader line to be specified. Notice the top note in Figure 15.51 has three line segments. The line was created using this option and adjusting the allowable number of line segments.

First Angle—This option will set the angle for the first leader line segment to a specified angle. This option should be used to achieve parallel leader lines similar to lower two leader lines in Figure 15.51.

Second Angle—This option sets the second angle for leader lines with multiple line segments.

Exit Options—This option will exit the option prompt and restore the MLEADER prompt.

Adding Additional Leaders

Additional leader lines can be added to an existing leader line using the **Mleaderedit** command. This command is started using the **Add Leader** button found on the **Multileader** toolbar or dashboard. The process to add two additional leader lines to an existing leader line is:

```
Click the Add Leader button.
Select a multileader: Select the existing leader line to receive
     additional leader lines
1 found
Specify leader arrowhead locations: (specify the location of the
     first leader line to be added)
Specify leader arrowhead locations: (specify the location of the
     second leader line to be added) Specify leader arrowhead
     locations: ENTER
```

Figure 15.52 shows the additional leader lines.

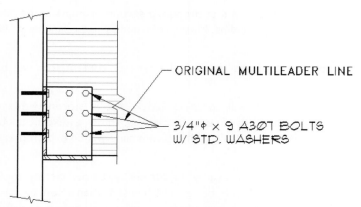

Figure 15.52   *Additional leader lines can be added to an existing multileader line using the **Add Leader** button. Once the command is entered, a prompt is provided to specify the original leader line followed for prompts for where to locate additional leader arrowheads. In this detail, the upper and lower leader lines were added.*

Removing Multileader Lines

The process for removing unwanted multileader lines is similar to the process for adding Mleader lines. Leaders are removed using the Mleaderedit command found on the **Multileader** toolbar or dashboard. Remove multileader lines using the following sequence:

```
Click the Remove Leader button.
Select a multileader: Select the existing leader line group to
     have a leader line removed
1 found
Specify Leaders to remove: (Select the line to be removed).
Specify Leaders to remove: ENTER
```

Aligning Multileader References

Many complicated drawings use keyed notes to keep the drawing as uncluttered as possible. Keyed notes are notes that are written as general sheet notes and represented on the drawings with a reference symbol containing a letter or number. The Mleaderalign command can be used to place key symbols in alignment. This command gives the technician the freedom to insert reference symbol into the drawing and then to make them look pretty once all of the material has been specified. Figure 15.53 shows an example of aligned keyed notes. Start the command by selecting the **Align Multileaders** button from the **Multileader** toolbar or dashboard. Symbols can be aligned using the following command sequence:

```
Click the Align Multileaders button.
Select multileaders: Select the existing leader lines using any
     selection method.
1 found, 4 total
Select multileaders: ENTER
Current mode: Use current spacing
Select multileader to align to or [Options]: Select the leader
     line to serve as the base.
Specify direction: Select a point to indicate an alignment line
     for the new location.
```

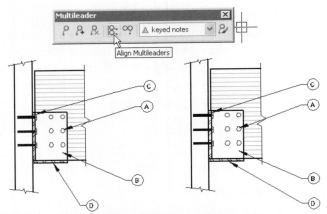

Figure 15.53 *The reference symbols for keyed notes can be inserted into a drawing using the* ***Multileader*** *command. The left portion of this figure shows the symbols placed in the drawing with little consideration for placement. The right side of the drawing shows the relocation of the symbols using the* ***Mleaderalign*** *command. The corresponding note for each symbol has been omitted from this drawing for space considerations.*

Collecting Multileaders

In addition to easily providing keyed notes, AutoCAD allows note symbols to be collected into groups using the Mleadercollect command. Figure 15.54 shows an example of collected keyed notes. Start \ command by selecting the **Collect Multileaders** button from the **Multileader** toolbar or dashboard. Symbols can be collected using the following command sequence:

```
Click the Collect Multileaders button.
Select multileaders: Select the leader lines associated with
     symbol A.
1 found
Select multileaders: Select the leader lines associated with
     symbol B.
1 found, 2 total
Select multileaders: Select the leader lines associated with
     symbol C.
1 found, 3 total
Select multileaders: Select the leader lines associated with
     symbol D.
1 found, 4 total
Select multileaders: ENTER
Specify collected multileader location or [Vertical/ Horizontal/
     Wrap]
<Vertical>: (Specify the direction of the collection) H ENTER
Specify collected multileader location or [Vertical/ Horizontal/
     Wrap]
<Horizontal >: (Specify the start point of the selection).
```

As the second point is specified, the symbols will be collected based on the location and placed in the order in which they were selected.

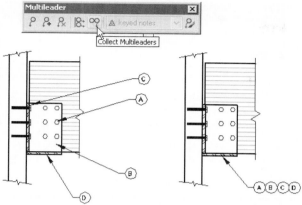

Figure 15.54   *The left portion of this figure shows the reference symbols for keyed notes placed in the drawing using the Multileader command. The right side of the drawing shows the relocation of the symbols using the **Mleadercollect** command. The corresponding note for each symbol has been omitted from this drawing for space considerations.*

CREATING MULTILEADERS WITH STYLE

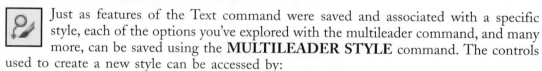

Just as features of the Text command were saved and associated with a specific style, each of the options you've explored with the multileader command, and many more, can be saved using the **MULTILEADER STYLE** command. The controls used to create a new style can be accessed by:

- Clicking the **Multileader Style** button from the **Multileader** toolbar.
- Clicking the **Multileader Style** button from the **Multileader** dashboard.
- Selecting **Multileader Style** from the **Format** menu.
- Typing **MLS** ENTER at the command prompt.

Each option will produce the **Multileader Style Manager** dialog box shown in Figure 15.55.

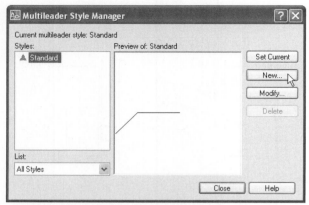

Figure 15.55 *Clicking the **Multileader Style** button displays the **Multileader Style Manager** dialog box that is used to create new, and modify existing, Mleader styles.*

CREATING A NEW STYLE

Clicking the **New** button on the **Multileader Style Manager** will display the **Create New Multileader Style**. A display similar to Figure 15.56 is used to name a style and start the process of defining the qualities of the new style.

Figure 15.56 *Clicking the **New** button on the **Multileader Style Manager** will display the **Create New Multileader Style** display.*

Naming a Style

As with naming other features, the name given to a multileader style should be descriptive of the contents. The **New Style Name** edit box is used to define the style name. For this example, the new style name will be STRAIGHT. Additional styles that should be considered include SPLINE and WELDING. Other styles may be desired for different styles of leader terminators.

Defining a Starting Point

The **Start with** edit box is used to define a starting point for defining the new style. By default, the existing style of STANDARD will be used as a base. As additional styles are defined, they will be added to this list to serve as a base for additional styles.

Additional Settings

Once the two major settings are set, click the **Annotative** button to make future leaders annotative. Options now consist of **Continue**, **Cancel** and **Help**. Clicking the **Cancel** button will end the process and return the drawing display. Selecting **Help** will provide command-specific help. Clicking the **Continue** button will display the **Modify Multileader Style** dialog box shown in Figure 15.57.

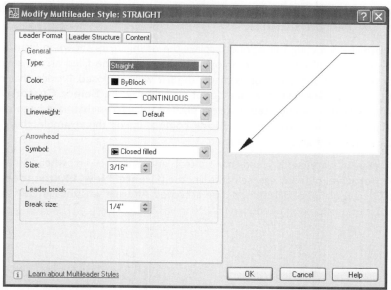

Figure 15.57 *Clicking the **Continue** displays the controls for defining a multileader style.*

DEFINING A STYLE

The three tabs of the **Modify Multileader Style** dialog box are used to set the defining qualities of a multileader style. Tabs for defining and modifying an existing style include **Leader Format**, **Leader Structure**, and **Content**. A fourth area is the preview box that adjusts the display as controls are adjusted.

Leader Format Tab

The **Leader Format** tab can be seen in Figure 15.57. It is divided into the area of **General**, **Arrowhead**, and **Leader break**.

General As the name implies, the **General** tab is used to define the general qualities of a leader line. Options include Type, Color, Linetype, and Lineweight.

- **Type**—Controls the shape of the leader line. Options include Straight, Spline, and no leader line. For this example the Straight method will be used.

- **Color**—Allows the color of the leader line to be defined. For this example, the by **ByLayer** option will be used.

- **Linetype**—Allows the linetype of the leader line to be set. Typically the settings will be either continuous or by layer. For this example the **ByLayer** option will be used.

- **Lineweight**—Allows the lineweight of the leader line to be set. Selecting this edit box will display the lineweight menu allowing the lineweight of the leader to be controlled. Use the lineweight that has been selected for thin lines for your drawing template. For this example, the **Default** setting will be used.

Arrowhead   This portion of the **Leaders** tab controls the shape and size of the terminator at the object end of the leader line.

- **Symbol**—Selecting this option displays the terminators shown in Figure 15.58 that can be used to replace the filled arrowhead. To replace the filled arrow, select the desired terminator. As the selection is made, the list will be closed. Notice that the last option in the list is **User Arrow**. This option will display the **Select Custom Arrow Block** dialog box that can be used to select predefined arrows that have been saved as a block. Chapter 18 will introduce the **Block** command. For this example, the default value of **Closed Filled** arrow will be used.

- **Size**—This option is used to control and set the arrowhead size. Because the **Annotative** button was selected in the **Create New Multileader Style** display, the default value of 3/16" should be accepted. Alternative sizes can be selected or typed into the edit box as needed.

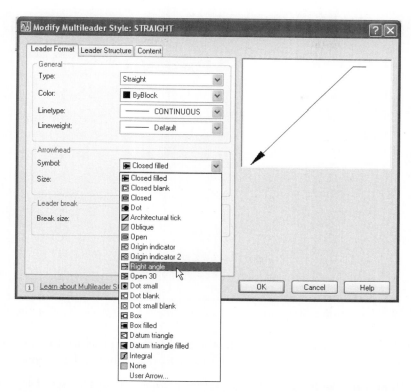

Figure 15.58 *Clicking the **Symbol** edit arrow displays options for leader line terminators.*

Leader break The final area of the **Leader Format** tab to be explored is the leader break size. This option controls the display of the break between the leader line and the text, if a dimension is used. The default setting of 1/8" is acceptable for most uses and will be accepted for this example.

Leader Structure Tab

This portion of the display defines qualities that will govern the shape of the leader. The tab can be seen in Figure 15.59. Major areas of the tab include the **Constraints**, **Landing Settings**, and **Scale**.

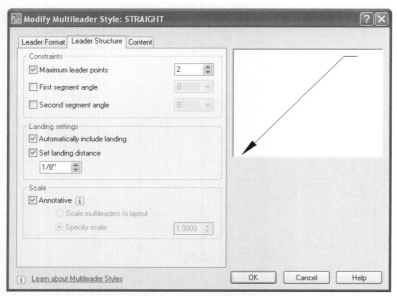

Figure 15.59 *The Leader Structure tab provides controls for defining the shape of the leader.*

Constraints The **Constraints** portion of the lead settings provides for the control of the leader segments and the angle that will be used to place the leader segments.

- **Maximum leader points**—This setting provides options for specifying how many line segments will be used as the leader line is placed. The default setting of 2 is generally fine for referencing most notes to a drawing. In the default setting of 2, you will be prompted to place text after specifying the two lines formed by the selected points. A higher number of segments are often required for placing welding symbols.

- **Angle Constraints**—Settings are available for the first and second angle settings, allowing the angle of each segment of the leader line to be controlled. By default, these values are **OFF**, allowing the angle to be set by the cursor. Default values for each option include 90, 45, 30, 15, and 0 degrees. Selecting an angle provides uniformity in appearance, but it might make placing text more challenging.

Landing Settings This area of the dialog box allows the tail of the leader line to be controlled. Options include:

- **Automatically include landing**—By default this option is active so that a horizontal landing will be provided with the leader line segment.

- **Set landing Distance**—With this option active, the edit box is activated so that a length for the tail can be specified. Use the arrows to scroll through the options or enter the desired size in the edit box. A size ranging from .125" to .25" is common for most uses, and a length of about 1" can be set for leader lines to be used with welding or elevation symbols.

Scale This option contains settings to control annotative and manual scale settings. With Annotative in the **ON** setting, other options in this area will be inactive. For most uses, take advantage of the use of the Annotative feature. With Annotative in the **OFF** setting, two options are available.

- **Scale multileaders to layout**—Layouts are used to display multiple drawings on a sheet at varied scales. Selecting this option will adjust the scaling of leader lines to the layout where they are displayed. Chapter 23 will introduce the creation and use of layouts.

- **Specify Scale**—This option controls the scaling for leaders that are displayed in model space.

Content Tab

This tab contains control areas for the text and for the connection of the leader line to the text. The **Multileader type** setting is used to define what will be placed at the end of the leader line tail. The options in this setting include **Mtext, Block** or **none**. Each setting functions in the same manner as when applied to the creation of a single leader line earlier in this chapter. In the dialog box, the selection will affect the display of other settings.

- In the default setting of **Mtext,** the display of the **Contents** tab will be similar to Figure 15.60.

- If the **None** option is selected, no text will be placed with the leader line, and the options in the **Text options** and **Leader connections** portions of the tab will be removed from the display.

- If the **Block** option is displayed, the **Contents** tab will be similar to Figure 15.61.

With the **Mtext** option active, the **Text options** and **Leader Connection** areas of the tab are active. Options include:

Controlling Mtext Options This area of the tab controls the appearance of text that will be placed with the Mleader command. Options include:

- **Default Text**—This option allows default text such as "TOP OF WALL" to be placed by the leader line. Clicking the [...] button will display the **Mtext In Place Editor** that can be used to define the text to be used.

- **Text style**—This option allows a predefined text style to be used. Text styles defined with the Text or Mtext commands can be selected with this option, or the command can be exited,\ and a new style created.

- **Text angle**—This setting controls the rotation angle of text placed with the Mleader command. Options included:

  - **Always Right-reading**

  - **As inserted**

  - **Keep horizontal**

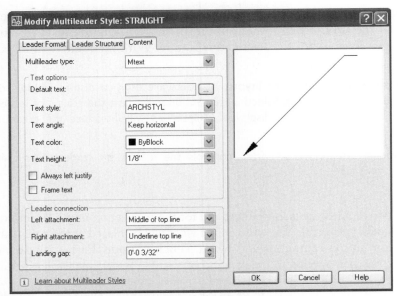

Figure 15.60 *The **Content** tab contains the settings for controlling the text placed with the **Mleader** command.*

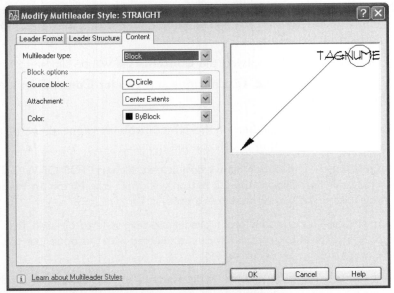

Figure 15.61 *The display on the **Content** page is altered based on the selection in the **Multileader type** edit box. With **Block** as the active option, controls for **Mtext** will be hidden, and only the controls for blocks will be displayed.*

- **Text color**—Selecting this setting allows a specific color to be assigned to the leader rather than using the color associated with the layer the leader will be placed on. Generally it's best to keep this setting as **By Layer**.

- **Text height**—This setting allows the height of the text placed with the Mleader to be controlled. Typically text is 1/8" high.

- **Always left justify**—With this setting active, text will always be left justified.

- **Frame text**—With this option active, a box will be placed around the text similar to the note in Figure 15.62.

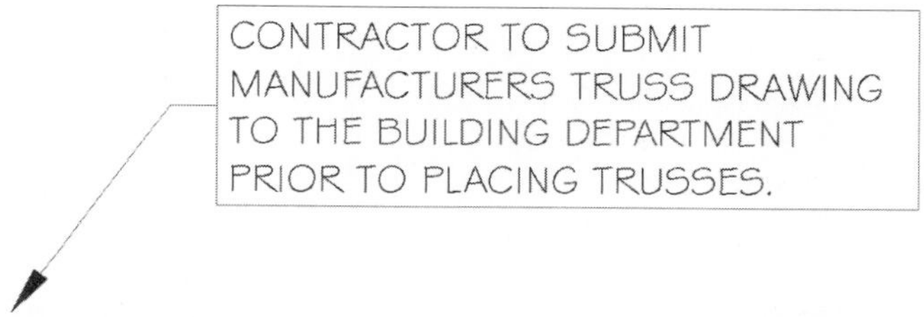

CONTRACTOR TO SUBMIT
MANUFACTURERS TRUSS DRAWING
TO THE BUILDING DEPARTMENT
PRIOR TO PLACING TRUSSES.

Figure 15.62 *A frame will be placed around the text added to the Mleader if the Frame text feature is active.*

Leader Connection   The settings in this area of the tab control the relationship of the Mleader to the text and provide options for underlining the text. Most offices prefer that when the leader is on the left side of the text, the Mleader tail should align with the middle of the top line of text. When the leader is on the right side of the text, the leader should be placed so that the leader tail aligns with the center of the bottom line of text.

Controlling Blocks inserted with Mleaders   If **Block options,** in the **Multileader type** area of the tab is selected, the **Content** display will resemble Figure 15.61. Options are the same as when the Mleader was created at the keyboard. Figure 15.54 shows Mleaders created using the circle option of the Block options.

Editing Mleaders

Selecting the **Modify** button in the **Multileader Style Manager** allows each of the Mleader features that were placed in a new text style to be altered. Once the **Modify** button is selected, each of the features can be altered using the exact methods by which they were created.

INCORPORATING TEXT INTO TEMPLATE DRAWINGS

Much of the text used to describe a drawing can be standardized and placed in a template drawing. Figure 15.63 shows an example of notes that an office includes as standard notes for all site plans. These notes can be typed once using the Mtext command and saved so that they do not have to be retyped with each use. The notes can then be either inserted into every site plan or nested in the template drawing on the SITE-ANNO layer. The layer can be frozen, thawed when needed, and then either the **Ddedit** or the **Properties** command can be used to make minor changes based on specific requirements of the job. Notice that several notes and each of the schedules include the text "xxx." These are notes that have been set up as blocks and stored with attributes that can be altered for each usage. Attributes will be discussed in later chapters.

GENERAL NOTES:
1. ENGINEER NOT RESPONSIBLE FOR LAND SURVEY OR TOPOGRAPHY.
2. ALL EXTERIOR SIGNS TO COMPLY W/ COUNTY SIGN ORDINANCE.
3. LOWER LEVEL FRAMED IN STEEL.
4. ALL STEEL COLUMNS TO BE 3" DIA. X 60" STEEL COLUMN FILLED W/ CONC.
   EMBEDDED 24" INTO GRADE.
5. SQUARE FOOTAGE OF LOT--00,000 SQ. FT.
6. SQUARE FOOTAGE OF STRUCTURE--00,000 SQ. FT. APPROXIMATELY.
7. 00.00% OF THE LOT IS COVERED.

UTILITY NOTES:
1. BALANCE OF WATER LINES TO BE SHOWN ON FLOOR PLANS.
2. CLEAN OUT @ ****

PLANTER NOTES:
1. ALL PLANTERS TO BE CONSTRUCTED W/ A SEPARATE PERMIT.
2. PLANTER ON WEST SIDE ALONG FLOYD MILLER BLVD. TO BE
   44' X 7.5' X 3' CONCRETE BLOCK W/ SIMILAR PLANTER IN THE.
   SOUTHWEST CORNER BY THE STAIRWAY.

PARKING NOTES:
1. PARKING TO BE 4" MIN. CONC. W/ 6" X 6"--4 X 4 WWM
   OVER 4" COMPACTED GRAVEL.
2. EVERY TWO PARKING SPACES TO HAVE A 6" X 6" WHEEL STOP W/
   (2) 1/2" x 12" STEEL DOWELS @ EA. WHEEL STOP.
3. PARKING SPACES TO BE MARKED WITH 4" WIDE WHITE STRIPES.

PARKING SPACES		
FULL-SIZE	9' X 20'	XX
COMPACT	7'-6" X 15'	XX
HANDICAPPED	12' X 20'	XX

	OCCUPANCY OF THE STRUCTURE	MAXIMUM ALLOWABLE FLOOR AREA (SQ. FT.)	TOTAL OCCUPANTS (FOR BLDG.)
FIRST FLOOR	XX	XX,XXX	XX
SECOND FLOOR	XX	XX,XXX	XX
THIRD FLOOR	XX	XX,XXX	XX

Figure 15.63 *Many offices have standard notes that are used for all site plans. These notes can be typed once and saved as a separate drawing or stored as a template so that they do not have to be retyped with each use.*

Many drawings, such as sections, contain basically the same notes that are displayed as local notes. Although sections for a tilt-up structure will be radically different from a section for a steel rigid frame, a template sheet can be developed for each type of construction. The actual drawing might look different, but the notes to specify standard materials are similar for most buildings using the same type of construction. Local notes can also be placed in the template drawing. These notes can be stored, moved to the desired drawing, thawed, moved into the needed position, and edited to greatly increase drafting efficiency.

CREATING TABLES

Tables are an important part of any set of drawings. The command offers many options and tools to insert or link information form software other than AutoCAD into an AutoCAD drawing. Rather than overwhelm you with all of the features, this chapter will introduce basic portions of the command, and assume that you'll rely on the Help menu for some of the more specialized features. Productivity in creating and editing data in tables is increased using the TABLE command. Access the command using one of the following methods:

- Click the **Table** button on the **Draw** toolbar.
- Click the **Table** button on the **Table** dashboard.
- Type **TABLE** at the command line.

Each method will open the **Insert Table** dialog box shown in Figure 15.64.

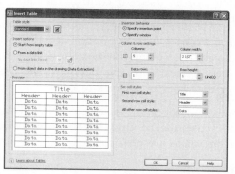

Figure 15.64 *The **Table** command can be used to organize information in construction documents. Select the **Table** icon from the **Draw** toolbar or the **Table** dashboard to open the **Insert Table** dialog box.*

EXPLORING THE INSERT TABLE DIALOG BOX

The Insert Table dialog box provides tools to easily create a table. Key areas of the display include controls for the style, insertion options, insertion behavior, rows and columns, cells, and the preview window.

Table Style

This option of the dialog box allows an existing table style to be used for a table that is being created. Currently only the **Standard** style is available, but as future tables are developed, their qualities can be saved in a style. Selecting the **Launch the Table Style dialog** button will show the **Table Style** display that can be used to define a new table style. Creating new styles will be explored shortly.

Insert Options

This set of options controls the methods that will be used to insert a table into a drawing. Options include:

- **Start from empty table**—This option can be used to create an empty table. Data in the table is placed once the table is created.

- **From a data link**—This option should be used if data from an external spreadsheet is to be placed in the table. Tables created in Excel are common types of material that can be added to AutoCAD tables.

  - **Launch Data Link Manager**—This edit box is inactive until the **From a data link** option is activated. When active, the **Launch Data Link Manager Dialog** button is active, and can be used to display the **Launch Data Link** dialog box.

- **Start from Data Extraction**—Selecting this option will allow the **Data Extraction** wizard to be launched.

Insertion Behavior

This portion of the dialog box is used to control the table insertion point within an AutoCAD drawing. Options in this area include:

- **Specify Insertion Point**—With this option active, the upper-left corner of the table will be used as the insertion point. The insertion point can be specified using the cursor or by entering coordinates. With this option active, the **Column width** setting is activated, allowing the width of columns to be set.

- **Specify window**—This option is used to control the location and the size of the table. With this option active, the controls for **Column width** and **Data rows** become inactive because the values are now controlled by the size of the window.

Column and Row Settings

This portion of the dialog box is used to control the display of cells and rows. These controls are active when the **Start from empty table** option is active. Options in this area include:

- **Columns**—This option controls the number of columns to be displayed in the table. A column icon is provided to illustrate the direction of a column. If the **Specify window** option is active and the width of columns has been defined, the column width is controlled by the number of columns and the size of the window.

- **Column width**—This option controls the width of each column. If the **Specify window** option is active, the column width is controlled by the number of columns and the size of the window.

- **Data Rows**—This option controls the number of rows to be displayed in the table. A row icon is provided to illustrate the direction of a row. If the **Specify window** option is active and the row height has been defined, the number of rows is controlled by the size of the window.

- **Row height**—This option controls the height of each row based on the text height and the cell margin. Both of these features are set as the table style is created. If the **Specify window** option is active, the height of each row is controlled by the number of rows to be created and the size of the window.

Set Cell Styles

This portion of the dialog box allows the cell style for rows to be defined. Three settings are provided for controlling the text in each row, and three options are provided for each of the row options. Row options include:

- **First row cell style**—By default the top row of the table will be displayed as titles.

- **Second row cell style**—By default the second row of the table will be displayed as headers.

- **All other row cell styles**—By default all rows of the table will be displayed as data.

Preview

The final portion of the display is the preview display. The preview area reflects the current style settings. Since no style has been created, the **Standard** setting is currently displayed.

CREATING A SIMPLE TABLE STYLE

The features that have just been explored can be used to create a simple table. The **Table Style** command can be used to create tables with varied features. Use the following steps to create a simple table:

1. Use the **Insert Table** dialog box opens as shown in Figure 15.64 to adjust the column and row settings.

2. Select **OK** to close the dialog box.

3. At the prompt for an insertion point, select the desired location in the drawing for the table placement.

4. Use the **Text Formatting** toolbar to set any text variables that may be desired. For this example the font was altered.

5. Type in the title for the table and press TAB to maneuver around and enter the desired header labels. Figure 15.65 shows an example of a door schedule created using the **Table** command.

DOOR SCHEDULE			
SYMBOL	SIZE	QUANITY	TYPE
1	3'-0" 6'-8"	1	R.P. W/ SIDE LITES
2	2'-8" × 6'-8"	1	S.C. / S.C.
3	2'-8" × 6'-8"	1	HOLLYWOOD
4	2'-8" × 6'-8"	5	H.C. 6 PANEL
5	2'-6" × 6'-8"	7	H.C. 6 PANEL
6	2'-4" × 6'-8"	4	H.C. 6 PANEL
7	6'-0" × 6'-8"	2	12 LITE FRENCH
8	6'-0" × 6'-8"	2	BI- FOLD
9	5'-0" × 6'-8"	3	BI- PASS
10	9'-0" × 7'-0"	3	OVERHEAD

Figure 15.65 *The table command can be used to create door, window, finish, hardware, and other types of tables for organizing information on construction drawings. The Style command was used to set the text font for the entire table. Once the table was created with uniform column width, the size of some columns was reduced to save space.*

ADDING A COLUMN/ROW TO A TABLE

Once a table has been created, columns or rows can still be added to the table. Use the following procedure to add a new column to an existing table.

1. Click on the column/row cell next to where you would like the new cell to reside. The cell will highlight with grips activated.

2. Right-click and select **Columns** or **Rows**.

   • If you select **Columns**, you can specify if you want the new column to be inserted to the right or left of the column you highlighted. If the **Delete** option is selected, the highlighted column will be removed.

   • If you select **Rows**, you can specify if you want the new row to be inserted above or below the row you highlighted. If the **Delete** option is selected, the highlighted row will be removed.

3. Make your selection, and the new column or row will be automatically added.

INSERTING A BLOCK IN A TABLE CELL

A useful feature in the **Table** command is the ability to insert blocks into a table cell. Blocks are repetitive symbols that will be introduced in chapter 18. This can prove to be extremely efficient in creating a legend, which displays all the symbols used in the drawing set. Creating a legend and updating it to meet project requirements creates standardization within the team. Use the following steps to insert a block into a table cell:

1. Select the desired cell; the cell will highlight with grips activated.

2. Right-click and select **Block** from **Insert** on the shortcut menu.

3. The **Insert a Block in a Table Cell** dialog box opens; scroll down the **Name** field and select the block or click the **Browse** button to select a block.

4. Adjust the Cell alignment and scale factor or keep the default **AutoFit** and **Rotation angle**. Select **OK** to close the dialog box and return to drawing session. The block will automatically be inserted in the table. Figure 15.66 shows the **Insert a Block in a Table Cell** dialog box.

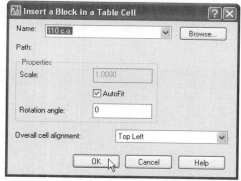

Figure 15.66   *The cell alignment, scale, and rotation angle can be adjusted in the* **Insert a Block in a Table Cell** *dialog box.*

MODIFYING TABLE PROPERTIES

Table properties can be modified using the **Properties** palette or by using the **Text Formatting** toolbar. To access the **Properties** palette, select the individual cell, right-click, and select **Properties** from the shortcut menu. The **Text Formatting** toolbar is accessed by double-clicking the select button while in an individual cell. Pressing the TAB or arrow keys moves the cursor across the cells when modifying the data. The grip command allows the table location, column width, and row height to be modified. The column width can be changed without modifying the overall width of the table by pressing CTRL while moving the grips. Data can also be imported from Microsoft Excel into AutoCAD as a Table object. To paste a table from Microsoft, you need to copy it to your clipboard. In AutoCAD, select **Paste Special** from the **Edit** menu. The **Paste Special** dialog box will open and prompt you to paste the table as a Microsoft Works Sheet or Chart, Picture, Bitmap, AutoCAD Entities, Image Entity, or Text. Equally, if tables are created within AutoCAD, it's possible to export the table data as a CSV format.

CREATING A TABLE STYLE

The **Tablestyle** command allows common table features to be set and saved as templates for future projects. Utilizing the **Table Styles** command will save valuable time and create CAD uniformity. Instead of selecting the default "standard" table, you'll be able to select a predefined table with all the settings already in place. Using the **Table Style** command is similar to using the **Text style** or **Mleader style** commands. A display similar to Figure 15.55 is displayed providing the options of **Set Current**, **New**, or **Modify**. Because you currently only have the **Standard** style, click the **New** button. This will display the **Create New Multileader Style** dialog box. This display is similar to the display used to name multileader styles shown in Figure 15.56. Assign a name that will clearly represent the table style, and then click the **Continue** button.

Once **Continue** is selected, the **New Table Style (new name)** dialog box is displayed and will appear similar to Figure 15.67. Within this dialog box, the cell and border properties can be adjusted using the **General**, **Text,** and **Borders** tabs. Cell properties include text style, text height, text color, fill color, and alignment. Border properties include grid, visibility, lineweight, and color.

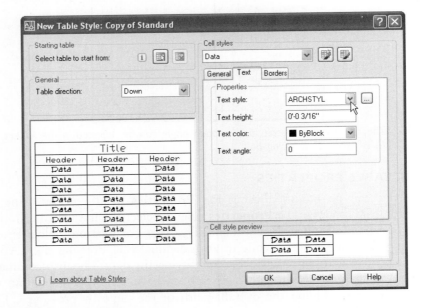

CHAPTER 15 EXERCISES

DIRECTIONS

Use capital letters to complete the follow exercises unless noted.

1. Start a new drawing and type the following text using left justified text with a height of .125" and a TrueType font:

All sheathing should be 1/2" standard grade 32/16-ply interior type with exterior glue. Lay perpendicular to rafters and stagger all joints. Nail with 8d's @ 6" o.c. at edge, blocking and beams, and 8d's @ 12" o.c. @ field. Use common wire nails.

Save the drawing as **E-15-1.**

2. Open a new drawing and make ROMANS the active font, with an obliquing angle of 10° and a height of 0.125":

   Use 8 ×8 ×16 grade "A" concrete block units with a triple score. Use # 5 diameter rebar @ 24" o.c. each way solid-grout all steel cells.

   Save the drawing as **E-15-2.**

3. Open a new drawing and make CityBlueprint the active font and a height of 0.125". Use a rotation angle of 35°. Write the following note:

   5 1/8" ×13 1/2" exposed glu-lam ridge beam f:2200

   Save the drawing as **E-15-3**.

4. Start a new drawing and make ROMANT the active font with a height of .125". Use a width factor of 1.25. Use the center option to type the following text:

   ATTENTION ALL CADD

   OPERATORS.

   DUE TO

   BUDGETARY PROBLEMS

   YOUR MOM DOES NOT WORK HERE.

   PLEASE CLEAN UP

   YOUR OWN MESS.

   Save the drawing as **E-15-4**.

5. Start a new drawing and create a style for text using **ROMANS** as the font with a height of 0.125". Use a width factor of 1.00. Create a second style named **TITLES** using **ROMANT** as the font, with a height of 0.25 and a width factor of 1.5. Assign each style a layer named for the style and provide a separate color for each. Use **Mtext** to create the following text:

   INSULATION NOTES

   Insulate all exterior walls with 5 1/2" high-density fiberglass batt insulation, R-21 min.

   Insulate all flat ceiling joists with 12" R-38 batts (no paper facing required).

   Insulate all vaulted ceiling with 10 1/4" high-density, paper-faced fiberglass R-38 min. with 2" air space above.

Insulate all wood floors with 8" fiberglass batts, R-25 with paper face. Install plumbing on heated side of insulation.

CAULKING NOTES

Caulk the following openings with expanded foam or backer rods. Elastomatic, copolymer, siliconized acrylic latex caulks may also be used where appropriate.

Any space between window and door frames.

Between all exterior wall sole plates and plywood sheathing.

On top of rim joist prior to plywood floor application.

Wall sheathing to top plate.

Joints between wall and foundation.

Joints between wall and roof.

Joints between wall panels.

Save the drawing as **E-15-5**.

6. Open drawing **E-15-5** and make the following changes to the insulation notes.

   Change wall insulation to reflect 6" fiberglass batt insulation, R-19 min.

   Change all flat ceiling to 12" R-38 batts with paper facing.

   Change all vaulted ceiling to 10" paper-faced fiberglass R-30 min. with 2" air space above.

   Change wood floors to 6" fiberglass batts, R-19 with paper face.

   Save the drawing as **E-15-6**.

7. Open drawing **E-11-9**. Use the Text command with an architectural style font and label the drawing. Complete the drawing and add all hatch patterns. Place the required text box on a layer separate from other text. Use the **Mleader** command with straight leaders and terminator, other than the filled arrowheads. Save the drawing as **E-15-7**.

8. Open drawing **E-13-10** and label the materials that have been drawn using a suitable font, either the Text or Mtext command, and attach the text to the proper material using spline leader lines. Change the text to use **Qtext** to display the text before saving the drawing. Create a separate layer for all text and save the drawing as **E-15-8**.

9. Open drawing **E-13-9**. Create separate layers, styles, and fonts for titles and text and label the drawing using Mtext. Use spline leader lines to connect text to the object being described. Use arrowheads other than filled arrows. Save the drawing as **E-15-9**.

10. Open drawing **FLOOR 14** and design a 2" wide title block along the right side of the page. Design space for page number, date, revision date, client information, and designer information. Since you are the designer, provide a company name and logo

and your name, address, and phone number. Create new layers for the title block lines and text that will remain constant and a separate layer for text that will vary for each job. Use a minimum of three different text fonts. Save the drawing as **FLOOR 14**.

11. Start a new drawing and type the following note: PROVIDE A 5° +/–. 5° BEND AT ALL #6 DIA. STEEL REINFORCING. STEEL TO BE WITHIN 1% OF SPECIFIED LENGTH. Save it as **E-15-11**.

12. Start a new drawing and create a line of text and make a copy of the text. Mirror one of the lines using a variable of 1 and one of the lines with a variable of 0. Save the drawing as **E-15-12**.

13. Start a new drawing and use the **Mtext** command to make a list of a minimum of three guidelines that describe the creation of multiline text. Use a second text style and justification point and type a second list of guidelines for describing text created with the **Text** command. Save the drawing as **E-15-13**.

14. Open drawing **E-15-13** and make a copy of each set of text. Edit the new text to add additional words in the existing sentences and change the color, style and width of text. Save the drawing as **E-15-14**.

15. Open a new drawing and create a window schedule showing at least 5 different windows. Create columns for symbol, size, window type, and quantity. Save the drawing as **E15.15**.

16. Open **E15.15** and edit the sizes of two of the existing windows. Edit the table to have two additional windows. Save the drawing as **E15.16**.

17. Open **E15.16** and save the table as part of a template drawing. Modify at least three of the text qualities. Save the drawing as **E15.17**.

18. Open any existing drawing and label at least five different features. Make a copy of this object and rename the features using keyed note symbols with a letter in a symbol. Save the drawing as **E15.18**.

19. Open drawing E15.18 and align the key symbols. Save the drawing as **E15.19**.

20. Create a table for a door, window, or finish schedule. Make up the data to fill the columns. Your table must contain at least a title row, a header row, and a minimum of 5 rows of data. Assign a minimum of 4 columns. Save the drawings as **E15.20**.

CHAPTER 15 QUIZ

DIRECTIONS

Answer the following questions with short complete statements. Type your answers using a word processor.

1. Place your name, chapter number, and the date at the top of the sheet.

2. Type the question number and provide the answer.

Warning: Some of the questions have not been covered in the reading material and will require the use of the help menu. You may also have to do some exploring to answer the questions.

1. What text option will allow text to be placed between two selected points at any angle?

_____

_____

2. What text option can be used to place multiple lines of text in a column?

_____

_____

3. If text is to be 0.125 high on a drawing that is to be plotted at a scale of 1"=20'–0", what is the easiest method to ensure that the text will be the correct height when plotted?

_____

_____

4. Explain the difference between rotation angle and oblique angle.

_____

_____

5. What is the difference between a style and a font?

_____

_____

6. What is the process for producing a degree symbol using the TEXT command?

_____

_____

7. What command is used to correct errors in text?

_____

_____

8. What effect will setting the style height option have on future text?

_____

_____

9. Explain how to correct the following problem: VERIFY EXATC hEIGHT AT JOBSITE.

_____

_____

10. When are lowercase letters appropriate for text on construction drawings?

_____

_____

11. Write the required entry to correctly specify three bolts that are three quarters of an inch in diameter at six inches on center. Include all special characters required if the style is set to be Romans.

_____

_____

12. A beam is located in an opening in a wall that is at 45° to horizontal. How should the text be oriented to describe the beam?

_____

_____

13. List and describe the two types of notes generally found on construction drawings.

_____

_____

14. What should be considered when choosing a font?

_____

_____

15. What are the effects of setting the text height in the **Style** command as opposed to setting the height in the **Text** command?

_____

_____

16. When would **Qtext** be used?

_____

_____

17. What is **Mtext** best suited for?

_____

_____

18. What should be the value for **Mirrtext** if you want to mirror the text with the object?

_____

_____

19. Your instructor wants you to create a body of text that starts in the upper right corner of a template drawing. The area for the notes will be 4" wide and be 1/2" down and 4 1/2" from the border. What are the commands and options, and what is the proper sequence for placing these notes?

_____

_____

20. What is the process required to save a paragraph created using **Mtext**?

_____

_____

21. A paragraph of multiline text has been created using ROMANS, but it should be in Italics. How can this be fixed?

_____

_____

22. Explain the major differences between **Ddedit** and **Properties** on multiline text.

_____

_____

23. What does the .CUS extension designate?

_____

_____

24. What is the advantage in placing stock notes on a template drawing?

_____

_____

25. A client has decided to change from floor joist to open-web trusses after all of the written specifications have been added to the drawing. How can this change be easily included?

_____

_____

26. List the two ways table properties can be modified.

_____

_____

27. What would be the benefit of creating a table style?

_____

_____

28. List three different types of data in a drawing file that you might want to create a field for.

_____

_____

29. How does the border offset factor affect the background color for MTEXT?

_____

_____

30. How can color be added to a text background?

_____

_____

Introduction to Dimensions

INTRODUCTION

In addition to the visual representation and the text used to describe a feature, dimensions are needed to describe the size and location of each member of a structure. Figure 16.1 shows a floor plan for a dormitory and the dimensions used to describe the location of structural members. In this chapter you will be introduced to:

- Basic principles of dimensioning
- Guidelines for placing dimensions on plan views and on drawings showing vertical relationships

Commands to be introduced in this chapter include:

- Dimlinear
- Dimaligned
- Dimordinate
- Dimradius
- Dimjogged
- Dimjogline
- Dimarc (Arc length)
- Dimdiameter
- Dimcenter (Center Mark)
- Dimangular
- Dimbaseline
- Dimcontinue
- Quick Dimension
- Dimbreak
- Dimspace

Note: Many of the commands used to place dimensions are listed with two names. Information about the command **Dimlinear** is listed under **D** for **Dimlinear** in the **Help** menu. The same command is listed on the toolbar and drop-down menu as **Linear**. Although most of the dimensioning commands can be accessed faster by toolbar, each command in this chapter will be referred to by its menu name and not its toolbar name.

Several methods will be used to enter the dimension commands. Because so many options are available to place dimensions, this chapter will examine only the default values for dimensioning. Chapter 17 will introduce methods for creating different styles of dimensions and describe how to adjust the variables that control the size and placement of the dimensions. Options that do not apply to the construction industry will be skipped entirely. Feel free to explore these options on your own, or consult the **Help** menu.

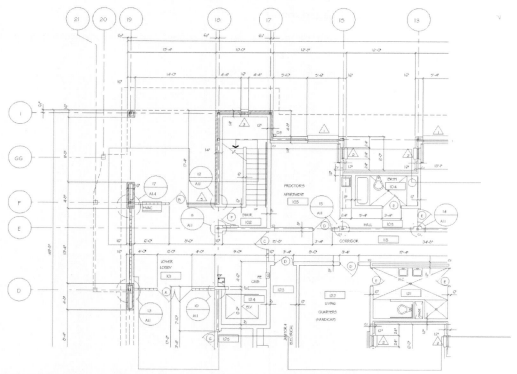

Figure 16.1   *Dimensions are used throughout construction drawings to show the location of each system to be installed. (Courtesy G. William Archer, AIA, Archer and Archer, P.A.)*

BASIC PRINCIPLES OF DIMENSIONING

To make better use of AutoCAD's potential, you need to understand several basic dimensioning concepts and terms before you explore the dimension commands. These principles include linetypes, placement of dimension features, and locating exterior and interior drawing features.

DIMENSIONING COMPONENTS

Dimensioning features include extension and dimension lines, text, and line terminators. Each component was introduced in an earlier chapter, and each feature is created using a dimensioning command. Common dimension features can be seen in Figure 16.2. Remember that sizes are given as a reference only and will be automatically determined by AutoCAD when using the 2D Annotation workspace.

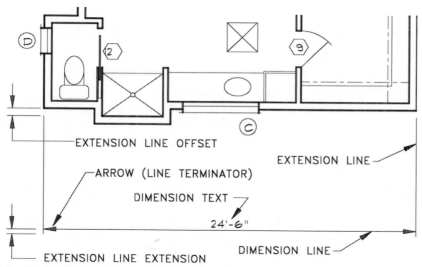

Figure 16.2 *Dimensions are composed of the dimension text, a dimension line, extension lines, and arrows (line terminators).*

Extension Lines

Extension lines are thin lines that extend out from the object being described and set the limits for the dimensions. Extension lines are usually offset about 1/8" from the object, and they extend past the last dimension line 1/8". The offsets and extensions can be seen in Figure 16.2. Two different types of linetypes may be used for extension lines.

- Solid lines are used to dimension to the exterior face of an object, such as a wall or footing.

- Centerlines are used to dimension to the center of objects. Figure 16.3 shows examples of each.

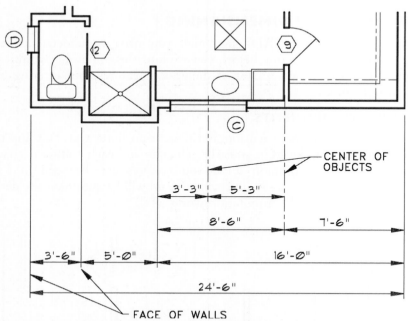

Figure 16.3 *Continuous extension lines are used to dimension to the edge of a surface; centerlines are used to dimension to the center of objects. Professionals often use a continuous line in place of a centerline to save time.*

Dimension Lines

Dimension lines are thin lines used to show the extent of the object being described. Exact location will vary with each office, but dimension lines should be placed in such a way as to leave room for notes yet still close enough to the features being described so that clarity will not be hindered. Guidelines for placement will be discussed later in this chapter.

Text

The text for architectural dimensions is placed above the dimension line and centered between the two extension lines. On the left and right sides of the structure, text is placed above the dimension line and rotated so that the text can be read from the right side of the drawing page. Examples of each placement can be seen in Figure 16.4. On objects placed at an angle other than horizontal or vertical, dimension lines and text are placed parallel to the oblique object, as seen in Figure 16.5. Often, not enough space is available for the text to be placed between the extension lines when small areas are dimensioned. Although options vary with each office, several alternatives for placing dimensions in small spaces can be seen in Figure 16.6.

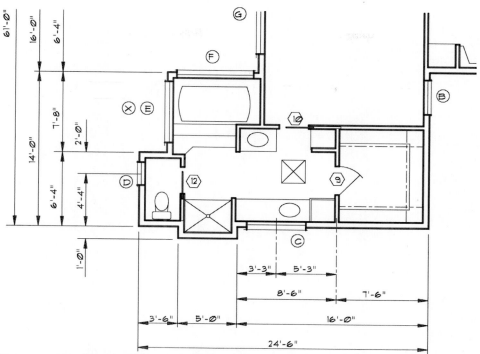

Figure 16.4 *Text should be placed above the dimension line and read from the bottom or right side of the drawing.*

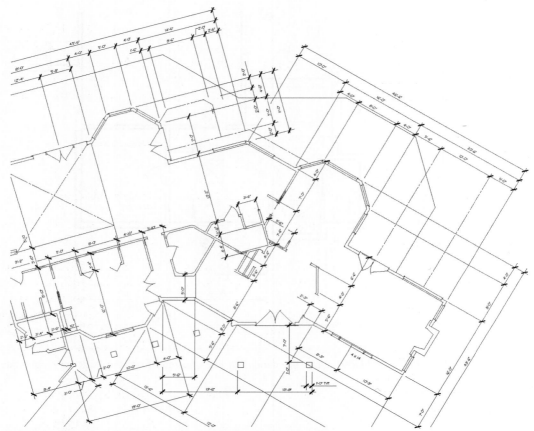

Figure 16.5 *When objects on an angle are dimensioned, dimension lines and text are placed parallel to the inclined surface. (Courtesy Residential Designs.)*

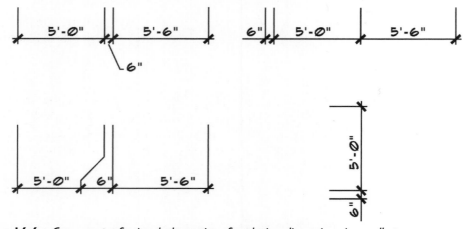

Figure 16.6 *Common professional alternatives for placing dimensions in small spaces.*

Terminators

The default method for terminating dimension lines at an extension line is with an arrowhead. Many offices use a tick mark or dot in an effort to duplicate their manual drawing styles. A third option is a thickened tick mark, which offers good contrast between lines, closely resembles its manual counterpart, and can be plotted quickly. All four options can be seen in Figure 16.7. AutoCAD also offers several other options that will be introduced in the next chapter.

Figure 16.7   *Alternatives for terminating dimension lines.*

DIMENSION PLACEMENT

Construction drawings requiring dimensions typically consist of plan views, such as the floor, foundation, and framing plans, and drawings showing vertical relationships, such as exterior and interior elevations, sections, details, and cabinet drawings. Each type of drawing has its own set of challenges to overcome in placing drawings. No matter the drawing type, dimensions should be placed on the * ANNO DIMS layer. The * represents the letter of the proper originator such as **A** (architectural) or **S** (structural), ANNO represents annotation, and DIMS represents dimensions.

> **Note:** Each of the examples that follow in this chapter have been created by professional designers and architects by altering the dimension style. This chapter will introduce each of the options for placing dimension in their default settings. Chapter 17 will introduce methods for altering the dimension qualities.

PLAN VIEWS

A discussion about placing dimensions in plan view can be divided into the areas of interior and exterior dimensions. Whenever possible, dimensions should be placed outside the drawing area.

Exterior Dimensions

Most offices start by placing an overall dimension on each side of the structure that is approximately 2" from the exterior wall. Moving inward, with approximately 1/2" between lines, are dimension lines used to describe major jogs in exterior walls, the distance from wall to wall, and the distance from wall to window or door to wall. Examples of placing these four dimension lines can be seen in Figure 16.8.

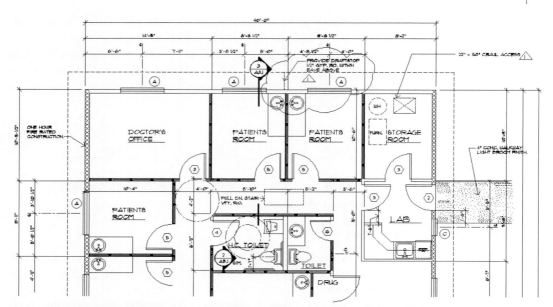

Figure 16.8 *Floor and framing plans are dimensioned by placing an overall dimension, dimensions to locate major jogs, wall-to-wall dimensions, and wall-to-opening dimensions. (Courtesy Scott R. Beck, Architect.)*

Two different systems are used to represent the dimensions between exterior and interior walls. Architects tend to represent the distance from the edge to edge of walls as seen in Figure 16.9. Engineering firms tend to represent the distance from exterior edge to center of interior wood walls using methods shown in Figure 16.10. Concrete walls are dimensioned to the edge by both disciplines. Concrete footings normally are dimensioned to the center. Examples of dimensioning concrete footings and walls can be seen in Figure 16.11.

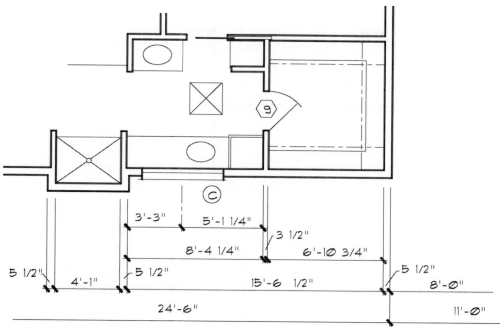

Figure 16.9 *Some architectural firms dimension from edge to edge of interior walls.*

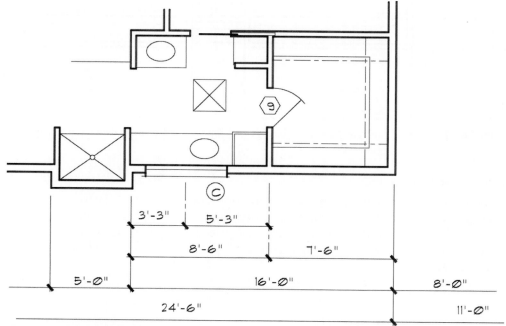

Figure 16.10 *Engineering firms often dimension to the center of interior walls.*

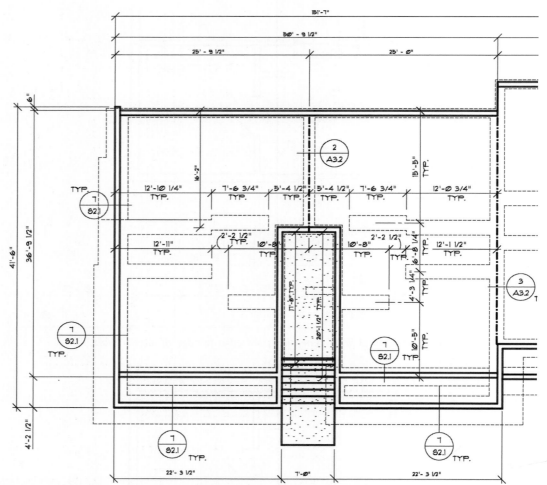

Figure 16.11 *Interior concrete piers and footings are dimensioned to their centers. (Courtesy Scott R. Beck, Architect.)*

Interior Dimensions

The two main considerations in placing interior dimensions are clarity and grouping. Dimension lines and text must be placed so that they can be read easily and so that neither interferes with other information that must be placed on a drawing. Information should also be grouped together, as seen in Figure 16.12, so that construction workers can find dimensions easily.

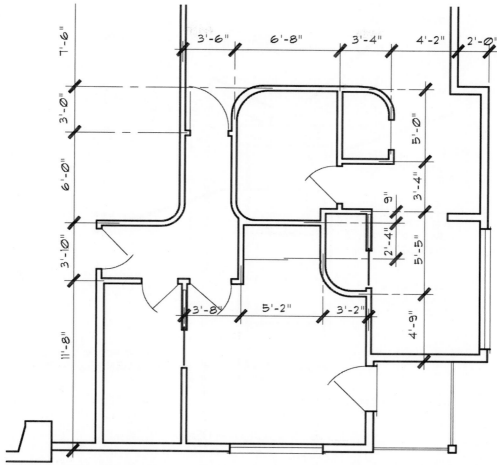

Figure 16.12   *Dimensions placed on the inside of a drawing should be grouped together and be clearly visible. (Courtesy Residential Designs.)*

VERTICAL DIMENSIONS

Unlike the plan views, which show horizontal relationships, the elevations, sections, and details require dimensions that show vertical relationships, as seen in Figure 16.13. Typically, these dimensions originate at a line that represents a specific point, such as the finish grade, a finish floor elevation, or a plate height.

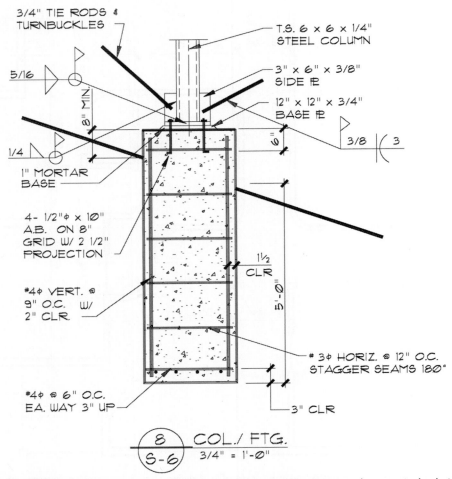

Figure 16.13 *Elevations, sections, and details each require dimensions to show vertical relationships.*

ACCESSING DIMENSIONING COMMANDS

Several methods are available for accessing the commands that control dimensions. To access the commands, use one of the following methods:

- Click the appropriate button on the **Dimension** toolbar.

- Select the appropriate option from the **Dimension** menu.

- Enter the appropriate alias at the Command prompt. (When entered at the Command prompt, a D is added as a prefix to the command name. For instance, the **Linear** command on the toolbar would be **DLI** at the Command prompt.)

Each of the options displayed on the menu and toolbar will be presented in this chapter. Because the toolbar is so much easier than mastering twelve names, the commands will be introduced assuming use of the toolbar. Dimension aliases will be noted as each command is introduced.

DIMENSIONING OPTIONS

Twenty-one methods are available from the default dimension toolbar for controlling and placing dimensions on a drawing. Each command is listed in the **Help** menu with a prefix of **Dim,** but the prefix is omitted when the selection is made from the toolbar or menu. Each can be seen on the Dimension toolbar shown in Figure 16.14. Commands include:

- The **Linear** and **Aligned** commands are used for providing linear dimensioning.

- **Diameter, Radius, Arc Length, Jogged, Angular,** and **Center Mark** are used to describe circular features.

- The **Baseline, Continue, Quick Dimension,** and **Ordinate** are used to place dimensions and can be combined with other commands.

- **Dimension Space, Dimension Break,** and **Jogged Linear** are used to alter dimensions that are already placed into the drawing.

- The **Tolerance** command is used to access geometric dimensioning and tolerance symbols. These symbols are rarely used and will not be covered in detail. The **Inspection** command is used to place specific dimensions for manufactured parts and is not relevant to construction drawings.

- **Dimension Style** can be used to access the **Dimension Style Manager** dialog boxes that are used to save dimensioning styles. Styles are a combination of values that control how the dimensions will be placed. Styles will be discussed in chapter 17.

- **Dimension Edit** and **Dimension Text Edit** are used to modify dimension qualities once they have been placed into the drawing. These features will be examined in chapter 17.

 Note: The following portion of this chapter will introduce each dimensioning method found on the **Dimension** toolbar. Chapter 17 will introduce methods of altering the features of each dimensioning method.

Figure 16.14 *The* **Dimension** *Toolbar found in the AutoCAD Classic workspace provides 21 methods for placing and altering dimensions. These same commands can be found in the dimension pallet of the 2D Annotation workspace. The toolbar will need to be expanded to display all of the buttons.*

LINEAR DIMENSIONS

 Linear dimensions are used to dimension straight surfaces common to construction features. To start the command, use one of the following methods:

- Click the **Linear Dimension** button on the **Dimension** toolbar.
- Click the **Linear Dimension** button on the **Dimension** dashboard.
- Select **Linear** from the **Dimension** menu.
 - Type **DLI** ENTER at the Command prompt.

Each method will produce the Command prompt:

```
Click the Linear button.
Specify first extension line origin or <select object>:
```

AutoCAD now offers two options to place a dimension. The default method is to specify two points to define the dimension. A second method allows selection of the object to be dimensioned. Once the object is selected, the end points are automatically defined.

Select Extension Line Origin

The start and end of a line or distance to be dimensioned can be specified. The command sequence for specifying each endpoint is as follows:

```
Specify first extension line origin or <select object>: (Select
    point.)
Specify second extension line origin: (Select point.)
Specify dimension line location or [Mtext/ Text/ Angle/
    Horizontal/ Vertical / Rotated]: (Select dimension line
    location.)
```

This option requires you to select the desired locations of the extension and dimension lines. Be sure to activate OSNAP settings before selecting the locations. As the second extension line location is selected, the dimension and extension lines can be dragged into position to allow for visual inspection. The coordinate display can also be useful in spacing dimension lines. The dimension location should be selected to allow room for all future notes. Place the dimension lines in the desired location and click the select button. The process can be seen in Figure 16.15.

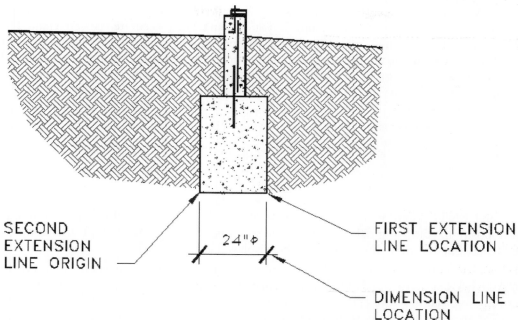

SECOND
EXTENSION
LINE ORIGIN

24"ϕ

FIRST EXTENSION
LINE LOCATION

DIMENSION LINE
LOCATION

Figure 16.15   *Horizontal and vertical dimensions can be placed using the **Linear** command. The dimension is placed by selecting the starting location for each extension line and then selecting the location for the dimension line. As the location for the second extension line is selected, the dimension is displayed and can be dragged into position.*

Mtext   Selecting the **Mtext** option allows the dimension value to be replaced by a body of text that is placed in a similar manner as with the **Text** option. Typing **M** ENTER at the dimension prompt will produce the **Text Formatting** toolbar and a display similar to Figure 16.16 with the text displayed in a highlighted edit box.

- With the cursor preceding the text, new text will be written followed by the dimension.

- Using the cursor to remove the existing text will remove the default dimension value and allow new text to be placed.

Entering the desired notation and selecting **OK** will remove the text editor and resume the dimension command.

Figure 16.16   *Type **M** ENTER at the Command prompt to display the **Text Formatting** toolbar.*

Text / No Text   One of the strengths of AutoCAD is also one of its weaknesses. AutoCAD is extremely accurate, but if you haven't been accurate as you created the drawing, the text assigned to the dimension will not produce the desired results. Choosing the **Text** option before the dimension line location is selected allows you to override the dimension text value by entering the value by keyboard. Use the following sequence to change the value from 3 1/2" to 4":

```
Specify dimension line location or [Mtext/ Text/ Angle/
    Horizontal/ Vertical/ Rotated]: T ENTER
Enter dimension text <3.5>: 4 ENTER
Specify dimension line location or [Mtext/ Text/ Angle/
    Horizontal/ Vertical/ Rotated]: (Select the dimension
    location.)
Dimension text = 3.5 ENTER
```

Pressing ENTER returns to the placement of the dimension and allows it to be dragged into position. The original dimension is still displayed at the end of the command sequence, but the new dimension will be displayed in the drawing.

Instead of entering a value when prompted for the new dimension text, pressing SPACEBAR and then ENTER will display the extension lines, dimension lines, and terminators, but no text. This option can be useful for specifying information about a specific area or product as seen in Figure 16.17.

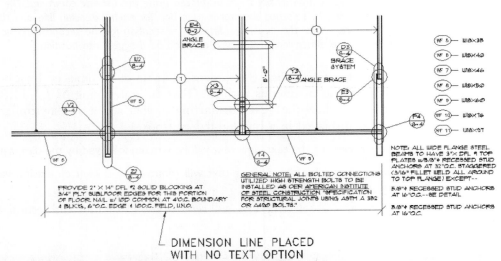

Figure 16.17   *Once the **T** option is selected, instead of entering a text value, press SPACE followed by ENTER to place a dimension line with no dimension.*

Angle   Type **A** ENTER at the line location prompt to alter the angle used to display the text. Although this is not often done, it could be especially helpful to make a dimension for a small space stand out. Figure 16.18 shows an example of text rotated at 45° to the dimension line. The command sequence, once the two endpoints are selected, is as follows:

```
Specify dimension line location or [Mtext/ Text/ Angle/
     Horizontal/ Vertical/ Rotated]: A ENTER
Specify angle of dimension text: 45 ENTER
Specify dimension line location or [Mtext/ Text/ Angle/
     Horizontal/ Vertical/ Rotated]: (Select the dimension
     location.)
Dimension text = 20'-6"
```

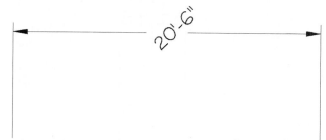

Figure 16.18   *The text angle can be rotated from the dimension line using the **A** option. Although not a common method of dimensioning, rotated text can be used to help draw attention to the dimension.*

Horizontal and Vertical   Typing **H** ENTER or **V** ENTER option at the prompt allows the specified dimension direction to be entered. AutoCAD automatically places a vertical dimension if two vertical endpoints are selected or a horizontal dimension line if two horizontal endpoints are selected.

Rotated   Typing **R** ENTER at the prompt rotates the extension and dimension lines to a specified angle from the desired surface. This option places the dimension in a rotated position and can be used to draw attention to a dimension describing a small space. Use this option to place the extension and dimension lines for referencing a note to a specific area. Figure 16.19 shows an example of the rotated option. The sequence to place the lines at 20° to a vertical surface is as follows:

```
Specify dimension line location or [Mtext/ Text/ Angle/
     Horizontal/ Vertical/ Rotated]: R ENTER
Specify angle of dimension line <0>: 20 ENTER
Specify dimension line location or [Mtext/ Text/ Angle/
     Horizontal/ Vertical/ Rotated]: (Select the dimension
     location.)
Dimension text = 6"
```

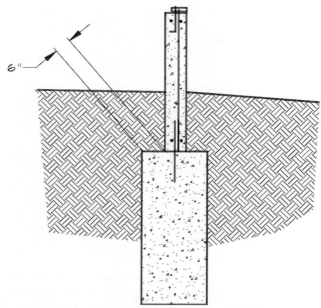

6"

Figure 16.19 *The extension lines and dimension lines can be rotated from the surface to be dimensioned using the* **R** *option.*

Select Object Option

Up to this point, each of the options discussed was made available once the two origins for the extension line endpoints were selected. By choosing the **Select Object** option, a specific line can be selected, and each endpoint will be determined automatically. The command sequence is as follows:

```
Click the Linear button.
Specify first extension line origin or <select object>: ENTER
```

Once ENTER is pressed, the crosshairs are changed to a pick box and a specific object can be selected. Once a surface is selected, the command continues as with any other option.

If the area to be dimensioned is smaller than the length required to place the arrows and text, each will be placed outside the extension lines. Methods of altering placement will be discussed in chapter 17.

ALIGNED DIMENSIONS

This method of placing dimensions is useful when the object being described is not parallel to the drawing borders. An example of aligned dimensions can be seen on the foundation plan shown in Figure 16.20. The command sequence is similar to the **Linear** command. To start the **Aligned** command, use one of the following methods:

- Click the **Aligned** button on the **Dimension** toolbar.
- Click the **Aligned** button on the **Dimension** dashboard.

- Select **Aligned** from the **Dimension** menu.

- Type **DAL** ENTER at the Command prompt.

The command sequence is:

```
Click the Aligned button.
Specify first extension line origin or <Select object>: (Select
    first point.)
Specify second extension line origin: (Select second point.)
Specify dimension line location or [Mtext/ Text/ Angle]:
Dimension text <45'-0">:
```

Once a location is selected for the dimension line, the line will automatically be placed parallel to the surface defined by the endpoints. Selecting the **Mtext** or **Text** option allows the default text value to be altered. Selecting the **Angle** option allows the angle of the text to be altered. Each of the options is the same as the corresponding option for the **Linear** command. Selecting the **Select Object** option automatically places the extension lines at the ends of the selected line and allows the dimension to be dragged into position.

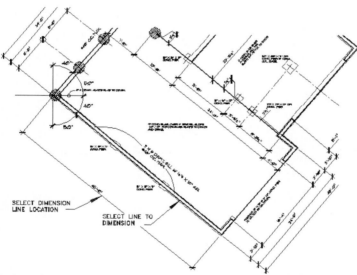

Figure 16.20 *When a surface is not horizontal or vertical, the **Aligned** command can be used to place dimensions. The process is similar to using the **Linear** command.*

ORDINATE DIMENSIONS

Ordinate dimensioning consists of placing dimensions without the use of dimension lines and arrows. An example of ordinate dimensions can be seen in Figure 16.21. Ordinate dimensions are occasionally used in placing vertical dimensions on details and sections. To start the **Ordinate** command, use one of the following methods:

- Click the **Ordinate** button on the **Dimension** toolbar.
- Click the **Ordinate** button on the **Dimension** dashboard.
- Select **Ordinate** from the **Dimension** menu.
- Type **DOR** ENTER at the Command prompt.

The command sequence is:

```
Click the Ordinate button.
Select feature location: (Select the surface to be dimensioned.)
```

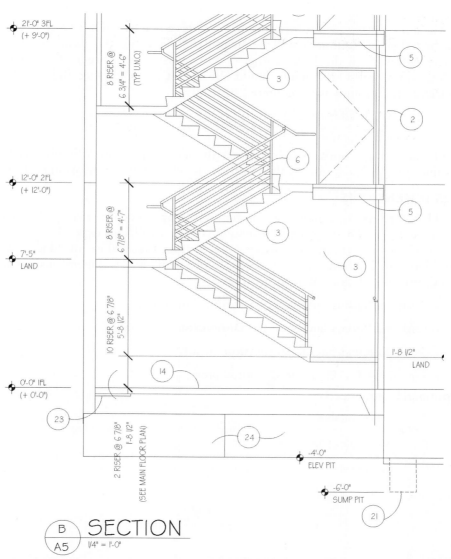

21'-0" 3FL
(+ 9'-0")

8 RISER @
6 3/4" = 4'-6"
(TYP U.N.O.)

12'-0" 2FL
(+ 12'-0")

8 RISER @
6 7/8" = 4'-7"

7'-5"
LAND

10 RISER @ 6 7/8"
5'-8 1/2"

0'-0" 1FL
(+ 0'-0")

2 RISER @ 6 7/8"
1'-8 1/2"
(SEE MAIN FLOOR PLAN)

1'-8 1/2"
LAND

-4'-0"
ELEV PIT

-6'-0"
SUMP PIT

B / A5   SECTION   1/4" = 1'-0"

Figure 16.21 *Placement of dimensions using the **Ordinate** command. All dimensions are placed using only a single extension line to represent each finish floor. All dimensions are referenced from the main floor level. (Courtesy Russ Hanson, HDN Architects, PC.)*

AutoCAD is prompting for the location that will serve as a base. On a section, this is often the finished floor elevation. It will be helpful to move the 0,0 location of the world coordinate system from the lower corner of the drawing to a corner of the object being dimensioned. The ORTHO mode should also be set to **ON**. Once the initial edge has been selected, the prompt will be altered to read:

```
Specify leader endpoint or [Xdatum/ Ydatum/ Mtext/ Text/ Angle]:
        (Select the endpoint of the extension line.)
```

The leader is a single horizontal line when the **Y** datum ordinate is used and a single vertical line for the **X** datum. As the endpoint is selected the dimension will be placed.

RADIUS DIMENSIONS

The **Radius** command is useful for many engineering and architectural applications. The way the information is placed will vary depending on the size of the circle or arc to be dimensioned and settings that will be introduced in the next chapter. Figure 16.22 shows each option for representing radius dimensions. To access the command, use one of the following methods:

- Click the **Radius** button on the **Dimension** toolbar.

- Click the **Radius** button on the **Dimension** dashboard.

- Select **Radius** from the **Dimension** menu.

- Type **DRA** ENTER at the Command prompt.

The command sequence is:

```
Click the Radius button.
Select arc or circle: (Select desired arc or circle.)
Dimension text <10'-6">:
Specify dimension line location or [Mtext/ Text/ Angle]: (Select
        location for dimension.)
```

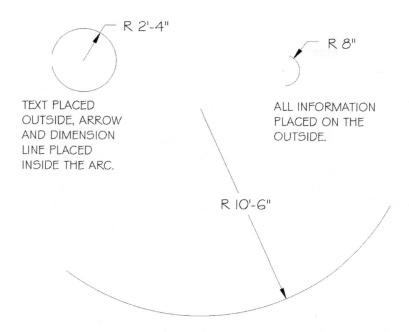

TEXT PLACED
OUTSIDE, ARROW
AND DIMENSION
LINE PLACED
INSIDE THE ARC.

ALL INFORMATION
PLACED ON THE
OUTSIDE.

ALL INFORMATION PLACED INSIDE OF THE ARC.

Figure 16.22 *The **Radius** command allows a radius to be placed using three different techniques.*

When you select the circle or arc to be dimensioned, the default dimension is dragged as you move the cursor. Entering **M**, **T**, or **A** allows the text or angle value to be altered. The **Mtext**, **Text**, and **Angle** options function exactly as discussed with other commands. Once the desired **Text** and **Angle** prompts have been responded to, the prompt will be returned to locate the dimension line.

JOGGED DIMENSIONS

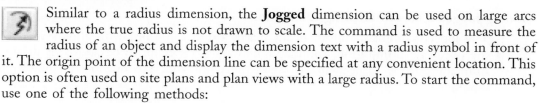

Similar to a radius dimension, the **Jogged** dimension can be used on large arcs where the true radius is not drawn to scale. The command is used to measure the radius of an object and display the dimension text with a radius symbol in front of it. The origin point of the dimension line can be specified at any convenient location. This option is often used on site plans and plan views with a large radius. To start the command, use one of the following methods:

- Click **Jogged** from the **Dimension** toolbar.
- Click **Jogged** from the **Dimension** dashboard.
- Select **Jogged** from the **Dimension** menu.
- Enter **DJO** ENTER at the Command prompt.

The command sequence is:

```
Click the Jogged button.
Select arc or circle: (Select desired arc or circle.)
Dimension text <40'-0">:
Specify center location override: (Select desired start point for
    radius.) Specify dimension line location or [Mtext/ Text/
    Angle]: (Select location for dimension.)
```

When you select the circle or arc to be dimensioned, the default dimension is dragged as you move the cursor. Entering **M**, **T**, or **A** allows the text or angle value to be altered. The **Mtext, Text**, and **Angle** options function exactly as discussed with other commands. Once the desired **Text** and **Angle** prompts have been responded to, the prompt will be returned to locate the dimension line. Figure 16.23 shows an example of the use of the **Jogged** command.

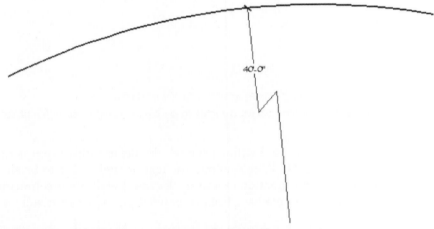

Figure 16.23 *Placing radius dimensions using the Jogged command.*

Adding Jogged Lines to Existing Dimensions

 The Jogged command automatically places a dimension line with a jog symbol in the line. The **Dimjogline** command allows a jog symbol to be added or deleted from an existing dimension. To start the command, use one of the following methods:

- Click **Dimjogline** from the **Dimension** toolbar.
- Click **Dimjogline** from the **Dimension** dashboard.
- Select **Jogged Linear** from the **Dimension** menu.
- Enter **DJL** ENTER at the Command prompt.

Each entry method will produce the following prompt:

```
Select dimension to add jog or [Remove]: (Select the existing
    dimension to receive the jog.)
Specify jog location (or press ENTER): (Select the desired
    location for the jog.)
```

As the location is specified, the jogged line is added and the command line is restored. The command also allows an existing jog to be removed using the following sequence:

```
Select dimension to add jog or [Remove]: R ENTER
Select jog to remove: (Select the jog to be removed.)
```

As the jog is specified, the jogged line is removed, and the command line is restored.

ARC LENGTH

 The **Arc Length** command can be used to specify the placement of the dimension line and to determine the direction of the extension lines. It is useful for placing dimensions of arcs on a site plan. To start the command, use one of the following methods:

- Click **Arc Length** from the **Dimension** toolbar.
- Click **Arc Length** from the **Dimension** dashboard.
- Select **Arc Length** from the **Dimension** menu.
- Type **DAR** ENTER at the Command prompt.

The command sequence is:

```
Click the Arc Length button.
Select arc or polyline arc segment: (Select desired arc.)
Dimension text <70'-0">:
Specify arc length dimension location or [Mtext/ Text/ Angle/
    Partial/ Leader]: (Select location for dimension.)
```

Selecting an arc will display the dimension and allow it to be dragged into position. Once a location is selected, the dimension will be placed with an arc symbol preceding the text. The **Mtext**, **Text**, and **Angle** options function in the same manner as with other dimensioning commands. Selecting the **Partial** option allows two points to be selected to define a partial arc length. If the **Leader** option is chosen, a leader is added to reference the dimension to the selected arc if the arc to be dimensioned is greater than 90°. The leader is drawn so that it is pointing towards the center of the arc being dimensioned. If the selected arc is less than 90°, the **No Leader** option is displayed to cancel the **Leader** option before the leader is created. Figure 16.24 shows examples of the use of the **Arc Length** command.

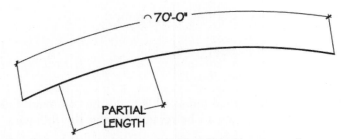

Figure 16.24 *Dimensioning arc lengths using the **Arc Length** command.*

DIAMETER DIMENSIONS

 The **Diameter** command will allow you to dimension the diameter of circular objects. To start the command, use one of the following methods:

- Click the **Diameter** button on the **Dimension** toolbar.
- Click the **Diameter** button on the **Dimension** dashboard.
- Select **Diameter** from the **Dimension** menu.
- Type **DDI** ENTER at the Command prompt.

Two methods of placing the diameter specifications on a drawing can be seen in Figure 16.25. The default method is to use a leader line to display the dimension. The command sequence is as follows:

```
Click the Diameter button.
Select arc or circle:
Dimension text <24.00>:
Specify dimension line location or [Mtext/ Text/ Angle]: (Specify
      a point or enter an option.)
```

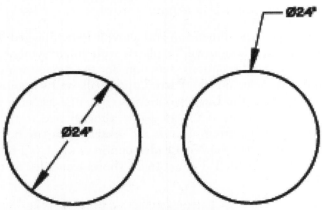

Figure 16.25 *The **Diameter** command allows the diameter to be placed using two different methods.*

When you select the circle to be dimensioned, the default dimension is dragged as you move the cursor. Depending on the size of the circle, the dimension toggles between being placed inside or outside the circle. Placing the dimension inside or outside the circle depends entirely on the size of the circle and the settings that will be introduced in chapter 17. Typing **M** ENTER, **T** ENTER, or **A** ENTER allows the text or angle value to be altered. Mtext, Text and Angle function exactly as discussed earlier for the **Linear** command. Once the desired prompts have been responded to, the prompt will be returned for locating the dimension line.

CREATING A CENTER MARK

The **Center Mark** command will provide a center mark for a circle or an arc. To start the command, use one of the following methods:

- Click **Center Mark** on the **Dimension** toolbar.

- Click **Center Mark** on the **Dimension** dashboard.

- Select **Center Mark** from the **Dimension** menu.

- Type **DCE** ENTER at the Command prompt.

        The command sequence is as follows: *Click the* **Center Mark** *button.*
        Select arc or circle: *(Select desired circle to receive center*
            mark.)

The mark will be placed once the circle is selected. Chapter 17 will introduce methods of altering the size of the mark.

ANGULAR DIMENSIONS

The **Angular** command can be used to dimension angles formed by selecting two straight nonparallel lines, an arc, a circle, and another point or three points. To start the command, use one of the following methods:

- Click the **Angular Dimension** button on the **Dimension** toolbar.

- Click the **Angular Dimension** button on the **Dimension** dashboard.

- Select **Angular** from **Dimension** menu.

- Type **DAN** ENTER at the Command prompt.

Each choice will produce the following prompt:

        Click the **Angular** *button.*
        Select arc, circle, line, or <specify vertex>:

Choosing Two Lines

One of the most common situations for dimensioning angles is describing the angle formed between two intersecting lines. Respond to the prompt by selecting a line. The command sequence continues with:

```
Select arc, circle, line, or <specify vertex>: (Select first
    line.)
Select second line: (Select second line.)
Specify dimension arc line location or [Mtext/ Text/ Angle/
    Quadrant]: (Select dimension location.)
```

As the dimension arc line location (Mtext/ Text/ Angle/ Quadrant) prompt is displayed, you will be able to choose the location of the dimension placement relative to the angle. Figure 16.26 shows alternatives for placing the angle dimensions. Each of the first three options function as they do with other dimensioning options. The quadrant option specifies the quadrant that the dimension should be locked to. When quadrant behavior is on, the dimension line is extended past the extension line when the dimension text is positioned outside of the angular dimension.

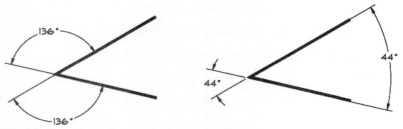

Figure 16.26 *Alternatives for placement of dimensions using the Dimangular command.*

Angles Based on an Arc

When you respond to the initial angle prompt by choosing an arc, dimensions can be placed to describe the tangent points of an arc. This could be especially useful when labeling subdivision maps or other large parcels of land. The command sequence is the same as the sequence used to dimension an angle based on two lines. As you select the second line, the angle is displayed. Depending on the location of the cursor, the display can be located in one of four positions. The **Text** and **Angle** options are the same as with other dimensioning commands. Figure 16.27 shows an example of how text will be placed.

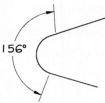

Figure 16.27 *Dimensioning of an arc using the Dimangular command.*

Angles Based on a Circle

When you respond to the initial angle prompt by choosing a circle, dimensions can be placed to describe angular patterns formed within a circle. This type of dimensioning is often used in specifying steel placement in circular columns and footings. Options for placing the dimension line can be seen in Figure 16.28. The command sequence is as follows:

```
Click the Angular button.
Select arc, circle, line, or <specify vertex>: (Select desired
    object.)
Specify dimension arc line location or [Mtext/ Text/ Angle/
    Quadrant]: (Select location.)
```

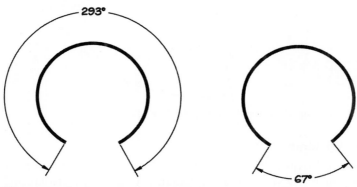

Figure 16.28   *Placement of angular dimensions using the circle option.*

Describing Three Points

When you respond to the original angular prompt with ENTER, you will be allowed to select three different points that can be used to describe an angle. This option can be used to describe the center point of bolts around a center point in a steel connector strap. The two options for placing the angle can be seen in Figure 16.29. The command sequence is as follows:

```
Click the Angular button.
Select arc, circle, line, or <specify vertex>: ENTER
Specify angle vertex: (Select a circle center point.)
Specify first angle endpoint: (Select a point.)
Specify second angle endpoint: (Select a point.)
Specify dimension arc line location or [Mtext/ Text/ Angle/
    Quadrant]: (Select desired text location.)
```

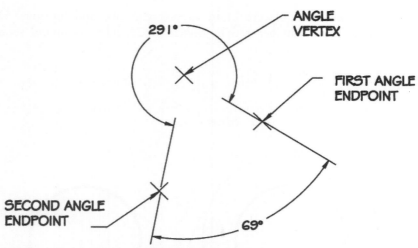

Figure 16.29 *Placement of dimensions using the three-point option.*

Describing the Quadrant

The Quadrant option provides control over which one of four possible angles will be specified for an angle. By specifying a quadrant, it allows you to ensure that the correct angle is dimensioned. Use the following command sequence to reference a dimension to the desired quadrant:

```
Click the Angular button.
Select arc, circle, line, or <specify vertex>: ENTER
Specify angle vertex: (Select a circle center point.)
Specify first angle endpoint: (Select a point.)
Specify second angle endpoint: (Select a point.)
Specify dimension arc line location or [Mtext/ Text/ Angle/
    Quadrant]: (Select desired quadrant.)
Specify dimension arc line location or [Mtext/ Text/ Angle/
    Quadrant]: (Use the Select button to accept the dimension,
    or edit with the Mtext or Text options.
```

BASELINE DIMENSIONS

Occasionally, when the spacing of objects is extremely critical, the baseline system of placing dimensions can be used. Baseline dimensions (also called datum dimensions) assume one edge to be perfect and reference all dimensions back to that surface. Figure 16.30 shows an example of baseline dimensions. Although this type of dimensioning is not used in residential design, it is common in some areas of commercial construction where a high degree of accuracy must be achieved. The **Baseline** command sequence must be started using an existing linear, angular, or ordinate dimension. Once the first dimension has been placed, start the **Baseline** command by using one of the following methods:

- Click the **Baseline** button on the **Dimension** toolbar.

- Click the **Baseline** button on the **Dimension** dashboard.

- Select **Baseline** from the **Dimension** menu.

- Type **DBA** ENTER at the Command prompt.

AutoCAD will use the first extension line from the existing dimension as the base point for other dimensions. The command sequence is as follows:

```
Click the Linear button.
Specify first extension line origin or <select object>: (select
    point)
Specify second extension line origin (select point)
Specify dimension line location or [Mtext/ Text/ Angle/
    Horizontal/ Vertical/ Rotated]: (Select dimension line
    location)
Command: (Click the Baseline button.)
Specify a second extension line origin or [Undo/Select] <Select>:
    (Select base for second point.)
Specify a second extension line origin or [Undo/Select] <Select>:
    (Select base point for the next dimension.)
Specify a second extension line origin or [Undo/Select] <Select>:
    (Select base point for the next dimension.)
Specify a second extension line origin or [Undo/Select] <Select>:
    ENTER
Select base dimension: ENTER
```

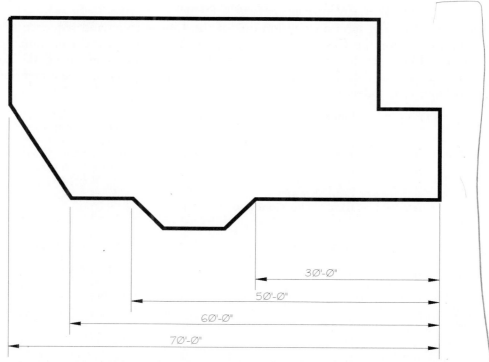

Figure 16.30   *Placement of dimensions using the **Linear** and **Baseline** commands. The dimension on the right was placed using **Linear** with the balance of the dimensions placed using **Baseline**.*

CONTINUOUS DIMENSIONS

A series of dimensions is needed to dimension a structure. In Figure 16.1, dimension lines extend from the exterior walls, to interior walls, and finally to the exterior wall on the opposite side. These dimensions can be placed with the **Continue** command. The command sequence starts by using the **Linear** command to place the first dimension. The **Continue** command must be based on a linear, angular, or ordinate dimension.

```
Click the Linear button.
Specify first extension line origin or <select object>: (Select
    point)
Specify second extension line origin (Select second point)
Specify dimension line location or [Mtext/ Text/ Angle/
    Horizontal/ Vertical/ Rotated]: (Select dimension line
    location)
```

To start the **Continue** command, use one of the following methods:

- Click the **Continue** button on the **Dimension** toolbar.
- Click the **Continue** button on the **Dimension** dashboard.
- Select **Continue** from the **Dimension** menu.
- Type **DCO** ENTER at the Command prompt.

AutoCAD uses the second extension line of the **Linear** command sequence as the first extension line for the **Continue** sequence. Press ESC to end the **Continue** command sequence. Figure 16.30 shows an example of placing dimensions using the **Continue** command with the default settings. The command sequence to place the dimensions across the bottom is as follows:

```
Click the Linear button.
Specify first extension line origin or <select object>: (Select
    first point)
Specify second extension line origin (Select second point)
Specify dimension line location or [Mtext/ Text/ Angle/
    Horizontal/ Vertical/ Rotated]: (Select the desired
    dimension location.)
Command: (Click the Continue button.)
Specify a second extension line origin or [Undo/Select]: <Select>
    (Select second point.)
Specify a second extension line origin or [Undo/Select]: <Select>
    (Select next point.)
Specify a second extension line origin or [Undo/Select]: <Select>
    (Select next point.)
Specify a second extension line origin or [Undo/Select]: <Select>
    ENTER
Select continued dimension: ENTER
```

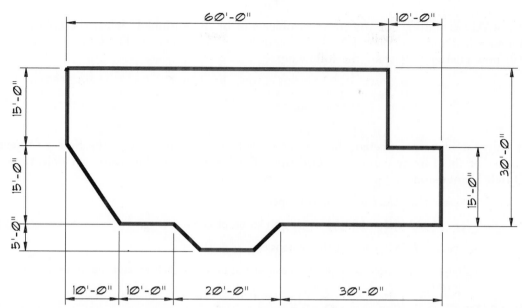

Figure 16.31 *Placement of dimensions using the **Linear** and **Continue** commands. The 10'–0"*
*dimension on the bottom left was placed using the **Linear** command. All of the other*
*horizontal dimensions were placed using the **Continue** command. The **Continue***
command uses the second extension line of the preceding sequence as the first exten-
*sion line of the **Continue** sequence.*

QUICK DIMENSIONS

You've been exposed to several different methods for placing dimensions throughout this chapter. Each works well for a specific need. AutoCAD provides an additional method for placing dimensions that combines features of many of the previous options. As the name implies, the **Quick Dimension** command allows a single dimension or a series of dimensions to be placed quickly. **Quick Dimension** excels in placing strings of continuous or baseline dimensions with very few prompts required. To start the command, use one of the following methods:

- Click the **Quick Dimension** button on the **Dimension** toolbar.
- Click the **Quick Dimension** button on the **Dimension** dashboard.
- Select **Quick Dimension** from the **Dimension** menu.
- Type **QDIM** ENTER.

Each method will produce the following prompt at the command line:

```
Click the Quick Dimension button.
Associative dimension priority = Endpoint
Select geometry to dimension: (Select object or objects to be
    dimensioned.)
```

AutoCAD is waiting for you to select objects to be dimensioned. Select the object to be dimensioned and then press ENTER to end the selection process. The Command prompt will now change to display the following:

```
Specify dimension line position, or [Continuous/ Staggered/
    Baseline/ Ordinate/ Radius/ Diameter/ datumPoint/ Edit/
    seTtings] <Continuous>: (Select the desired dimension
    location.)
```

If only one object is selected, the command will function similarly to the **Select Objects** option of the **Linear** command. Use the following steps to dimension an object using **Quick Dimension**.

- Select the object to be dimensioned.

- Specify a location for the dimension to be placed.

- Press ENTER to accept the dimension.

- When the dimension is in the desired position, left-click to add the dimension to the drawing.

Quick Dimension Placement Options

The best feature of **Quick Dimension** can be seen when multiple objects for dimensioning are selected. Once the selection set is completed and ENTER is pressed, dimensions to describe each feature will be displayed. Options include:

```
Specify dimension line position, or [Continuous/ Staggered/
    Baseline/ Ordinate/ Radius/ Diameter/ datumPoint/ Edit/
    seTtings] <Continuous>
```

Using the Continuous Option   The **Continuous** option of **Quick Dimension** places multiple dimensions based on selecting one location point. Figure 16.32 shows an object dimensioned using the Continuous option. The command sequence is:

```
Click the Quick Dimension button.
Associative dimension priority = Endpoint
Select Geometry to dimension: (Select object to be dimensioned.)
1 found
Select Geometry to dimension: (Select next object to be
    dimensioned.)
1 found, 2 total
Select Geometry to dimension: (Select next object to be
    dimensioned.)
1 found, 3 total
Select Geometry to dimension: (Select next object to be
    dimensioned.)
1 found, 4 total
Select Geometry to dimension: (Select next object to be
    dimensioned.)
1 found, 5 total
Select Geometry to dimension: ENTER
```

```
Specify dimension line position, or [Continuous/ Staggered/
    Baseline/ Ordinate/ Radius/ Diameter/ datumPoint/ Edit/
    seTtings <Continuous>: ENTER
```

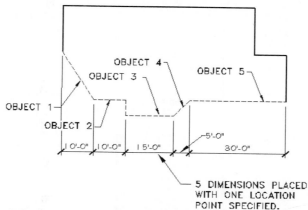

Figure 16.32 *The best features of* **Quick Dimension** *can be seen when multiple objects are selected to be dimensioned. In this example, five objects have been selected to be dimensioned using the* **Continuous** *mode. A dimension to describe each feature is placed by identifying one location point.*

Using the Baseline Option The **Baseline** option of **Quick Dimension** combines the features of the **Baseline** dimension option and the Continuous process to place multiple dimensions. The command functions in a method similar to the **Continuous** option in that each object must be selected, followed by specifying the placement method (in this case Baseline), and then the location is identified. Figure 16.33 shows an example of placing dimensions using the **Baseline** option of **Quick Dimension**. The command sequence is as follows:

```
Click the Quick Dimension button.
Associative dimension priority = Endpoint
Select Geometry to dimension: (Select object to be dimensioned.)
1 found
Select Geometry to dimension: (Select next object to be
    dimensioned.)
1 found, 2 total
Select Geometry to dimension: (Select next object to be
    dimensioned.)
1 found, 3 total
Select Geometry to dimension: (Select next object to be
    dimensioned.)
1 found, 4 total
Select Geometry to dimension: (Select next object to be
    dimensioned.)
1 found, 5 total
Select Geometry to dimension: ENTER
```

```
Specify dimension line position, or [Continuous/ Staggered/
    Baseline/ Ordinate/ Radius/ Diameter/ datumPoint/ Edit/
    seTtings <Continuous>: B ENTER
Specify dimension line position, or [Continuous/ Staggered/
    Baseline/ Ordinate/ Radius/ Diameter/ datumPoint/ Edit/
    seTtings] <Baseline>: (Select location to place the
    dimension string.)
```

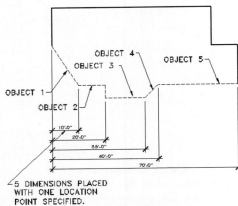

Figure 16.33 *The **Baseline** option of the **Quick Dimension** command functions similarly to the **Continuous** option in that each object must be selected, followed by the location.*

Using the Staggered Option   The **Staggered** option of **Quick Dimension** can be used to place staggered dimensions. Once objects to be dimensioned are selected, type **S** ENTER when prompted for an option. Figure 16.33 shows an example of staggered dimensions.

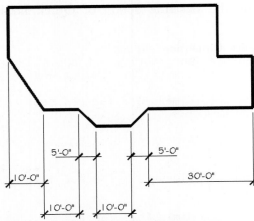

Figure 16.34 *Typing **S** ENTER when prompted for an option allows staggered dimensions to be placed.*

Using the Ordinate Option   Typing **O** ENTER at the prompt allows ordinate dimensions to be placed using methods for selection similar to other **Quick Dimension** options. The **Radius** and **Diameter** options of **Quick Dimension** allow you to select and dimension multiple circles or arcs of various sizes in one command sequence rather that having to repeat a dimension sequence for each object.

Using the DatumPoint Option   The **datumPoint** option will place dimensions in a pattern similar to the Ordinate option. The major difference between the two options is that **datumPoint** allows any point to be set as the base point. This option could be used to establish a finish floor level as 00, and allow dimensions to be placed above and below the base point (the finish floor). Figure 16.34 shows an example of dimensions placed using the **datumPoint** option. The option is started by typing **P** ENTER at the option prompt. Once objects are selected for dimensioning, the command sequence to place **datumPoint** dimensions is as follows:

```
Select Geometry to dimension: ENTER
Specify dimension line position, or [Continuous/ Staggered/
     Baseline/ Ordinate/ Radius/ Diameter/ datumPoint/ Edit/
     seTtings] <Continuous>: P ENTER
Select new datum point: (Select the point to be the base of the
     dimensions.)
Specify dimension line position, or [Continuous/ Staggered/
     Baseline/ Ordinate/ Radius/ Diameter/ datumPoint/ Edit/
     seTtings] <Ordinate>: (Select the location to place the
     dimension string.)
```

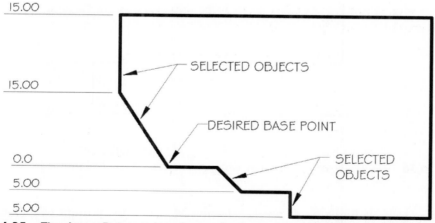

Figure 16.35   *The **datumPoint** option allows ordinate dimensions to be placed using any point to be set as the base point.*

USING THE DIMBREAK COMMAND

The **Dimbreak** command allows crossing dimensions that have been placed in a drawing to be edited. AutoCAD will provide prompts to choose which of the two dimensions will be continuous and which dimension will be broken. Start the command using one of the following methods:

- Click the **Dimbreak** button on the **Dimension** toolbar.
- Click the **Dimbreak** button on the **Dimension** dashboard.
- Select **Dimension Break** from the **Dimension** menu.
- Type **DIMBREAK** ENTER at the Command prompt.

Each method will produce the following prompt:

```
Select a dimension or [Multiple]: Select the extension line to
    be broken.
Select object to break dimension or [Auto/ Restore/
    Manual]<Auto>: Select the extension line to be continuous.
```

As the second line is selected, the first line will be broken to allow the second line to continue through to the object. Figure 16.36 shows the use of the **Dimbreak** command.

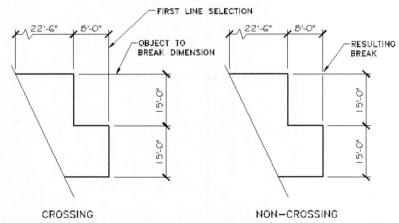

Figure 16.36 *The Dimbreak option allows crossing dimensions to be broken to add clarity to a drawing.*

In addition to providing a single break, selecting **M** at the first prompt will allow multiple dimensions to be selected to add or delete breaks. If the multiple option is selected the following prompt will be displayed:

```
Enter an option [Break/ Restore] <Break>:
```

Command options will produce the following results:

- **Auto** will place a dimension break at each intersection point of the selected dimension.

- **Break** will automatically place breaks at all intersection points of the selected dimensions.

- **Manual** will place a break, once two points on a dimension or extension line have been specified.

- **Restore** will remove a break from the selected dimensions.

ALTERING DIMENSION SPACING

The **Dimspace** command can be used to alter the spacing between parallel linear and angular dimensions. Start the command using one of the following methods:

- Click the **Dimspace** button on the **Dimension** toolbar.

- Click the **Dimspace** button on the **Dimension** dashboard.

- Select **Dimension Space** from the **Dimension** menu.

- Type **DIMSPACE** ENTER at the Command prompt.

Each method will produce the following prompt:

```
Select base dimension: Select a dimension to serve as a base
    location for other dimension to be selected.
Select dimensions to space: (Select parallel dimensions that are
    to be relocated. Press ENTER when the selection set is
    complete.) ENTER
Enter value or [Auto] <Auto>: (Enter the desired spacing.)
    .5 ENTER
```

Options include entering a distance or using the Auto setting. The NCS requires a minimum distance of 0.375" between dimension lines, but a distance of 0.5" is preferred if space allows. Once the distance is entered at the prompt, the selected dimensions will be relocated. If the Auto option is selected, the spacing between dimension lines will be assigned to be twice the height of the dimension text.

CHAPTER 16 EXERCISES

Complete the following exercises using the 2D Drafting Annotation workspace. Assign a scale appropriate to the drawing. Unless noted, assume a scale of 1/4" = 1'-0" for any plan views and a scale of 1/2" = 1'- 0".

Warning: Most of the dimensions that you place in the following drawings will not appear as they do in the examples of this chapter. *Don't panic, they're not supposed to.* Variables that affect most of the dimensions in this chapter have been altered to show you the proper method to use the dimensions. Complete the drawings as described. The next chapter will teach you what variables to adjust need to be altered and explain the process to make the needed adjustments.

1. Start a new drawing and set the units to architectural with 1/16" accuracy. Draw a hexagon with a horizontal top and base inscribed in a 4" diameter circle. Use the **Linear** command to provide overall dimensions. Save the drawing as **E-16-1**.

2. Open drawing **E-16-1**. Change the hexagon to a layer titled BASE. Change the dimensions to a layer titled DIMEN and turn the layer OFF. Create a layer titled ANGLE and make it current. Provide dimensions to indicate the angle of the sides. Save the drawing as **E-16-2**.

3. Open drawing **E-9-5** and create a layer titled DIM1. Rescale the drawing by a factor of .25 and completely dimension the drawing so that the metal hanger can be fabricated. Save the drawing as **E-16-3**.

4. Open drawing **E-10-9** and create a layer titled DIM1. Change the lines representing the one-story footing to .5 wide polylines. Place the two-story footing on a new layer and set that layer to OFF. Rescale the drawing using a scale factor of .25. Completely dimension the one-story footing and save the drawing as **E-16-4**.

5. Start a new drawing and draw the following object. Starting at 2,2:

   B. 1.5,0
   C. 0,1
   D. 1.25,0
   E. .75,.75
   F. 1.5,0
   G. 0,.75
   H. 1.5,0
   J. 0,1.75
   K. −2,0
   L. 0,−.75
   M. −2,0
   N. 0,−1.25
   P. −2.5,0
   Q. .5,−1.5
   R. AND THEN CLOSE.

   Change the line segments to polylines. Completely dimension the drawing, using the **Continue** command where possible. Save the drawing as **E-16-5**.

6. Open drawing **E-16-5** and create a new layer for the existing dimensions. Change the dimensions to the new layer that was just created, and set it to **OFF**. Create another layer for dimensions and completely dimension the object using baseline dimensioning. Assume the lower-left corner as the starting point for horizontal and vertical dimensions. Save the drawing as **E-16-8.**

2. Open the architectural template and draw an object similar to Figure 16.30 Create a layer for placing dimensions. Dimension every surface using the **Continuous** option of **Quick Dimension**. Save the drawing as **E-16-7**.

3. Open drawing **E-16-7** and create a new layer for dimensions. Dimension each surface using the **Baseline** option of **Quick Dimension**. Save the drawing as **E-16-8**.

4. Open drawing **E-16-8** and create another layer for dimensions. Dimension each surface using the Staggered option of **Quick Dimension**. Save the drawing as **E-16-9**.

CHAPTER 16 QUIZ

DIRECTIONS

Answer the following questions with short complete statements. Type your answers using a word processor.

1. Place your name, chapter number, and the date at the top of the sheet.

2. Type the question number and provide the answer.

 Warning: Some of the questions have not been covered in the reading material and will require the use of the help menu. You may also have to do some exploring to answer the questions.

1. What four components comprise dimensioning?

_____

_____

2. How should text be placed on architectural drawings to ease the job of the print reader?

_____

_____

3. List four options typically used for the intersections of dimension and extension lines.

_____

_____

4. Describe how to place dimensions relative to a concrete wall.

_____

_____

5. List four groups of commands that relate to placing dimensions on a drawing.

_____

_____

6. Describe two methods of locating extension lines for **Linear** dimensions.

_____

_____

7. List four methods to describe angular dimensions.

_____

_____

8. Open one of the drawing exercises that were completed for this chapter and provide the name of the layer that was added to the drawing base by AutoCAD.

_____

_____

9. List three methods of placing linear dimensions.

_____

_____

10. Explain the difference between aligned and rotated dimensions.

_____

_____

11. Should the default text value for dimension text always be accepted? Explain your answer.

_____

_____

12. What is the effect of pressing ENTER at the first **Linear** Command prompt rather than selecting a point?

_____

_____

13. Should a dimension be placed inside or outside a circular object?

_____

_____

14. What is the process for placing a 6" center mark in the center of a circular steel tube?

_____

_____

15. Explain the options for **Ordinate** dimensioning.

_____

_____

Placing Dimensions on Drawings

INTRODUCTION

In chapter 16 you were introduced to dimension requirements of different drawings and the methods of describing linear, angular, circular, continuous, and baseline dimensions. Each system was described using the default values. If all of your drawings could fit on a computer screen, everything would be fine. To allow for drawings of various sizes, the spacing of extension and dimension lines and the size of text and arrows must be altered. In this chapter you'll be introduced to:

- Altering dimension variables
- Establishing dimension styles
- Editing existing dimensions

Commands to be explored include:

- **Dimstyle** (Dimension style)
- **Flip Arrows**
- **Initial Length**
- **Dimoverride** (Dimension override)
- **Dimedit** (Dimension edit)
- **Dimtedit** (Dimension text edit)
- **Dimension Update**

CONTROLLING DIMENSION VARIABLES WITH STYLES

As text was explained, you were introduced to the **Style** command. **Style** allows you to group variables such as the font, text height, width, oblique angle, and orientation so that several different styles of text can easily be created within one drawing. AutoCAD also allows dimension styles to be created within a drawing to meet the needs of various situations. To save time, individual styles as well as each dimensioning variable should be set on template drawings on the ANNO DIMS layer.

INTRODUCTION TO DIMENSION STYLES

 A dimension style is a set of dimension variables that control various aspects of placing the extension lines, dimension lines, and annotation in relation to these lines and line terminators. The type of drawing you are working on will affect how the variables are to be set.

- Dimensions that are placed on a site plan are often written in engineering units.

- Dimensions that locate the property lines are placed with no extension or dimension lines.

- Dimensions placed to locate utilities in an easement are placed using baseline dimensions expressed in feet and inches.

- Dimensions written on architectural drawings are written in feet and inches using continuous dimensions. Dimensions on elevations and sections can be placed using ordinate dimensioning.

You can create styles that incorporate the requirements for each of these dimensions and save them in a template drawing. You will be introduced to creating various styles later in this chapter.

Before a style can be created, you must have a thorough understanding of the dimension variables. Dimension variables are the qualities that control how dimensions will appear. You will be introduced to dimensioning variables through the default style of **Standard**. Later you'll be introduced to methods of creating specific styles to meet the needs of each drawing type.

CONTROLLING DIMENSION VARIABLES

Dimension variables are altered using the **Dimension Style Manager** dialog box, displayed in Figure 17.1. To access the dialog box, use one of the following methods:

- Select **Dimension Style** on the **Dimension** toolbar.

- Select **Dimension Style** on the **Dimension** dashboard.

- Select **Dimension Style** from the **Format** menu.

- Select **Dimension Style** from the **Dimension** menu.

- Type **D** ENTER at the Command prompt.

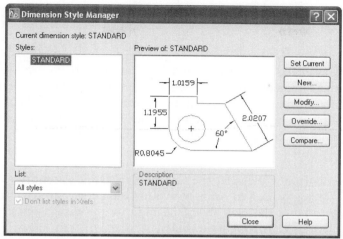

Figure 17.1 *Dimension variables are created and altered using the **Dimension Style Manager** dialog box. Access the dialog box by clicking the **Dimension Style** button on the **Dimension** toolbar and Dimension dashboard, by selecting **Dimension Style** from the **Format** menu, by selecting **Style** from the **Dimension** menu, or by typing D ENTER at the Command prompt.*

DIMENSION STYLE MANAGER

Setting a dimension style is similar to the methods used to set the styles for Text and Mleaders. The main areas of the **Dimension Style Manager** include the **Styles** box, a **Preview** window, and the **Set Current**, **New**, **Modify**, **Override**, and **Compare** buttons.

Current Dimension Style

Notice in Figure 17.1 that the current style is named **Standard**; this style features numbers displayed with four-place decimal units. As styles are created, the name of each style will be listed in the **Styles** box. The current style will be highlighted. Once multiple styles have been created, a different style can be made current by highlighting the name of the style and then selecting the **Set Current** button.

List

The **List** box displays the options that control the display of dimension styles. Options include **All Styles** and **Styles in Use**.

- With **All Styles** selected, all of the dimension styles contained in the drawing will be displayed.

- With **Styles in Use** selected, the dialog box will display only dimension styles that are referenced by dimensions in the drawing.

- Just below the **List** box is the **Don't list styles in Xrefs** check box. XREF drawings will be discussed in future chapters. With this box inactive, dimension styles contained in referenced drawings will be listed. When active, AutoCAD suppresses the display of dimension styles in externally referenced drawings in the **Styles** list.

Set Current

Five buttons are available for creating and altering the dimension style. Each will be introduced now and explored in detail throughout the chapter. The **Set Current** option has already been briefly introduced. Once a style is created, you can make it the current style by highlighting the style name and picking the **Set Current** button. Each of the settings of the drawing style will be applied to the new dimensions.

New

Selecting the **New** button will display the **Create New Dimension Style** dialog box shown in Figure 17.2. The box is used to name new dimension styles, to select the style that will be used as a base to build the new style, and indicate the dimension types that the new style will be applied to. Components include:

New Style Name—By default, AutoCAD will assign **Copy of Standard** as the new style name. With the name highlighted, change the style name to provide a useful description.

Start With—This list box determines what will be used as the base to form the new style. Rather than starting from scratch, AutoCAD will use the **Standard** style as a base and allow these options to be adjusted to meet the drawing needs. As new styles are created, each style name will be listed, allowing it to be selected as the base for a new style.

Annotative—This feature is used to assign annotative features to dimensions.

Use for—Selecting the **Use for** edit arrow will display a list of options. This allows a style to be created that will be applied only to certain types of dimensions. This option could be used if you would like to use the **Standard** style with black text as a base but allow all radius dimensions to be red. Selecting **Standard** as the **Start With** option, and **Radius Dimensions** as the **Use for** selection will begin the process of creating a sub-style of the **Standard** style. With these options active, each time the **Standard** style is used, radius dimensions will be red, and all other dimensions will be black.

Continue—Selecting the **Continue** button will display the **New Dimension Style** dialog box. The display will be explored in the next section of this chapter.

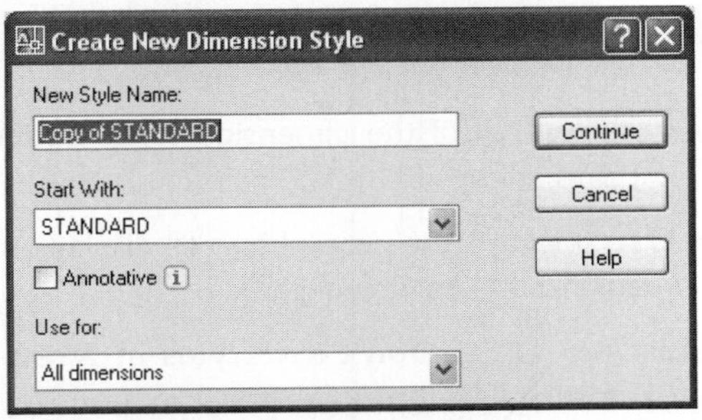

Figure 17.2 *Selecting the **New** button will display the **Create New Dimension Style** dialog box.*

Modify

Selecting the **Modify** button allows an existing style to be modified. As the **Modify** button is selected, the **Dimension Style Manager** dialog box is removed and the **Modify Dimension Style** dialog box is displayed. The name of the current style will also be displayed in the Title bar. The dialog box is identical to the **New Dimension Style** dialog box. Methods of altering an existing style will be explored later in this chapter.

Override

Selecting the **Override** button will produce the **Override Dimension Style** dialog box. The box is identical to the **New Dimension Style** dialog box. This option allows a temporary override to a dimension style. Methods of overriding an existing style will be explored later in this chapter.

Compare

Selecting this option displays the **Compare Dimension Styles** dialog box. The contents of the display will depend on the styles being compared. This dialog box can be very helpful in comparing the differences between two styles or in displaying all of the properties of one style.

CREATING A NEW STYLE

Selecting the **Create New Dimension Style** dialog box displays the **New Dimension Style** dialog box shown in Figure 17.3. The exact title will vary depending on the name of the style being created. The dialog box contains seven tabs for controlling the dimension style properties.

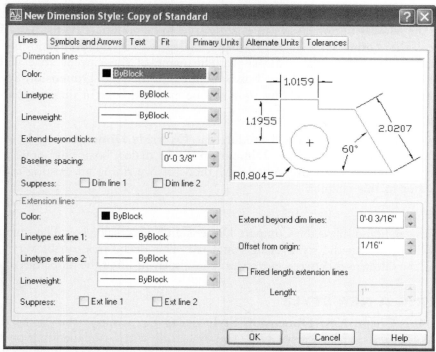

Figure 17.3 *Selecting the **Continue** button in the **Create New Dimension Style** dialog box will display the **New Dimension Style** dialog box and allow the properties of a dimension style to be set.*

LINES TAB

The **Lines** tab contains two major areas for adjusting the lines to be used with the new style. These are the **Dimension Lines,** and **Extension Lines** display windows. A display window is part of each tab and provides a visual representation of how the dimension properties will be displayed in the drawing area using the current settings. Each time a property is modified, the display will be updated to reflect the new property setting.

Dimension Line Controls

The **Dimension Lines** area contains settings to control **Color, Linetype, Lineweight, Extend beyond ticks, Baseline spacing**, and **Suppress.** These controls are used to adjust the properties of the dimension line only.

Color The **Color** list displays and sets the color for dimension leader lines. Selecting the **Color** edit arrow displays the seven basic colors. Selecting the **Other** option displays the **Select Color** dialog box. Although a different color can be set for each aspect of a dimension, the use of layers is a common method of controlling the color of dimension lines. For best results, select the **Color** of **ByLayer**. This setting indicates that the current layer setting will be used to describe the selected object. Setting a color here will allow the dimension to be a different color than other objects on the same layer.

Linetype   Selecting the **Linetype** edit arrow displays the linetype list. This option will allow any linetype that is loaded into the drawing to be used as the linetype to create dimension lines. No matter how may options you have, the continuous linetype controlled by the LAYER command is best for dimension lines. For best results, set the **Linetype** to **ByLayer**.

Lineweight   Selecting the **Lineweight** edit arrow displays the lineweight list, allowing the lineweight of dimension lines to be set. The lineweight for dimension lines should be thin lines controlled by the LAYER command. For best results, set the **Lineweight** to **ByLayer**.

Extend Beyond Ticks   By default, this option is inactive. It will become active once the arrowhead is altered using the **Symbols and Arrows** tab. This option controls how far the dimension line will extend past the extension line. Figure 17.4 shows examples of adjusting the extension. Most architectural offices set the extension value to equal the height of the text to be used. Once the arrowhead has been changed, come back and set this value to 1/8".

Figure 17.4   *The default placement of the dimension line extension on the left is often altered so that the dimension line extends past the extension line.*

Baseline Spacing   Selecting the **Baseline spacing** edit box allows the distance between lines of dimensions to be set. AutoCAD refers to this increment as the baseline setting. The default setting for architectural units is 0'–0 3/8". Most professionals use a distance of 1/2". Figure 17.5 shows an example of the baseline spacing.

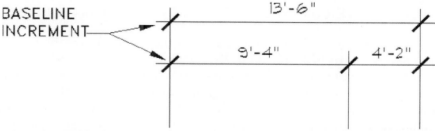

Figure 17.5   *The **Baseline spacing** setting controls the spacing between stacked lines of dimensions.*

Suppress This option works as a toggle that allows parts of the dimension line to be omitted. The **Dim Line1** and **Dim Line2** check boxes allow the first or second portion of the dimension line to be omitted from the drawing. The first or second dimension line is relative to the selection of the first and second extension line. These options are generally not set as active for construction drawings.

Extension Line Controls

This area has features similar to those in the **Dimension Lines** area, except these settings control each aspect of the extension lines in relation to the object being dimensioned.

Color This option can be used to set the color of the extension lines. Selecting the **Color** edit arrow displays the color list. Although a different color can be set for extension lines, the use of layers is the best method of controlling the color of extension lines. For best results, select the **Color** of **ByLayer**.

Controlling Linetype AutoCAD provides options for controlling the linetype used for each extension line. Options include **Linetype ext line 1** and **Linetype ext line 2**. Each is used to determine the type of line to be used to represent the extension line. The first or second extension line is relative to the selection of the first and second points used to determine what will be dimensioned. Selecting the **Linetype** edit arrow displays the linetype list. This option will allow any linetype that is loaded into the drawing to be used as the linetype to create extension lines. This option allows a continuous line to be used for one extension line and a centerline to be used for another. Options for using this feature include:

- Creating one style for continuous extension lines, and a separate style for centerlines.

- Setting both extension lines as continuous, and using **Properties** to override the **Linetype** setting. This option will be explored later in this chapter as editing dimensions is explored.

For now set both extension lines using the **ByLayer** setting.

Lineweight Selecting the **Lineweight** edit arrow displays the lineweight list allowing the lineweight of extension lines to be set. Lineweights for extension lines should be thin lines controlled by the LAYER command. For best results, set the **Lineweight** to **ByLayer**.

Suppressing Extension Lines This option allows for the suppression of the first and second extension lines, as shown in Figure 17.6. By default, extension lines will be displayed at each end of the dimension line. This option works well for establishing overall sizes. The first extension line can be omitted by selecting the **Ext Line1** box. With this option active, as the location for extension lines is selected, the first location will not receive an extension line. Selecting both boxes suppresses the extension lines at each end of the dimension line. This is useful if a dimension line is to be placed between two existing extension lines. This option also works well when the object is being used as the extension lines. This is typically done when dimensioning between floor and ceiling levels on sections or elevations. Figure 17.7 shows examples of suppressing extension lines.

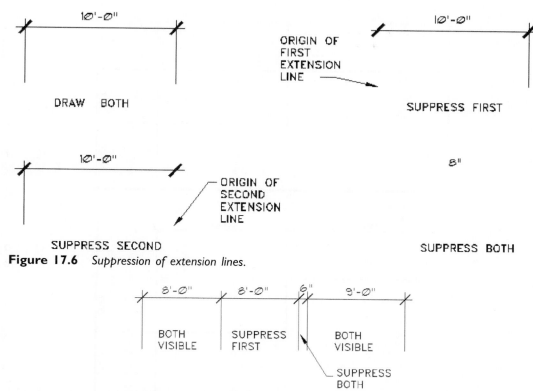

Figure 17.6   *Suppression of extension lines.*

Figure 17.7   *Practical applications for suppressing extension lines.*

Extend Beyond Dim Lines   The **Extend beyond dim lines** option allows the distance that the extension line extends past the dimension line to be controlled. The default value will depend on the current precision that was selected in the **Units** setup. An extension offset of 1/8" is common on many professional drawings. The effect of setting **Extend beyond dim lines** can be seen in Figure 16.2.

Offset from Origin   The **Offset from origin** option is used to control the gap between the extension lines and the object being dimensioned. A value of 1/8" is common in many offices. The effect of setting the **Offset from origin** is shown in Figure 16.2.

Fixed Length Extension Lines   This option allows the length of the extension lines to be controlled. This is helpful when placing dimensions on an **L** shaped structure. Figure 17.8 shows the contrast of toggling this option **ON** or **OFF**. If the option is set to ON, the

Length option will now be activated. The length entered in the edit box will control the length of the short extension line.

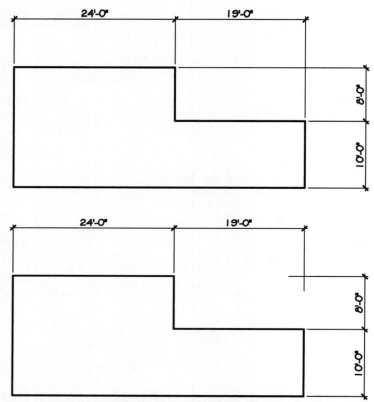

Figure 17.8 *The Fixed Length Extension Lines option allows the length of extensions to be controlled.*

SYMBOLS AND ARROWS

The **Symbols and Arrows** tab shown in Figure 17.9 contains controls for **Arrowheads, Center marks, Arc length,** and **Radius** dimensions. The display window presents a visual representation of how the dimension properties will be displayed in the drawing area using the current settings. Each time a property is modified, the display will be updated to reflect the new property setting.

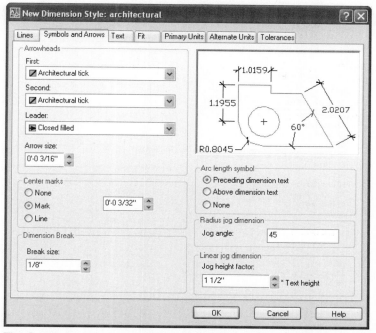

Figure 17.9 *The **Symbols and Arrows** tab provides options for fine-tuning dimension features.*

Arrowheads

This area controls the type of dimension line terminator to be used. The default selection for the first and second terminator, the closed filled arrow, is displayed in the edit box. By default, AutoCAD uses the terminator selected for the first dimension line used for the second line. The type of terminator can be selected by selecting the button or the edit arrow on the right side of the **First** list box. Selecting the button or the down arrow displays the options list shown in Figure 17.10. Making the selection of **Architectural Tick** automatically alters the display of each arrow to tick marks. You can alter the Second terminator to be different from the first by using the list for second option. This is rarely done on professional drawings, and you should not decide to start a new trend!

 Note: Once you've set the terminators, go back to the **Lines** tab and set the **Extend beyond Ticks** setting to 1/8".

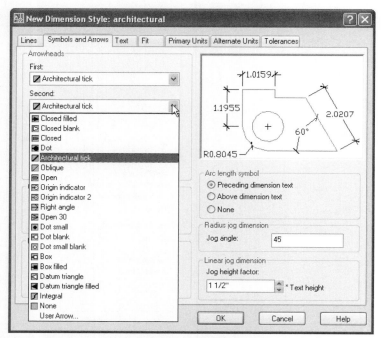

Figure 17.10 *Selecting the button or the down arrow displays the terminator list.*

User Arrow Most architectural and engineering offices use the **Oblique, Architectural Tick,** or **User Arrow** for a terminator. The user-defined tick (User Arrow) can be created using the **Block** command (discussed in chapter 18). Once the desired block is drawn, it can be named and saved to the drawing base. Selecting **User Arrow** displays the **Select Custom Arrow Block** dialog box. The name that was assigned to the block can be assigned in the **Select Custom Arrow Block** dialog box. The display window will now display the name of the block rather than showing an icon.

Leader The **Leader** option is used to control the type of the arrowhead used on leader lines.

Arrow Size The **Arrow size** edit box is used to determine the size of the arrowhead for leader lines relative to other dimensioning features. Selecting a tick value that is the same height as the text is a safe guideline to follow. If 1/8" high text is to be used, change the value to .125.

Center Marks

The **Center Marks** box is used to control how and when the center mark for radial dimensions will be displayed. The center mark is drawn only if the dimensions for **Center, Radius,** or **Diameter** are placed outside the circle or arc being described. Figure 17.11 shows the options for placement.

- The **Mark** setting displays the center mark using the **Center Size** value.

- Selecting the **Line** option displays centerlines that project beyond the limits of the circle.

- Selecting the **None** radio button will display no markings for the center point of a circle or arc.

Size   The **Size** edit box is used to control the size of the center mark. The setting controls the size of one-half of the center mark. Setting a size value of .125" will produce a center mark with a total length of .25". Many professionals adjust the default setting of 1/16" to 1/8".

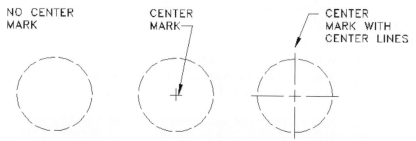

Figure 17.11   *Options for displaying center marks include none, mark and line.*

Dimensions Break

This setting controls the gap width of dimension breaks. The **Break size** edit box displays and sets the size of the gap used for dimension breaks.

Arc Length Symbol

The **Arc Length Symbol** area controls the display of the arc symbol in an arc length dimension. (See Figure 16.23.) Options include:

- **Preceding Dimension Text**—This toggle switch places arc length symbols before the dimension text.

- **Above Dimension Text**—This toggle switch places the arc length symbol above the dimension text.

- **None**—This toggle switch controls the suppression of the display of arc length symbols.

Radius Dimension Jog controls the display of the zigzag used in a radius dimension placed with the **Jogged** command. The **Jog angle** determines the angle of the transverse line that connects the extension and dimension lines of a radius dimension. See Figure 17.12.

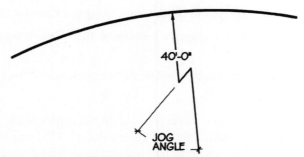

Figure 17.12 *The **Jog angle** setting controls the angle of the transverse line in a dimension placed using the **Jogged** command.*

TEXT TAB

The text used with dimensioning can be controlled using the **Text** tab. The tab contains controls for **Text appearance, Text placement, Text alignment,** and the image box. Each of these elements can be seen in Figure 17.13.

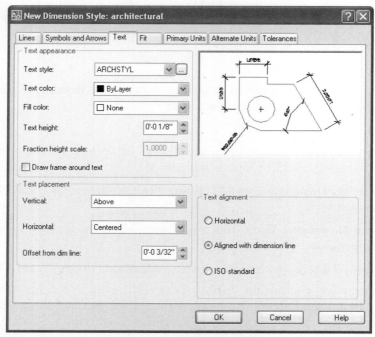

Figure 17.13 *The **Text** tab is used to set the appearance, placement, and alignment of the text that is used with dimensions.*

Text Appearance

The **Text Appearance** portion of the **Text** tab contains controls for the dimension text style, color, height, size, and frame.

Text Style   The **Text style** edit box displays the name of the current text style. It also allows one of the defined text styles to be used for dimension text. Selecting the **[...]** (ellipsis) button beside the edit box displays the **Text Style** dialog box, which allows a new text style to be created. For this example, the ARCHITECTURAL style created as the style for Text will be used.

Text Color   The **Text color** box displays and sets the color for dimension text. Selecting the **Color** edit arrow displays the seven basic colors. Selecting the **Other** option displays the **Select Color** dialog box. Assigning colors to dimension text by layer is generally the preferred method.

Text Height   The **Text height** box displays the current text height and allows the height of the dimension text to be altered. The height entered here must be coordinated with the height that is entered in the **Text Style** dialog box. If a text height is entered in the **Text Style** dialog box, that height will override the text height entered here in the **Text** tab. For the value of the **Text** tab to be active, a text value of 0 must be entered in the **Text Style** dialog box. Dimension text that is 1/8″ (3 mm) high is typical.

Fraction Height Scale   The **Fraction height scale** option sets the scale of fractions relative to the dimension text. In its current state, the option is inactive. For this option to be active, the **Precision** must be set on the **Primary Units** tab. Once a precision has been set, the scale factor can be altered. With the default value of 1.0000, the height of fractions will be the same as the dimension text height.

Framing Text   The **Draw frame around text** check box is currently inactive. Selecting this option will place a box around the text. With the current settings, selecting the box will do nothing. Once values are altered in the **Text Placement** area, selecting this option box will display a box around a dimension similar to Figure 17.14.

FRAME AROUND TEXT

Figure 17.14   *Once values are altered in the **Text Placement** area, selecting the **Draw frame around text** check box will display a box around a dimension. This option can be used to draw attention to a dimension or to distinguish between new and existing construction.*

Text Placement

The **Text Placement** area of the **Text** tab controls the vertical and horizontal placement of the dimension text, as well as the offset distance.

Vertical   This option controls the vertical justification of dimension text relative to the dimension line. Options include **Centered, Above, Outside,** and **JIS.** Figure 17.15 shows the two options that relate to construction drawing. The **Outside** or the **JIS** (Japanese Industrial Standards) are not used on construction drawings.

> **Centered**—This option breaks the dimension lines so that text is centered between the dimension lines.

> **Above**—This option is often used for architectural and engineering drawings. The option places the dimension text above the dimension line. The distance from the dimension line to the bottom of the text is controlled by the **Offset** from dim line option.

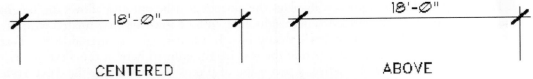

Figure 17.15   *Two common settings of the* **Vertical Text Placement** *list.*

Horizontal   The **Horizontal** option allows the location of text along the dimension line to be set. The edit box can be used to control where the text is placed. Selecting the down arrow allows the options to be altered. Selecting an option makes it current and removes the list. Examples of each option can be seen in Figure 17.16. Options include:

> **Centered**—The default value places text centered along the dimension line.

> **At Ext Line 1**—This option places the text close to the first extension line.

> **At Ext Line 2**—This option places the text close to the second extension line.

> **Over Ext Line 1**—This option places the text parallel and above the first extension line.

> **Over Ext Line 2**—This option places the text parallel and above the second extension line.

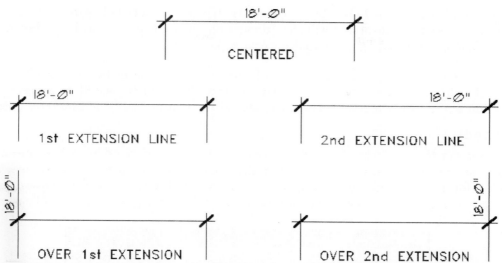

Figure 17.16   *Placement of the dimension text using the **Horizontal Text Placement** list.*

Offset from the Dimension Line   The **Offset from dim line** edit box displays the current gap between the text and the dimension line. The setting also allows the gap between the dimension line and the text to be altered. In construction drawings, the text is placed above the dimension line. The gap value should be set to equal half of the text height value.

Text Alignment

The **Text Alignment** area of the **Text** tab can be used to control the orientation of dimension text to the dimension line. Choosing one of the inactive options alters the selection. Options include the default option of **Horizontal, Aligned with dimension line**, and **ISO Standard**. Figure 17.17 shows an example of each type of text placement.

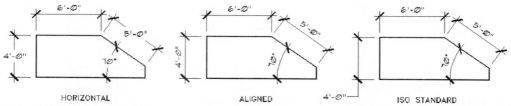

Figure 17.17   *The **Text Alignment** area of the **Text** tab can be used to control the orientation of dimension text to the dimension line. Most construction drawings use aligned dimensions.*

Horizontal   With this box active, dimension text is placed horizontal, regardless of the angle of the dimension line. The option should be set to **OFF** for most construction drawings so that the text remains parallel to the dimension line.

Aligned with Dimension Line This option is typically associated with construction drawings. Selecting this option will align dimension text to be parallel with the dimension line. This option should be activated for cost applications.

ISO Standard This option aligns text with the dimension line if the space between the extension lines is wide enough to display the text. Text will be placed in a horizontal position outside the extension lines if the text will not fit between the extension lines.

FIT TAB

The **Fit** tab provides controls for the placement of text in relationship to the dimension line. The tab can be used to set the values that control the placement of text, arrows, and leaders for dimension placement. Figure 17.18 shows the **Fit** tab of the **Dimension Style** dialog box. Figure 17.19 shows common text placements that might be seen on a floor plan.

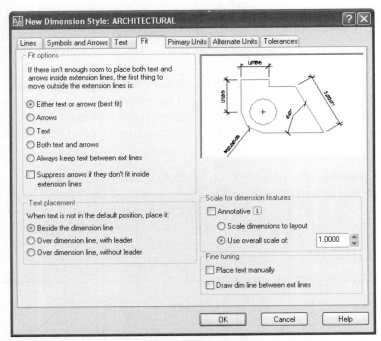

Figure 17.18 *The **Fit** tab of the **Dimension Style** dialog box.*

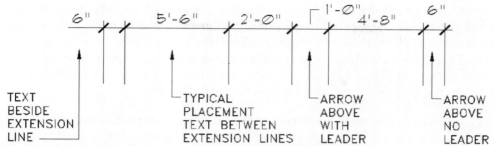

Figure 17.19 *Common dimension text placements used on floor plans.*

Fit Options

This portion of the **Fit** tab controls the placement of text and arrowheads if there is not enough space to place both inside the extension lines. If you are using Architectural Ticks, you'll have more space between the extension lines than if you use arrows. Figure 17.20 shows examples of placement options.

 Note: Remember you're choosing what you want to move when space it tight. Always try and keep the text in the default position.

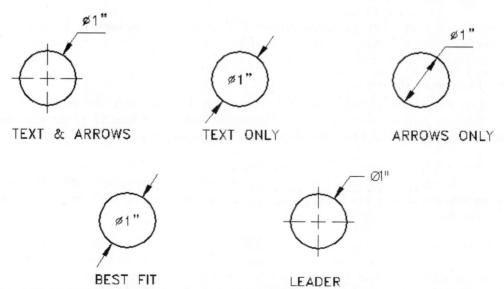

Figure 17.20 *Examples of **Fit** options. Notice that depending on the size of the space to be dimensioned, some options produce the same results.*

Either the Text or the Arrows, Whichever Fits Best   Selecting the default option will display the arrows and the text inside the extension lines when space is available. If sufficient space is not available between the extension lines, the text is placed between the extension lines and the arrows are placed outside the extension lines. If no space is available for text, only the arrows are placed between the extension lines. If space is not available for either text or arrows, both will be placed outside the extension lines.

Arrows   This option displays the arrows and the text inside the extension lines when space is available. If sufficient space is not available between the extension lines, the arrows will be placed outside the extension lines.

Text   This option also displays the arrows and the text inside the extension lines when space is available. If sufficient space is not available between the extension lines, the arrows are placed between the extension lines and the text is placed outside the extension lines. If no space is available for text, both will be placed outside the extension lines.

Both Text and Arrows   This option always displays the arrows and the text outside the extension lines. This option works well with small dimensions placed at the end of a dimension string.

Always Keep Text Between Ext Lines   This option always displays the text between the extension lines. This option works well with small dimensions placed in the middle of a dimension string.

Suppressing Arrows   The **Suppress arrows if they don't fit inside the extension line** option keeps the text between the extension lines and removes arrowheads when space is not available.

Text Placement

The **Text Placement** portion of the **Fit** tab controls the placement of text when it is not in the default position. Options include **Beside the dimension line**, **Over the dimension line, with a leader**, and **Over the dimension line, without a leader**.

Beside the Dimension Line   This option displays the text and the arrows outside the extension lines. This option works well in combination with using the **Either the text or the arrows, whichever fits best** option of the **Fit Options** area with small dimensions placed at the end of a dimension string.

Over the Dimension Line, with a Leader   This option displays the text above the normal position of text and to the side of the dimension line and uses an arrow to reference the text to the dimension line. This option works well in combination with using the **Either the text or the arrows, whichever fits best** option of the **Fit Options** area with small dimensions placed in the middle of a dimension string, when several small dimensions will be placed near each other.

Over the Dimension Line, without a Leader   This option displays the text above the normal position of text above the dimension line but does not use an arrow to reference the text to the dimension line. This option works well in combination with using the **Either the text or the arrows, whichever fits best** option of the **Fit Options** area with small dimensions placed in the middle of a dimension string.

Scale for Dimension Features

Up to this point of planning dimensions, each dimension will only be legible on a 12" × 9" screen. This area of the **Fit** tab sets the overall scale value or the paper space scaling for all dimensions. Options include **Annotative, Scale dimensions to layout** (paper space) and **Use overall scale of.**

Annotative   Activating the **Annotative** button will assign annotative features to each component of a dimension. With the option **ON,** the scale of annotative features will be determined by the scale that is assigned to the workspace.

Scale Dimensions to Layout   Selecting the **Scale dimensions to layout** (paper space) box will apply a scaling factor to dimensions plotted in paper space. The option sets a scale factor based on the scaling between the current model space viewport and paper space.

Use Overall Scale of   The **Use overall scale of** edit box allows the scale of all dimensioning features to be increased or decreased based on the scale a drawing is to be plotted at. You've been drawing in model space with actual sizes so far. Text that is to be 1/8" high when plotted at a scale of 1/4"=1'–0" will need to be set to 6" in height. To have legible arrows, offsets, and text you'll have to convert every variable. If you actually have a life, you might just want to set the scale factor to 48. This will control all dimension variables and leave you a little free time. To find the desired scale factor, take the reciprocal of the drawing scale. A scale of 1/4"=1'–0" is 12/.25, with a scale factor of 48. Other common scale factors include the following:

ARCHITECTURAL VALUES		ENGINEERING VALUES	
Plotting Scale	**Scale Factor**	**Plotting Scale**	**Scale Factor**
3/4"=1'–0"	16	1"=1'–0"	12
1/2"=1'–0"	24	1"=10'–0"	20
3/8"=1'–0"	32	1"=100'–0"	1200
1/4"=1'–0"	48	1"=20'–0"	240
3/16"=1'–0"	64	1"=200'–0"	2400
1/8"=1'–0"	96	1"=30'–0"	360
3/32"=1'–0"	128	1"=40'–0"	480
1/16"=1'–0"	182	1"=50'–0"	600
		1"=60'–0"	720

Fine Tuning

The final portion of the **Fit** tab allows minor text adjustments to be made: Options include **Place text manually** and **Draw dim lines between ext lines**.

Place Text Manually   Activating this check box allows the dimension text location to be specified as you place each dimension regardless of the justification. With this box active, AutoCAD ignores other justification settings and allows you to specify where the text will be placed. Adjusting the variable can be useful on radial dimensions, as shown in Figure 17.21.

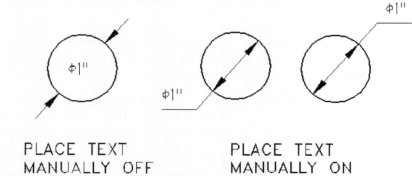

PLACE TEXT
MANUALLY OFF

PLACE TEXT
MANUALLY ON

Figure 17.21   *With the **Place text manually** check box active, AutoCAD ignores other justification settings and allows you to specify where the text will be placed.*

Draw Dimension Line   When a small space is dimensioned the text will be placed outside the extension lines. Selecting the **Draw dim line between ext lines** option will place a dimension line between the extension lines even when the text and arrows are outside the extension lines. Figure 17.22 shows the effect of using this option on a radial dimension.

OFF

ON

Figure 17.22   *Selecting the **Draw dim line between ext lines** option will place a dimension line between the extension lines even when the text and arrows are outside the extension lines.*

PRIMARY UNITS TAB

A dimension of 14'–0" is considered a primary dimension unit. Most construction-related dimensions consist of primary units, with no other information required. Occasionally a tolerance might be added to the primary units. This portion of the **Modify Dimension Style** dialog box controls the display of the primary measurement units and any prefixes or suffixes for the dimension text. Selecting the **Primary Units** tab displays a tab similar to Figure 17.23. This tab is used to set the variables that control the appearance of text for dimension and leader lines. The tab includes the areas of **Linear Dimensions**, **Measurement Scale**, **Zero Suppression**, and **Angular Dimensions**.

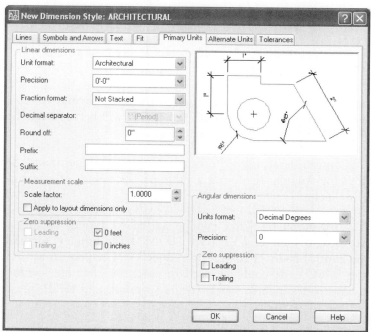

Figure 17.23 The **Primary Units** tab.

Linear Dimensions

This portion of the **Primary Units** tab displays and sets the format and precision for linear dimensions. Key elements include the **Unit format**, **Precision**, **Fraction format**, **Decimal separator**, **Round off**, **Prefix**, and **Suffix**.

Unit Format Even though the units of the drawing are set to Architectural, the default **Unit format** setting is decimal. Select the down arrow to display the list, which includes Scientific units, Decimal units, Engineering units, Architectural units, Fractional units, and Windows Desktop. The **Architectural** units will meet the needs for most drawings, although **Engineering** units are used on site-related drawings.

Precision Selecting the down arrow of the **Precision** edit box displays the precision of the dimension. Options for the Architectural format range from 0'–0" through 0'–0 1/256" For most dimensions, a setting of 0'–0" will provide needed accuracy. Most engineers and architects determine the fractions based on calculations rather than relying on drawing accuracy. If a fraction needs to be added to the dimension, it can be added by editing the text once the dimension is displayed.

Fraction Format The **Fraction format** edit box sets the format for the display of fractions. Options include **Diagonal, Horizontal**, and **Not Stacked**. Each option can be seen in Figure 17.24.

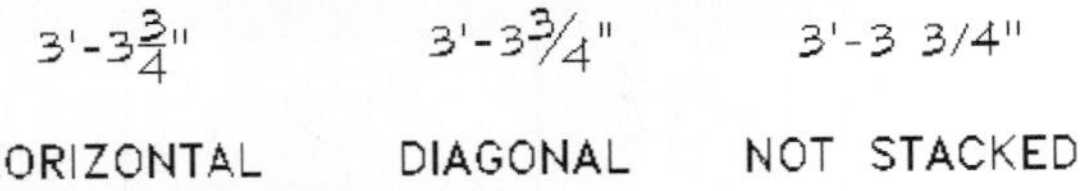

$$3'-3\tfrac{3}{4}" \qquad 3'-3\,^3\!/_4" \qquad 3'-3\ 3/4"$$

HORIZONTAL   DIAGONAL   NOT STACKED

Figure 17.24   *The **Fraction format** options.*

Decimal Separator This option is inactive when the **Unit format** is set to **Architectural** units. If the **Unit format** is set to **Decimal**, the option will be activated. Selecting the edit arrow displays the period, comma and space options.

Round Off This option determines the value of rounding measurements for all dimensions types except for Angular. Entering a value of 1/8 rounds all measurement values to the nearest eighth inch.

Prefix and Suffix A prefix or suffix can be added by typing in the appropriate box. Common suffixes might include **TYP** (typical), **MM** (millimeter), or **MAX** (maximum). You may use the control codes that were introduced in Chapter 16 as prefixes or suffixes. For example, the **%%C** code could be used to enter the diameter symbol as a prefix or suffix. Other control codes include **%%D** (degree symbol) and **%%P** (plus/minus symbol).

Measurement Scale

The **Measurement Scale** area of the **Primary Units** tab provides options for setting the scale factor and a method for limiting the scale values.

Scale Factor The **Scale** box specifies the scale factor of a dimension without affecting the component, angles, or tolerance values. The option affects all linear distances except Angular. Alter the scale factor by using the up/down arrows or by highlighting the existing value and then typing a new value. A scale factor of 2 doubles the value of the dimension that is entered on the drawing. With a scale factor of 2, a dimension of 8" will be written when a 4" long object is dimensioned. For most architectural drawings the default value of 1 does not need to be altered.

Apply to Layout Dimensions Only The **Apply to layout dimensions only** check box will apply a scaling factor to dimensions plotted in paper space. The layout scale factor will be adjusted to reflect the zoom scale factor for objects created in model space viewports.

Zero Suppression

This area of the **Primary Units** tab controls the display of zeros for leading, trailing, feet and inches. When a check mark is in the box, it is activated.

Leading   With this box active, the zero is suppressed preceding a decimal point in units less than one. Seven tenths will be written .7 rather than 0.7. This option will not be available when Architectural units are current.

Trailing   This box controls the display of zeros after a decimal point. With the box active, no zeros will be displayed after the decimal. Seven tenths will be written .7 rather than .70. This option will not be available when **Architectural units** are current.

0 Feet   This option is only available when Architectural units are current. A dimension less than one foot can be displayed as 0'–8" or 8". Professional practice is mixed, and the setting for the **0 Feet** box will vary from office to office. Selecting the **0 Feet** box causes the zero representing feet to be suppressed when the dimension is less than twelve inches.

0 Inches   This option is only available when Architectural units are current. Most professionals display the zero after the foot symbol. Writing 5'–0" is preferred to writing 5'. With this box active, zeros will be suppressed. This box should be inactive.

Controlling the Display of Angles

The **Angular Dimensions** area controls the format of angular dimensions. Options include **Units format, Precision,** and **Zero Suppression.** Options are similar to their linear counterparts.

Units Format   The **Units format** for angular dimensions includes the options of Decimal Degrees, Degrees/Minutes/Seconds, Gradians, and Radians. The Decimal Degrees option will meet the needs for most architectural drawings. The Degrees/Minutes/Seconds option will work best for most site-related drawings.

Precision   The Precision list will vary depending on the **Units format** setting. When units are measured in decimals, the precision will be measured in decimal places from zero to eight places.

Zero Suppression   This area suppresses the leading or trailing zero of angular dimensions. With the **Leading** option activated, 0.75 will be displayed as .75. The **Trailing** option, when activated, suppresses the zeros that follow the decimal. .7500 will be displayed as .75.

ALTERNATE UNITS TAB

This portion of the **Modify Dimension Style** dialog box is not used to create a style for most construction drawings. As you display the tab, each option is inactive. Selecting the **Display alternate units** check box activates the options. The **Alternate Units** tab is shown in Figure 17.25. Major components include the **Alternate Units, Zero Suppression,** and the **Placement** areas. Features of this tab function in a manner similar to their counterparts on the **Primary Units** tab. One option that should be considered on this tab is the ability to create and display metric units behind the architectural primary units. Selecting **Decimal** as the **Unit format**, a multiplier for all units of 25.4, and a suffix of mm will display dimensions similar to Figure 17.26.

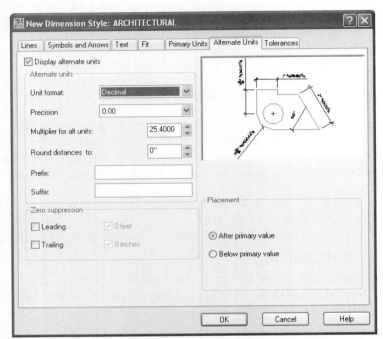

Figure 17.25 *The **Alternate Units** tab options are inactive as it is first displayed. Selecting the **Display alternate units** check box activates this tab, allowing alternate units to be displayed behind the primary units.*

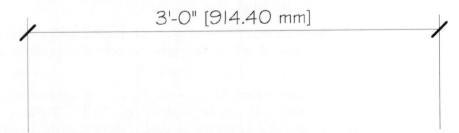

Figure 17.26 *Selecting **Decimal** as the alternate unit format, a multiplier for all units of 25.4, and a suffix of mm will display architectural units followed by metric dimensions.*

TOLERANCES TAB

A tolerance is a range that is allowed to compensate for human error. If a note specifies that bolts in a steel plate are to be 3" o.c., and the holes are placed at 3 1/16" o.c., should the plate be scrapped? A tolerance specification will define what is an acceptable range of variation. Most areas of construction do not use a tolerance zone, although some steel and concrete detailing might require tolerances. The **Tolerances** tab shown in Figure 17.27, controls the display and precision of the tolerance. Major components include the **Tolerance format, Tolerance alignment, Zero suppression**, and the **Alternate unit tolerance** areas. Features of this tab function in a manner similar to their counterparts on the **Primary Units** tab. With the default **Method** value of **None**, most of the box is inactive. Once a method of tolerance specification is selected, the balance of options on the tab will be activated. Figure 17.28 shows examples of the available tolerance methods.

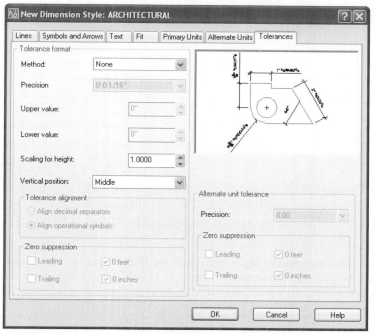

Figure 17.27 The **Tolerances** tab of the **Modify Dimension Style** dialog box controls the display and precision of the tolerances.

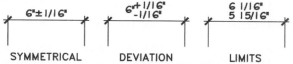

Figure 17.28 Tolerance methods available in AutoCAD include symmetrical, deviation, and limits.

COMBINING VARIABLES TO CREATE A STYLE

If you're feeling overwhelmed by styles, variables, and values, don't be discouraged. Remember, most of these values will be set on a template drawing and left alone. Throughout this chapter comparisons have been made using the **New Dimension Style** dialog box and its tabs. This section will help you combine these values to create a style. Dimensioning variables can be combined to create a particular style in much the same way that options are combined with **Text** and **Multileaders** to create different text styles. Styles can be created, saved, modified and quickly loaded by using the **Dimension Style Manager** dialog box shown in Figure 17.1.

CREATING DIMENSIONING STYLES

Dimensioning styles allow you to save groups of settings for dimension variables. If an office does both architectural and civil engineering projects, two distinct dimensioning styles could be created and saved in template drawings to speed the dimensioning process. Even in a firm that does only architectural drawings, varied styles can be useful to group option settings. For instance, on a multilevel structure, some dimensions will be shown on each floor plan and the framing and foundation plans. These might be placed in a style titled ARCHBASE that uses a certain size, color and style of text with two extension lines. Styles might be created for dimensions that will be used to dimension to the center of interior walls and have the left or right extension line suppressed, with another style created for a style that has both extension lines suppressed. Once created, these styles can be quickly loaded and interchanged.

Use the following steps to create a dimension style suitable for a plan view using annotative features.

1. Select the **Dimension Style** button on the **Dimension** toolbar, select **Dimension Style** from the **Format** menu or type **D** ENTER a the Command prompt. Each method will display the **Dimension Style Manager** dialog box similar to Figure 17.1.

2. Select the **New** button. This will display a **Create New Dimension Style** dialog box similar to Figure 17.2.

3. Enter the desired style name. In this case **PLAN VIEW BASE** will be used.

4. Select an existing style to use as a base. In this case the **Standard** style will be used.

5. Click the **Annotative** button so that this feature is active.

6. Select what dimensions that this style will be applied to in the **Use for** edit box. In this case, **All Dimensions** will be used.

7. Select the **Continue** button. This will display a **New Dimension Style: PLAN VIEW BASE** dialog box. Proceed through each tab and set the options to meet your specific situation. The next section will recommend values that you might want to consider using.

8. Once each option has been selected, click the **OK** button. This will return the **Dimension Style Manager** dialog box and show the name of the new style that has just been created.

9. Select PLAN VIEW BASE from the **Styles** box.

10. Select the **Set Current** button.

11. Select the **Close** button.

This process will create the PLAN VIEW BASE style and set it as the current dimension style.

COMMON STYLE SETTINGS

Although offices often have individual dimensioning standards, several common options can be set to meet most needs. Use the **Standard** style as a base to create the desired settings to update your template:

Lines Tab

> Color and Lineweight: **By Layer**
>
> Extend beyond ticks: **1/8"**
>
> Baseline spacing: **1/2"**
>
> Extend beyond dim lines: **1/8"**
>
> Offset from origin: **1/8"**

Symbols and Arrows Tab

> Arrowheads: **Architectural Tick**
>
> Leader: **Closed filled**
>
> Arrow size: **3/16"**
>
> Center Mark: **Line**
>
> Center Mark Size: **1/16"**

Text Tab

> Text style: *Stylus BT*
>
> Text color: *By Layer*
>
> Text height: *1/8"*
>
> Offset from dim line: *1/16"*
>
> Vertical: *Above*
>
> Horizontal: *Centered*
>
> Aligned with dimension line: *ON*

Fit Tab

Fit Options: **Arrows**

Text placement: **Beside**

Annotative: **ON** (If you decide not to use the annotative feature, use an overall scale of: **48** [for 1/4"=1'-0" scale].)

Draw dim line between ext lines: **ON**

Primary Units Tab

Unit format: **Architectural**

Precision: **0'–0"** (varies based on drawing)

Fraction format: **Not Stacked**

Zero suppression: 0 feet **ON** / inches **OFF**

Following the guidelines to create PLAN VIEW BASE will produce dimensions similar to Figure 17.29a. The dimension style works well for the overall dimensions, but it is somewhat limited when an object has a jog in it. If the style is used to place dimension A, the desired results will be obtained, but two extension lines will be placed on the left side of space A. The style will reproduce with the desired results if you're careful to always use OSNAP when you select where the extension lines will be placed. Failure to use OSNAP might produce results similar to Figure 17.29b. Even if OSNAP or Quick Dimension is used to place the dimensions for space B and C, the desired results will not be achieved. The extension line for B will extend to be within 1/8" of surface B, leaving no offset for surface A. The same problem will occur at B as surface C is dimensioned.

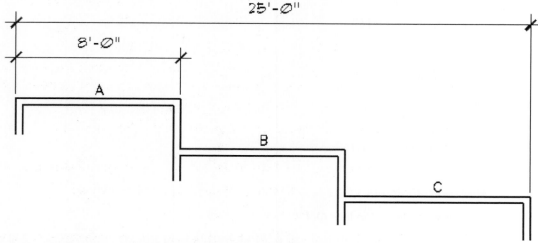

Figure 17.29a *Using the PLAN VIEW BASE style will produce dimensions suitable for many architectural applications.*

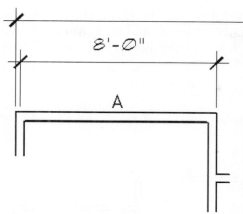

Figure 17.29b *Failure to use OSNAP might produce extension lines that do not align.*

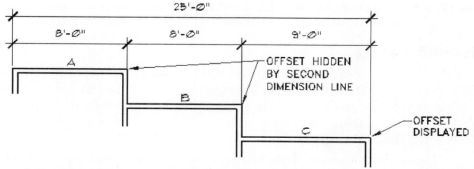

Figure 17.29c *Even if OSNAP or* **Quick Dimension** *is used to place the dimensions for space B and C, the desired results will not be achieved. The extension line for B will extend to be within 1/8" of surface B, leaving no offset for surface A. The same problem will occur at B as surface C is dimensioned.*

CREATING A NEW STYLE BASED ON AN EXISTING STYLE

A better method of placing the interior dimensions is to create a style that will suppress one or both of the extension lines. This can easily be done using the following procedure:

1. Display the **Dimension Style Manager** dialog box and select the **New** button.

2. In the **Create New Dimension Style** dialog box, enter a style name of INTERIOR.

3. Use **PLAN VIEW BASE** as the **Start With** style and use it with All Dimensions.

4. Click **Continue** to advance to the **New Dimension Style** dialog box.

5. Use the **Lines and Arrows** tab to suppress Line 1.

6. Click the **OK**, **Set Current**, and **Close** buttons to set the new style as the current style and return to the drawing area.

Using the PLAN VIEW BASE and the INTERIOR styles will allow dimensions to be placed accurately for the overall length and for spaces A, B, and C. Figure 17.29d shows the result of using the PLAN VIEW BASE and the INTERIOR styles.

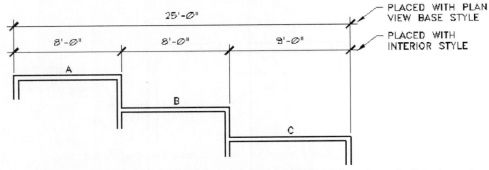

Figure 17.29d *Using the **PLAN VIEW BASE** and the **INTERIOR** styles will allow dimensions to be placed accurately for the overall length and for spaces A, B, and C.*

MODIFYING A STYLE

Once a style has been created, you might find that it is not all you hoped it would be. The **Modify** option in the **Dimension Style Manager** dialog box allows an existing dimension style to be altered. The process for altering an existing style is similar to the process used to create a new style using an existing style as a base. An existing style can easily be modified using the following procedure:

1. Display the **Dimension Style Manager** dialog box and select the style to be modified from the **Styles** display box.

2. With the desired style highlighted, select the **Modify** button. This will display the **Modify Dimension Style** dialog box. The box is exactly the same as the **New Dimension Style** dialog box.

3. Use the necessary tabs to alter the existing values. For the example, the **Symbols and Arrows** tab was used to alter the arrowheads.

4. Click the **OK** and the **Close** buttons to modify the existing style and return to the drawing area.

Using the **Modify** option on the INTERIOR style will alter the display to resemble Figure 17.30.

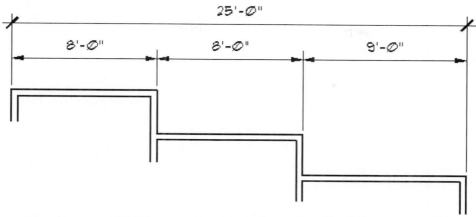

Figure 17.30 *Using the **Modify** option on the INTERIOR style will alter the display shown in Figure 17.29d to resemble this display. Remember that just because AutoCAD will allow something, doesn't make it acceptable. Don't mix arrowheads and tick marks on dimensions.*

OVERRIDING EXISTING STYLES

Using the **Modify** option of the **Dimension Style Manager** dialog box changes an entire style to the specified parameters. Using the **Override** option allows an existing style to stay intact while altering new features that are added to the style. Use the following steps to override an existing style:

1. Display the **Dimension Style Manager** dialog box and select the style to override from the **Styles** display box. In this example, the INTERIOR style will be used as the style to be altered.

2. With the desired style highlighted, select the **Override** button. This will display the **Override Current Style** dialog box. The box is exactly the same as the **New Dimension Style** dialog box.

3. Use the necessary tabs to alter the existing values. For the example, the **Lines and Arrows** tab was used to alter the arrowheads, and the **Text** tab was used to center the text between the dimension line.

4. Click the **OK** and the **Close** buttons to override the existing style and return to the drawing area.

Using the **Override** option on the INTERIOR style will alter the display to resemble Figure 17.31.

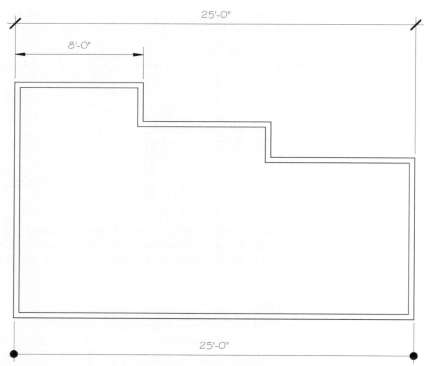

Figure 17.31 *Using the **Override** option on the INTERIOR style to alter the display to show dot arrowheads and centered text.*

COMPARING STYLES

The final option of the **Dimension Style** dialog box is to compare styles. This option can be used to list the properties of a style or to compare the properties of two dimension styles. The function performed depends on the settings of the **Compare** and **With** edit boxes. To list the properties of a style, select the name of the desired style in the **Compare** edit box, and select <**None**> in the **With** list. Entering a style name in the **Compare** edit box and a different style name in the **With** edit box will display a listing of the property differences similar to Figure 17.32. The list of properties and the list of differences can be copied to the Windows Clipboard. Once copied to the Clipboard, the listings can be pasted into other Windows applications.

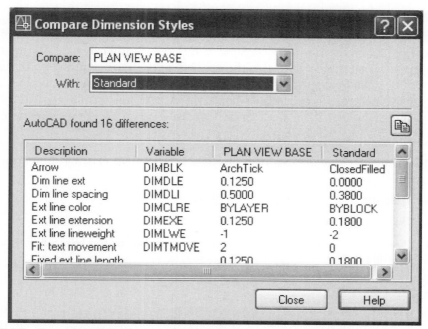

Figure 17.32 *Entering a style name in the* **Compare** *edit box and a different style name in the* **With** *edit box will display a listing of the property differences.*

RENAMING OR DELETING STYLES

A style can be renamed or deleted using the **Styles** portion of the **Dimension Style Manager** dialog box. Use the following procedure to rename a style:

1. Display the **Dimension Style Manager** dialog box and move the cursor to the **Styles** portion.

2. Highlight a style in the **Styles** portion of the dialog box.

3. With the style highlighted, right-click to produce a shortcut menu.

4. Select the **Rename** option and enter the desired new name.

5. Continue renaming other menu listings or select **Close** to remove the dialog box and return to the drawing area.

Similarly, deleting a dimension style is done in the same dialog box, using the following procedure:

1. Once the style to be deleted is highlighted, right-click and select the **Delete** option from the shortcut menu.

2. Continue deleting other styles or select **Close** to remove the dialog box and return to the drawing area.

APPLYING STYLES

Each dimension you create has a style. If you started to place dimensions prior to reading this chapter, the dimensions were created using the **Standard** style. Throughout this chapter, you've learned to create, modify, and override dimension styles. In addition to selecting the **Set Current** button in the **Dimension Style Manager** dialog box, two other methods are available to alter the dimension style. A new style can be applied by using the shortcut menu or by using the **Dim Style Control** box on the **Properties** toolbar.

Using the Shortcut Menu

Use the following steps to assign a new dimension style to an existing dimension with the shortcut menu:

1. Select the desired dimension and right-click.

2. Choose **Dim Style** from the shortcut menu.

3. Choose the name of the desired style to be applied to the selected dimension and press ENTER.

The process can be seen in Figure 17.33.

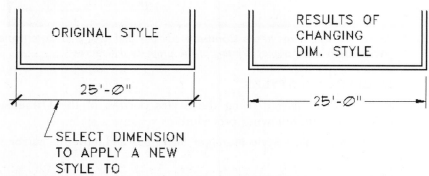

Figure 17.33 *A dimension style can be altered by activating the grips and then right-clicking to produce the shortcut menu. Select* **Dim Style** *from the shortcut menu, and then select the name of the desired style to be altered.*

Using the Dim Style Control Box

Use the following steps to assign a new dimension style to an existing dimension with the **Dim Style** edit box:

1. Select the desired dimension.

2. Display the **Dim Style Control** menu using the **Styles** toolbar.

3. Choose the name of the desired style to be applied to the selected dimension and press ENTER.

EDITING DIMENSION PLACEMENT

The placement of dimensions can be edited using grips or the editing commands. The editing commands can be selected from the shortcut menu or by the methods presented in earlier chapters. To use the shortcut menu, select a dimension with the cursor. Each of the grips will be activated. Right-click to obtain the editing menu. Notice in Figure 17.34 that a grip is displayed for the dimension text, at each intersection of the dimension and extension lines, and at what appears to be the end of the extension line. Closer examination will show the grips that appear to be at the end of the extension lines actually lie beyond the lines. These grips mark the definition points or **Defpoints** for the dimensions. AutoCAD automatically assigns definition points for every dimension to define the dimension location. Use the **Layers** button to examine the layers for the current drawing and you'll find that the program created a layer titled Defpoints to display the definition points. The layer will not be plotted. Dimensions can be edited using grips without knowing the location of the definition points. The Defpoints will automatically be displayed. If you alter dimensions using the edit commands, the Defpoints must be included in the selection set. Using the **Node** mode of **Object Snap** will aid in **Defpoint** selection.

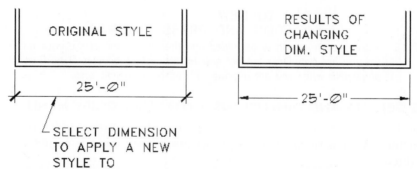

Figure 17.34 *A grip is displayed for the dimension text, at each intersection of the dimension and extension lines, and at the definition points (Defpoints) for the dimensions. AutoCAD automatically assigns definition points for every dimension to define the dimension location.*

STRETCHING DIMENSIONS

Dimensions can be stretched to alter their position using either grips or the **Stretch** command. The process of stretching a dimension can be seen in Figure 17.35. Use the following steps to alter the dimension position using grips:

1. Select the dimension to be stretched.

2. Make one of the grips on the dimension line the hot grip.

3. Drag the grip to the desired location. OSNAP will aid in accurate placement.

This process will stretch all features of the dimension. To move just the text, select the text grip. Make it the hot grip, and move the text to the desired location.

The **Stretch** command can be used to stretch dimensions using the same procedure that is used to move drawing objects. Use the **Crossing** option to select the dimension text to be relocated.

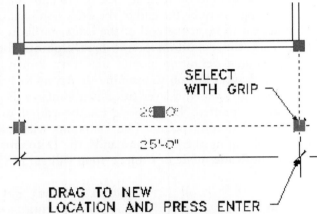

Figure 17.35 *Select the dimension to be edited and then make one of the grips on the dimension line the hot grip. Drag the hot grip to the desired location. OSNAP will aid in accurate placement when you are aligning with other dimension lines.*

EDITING OBJECTS AND DIMENSIONS WITH THE MODIFY MENU

Objects and dimensions can be edited using the editing commands contained on the **Modify** menu as well as the grip editing functions. As an object is edited, the dimensions are also edited.

Stretch

Dimensions can be stretched in the same way that drawing objects are stretched. If the dimension is selected with the object, as seen in Figure 17.36, both will be stretched, and the dimension text will be revised automatically.

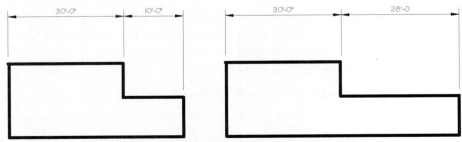

Figure 17.36 *Using the **Stretch** command to alter an object with dimensions.*

Extend

The **Extend** command sequence also can be used to extend a dimension line and adjust the text accordingly. Figure 17.37 shows the effects of the **Extend** command on dimensions.

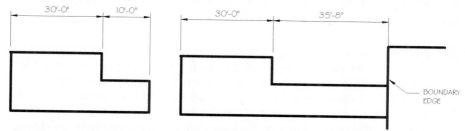

Figure 17.37 *The effect of the **Extend** command on an object with dimensions.*

Trim

The **Trim** command sequence can be used to shorten a dimension line and adjust the text accordingly. Figure 17.38 shows the effects of **Trim** on dimensions.

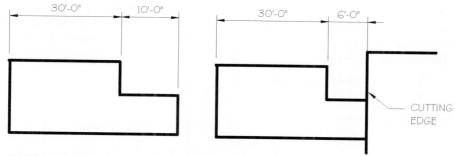

Figure 17.38 *The **Trim** command also affects an object, the dimension lines, and text.*

EDITING DIMENSION TEXT

Dimensions that have been placed on a drawing can be edited using the **Properties, Dimedit,** and **Dimtedit** commands.

EDITING DIMENSION TEXT USING PROPERTIES

 The dimension text can easily be edited using the **Properties** command. Use the following steps to alter dimension text using the **Properties** command:

1. Select the dimension to be edited.

2. Select the **Properties** button on the **Standard** toolbar.

3. Open the **Text** menu and choose **Text Override**.

4. Enter the desired dimension in the **Text Override** edit box.

5. Continue to alter properties or close the **Properties** dialog box.

Text entered in the **Text Override** edit box will override the measured dimension of the feature. The true measurement will continue to be displayed in the **Measurement** box of the **Properties** dialog box. Dimension text properties can also be altered by selecting the text, right-clicking and then selecting **Properties** from the shortcut menu. Other options for altering the location and text will be introduced later in this section.

Note: This feature can be especially helpful to dimension an object that must be a certain size, but you were unsure of the exact size when the part was drawn and dimensioned. If you know bolts are typically placed in steel plates at a 3" spacing, you can draw and dimension a detail prior to receiving the final calculation from the engineer. If the engineer determines that this application needs the bolts at 3 1/2" spacing, the dimension can be altered without editing the bolt locations on the drawing. The bolt locations will not be drawn to scale, but if the math works, so does the detail, and you've saved time.

EDITING DIMENSION TEXT USING DIMEDIT

The **Dimedit** command allows dimension text to be edited. The existing text can be edited, rotated, moved, or restored to its original position. The angle of the extension lines can also be altered from perpendicular to oblique. Start the command by selecting the object to be edited and then select the **Dimension Edit** button on the **Dimension** toolbar:

```
Click the Dimension Edit button.
Enter type of dimension editing [Home/New/Rotate/Oblique]:
```

Changing the Dimension Text

To alter existing text, select the **New** option:

```
Enter type of dimension editing [Home/New/Rotate/Oblique]:
    N ENTER
```

This will produce the **Text Formatting Editor** and allow the desired text to be entered. Remove the existing text, enter the desired dimension and then select the **OK** button to close the editor. The result of the command can be seen in Figure 17.39.

Figure 17.39 *The New option of the **Dimedit** command can be used to edit existing dimension text.*

Rotating Existing Text

Choosing the **Rotate** option from the menu allows existing text to be rotated. Selecting the existing text in Figure 17.40 to be rotated to a 45° angle would produce the following results.

Figure 17.40 *The Rotate option of the* **Dimedit** *command can be used to rotate existing dimension text.*

Altering the Extension Line Angle

The angle of existing extension lines can be altered using the Oblique option of **Dimedit**. Selecting Oblique from the menu and then entering a -75° angle value will produce the results shown in Figure 17.41.

Figure 17.41 *The Oblique option of the* **Dimedit** *command can be used to alter the oblique angle of existing extension lines.*

Returning Home

Using the default option of **Home** will return the text to its default position.

EDITING DIMENSION TEXT USING DIMTEDIT

The **Dimtedit** command is used to position the text relative to the dimension line. To start the command:

- Click the **Dimension Text Edit** button on the **Dimension** toolbar.

- Select **Align Text** from the **Dimension** menu.

- Type **DIMTED** ENTER at the Command prompt.

Any access method will prompt you to select a dimension to be edited. Once a dimension is selected, the following prompt will be displayed.

```
Specify new location for dimension text or [Left/Right/Center/
    Home/Angle]:
```

Choosing the **Left** option from the menu will move the text toward the left end of the dimension line. Choosing **Right** will move the text toward the right end of the dimension line. Choosing **Angle** will allow the text to be rotated, choosing **Center** will center the text on the dimension line, and choosing **Home** will return the text to its default position. Each option can be seen in Figure 17.42.

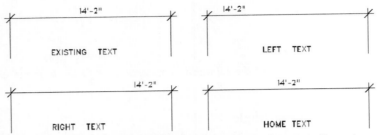

Figure 17.42 *The options of **Dimtedit** will alter the location of existing dimension text.*

UPDATING DIMENSIONS

The **Dimension Update** command allows existing dimensions to be selected and altered to a different style. The command is started by selecting the **Dimension Update** button on the **Dimension** toolbar or by selecting **Update** from the **Dimension** menu. The command duplicates the action of the **Dimension Text Edit** command but automatically applies the current dimension style. Use the following procedure to Update a dimension style:

1. Select the **Dimension Update** button.

2. Select the dimensions to update to the current dimension style.

3. Press ENTER.

WHAT IT ALL MEANS

More so than any area of AutoCAD, dimensioning seems to cause new users the most anxiety. You've read through what must seem like an unlimited number of commands and options, and you might be having a hard time getting all of this information organized. The good news is that once you set the variables using the dialog boxes and store them in template drawings, most of your work is done.

CHAPTER 17 EXERCISES

1. Open drawing **Floor 14**. Use the appropriate dialog box and a style suitable for dimensions with both extension lines, a style with one dimension line, and a style with no extension lines. Set the text offset from the dimension line for all styles to be 1/16", with a baseline increment of 1/2". Assign each style a different color. Mark circles with a 1/8" center mark and provide for centerlines. Use 1/8" for the feature offset and extension above line settings. Use 1/8" long tick marks and a 1/8" long tick extension. Set the text height as 1/8" with the text forced inside, and above the extension line. Set the zero suppression controls so that a zero will be shown following feet, but not preceding inch sizes. Save these styles and then save the drawing as **FLOOR14**.

2. Use appropriate dimension techniques to label the floor to plate height of drawing **E-13-10**. Save the drawing as **ELEV**. Assume plotting at a scale of 1/4" = 1'-0".

3. Use appropriate dimension techniques to label the base footings created for drawing **E-13-4**. Save the drawing as **FOOTINGS**. Assume plotting at a scale of 3/4" = 1'-0".

4. Use appropriate dimension techniques to label the two-story footing created for drawing **E-13-4**. Show the concrete as 8" above grade, with an 8" stem wall, a 7" × 15" footing that extends 18" below the finish grade. Save the drawing as **FOOT-INGS 2**. Assume plotting at a scale of 1/2" = 1'-0".

5. Open drawing **13-6**. Use baseline dimension methods with dot line terminators to locate each tree center. Place all dimensions separate from all other information. Save the drawing as **E-17-5**. Assume a plotting scale of 1" = 10'-0".

6. Open drawing **E-9-14** and provide the needed dimensions to describe the structure. Hatch the 8" wide wall with ANSI31. Save the drawing as **E-17-6**. Assume a plotting scale of 1/4" = 1'-0".

7. Open drawing **E-9-7** and provide the needed dimensions and notes to describe the foundation plan. Change the lines representing the beams to 3" wide lines. Label the main structural members to represent post and beam methods of your region. Save the drawing as **E-17-7**. Assume a plotting scale of 1/4" = 1'-0".

8. Start a new drawing and use the attached drawing as a guide to show the support for two 6 3/4" × 37 1/2" glu-lam beams. Establish separate layers for the beam, steel, text, and dimensions. Completely label and dimension the drawing. Set the required scale factors for plotting at a scale at 1"=1'-0".

   The beams will be supported on a 5/16" × 6 7/8" × 24" base plate with two 5/16" × 10" × 24" side plates welded to the base plate with a 1/4" fillet weld. The base plate will be supported by a 6" × 6" × 1/4" steel column and welded with a 5/16" fillet weld all around. Each beam will be bolted to the side plate with four 5/8" diameter machine bolts. Bolts will be 1 1/2" in from the top and sides and be 3" on center.

   A second connector plate will be located with the center of the 3" × 22" × 5/16" steel plate 8" down from the top of the beam. Two bolts will be used to join the plate to the beam. Bolts will be 2" from the plate end, and 3" on center. Save the drawing as **E-17-8**.

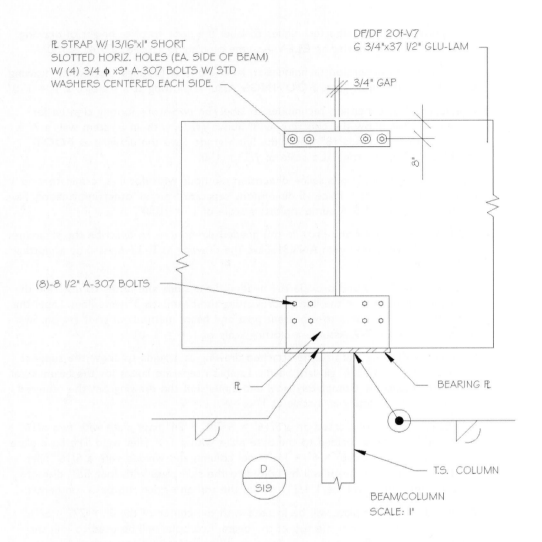

PL STRAP W/ 13/16"x1" SHORT
SLOTTED HORIZ. HOLES (EA. SIDE OF BEAM)
W/ (4) 3/4 φ x9" A-307 BOLTS W/ STD
WASHERS CENTERED EACH SIDE.

DF/DF 20f-V7
6 3/4"x37 1/2" GLU-LAM

3/4" GAP

8"

(8)-8 1/2" A-307 BOLTS

PL

BEARING PL

PL

T.S. COLUMN

D
S19

BEAM/COLUMN
SCALE: 1"

9. Start a new drawing and use the drawing on the next page as a guide to show the column-to-base-plate connection. Establish separate layers for the steel, concrete, text, hatch, and dimensions. Completely label and dimension the drawing. Set the required scale factors for plotting at a scale of 1 1/2"=1'–0".

The column will be a 6" × 6" × 3/8" TS welded to a 10" × 10" × 7/8" base plate with a 3/16" fillet weld all around. Use four 18" long anchor bolts set in an 8 1/2" square grid around the column. Support the base plate on 1" mortar. Save the drawing as **E-17-9**.

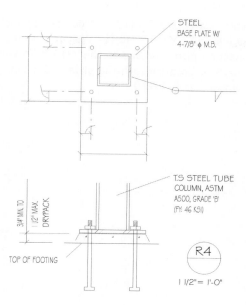

STEEL
BASE PLATE W/
4-7/8" φ M.B.

T.S STEEL TUBE
COLUMN, ASTM
A500, GRADE 'B'
(FY: 46 KSI)

3/4" MIN. TO
1/2" MAX.
DRYPACK

TOP OF FOOTING

R4

1 1/2"= 1'-0"

10. Start a new drawing and use the drawing on the next page as a guide to show the beam-to-beam connection. Establish separate layers for the steel, text, hatch, and dimensions. Completely label and dimension the drawing. Set the required scale factors for plotting at a scale of 1 1/2"=1'–0".

The support beam is a W18 × 76 (18 1/4" dp.) steel wide flange with a W18 × 46 (18" dp.) on the left and a W18 × 40 (17 7/8" dp.) on the right. Use two 4" × 12" × 5/16" steel connection plates welded to the W18 × 76 with 5/16" × 12" fillet each side. Use four machine bolts through the plate and beam @ 3" o.c., 1 1/2" from the edge. Save the drawing as **E-17-10**.

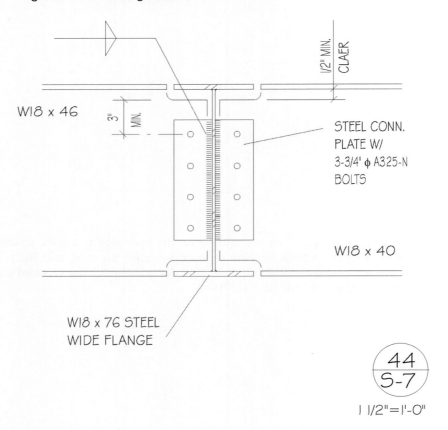

11. Use the attached drawing as a guide to draw the slab-on-grade plan. Establish separate layers for the concrete, text, and dimensions. Completely label and dimension the drawing. Save the drawing as **E-17-11**. Assume a plotting scale of 1/8" = 1'-0".

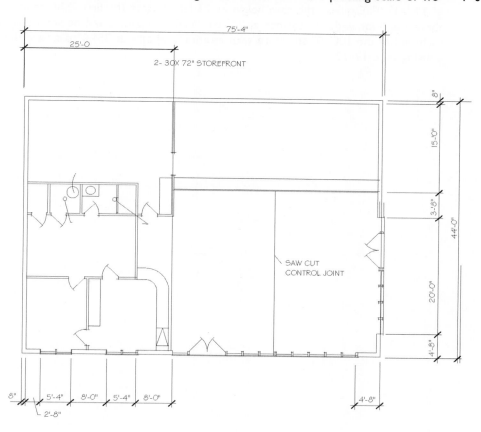

SLAB   PLAN
1/8" = 1'-0"

12. Start a new drawing and use the attached drawing as a guide to steel framing elevation. Establish separate layers for the steel, text, and dimensions. Completely label and dimension the drawing. Set the required scale factors for plotting at a scale of 3/8"=1'-0". Establish the eave height as 24 feet. Locate the first Z brace at 7 feet above the floor, with the balance at 72" o.c. Vertical supports will be set 12" in from each end of the 100'-0" structure with interior supports at 25'-0" o.c. Save the drawing as **E-17-12**.

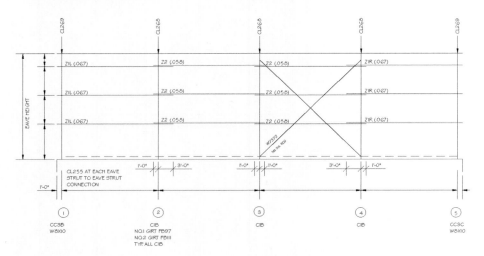

13. Start a new drawing and use the attached drawing as a guide to draw the wall and floor detail. Establish separate layers for the wood, concrete, steel, text, and dimensions. Completely label and dimension the drawing. Set the required scale factors for plotting at a scale of 1/2"=1'–0". Establish the upper floor 30" above the lower floor, and the bottom of the 4 × 10 that supports the lower floor is to be 18" clear to finish grade. Dimension the height of the rail as 42". Save the drawing as **E-17-13**.

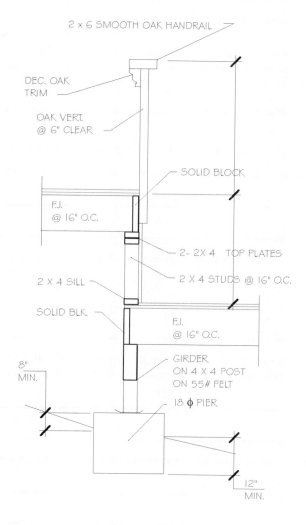

CHAPTER 17 QUIZ

DIRECTIONS

Answer the following questions with short complete statements. Type your answers using a word processor.

1. Place your name, chapter number, and the date at the top of the sheet.

2. Type the question number and provide the answer.

 Warning: Some of the questions have not been covered in the reading material and will require the use of the help menu. You may also have to do some exploring to answer the questions.

1. A group of dimensioning controls that can be saved for future use is called a

   _____ .

2. A _____ is used to control individual aspects of dimensioning.

3. What command can be typed to access a listing of dimensioning features?

   _____

   _____

4. How can styles be used in an engineering office that does both civil and structural projects?

   _____

   _____

5. What value controls the size of each dimension feature and how should this value be set for a drawing that will be reproduced at a scale of 1/4"=1'–0"?

   _____

   _____

6. What option will remove both extension lines from a dimension?

   _____

   _____

7. The spacing between two dimension lines is controlled by what option?

_____

_____

8. What does **Dimgap** control?

_____

_____

9. List two features that can be displayed to describe round objects.

_____

_____

10. List five options for terminating a dimension line that are suitable for an architectural office.

_____

_____

11. List and describe two features of the extension line as it relates to the dimension line and the object being described, and give the value that controls each.

_____

_____

12. What are the options for controlling the height of text?

_____

_____

13. Name the dialog box that can be used to make changes to dimensioning text.

_____

_____

14. What style will be used as a base when dimensioning styles are created?

_____

_____

15. What option will allow for changing an attribute without changing the entire style?

_____

_____

16. What does the **Dimtedit** command affect?

_____

_____

17. Describe how associative dimensions will affect editing.

_____

_____

18. How can the **Mleader** command be used to provide two arrows pointing to the same text?

_____

_____

19. What dimensioning command is used if the text is to be placed so that it is not parallel, perpendicular, or aligned with a surface?

_____

_____

20. What command can be used to change existing dimensions to the current style?

_____

_____

21. Sketch examples of right angle and closed terminators.

_____

_____

22. Explain the process for creating a user-defined terminator.

_____

_____

23. Explain the options to display a centerline in each circle to be dimensioned.

_____

_____

24. Give the proper scale factor for drawing a site plan to be plotted at a scale of 1/8"=1'–0" and explain how the value is determined.

_____

_____

25. Explain the process to place text above the dimension line.

_____

_____

26. List and explain the options for controlling zeros in dimensions.

_____

_____

27. Explain the difference between associative and normal dimensions.

_____

_____

28. Explain the options of the **Dimedit** command.

_____

_____

29. Explain the effect of selecting an extension line to be erased.

_____

_____

30. Explain the effect of stretching a window that is dimensioned 3'–0" long to a length of 6'–0".

_____

_____

Creating Blocks, Wblocks, and Dynamic Blocks

INTRODUCTION

You might have been wondering, while working through the first 17 chapters, why you haven't started drawing some really neat projects. Since chapter 2 you've had the basic drawing skills to draw most projects you could imagine. Subsequent chapters have helped you to develop skills needed to ease the drawing and design process but still have not equipped you to master a complicated project easily. This chapter will:

- Introduce you to creating blocks for repetitive use

- Explore methods of inserting blocks into a drawing

- Examine methods for editing blocks that have been inserted in a drawing

- Explore methods to redefine blocks that have been inserted in a drawing

- Examine the benefits of using wblocks in multiple drawings

- Explore methods of assigning parameters and actions to a block to create a dynamic block.

Commands to be introduced in this chapter include:

- **Block**

- **Insert**

- **Minsert** (Multiple inserts with array)

- **Base**

- **DesignCenter**

- **Bedit** (Block edit)

- **Block Definition Editor**

- **Refedit** (Reference edit)

- **Save Reference Edits**

- **Explode**

- **Wblock** (Write block)

- **ResetBlock**

- **Attsync** (Attribute synchronize)

- **Bauthorpalette** (Block authorize)

- **Bparameter** (Block parameter)

- **Bassociate** (Block associate)

- **Bsave** (Block save)

BLOCKS

Drawings are composed of symbols. To show a door on a floor plan, you don't draw a door; instead you use a line and an arc to represent a door. A block is a group of objects that are treated as one object. The objects used to represent a door symbol can be saved as a block and inserted throughout the drawing as one object. The **Block** command allows symbols that are used repeatedly to be created, saved, and inserted easily in the drawing in which they were created, which provides many benefits for the user. Tools such as **DesignCenter** allow blocks to be moved from one drawing to another to expand their use.

BENEFITS OF BLOCKS

Symbols that are used frequently can be saved as a block, which will greatly increase drawing speed, uniformity, and efficiency.

Speed

If you're drawing a typical residential floor plan, doors will need to be represented 19 or 20 times. Rather than draw each door, you can draw one door and save it as a block. As a block, a door can be inserted quickly and the size altered or the position rotated, as seen in Figure 18.1.

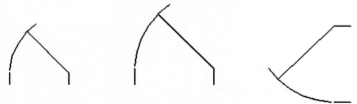

Figure 18.1 *Blocks can be inserted in a drawing, altered in size, and rotated to meet specific needs.*

Editing

Throughout the evolution of construction drawings, many changes and revisions will take place. Blocks can greatly aid in this revision process. If a client decides that all of the doors in a building are to be 6 inches wider, several hours might be required to change all of the doors on the various floor plans. If the doors are drawn as a block, the block can be redefined and all references to that block will be updated automatically.

Attributes

An attribute is text that is associated with a block. Many blocks, such as a door, require text to explain variations. Size, material, quantity, and supplier are examples of information that can be saved as an attribute with a door block. This text is called an attribute and can vary with each application of the block. Attributes can be displayed on the screen as text or they can be stored with the drawing but made invisible. Attributes associated with blocks can also be extracted from a drawing and transferred to a database form, such as a door schedule. Chapter 19 will explore adding attributes to blocks.

Efficiency

In addition to saving time and providing a quick method for revisions, the **Block** command combines the storage of groups of objects into one unit. In drawing a door, information must be stored about the start and end points of the original line and arc, the radius of the arc, the scale factor as the symbol is enlarged or reduced for various uses, and the location of each unit as it is displayed throughout the drawing. When the symbol is installed as a block, only a block reference is stored, saving valuable storage space.

USES FOR BLOCKS

Before a possible listing of blocks is considered, it should be remembered that the list is endless. Any symbol, groups of notes, or a portion of a drawing that is used repeatedly should be stored in a template drawing as a block. A sample listing for architectural blocks would include items for the site, floor, and foundation plans, elevations, and sections as listed below.

Site plans—North arrow, trees, shrubs, spas, pools, fountains, curbcuts and driveways, utility symbols, and general notes. Drawings for subdivisions might include structure shapes, sidewalks, and decks or patios.

Floor plans—Appliances, door and window symbols, heating equipment, heating and cooling symbols, electrical symbols, and plumbing symbols.

Framing plans—Joist and rafter direction indicators, section tags, and general notes.

Foundation plans—Piers, post and piers, beam and column symbols, general notes, and foundation details.

Elevations—Door styles, window styles, shutters, trim, eave molding, decorative posts, garage doors, gable wall vents, trees and shrubs, lighting fixtures, and people.

Sections—Structural shapes such as plates, sills, double sills, beams, columns and piers, foundation components, eave components, insulation, standard dimensions, and common notes.

In addition to drawing simple symbols as a block, you can combine several individual blocks to form a larger block. Figure 18.2 shows an example of a typical truss-to-plate intersection. The block could be inserted in a section drawing, mirrored, and then the truss shape completed by the **Fillet** and **Extend** commands, as seen in Figure 18.3.

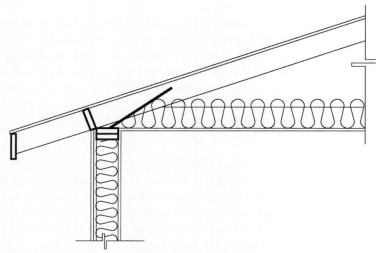

Figure 18.2   *Several objects can be combined to form a block.*

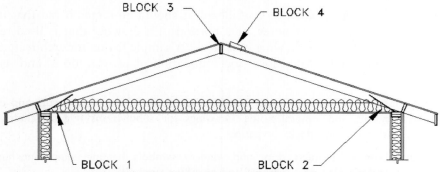

Figure 18.3   *Using the **Mirror**, **Fillet**, and **Extend** commands, this section was created by inserting and connecting blocks. Block 1 is the drawing shown in Figure 18.2. Block 2 is the mirror image of Block 1. Block 3 is a copy of one of the plates from block 1 that was rotated and saved as a separate block. Block 4 is a ridge vent that was saved from a previous drawing and inserted into this drawing.*

DRAWING OBJECTS FOR BLOCKS

Because a block is a group of objects that is saved as one object, there is no special method for drawing the block. The components that will comprise the block can be created using any of the drawing or editing commands. Some consideration should be given to the layer the block is created on and the size of the block while it is being constructed, to avoid problems when the block is inserted in the drawing.

The Effects of Layers on Blocks

A block will take on the characteristics of the layer on which it is created. A block can also consist of objects that have been drawn on several layers with different linetypes, lineweights and colors. Although a block is always inserted on the current layer, the block reference preserves information about the original layer, color, and linetype properties of the objects that are contained in the block. You can control whether objects in a block retain their original properties or inherit their properties from the current layer, color, linetype, or lineweight settings. If the block was created on one or more layers that do not exist in the current drawing, those layers will be created. There are, however, three exceptions to this guideline.

- Blocks that are created on the 0 layer will be placed on the current layer when placed in a drawing.
- Blocks drawn with the color created ByLayer are drawn with the color of the current layer, but will reflect the color of the layer into which the block is inserted.
- Blocks drawn with the linetype created ByBlock are drawn with the linetype of the current layer, but will reflect the linetype of the layer in which the block is inserted.

Block Size

Once the block is saved, it can be inserted throughout a drawing. When the **Insert** command is introduced later in this chapter, you will have an opportunity to alter the size of the block in the X and Y directions. Altering the size of the block during the insertion process can be easier if the size of the block is considered during creation. Some blocks, such as toilets, electrical symbols, and appliances, are a standard size. These blocks can be drawn at their actual size and inserted. Other blocks, such as doors, windows, tubs, and showers, vary in size with each application. Drawing these blocks in one-unit squares will allow the size to be adjusted each time it is inserted. Rather than having separate blocks for 4', 5', and 6' windows, one window can be created, and the X value controlling its length can be altered. You'll also have an opportunity to alter the size of existing blocks using the **Block Definition Editor**, and to create dynamic blocks. Dynamic blocks can be altered by updating the block definition prior to inserting the block so that the size, rotation angle and several other features can be altered.

The size of the unit can vary depending on the drawing the block will be inserted in. On floor plans, you might use 1" as the unit and scale up or down. Figure 18.4 shows a window created using a 12" unit, and the effects that can be achieved with one block.

Figure 18.4 *Blocks that vary in size should be created using one unit, such as a foot, so the size of the block can be adjusted easily each time the block is inserted. A window that is 1 unit long (a foot) can be inserted at a horizontal scale of 5.5 to produce a window that is 5'–6" long.*

CREATING A BLOCK

Figure 18.5 shows the drawing process for creating a toilet, which can then be saved as a block. Keep in mind that although the toilet has been drawn, it is not a block yet. Once the drawing is ready to be reproduced, start the **Block** command by using one of the following methods:

- Type **B** ENTER at the Command prompt.
- Click the **Make Block** button on the **Draw** toolbar.
- Select **Make** from the **Block** cascading menu on the **Draw** menu.

Each method will produce the **Block Definition** dialog box shown in Figure 18.6.

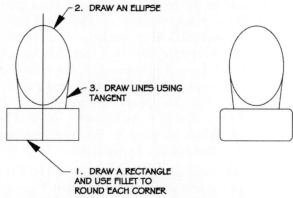

Figure 18.5 *A toilet can be created by drawing an ellipse and a rectangle and connecting the two with straight lines. The **Fillet** command was used to round each corner.*

Figure 18.6 *Start the* **Block** *command by typing* **B** *ENTER at the Command prompt, by clicking the* **Make Block** *button on the* **Draw** *toolbar, or by selecting* **Make** *from* **Block** *on the* **Draw** *menu.*

Key elements of using the **Block Definition** dialog box to create a block include: choosing a block name, selecting the objects to be in the block, selecting an insertion point, selecting block to be annotative, and saving the block.

Choosing a Name

Start the process by entering the name of the block to be created in the **Name** edit box. A name of up to 255 characters can be used to describe the block. The block name can contain letters, digits, and special characters such as $, -, and _. Any character not used by Microsoft Windows or AutoCAD, including blank spaces can be used. Use a name that will describe the contents. For instance, DOOR SWG 32 would be an appropriate name to describe a 2'–8" swinging door. If a name for an existing block is provided for the new block, the old block will be destroyed and the new block will take its place. Before it is vaporized, you will be given an alert message by AutoCAD warning that the block is about to be redefined.

Selecting the Objects

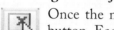

Once the name has been provided, click either the **Pick point** or **Select objects** button. Each option must be selected, and the order will not affect the outcome of the command. For this discussion the object will be selected first. Click the **Select objects** or **Quick Select** button to define the objects that will be in the block. (The use of the **Quick Select** dialog box will be explained shortly.) Clicking the **Select objects** button temporarily removes the dialog box from the screen and allows the objects that will make up the block to be selected. Objects to comprise the block can be selected using any object selection method. Once the desired objects have been selected, press ENTER to return the **Block Definition** dialog box. An image of the selected objects will be displayed in the **Preview icon** display beside the **Name** edit box.

Note: The **Blockicon** command can be used to display images of blocks stored in drawings created using AutoCAD 2000 or previous releases. The command can be entered at the Command prompt and will provide prompts for what blocks you would like to update. Pressing ENTER will update all blocks.

Using Quick Select

 If the **Quick Select** button is selected, a **Quick Select** dialog box similar to Figure 18.7 will be displayed. The **Quick Select** dialog box can be used to define the filtering criteria that will be used to select the objects for the block. Although simple blocks such as the toilet can easily be selected using a window, **Quick Select** can be used to select complex shapes or multiple objects. Notice that the current filter is defined as the entire drawing in the **Apply to** edit box. If there is no current selection set, the entire drawing will be used as a block; this option will be explored later. To select an object for the block, click the **Select Objects** button to the right of the **Apply to** edit box. This will temporarily remove the dialog box from the screen and allow the objects to be selected. The process is the same as used with the **Select objects** button in the **Block Definition** dialog box. Once the desired objects have been selected, pressing ENTER will return the **Quick Select** dialog box. Other key elements of the **Quick Select** dialog box include the following:

- **Object type** specifies the type of object to use to define the filter. The current choices are Multiple, and Block Reference. Since you're defining a block, choose Block Reference.

- The **Properties** box can be used to specify properties to be used as a filter. Objects to form the block can be defined by the color, lineweight, linetype, or any of the other listed properties.

- **Operator** controls the range of the selected property filter. Depending on the selected property, options can include **Equals**, **Not Equal**, **Greater Than**, **Less Than**, and **Select All**. Selecting the edit arrow displays a list to allow the operator value to be altered.

- **Value** defines the filter property value. If the selected property is a color, the value will specify a color. If the property is a layer, the 0 layer is the default value. Selecting the edit arrow displays a list to allow the value to be altered.

- The **How to apply** area of the dialog box specifies if the selected objects will form or be excluded from the selected set to form the block. Choosing **Exclude from new selection set** will create a block of all objects that do not match the selection filter.

 Note: To create a block such as a toilet, use **Select objects**. **Quick Select** can be used to select all of the walls of a floor plan to form a block without having to select each one of them. If a filter set is defined that will include all of the walls, or exclude everything except the walls, the block can be defined much faster than if each wall has to be selected.

Figure 18.7   *The **Quick Select** dialog box can be used to define the filtering criteria that will be used to select the objects for the block.*

Object Selection Options

In addition to methods for selecting objects to define the block, the **Objects** area in the **Block Definition** dialog box contains options to retain, convert, and delete the selected objects.

> **Retain**—Selecting **Retain** will keep the selected objects in their original state in the selection set. Once the object is selected to be a block, the original objects will remain in their current location.

> **Convert to block**—Selecting this option will remove the selected objects from the selection set and replace them with a previously defined block.

> **Delete**—Selecting this option will remove the objects that define the block from the drawing. This option works well if you create the toilet (or any other object) that is to become the block in the middle of the living room. With **Delete** active, once the object is selected to be a block, the original object will be removed from the drawing. It will still exist as a block in the drawing base, but it can't be seen until the block is inserted.

SPECIFYING THE BASE POINT

The base point of the block will align with a point called the insertion point. An insertion point will be designated as the block is inserted into a drawing. The base point can be selected by picking any point of the object or by entering X and Y coordinates. The **Pick point** button temporarily removes the dialog box from the screen and allows the base point to be selected. Figure 18.8 shows the effect of the base point of a block with the insertion point at the destination. The insertion point should be chosen using OSNAP based on the block's use. Blocks such as a toilet, a lavatory, or a sink will be inserted most easily using a center point. Other objects such as a sill or beam might best be located from an edge or intersection. Figure 18.9 shows some typical insertion points for common architectural features. The insertion point also can be used as a rotation point for the block as the block is inserted in its new location. Figure 18.10 shows how a block can be rotated about its insertion point. Other options will be discussed as the **Insert** command is explored. Using a center point or the lower left corner of a block will help to keep track of where to insert the block.

 Note: Later in this chapter you'll be introduced to methods of assigning and toggling through multiple insertion points, and other methods of altering the rotation.

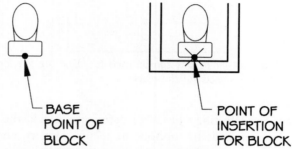

BASE
POINT OF
BLOCK

POINT OF
INSERTION
FOR BLOCK

Figure 18.8 *An insertion point defines the location of the block to be used as a handle.*

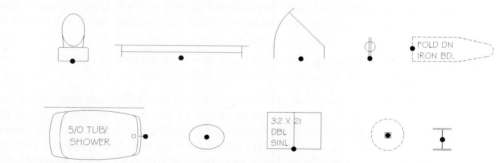

Figure 18.9 *Typical insertion points for common architectural features.*

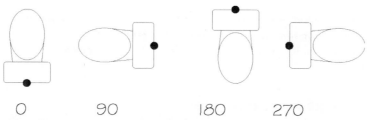

Figure 18.10   *Rotating a block around its base point.*

The **Annotative** check box allows the block to be annotative. An annotative block has the ability to have multiple scale factors assigned to the selected block. This will automate the process of scaling drawing objects. See chapter 23 for more on annotative objects.

The **Description** edit box allows a description to be assigned to the block. A block can be created without providing a description, but on a complex block, the description might help you distinguish it from a similar block.

Open in Block Editor

With this option active, the **Block Editor** dialog box will be displayed when **OK** is selected to end the block selection. The **Block Editor** is used to define dynamic behavior for a block definition. It adds parameters and actions, which define custom properties and dynamic behavior. The **Block Editor** contains a special authoring area in which you can draw and edit geometry as you would in the drawing area. The affects of dynamic blocks will be introduced later in this chapter once blocks have been explored.

SUMMARY OF STEPS IN CREATING A BLOCK

It's taken several pages to explain what a block is. Remember that a block is just a collection of objects that can be used over and over. Because they are so important to your drawing future, use the following steps as you create blocks:

1. Display the **Block Definition** dialog box by typing **B** ENTER at the Command prompt, by clicking the **Make Block** button on the **Draw** toolbar, or by selecting **Make** from **Block** on the **Draw** menu.

2. Provide a name that will define the block.

3. Select the drawing objects to be reproduced as blocks.

4. Select a base point to be used the block is inserted in the drawing.

5. Select modifiers if desired in the **Objects** and **Settings** portions of the dialog box.

6. Click **OK** to save the block to the drawing base.

The object has now been saved as a block definition and is ready to be reused throughout the drawing. If another toilet, or what ever object is used as a block, needs to be added to the drawing, you can add it to the drawing by inserting the block rather than redrawing the objects.

Note: You've just read through the process to create a block. As you create a block, it might leave you waiting for something exciting to happen. The creation of a block is nothing special. It's what you can do with it that will revolutionize your drawing process.

PREPARING BLOCKS FOR INSERTION

 Once a block has been created and saved, it can be placed in a drawing using the **Insert** command. To access the command, use one of the following methods:

- Click the **Insert Block** button on the **Draw** toolbar.
- Type **I** ENTER at the Command prompt.
- Select **Block** from the **Insert** menu.

Each method will produce an **Insert** dialog box similar to Figure 18.11.

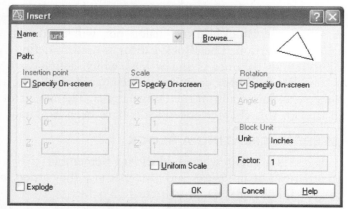

Figure 18.11 *Selecting the **Insert Block** button on the **Draw** toolbar, typing **I** ENTER at the Command prompt, or selecting **Block** on the **Insert** menu will display the **Insert** dialog box.*

Choosing a Block

The **Name** edit box is used to select the block to be inserted. The name of the block to be inserted can be typed in the edit box or selected from a list of blocks contained in the drawing. Selecting the arrow beside the edit box will display a listing of all blocks contained in the drawing. Selecting the **Browse** button will open the **Select Drawing File** dialog box. The dialog box is the same as the file selection dialog boxes that you've used as drawing files are opened. The dialog box allows a block or file to be selected for insertion. Inserting files will be introduced later in this chapter.

Specifying an Insertion Point

Activating the **Insertion point Specify On-screen** check box allows an insertion point to be used to specify where the block will be located. With the **Specify On-screen** box inactive, the X and Y coordinate locations (Z is the vertical reference for 3D) can be used to locate the insertion point. As the block was created, a base point was determined. Once the desired base parameters are adjusted, a prompt will be displayed to provide a location to place the original base point. Figure 18.12 shows the relationship of the block base point to the destination insertion point. Once the insertion point is selected, the block will be dragged into the indicated position, and the scale can be altered. The parameters to control the insertion can each be adjusted before removing the dialog box to insert the block, or at the Command prompt before the block is inserted.

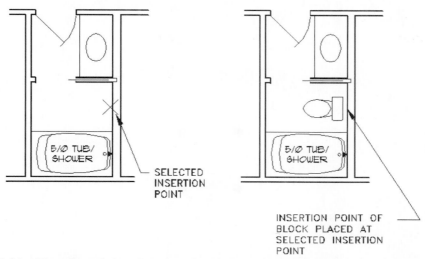

Figure 18.12   *The effect of the selected insertion point of a drawing and the insertion point of a block.*

Scale

This option allows you to define scale factors for the block to be inserted. With the **Specify On-screen** check box active, a prompt will be displayed at the Command prompt before the block is inserted. With the box inactive, the X, Y, and Z edit boxes are activated. Unless a different value is entered, the value entered for X will become the default value for Y. Any dimensions contained in the block will be multiplied by the scale factor as the block is inserted. If the toilet in Figure 18.12 were drawn as full size, the scale for inserting the toilet would be 1. Inserting a window created as a 1' block to be a 6' long window in a 6" exterior wall would require an X scale factor of 6, and a Y scale factor of .5. With the **Uniform Scale** check box active, the Y value will equal the X value. The effects of the scale factor can be seen in Figure 18.13. Using dynamic block features will also provide options for adjusting the scale of blocks.

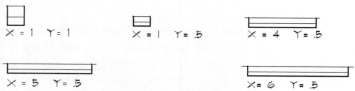

Figure 18.13 *The effect of X/Y scale factors on a block.*

Negative Scale Factors The location of a block can be altered by using negative X and Y values. Using negative values combines the **Insert** and **Mirror** commands. Figure 18.14 shows an example of negative value alternatives.

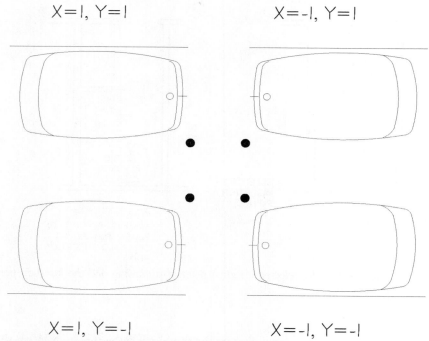

Figure 18.14 *Altering a block using negative scale factors.*

Rotation

The Rotation area allows you to supply a rotation angle. With the **Specify On-screen** box active, a prompt will be displayed at the Command prompt before the block is inserted. With the box inactive, the **Angle** edit box is activated and an angle can be entered to control the rotation of the block at insertion. Figure 18.15 shows an object rotated around its insertion point. Any angle can be provided for the rotation angle.

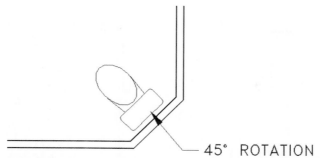

— 45° ROTATION

Figure 18.15 *Specifying an angle of rotation allows a block to be rotated relative to its insertion point.*

Explode

Blocks function as one object when they are inserted in a drawing. If the **Explode** check box is activated, as the block is inserted in the drawing, each object of the block will function as a separate object. This allows you to edit the block; however if you choose to modify the template block and update your drawing all exploded blocks will be excluded from the update.

INSERTING BLOCKS

Once the parameters of the **Insert** dialog box have been set, selecting the **OK** button will close the dialog box and allow a selection point to be specified. Selecting an insertion point will insert the block at the specified parameters.

Inserting Blocks with the Divide Command

The **Divide** command can also be used to insert blocks at repeated distances. You can insert blocks by dividing a line into the desired number of segments and then inserting the block at each marker. This can be done if the Block option is selected. Figure 18.16 shows the effect of aligning blocks with the object. The command sequence to divide a circle into 14 even spaces and place a block named SLOT is as follows:

```
Enter the number of segments or [Block]: B ENTER
Enter name of block insert: (Type name of desired block.) SLOT
    ENTER
Align block with object? [Yes/No]<Y>: ENTER
Number of segments: 14 ENTER
```

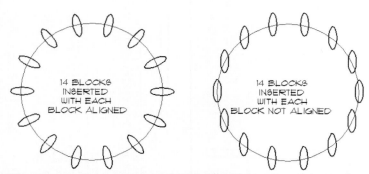

Figure 18.16   *Blocks can be inserted into a drawing at each marker using the **Block** option of the **Divide** command.*

WORKING WITH BLOCKS

Several other options can be used to control the use of blocks. Useful tools include the **Minsert** command for multiple inserts, using an entire drawing as a block, building blocks by using other blocks in a process known as nesting, exploding blocks, and controlling blocks with the DesignCenter.

MINSERT—MAKING MULTIPLE COPIES

The **Minsert** command combines the features of **Insert** and **Array**. The array pattern created with **Minsert** has the qualities of a block, except that it cannot be exploded. **Minsert** allows a block to be arrayed in a rectangular pattern typical of what might be found in the interior column supports for multilevel structures, or the supports of a post-and-beam foundation. Start the **Minsert** command by typing **MINSERT** ENTER at the Command prompt. Figure 18.17 shows an example of a multiple insert. The drawing was created using the following command sequence:

```
Command: MINSERT ENTER
MINSERT
Enter block name or [?]: <toilet> PIER ENTER
Specify insertion point or [Scale/X/Y/Z/Rotate/PScale/PX/PY/PZ/
    PRotate]: (Select the insertion point.)
Enter X scale factor, specify opposite corner, or [Corner/
    XYZ]<1>: ENTER
Enter Y scale factor, < use X scale factor>: ENTER
Specify rotation angle: <0>: ENTER
Enter number of rows (+m-) <1>: 3 ENTER
Enter number of columns ($\vert \vert \vert $ ) <1>: 6 ENTER
Enter distance between rows or specify unit cell (+m-): 96 ENTER
Specify distance between columns ($\vert \vert \vert )$: 48 ENTER
Command:
```

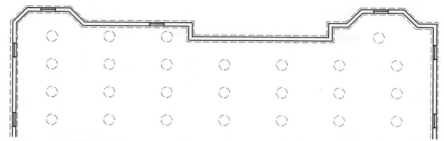

Figure 18.17   *The **Minsert** command allows blocks to be arrayed in a rectangular pattern.*

MAKING AN ENTIRE DRAWING A BLOCK

An entire drawing can be used to form a block. When the entire drawing is to be a block, an insertion point other than 0,0 might be desired. The **Base** command can be used to define a new base point. Access the command by typing **BASE** ENTER at the Command prompt or by selecting **Base** from **Block** from the **Draw** menu. Each method will produce the following prompt:

```
Command: BASE ENTER
BASE
Enter base point: (Select a point.)
```

The base point will now become the insertion point.

A better alternative to this method of using an entire drawing as a block will be introduced later in this chapter when wblocks are introduced, and in chapter 22 where externally referenced drawings (Xrefs) are explored.

NESTING BLOCKS

So far you've been introduced to creating blocks to represent common symbols such as a toilet, sink, door, or any number of repetitive drawing elements. Blocks can also be combined with other objects or blocks to form a nested block. A kitchen for a large apartment complex could be created as a block and inserted in each unit. The kitchen block will contain other blocks for the dishwasher, sink, stove, refrigerator, and pantry. Joining blocks of individual units can create entire buildings of the complex. Uses for nested blocks are almost endless for construction drawings. Figure 18.18 shows an example of nested blocks used to create a foundation plan.

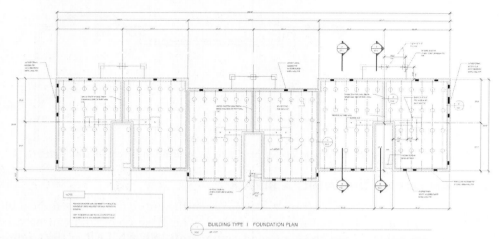

BUILDING TYPE 1 FOUNDATION PLAN

Figure 18.18 *Blocks for the vents, piers, and girders are combined to create a foundation for each unit. A block was then created for each unit and mirrored to form the adjoining unit. A block was then made of each pair of units, and then blocks of each pair were inserted to form the building foundation.*

Care must be taken in the use of nested blocks to avoid confusion in naming and layering. Use the following guidelines to avoid confusion:

- Assign properties individually rather than **ByBlock** if all instances of a block require the same color, layer, linetype and lineweight properties.

- To control the properties of each block using the properties of the layer the block is inserted on, draw the objects that form each block, including nested blocks, on the 0 layer. Set properties such as color, layer, linetype and lineweight properties ByLayer.

USING BLOCKS IN MULTIPLE DRAWINGS

One of the benefits of working with multiple drawings is that you can move or copy blocks in one drawing to a different drawing. This can be done with opened or closed drawings. Use the following steps to copy blocks between two open drawings.

1. Select the block to be copied.

2. Right-click to display the shortcut menu.

3. Select **Copy**.

4. Move to the other drawing and make it the current drawing by selecting the title bar.

5. Select **Paste** from the shortcut menu or the **Edit** menu.

6. Insert the block in the desired location.

CONTROLLING BLOCKS WITH DESIGNCENTER

Although the full potential of **DesignCenter** will not be explored until chapter 21, it has features that are useful for working with blocks. The **DesignCenter** allows blocks to be inserted from drawings that are not open. If you work with consulting firms, you can copy a block from a drawing in their office, if you have the proper access to their network. For now, we'll work on copying a block from one drawing to another. Future chapters will provide information on working with networked machines and consulting firms using the Internet. To copy a block from a drawing file on your computer, display the **DesignCenter** by clicking the **DesignCenter** button on the **Standard** toolbar. The **DesignCenter** will resemble Figure 18.19. The exact appearance will depend on what drawings are opened. Click the **Folders** tab in the upper left corner. This will produce a display matching the **Standard File Selection** dialog box. Selecting the name of a stored drawing will provide access to the contents of that drawing for transfer to another file. Selecting the **Open Drawings** tab will produce a display of the contents of an active drawing similar to Figure 18.20. Double-clicking the **Blocks** listing will provide a listing of the blocks contained in the drawing similar to Figure 18.21. With both drawings displayed on the screen, highlighting the name of the desired block allows that block to be dragged into the new drawing.

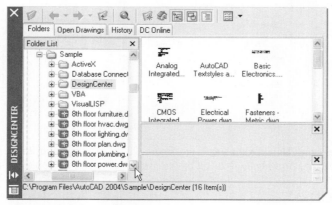

Figure 18.19 *The DesignCenter tool palette.*

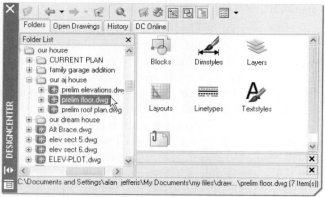

Figure 18.20   Selecting a drawing from the **Folders** tab list will allow the contents of an unopened file to be displayed. Selecting the **Open Drawings** tab lists the contents of an active file.

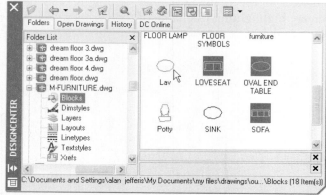

Figure 18.21   Selecting blocks from the list in Figure 18.20 will display a list of blocks in the selected drawing. One or all of the listed blocks can be copied to a new drawing.

USING HYPERLINKS

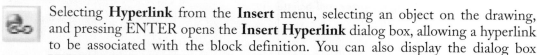

Selecting **Hyperlink** from the **Insert** menu, selecting an object on the drawing, and pressing ENTER opens the **Insert Hyperlink** dialog box, allowing a hyperlink to be associated with the block definition. You can also display the dialog box typing **HYPERLINK** ENTER at the Command prompt. Hyperlinks are created in AutoCAD drawings as pointers to associated files. Hyperlinks can be used for varied tasks such as launching a word processing program, opening a specific file, pointing to a named location in a file, or activating your Web browser and loading a specified HTML page. Hyperlinks can be used to point to locally stored files, files on a network drive, or files on the Internet. Cursor feedback is automatically provided to indicate when the crosshairs are over a graphical object that has an attached hyperlink. You can then select the object and use the **Hyperlink** shortcut menu to open the file associated with the hyperlink. Chapter 26 will also explore the **Hyperlink** command.

EDITING BLOCKS

Blocks can be edited by changing the description, by using the **Explode** command, or by using the **Redefine** option of the **Block** command.

EDITING BLOCK DEFINITIONS

A block description can be altered in the **Block Definition** dialog box using the following steps:

1. Display the dialog box by selecting the **Block Edit** button on the **Standard** toolbar, or typing **BEDIT** ENTER at the command prompt. Each will display the **Edit Block Definition** dialog box similar to Figure 18.22.

2. Select the edit arrow beside the **Name** box to display a listing of the current blocks.

3. Select the name of the block that will be altered and select the **OK** button. This will display the **Block Editor** display shown in Figure 18.23.

4. Make the required alterations to the block while it is displayed in the **Block Editor**. Typically, this might include adding objects to the block. The full editing power of the editor will be explored in the next section of this chapter.

5. Select the **Close Block Editor** to return to the drawing area.

6. Select **Yes** when prompted to accept the changes. This will close the editor, update each existing block, and close the **Block Editor**.

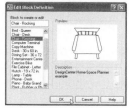

Figure 18.22   *The **Edit Block Definition** dialog box can be used to select a block to be edited. Scroll the list of blocks included in the drawing to select the block to be edited. Select the **OK** button to proceed.*

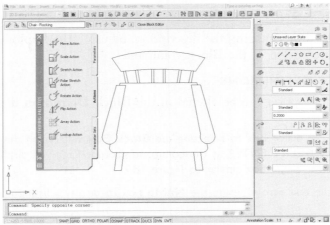

Figure 18.23 *Once **OK** has been selected in the **Edit Block Definition** dialog box, the **Block Editor** will be displayed. Make any changes to the block in the editor, and then select the **Close Block Editor** button when editing is complete.*

REDEFINING A BLOCK

The **Refedit** command provides an additional method to alter blocks. To start the command, use one of the following methods:

- Click the **Edit Reference in Place** button from the **Refedit** toolbar.

- Select **Xref and Block In-Place Editing** from the **Tools** menu, and then selecting **Edit Reference In-Place**.

- Type **REFEDIT** ENTER at the command prompt.

This command can be used to redefine externally referenced drawings (Xrefs) and blocks. Xrefs will be considered in a later chapter. Even if a block has been inserted in a drawing in several locations, it is possible to update all of the blocks without editing them individually. Blocks are edited by redefining a copy of one of the blocks, then saving the edited copy of the block using the existing block name. It's also important to remember that this process will update ALL blocks of the specified name. There is no "I only wanted a few of them" option. Use the following sequence to redefine a block. For the following example, the STD TUB SHOWER block will be altered.

1. Click the **Edit Reference in Place** button from the **Refedit** menu. This will automatically start the **Refedit** command and produce the Command prompt: **Select reference**.

2. Select the block to be edited. Selecting a block to be altered will produce a **Reference Edit** dialog box similar to Figure 18.24 listing the selected block in the **Reference name** display. If the selected block belongs to a nested block, all of the references available for selection will be displayed in the **Reference Edit** dialog box.

3. Select the name of the block to be edited from the **Reference name** edit box. As the block is selected, AutoCAD blocks the reference file to prevent the file from being accessed by multiple users.

4. On the **Settings** tab, select the **Create unique layer, style and block names** check box.

5. Click the **OK** button. This will close the **Reference Edit** dialog box. All other occurrences of the block to be edited will be removed from the display until the editing process is complete.

6. Select the block to be edited and press ENTER. Other blocks appear faded and are not allowed to be altered.

7. Edit the block as desired.

8. Select the **Save Reference Edit** button on the **Refedit** toolbar. This will display a warning box indicating that all of the selected blocks are about to be updated.

9. Click the **OK** button in the warning box. The box will be closed and all of the selected blocks will be redisplayed to reflect the indicated changes.

Each copy of the selected block used throughout the drawing will now be updated to reflect the current changes.

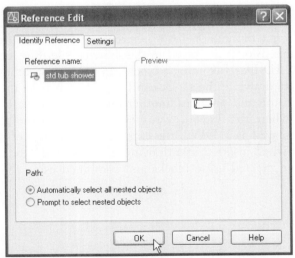

Figure 18.24 *The **Reference Edit** dialog box can be used to select and edit a block.*

EXPLODING BLOCKS

If the **Explode** option is activated in the **Insert** dialog box, each object in the block will function independently of other objects in the block. By default, as blocks are inserted in a drawing, all objects in the block will function as one object for editing purposes. Combining all of the objects of a block as one object is great if you want to

Move, **Rotate**, or **Copy** a block. Only one object has to be selected and the entire block will respond to the command. Using other editing commands such as **Erase** to remove one object also will cause the entire block to be erased. To erase just one object in the block, you must use the **Explode** command. To access the command, use one of the following methods:

- Click the **Explode** button on the **Modify** toolbar.

- Type **X** ENTER at the Command prompt.

- Select **Explode** from the **Modify** menu.

```
The command sequence is as follows:
Select the Explode button.
Select objects: (Select the desired blocks.)
Select object: (This prompt will continue until all of the
      desired blocks are selected.) ENTER
```

The block will be displayed with no apparent change. Individual components of the block can now be edited. Blocks with equal X, Y, and Z values will keep their shape when exploded. Blocks with unequal X, Y, and Z values may change shapes when exploded. Blocks that are inserted with **Minsert** can't be exploded.

WBLOCKS

Blocks are a great way to make repeated copies of a symbol throughout a drawing. A symbol that is used on a foundation for one structure can be used on the foundation of several other structures. Blocks are saved in the drawing file they are created in, but with the aid of the AutoCAD DesignCenter, a block can be used on other drawings. If a block is saved on a template drawing, the block can be used each time the template drawing is used. The **Wblock** command provides another option for blocks to be used in multiple drawings no matter where the drawing was created, by saving the block as a .DWG file. A wblock will be displayed as a .DWG file each time a directory of drawings is made, but it will act as a block when inserted in a drawing. Figures 18.25 and 18.26 show examples of details and sections created using blocks and combined using the **Wblock** command.

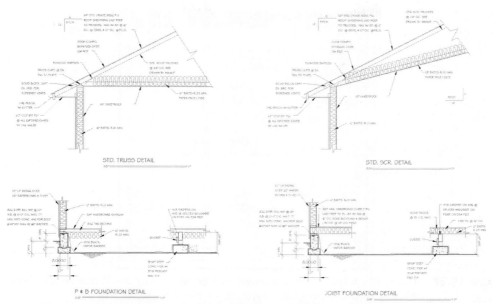

Figure 18.25 *Common details can be stored using the **Wblock** command.*

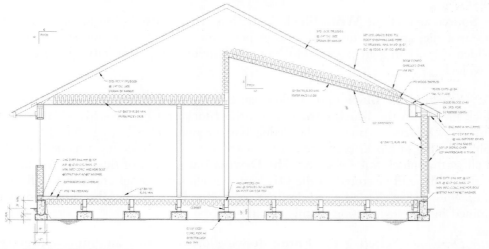

Figure 18.26 *Blocks saved using the **Wblock** command can be combined to make other drawings.*

CREATING WBLOCKS

Typing **W ENTER** at the Command prompt will start the command and display a **Write Block** dialog box similar to Figure 18.27. The dialog box is used to create and save blocks to a file for future use. The two major areas of the **Write Block** dialog box include the **Source** and **Destination** areas. The areas that are active will depend on the selections made in the **Source** portion of the dialog box.

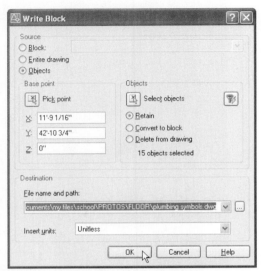

Figure 18.27 *Typing **W** ENTER at the Command prompt will display the **Write Block** dialog box allowing blocks, drawings, or objects to be saved for future use.*

Source

The **Source** area of the **Write Block** dialog box selects existing blocks or objects, writes them as a file, and specifies an insertion point for when they are inserted. Options are provided for selecting an existing block or an entire drawing, or specifying objects to form the wblock.

Block Selecting the **Block** option allows an existing block to be selected to be stored as wblock file. With the **Block** setting active, the **Block** edit box is activated and the **Base point** and **Objects** portions of the dialog box are inactive. Selecting the edit arrow will display a listing of blocks contained in the current drawing. Selecting a name from the list will allow that block to be inserted. The **Destination** portion of the block displays where the wblock file will be stored. If the **OK** button is clicked, the selected block will be saved as a drawing file in the indicated destination. Methods of altering the destination will be examined in the next section of this chapter.

Entire Drawing Selecting the **Entire drawing** option allows an entire drawing to be saved as a wblock. This option will have the same results as when **Saveas** is used. The benefit of using this option is that unreferenced objects such as blocks, layers, and linetypes that were not used will be removed from the drawing file. This is an excellent method for reducing the size of a drawing and for saving a drawing at the end of a work session.

Objects Using the default option of **Objects** allows drawing objects to be selected to form the wblock. The process is very similar to the procedure used to create a block. With the **Objects** button active, the **Base point** and **Objects** portions of the dialog box are activated so that objects and a base point can be selected. The order in which the base point or the objects are selected can be altered. Each of these functions is exactly the same as their counterparts in the **Block** command. As each button is selected, the dialog box is removed, allowing for a selection to be made. Once the selection is made, the dialog box is restored, allowing for other options to be altered.

Destination

The box allows the name, location, and units of the drawing file to be specified.

File Name and Path   The **File name and path** display will vary depending on which of the three **Source** options is active. If the **Block** option is active, the name of the block entered in the **Block** edit box will be displayed in the **File name and path** edit box. If the **Entire drawing** box is active, the current name of the drawing will be displayed in the **File name and path** edit box. With the **Objects** option active, the default name of **new block.dwg** will be displayed. Best results will be achieved by providing a more descriptive name.

Because blocks stored using the **Wblock** command are treated as a drawing file, a storage location must be given. The **File name and path** edit box displays the current drive and path that will be used to store the drawing. Selecting the browse **(...)** button allows the drive and folder to be altered. Selecting the button produces a display similar to Windows Explorer. Once a folder name has been selected, the **Create Drawing** dialog box will allow for selection of a file name and a location for the file.

 Note: Great care should be taken in the naming and storing of wblocks. The name should clearly describe the contents of the drawing. Wblocks should be stored in folders and subfolders that will define the use. For instance, a name of PROTOS could be used for the folder. Subfolder names would include names such as SITE, FLOOR, ELEV, FND, or other common drawings to be used. Each subfolder would contain only wblocks for that specific type of drawing.

Summary of Steps in Creating a Wblock

Remember that a wblock is just a collection of objects that can be used over and over in a multitude of drawings. Use the following steps to create wblocks:

1. Display the **Write Block** dialog box by typing **W** ENTER at the Command prompt.

2. Choose the source of the block to be created.

3. If the wblock will be comprised of a pre-existing block, provide a name for the block.

4. Select the drawing objects to be reproduced as a block.

5. Select a base point to be used at the block is inserted in the drawing.

6. Select a name to designate the file and the drive, folder, and subfolder to be used as the destination.

7. Click the **OK** button to save the block as a drawing file.

The object has now been saved as a drawing file and is ready to be reused throughout the existing drawing or in a new drawing. If the toilet from Figure 18.8 needs to be added to the existing drawing or to a totally new drawing file, it can be added to the drawing using the **Insert** command.

INSERTING WBLOCKS

Once a symbol or drawing has been saved as a wblock, it can be used just as any other drawing. It can be retrieved using the drawing **Startup** dialog box as a new drawing session is started, or it can be inserted in an existing drawing using the **Insert** command. The **Insert** command will bring an existing drawing into the current drawing from the Command prompt. Each aspect of inserting a wblock is the same as for inserting a block.

EDITING WBLOCKS

One of the strengths of using an entire drawing as a wblock is that when it is inserted, it will move as one object. Insert a wblock of a floor/roof plan, and all of the components will move as one object. A wblock can be altered using the same commands used to edit a block.

WORKING WITH DYNAMIC BLOCKS

Up to this point, you've created and edited blocks and wblocks. They work great but they each have limitations. If you needed bathtubs that are 60 and 72 inches long or left or right entry, a block for each option would need to be created. AutoCAD allows one dynamic block to be created that is flexible enough to meet the needs of several blocks. A dynamic block reference can be edited in a drawing while you work. You can manipulate the objects that create the dynamic block reference through custom grips or custom properties. This allows you to adjust the block in-place as necessary rather than searching for another block to insert or redefining the existing one.

If a block reference for a 60" tub is inserted in a drawing, the size may need to be altered while you're editing the drawing. If the block is dynamic and defined to have an adjustable size, you can change the size by simply dragging the custom grip or by specifying a different size in the **Properties** palette. The block might also contain an alignment grip, which allows you to align the block reference easily to other geometry in the drawing.

EXPLORING DYNAMIC BLOCK TOOLS

A dynamic block can be created from scratch by drawing objects just as you would for a normal block, or you can add dynamic behavior to an existing block definition. Dynamic blocks are created using the **Block Editor**. The **Block Editor** was used earlier to edit block definitions. The **Block Editor** can also be used to add the elements that make a block dynamic. The editor is accessed by selecting the **Block Editor** icon from the **Standard** toolbar. Selecting the icon will display the **Edit Block Definition** dialog box. Figure 18.28 shows the **Edit Block Definition** dialog box with an existing block selected for editing. Selecting the **OK** button will display the selected block in the **Block editor as shown in** Figure 18.29.

Note: The names of several commands that control dynamic blocks were listed at the start of this chapter. These commands are executed in the background as you pick options from the **Block Editor** or one of the **Authorizing** palettes. Although these commands do not have to be selected to complete a dynamic block, they may be helpful in your research to expand your knowledge of blocks.

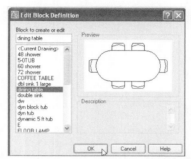

Figure 18.28 *Selecting the **Block Editor** icon from the **Standard** toolbar will display the **Edit Block Definition** dialog box. Select the name of an existing block and click **OK** to begin the process of creating a dynamic block.*

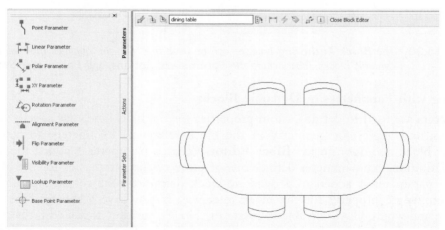

Figure 18.29 *The selected block has been placed in the **Block Editor** with the **Parameter** palette displayed.*

In addition to adding objects to the block, parameters and actions can be added to a block in the editor to make it dynamic. **Parameters** define custom properties for the dynamic block by specifying positions, distances, and angles for objects in the block. When a parameter is added to a block definition, custom grips and properties are automatically added to the block. You use these custom grips and properties to manipulate the block reference in the drawing. **Actions** define how the geometry of a dynamic block reference will move or change when the block reference is manipulated in a drawing. When you add actions to the block, you must associate them with parameters and usually geometry. Controls for **Parameters** and **Actions** can be found on the **Block Authoring** palettes shown in Figure 18.30.

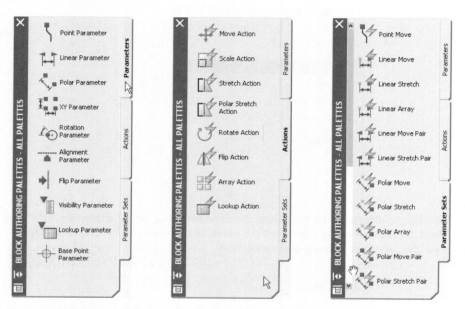

Figure 18.30 *The **Block Authoring** palettes can be used to define the qualities that will create dynamic blocks. Tabs include the **Parameters**, **Actions**, and **Parameter Sets**.*

Working with Parameters in Dynamic Blocks

Parameters are used to define custom properties for the block and may specify positions, distances, and angles for geometry in the block reference. Parameters are added to a dynamic block definition in the **Block Editor.** When a parameter is added to a dynamic block definition, the parameter defines one or more custom properties for the block. If a rotation parameter is added to a dynamic block definition, the rotation parameter will define an **Angle** property for the block reference. If you have a shower block in your drawing, and would like to be able to rotate the block's position as you edit, the parameter defines the axis whereby the block can be rotated. If a **Point** parameter is added to a dynamic block definition, the point parameter defines two custom properties for the block reference: Position X and Position Y relative to the base point of the block reference.

To be considered dynamic, a block definition must contain at least one parameter. Once a parameter is added to a dynamic block definition, grips associated with key points of the parameter will be added automatically. An action for the block definition and associate the action with a parameter must then be added.

Parameter Grips   A grip or custom property in the **Properties** palette is used to manipulate the block reference in a drawing. When the block reference is altered by moving a grip or changing the value of a custom property in the **Properties** palette, the value of the parameter that defines that custom property in the block is changed. When you change the value of the parameter, it drives the action that is associated with that parameter, which changes the geometry of the dynamic block reference. You can view the associated parameter description by floating the cursor over a dynamic block grip. The parameter description is displayed on the screen next to the dynamic block.

Assigning Values Sets   **Parameters** also define and limit the values that affect the dynamic block reference's behavior in a drawing. A parameter can be assigned a fixed set of values, minimum and maximum values, or incremental values. A linear parameter used with block may have values that are set at 15, 30, and 45 degrees. When the block reference is inserted into a drawing, the block can only be changed to one of these values. Adding a value set to a parameter allows limits to be placed on how the block reference can be manipulated in a drawing.

The following table lists and describes the types of parameters you can add to a dynamic block definition and the types of actions you can associate with each parameter.

Parameter	Action	Options	Description
Alignment	None	Perpendicular Basepoint	Automatically rotatesand aligns with objects.
Base	None	Location	Defines base points for dynamic blocks
Flip	Flip	Reflection line	Flips blocks about their reflection line.
Linear	Array Move Scale Stretch	Start point End point Chain Midpoint List	Limits grip movement to preset angle.
Lookup	Lookup	Action name Properties Audit Input Lookup	Defines custom properties that evaluate values from predefined lists and tables.
Point	Move Location Array	Location Chain	Defines an X and Y location in the drawing.
Polar	Move Polar stretch Scale Increment	Basepoint Endpoint Chain List Stretch	Shows the distance between two anchor points and displays an angle value. Both grips and the Properties palette can be used to change both the distance value and the angle.

Parameter	Action	Options	Description
Rotation	Flip	Base point Radius Base Angle Default Angle Chain List	Defines angles.
Visibility	None	Display	Toggles the visibility states of blocks. Shows the X and Y distances from the base points of blocks.
XY	Array Move Scale Stretch Increment	Base Point Endpoint Chain List	Toggles the visibility states of blocks. Shows x,y distances from the base points of blocks.

Using Actions in Dynamic Blocks

Actions define how the geometry of a dynamic block reference will be altered when the custom properties of the block reference are altered in a drawing. With the exception of Alignment, Base and Visibility, a dynamic block will contain at least one action. When an action is added to a dynamic block definition, it must be associated with a parameter, a key point on the parameter, and geometry. A **key point** is the point on a parameter that controls the associated action when edited. The geometry associated with an action is referred to as the selection set.

In Figure 18.31, the dynamic block definition represents a dining table with a linear parameter with one grip specified for its endpoint, and a stretch action associated with the endpoint of the parameter and the geometry for the right side of the block. The endpoint of the parameter is the key point. The geometry on the right side of the block is the selection set.

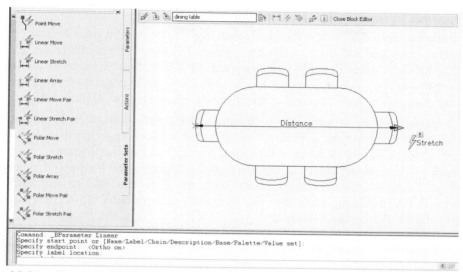

Figure 18.31 *A dynamic block with a linear parameter with one grip specified for its endpoint, and a stretch action associated with the endpoint.*

The types of actions and the parameters that can be associated with each type of action can be seen in the table below.

Action	Parameter	Description
Array	Linear, polar, XY	Selected objects can be arrayed.
Flip	Flip	Objects can be flipped about a mirror line.
Lookup	Lookup	Displays the Property Lookup table dialog box.
Move	Point, linear, polar, XY	Objects move in a linear or polar direction.
Polar Stretch	Polar	Stretches blocks by angle and distance.
Rotate	Rotation	Allows rotation around a select point.
Scale	Linear, polar, XY	Objects are resized in a linear, polar, or an XY direction.
Stretch	Point, linear, polar, XY	Objects can be stretched in a Point, linear, polar, or XY direction.

More than one action can be assigned to the same parameter and geometry. Two or more of the same type of actions should not be assigned to the same key point on a parameter if both actions affect the same geometry. Adding two or more of the same type of actions can result in unexpected behavior in the block reference.

Working with Parameter Sets in Dynamic Blocks

The **Parameter Sets** tab on the **Block Authoring** palette can be used to add commonly paired parameters and actions to a dynamic block definition. A parameter set is added to a block in the same manner used to add a parameter. The action included in the parameter set is automatically added to your block definition and associated with the added parameter. You must then associate a selection set (geometry) with each action.

When a parameter set is added to a dynamic block definition, a yellow alert icon is displayed next to each action. The icon indicates that a selection set needs to be associated with each action. Double click the yellow alert icon or use the **Bactionset** to associate the action with a selection set. Selecting the alert icon will display the **Property Lookup Table** dialog box. Lookup actions are associated with the data you add to this table, not a selection set. The following table lists the parameter sets provided on the Parameter Sets tab of the Block Authoring palette.

Parameter Set	Description
Flip	Adds a flip parameter with one grip and an associated flip action to the dynamic block definition.
Linear Array	Adds a linear parameter with one grip and an associated array action to the dynamic block definition.
Linear Move	Adds a linear parameter with one grip and an associated move action to the dynamic block definition.
Linear Move Pair	Adds a linear parameter with two grips and a move action associated with each grip to the dynamic block definition.
Linear Stretch	Adds a linear parameter with one grip and an associated stretch action to the dynamic block definition.
Linear Stretch Pair	Adds a linear parameter with two grips and a stretch action associated with each grip to the dynamic block definition.
Lookup	Adds a lookup parameter with one grip and a lookup action to the dynamic block definition.
Polar Array	Adds a polar parameter with one grip and an associated array action to the dynamic block definition.
Polar Move	Adds a polar parameter with one grip and an associated move action to the dynamic block definition.
Polar Move Pair	Adds a polar parameter with two grips and a move action associated with each grip to the dynamic block definition.
Polar Stretch	Adds a polar parameter with one grip and an associated stretch action to the dynamic block definition.
Polar Stretch Pair	Adds a polar parameter with two grips and a stretch action associated with each grip to the dynamic block definition.

Parameter Set	Description
Rotation	Adds a rotation parameter with one grip and an associated rotate action to the dynamic block definition.
Visibility	Adds a visibility parameter with one grip. No action is required with a visibility parameter.
XY Array Box Set	Adds an XY parameter with four grips and an array action associated with each grip to the dynamic block definition.
XY Move	Adds an XY parameter with one grip and an associated move action to the dynamic block definition.
XY Move Box Set	Adds an XY parameter with four grips and a move action associated with each grip to the dynamic block definition.
XY Move Pair	Adds an XY parameter with two grips and move action associated with each grip to the dynamic block definition.
XY Stretch Box Set	Adds an XY parameter with four grips and a stretch action associated with each grip to the dynamic block definition.

CREATING DYNAMIC BLOCKS

Before you create a dynamic block take time to plan the contents. Some things to consider include:

- Determine what the block will look like and how it will be used in a drawing.
- Decide which objects within the block will change or move when the dynamic block reference is manipulated.
- Decide how these objects will change throughout the drawing.
- Will the block need to be resized?
- When the block reference is resized, will additional geometry be displayed?

These factors will determine the types of parameters and actions that need to be added to the block definition, and how the parameters, actions, and geometry will work together.

Use the following steps to create dynamic blocks.

1. Draw the objects that will form the new block or use an existing block definition.

2. Open the **Authoring Palettes** and add the required parameters. See Figure 18.30.

3. Assign **Actions** to the parameters. When you add an action to the block definition, you will need to associate the action with a parameter and a selection set of geometry. This creates a dependency. See Figure 18.31.

4. Select the **Close Block Editor** button and then save the changes.

5. Insert the block and see if it responds as you expected.

CHAPTER 18 EXERCISES

1. Open the **FOOTINGS** drawing that was updated in chapter 17. Save the drawing as a block titled **FND1STORY**.

2. Open drawing **FLOOR14** and use the drawing to create a wblock titled **BASE14**.

3. Open drawing **E-17-7**. Save the drawing as a wblock titled **FNDPLN**. Insert **FNDPLN** in drawing **FND1STORY**, rescale the footing and save the entire drawing as **FNDDLT**.

4. Open drawing **FNDDLT** and use it to create a wblock titled **FNDPLN1**. Insert **FNDPLN1** in **BASE14**.

5. Use the following drawing as a guide to draw these symbols and save them as a wblock in a folder titled **\PROTO\FLOOR** using the designated block name. Create all blocks on the 0 layer and use 4" high text.

DESCRIPTION	SYMBOL	BLOCK NAMEBATH
60" × 32" tub	A	TUB60
72" × 32" tub	B	TUB72
19" × 16" lavatory	C	LAV
Toilet	D	POTTY
36" shower	E	SHOWER36
42" shower	F	SHOWER42
60" shower	G	SHOWER60
32" × 21" double sink with a garbage disposal on either side	H	DBLSINK
36" × 21" double sink with a garbage disposal on small side	J	XDBLSINK
43" × 21" triple sink with a garbage disposal in the middle	K	XXSINK
16" × 16" veggie sink	L	VEGSINK
Dishwasher	M	DW
Trash compactor	N	TC
Lazy susan	P	LS
Range	Q	RANGE
Double oven	R	DBLOVEN
18" pantry	S	PANUTILITY
Washing machine	T	WASH
Dryer	U	DRYER

DESCRIPTION	SYMBOL	BLOCK NAMEBATH
Built-in ironing board	V	IRON
21" × 21" laundry tray	W	LT
Water heater	X	WH
Hose bibb	Y	HB

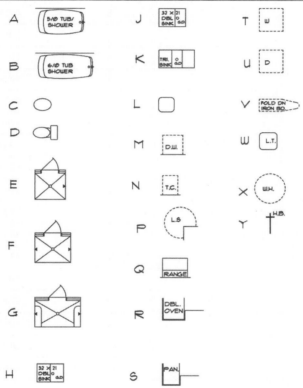

6. Use the following drawing as a guide to draw these symbols, and save them as a wblock in a folder titled **\PROTO\ELECT** using the designated block name. Use a 6" diameter circle for all electrical symbols with no text. Use an 8" diameter symbol with 4" high text for symbols that require text to be placed inside the symbol. Create all blocks on the 0 layer.

DESCRIPTION	SYMBOL	BLOCK NAMEBATH
110 c.o.	A	110 CO
Multi 110 c.o.	B	110 CO2
220 c.o.	C	220 CO
110 half-hot c.o.	D	110 HH
110 waterproof	E	110 WP
110-ground fault	F	110 GFI interrupter
Floor-mounted 110 c.o.	G	110F
Junction box	H	JBOX LIGHT FIXTURES
Ceiling-mounted light	J	CEIL LITE fixture
Wall-mounted light	K	WALL LITE fixture
Can light fixture	L	CAN LITE
Recessed ceiling light	M	RECESS LITE fixture
Wall-mounted spotlights	N	WALL SPOT
24" × 48" surface-mounted fluorescent fixture	P	24 FLUOR
48" × 48" surface-mounted fluorescent fixture	Q	48 FLUOR
48" shop fluorescent fixture	R	48 SHOP
48" track light	S	TRACK SWITCHES
Single-pole switch	T	SWITCH
Three-way switch	U	SWITCH3
Switch with dimmer	V	DIM SWITCH SPECIAL SYMBOLS
Vacuum	W	VACUUM
Intercom	X	INTERCOM
Phone	Y	PHONE
Stereo speakers	Z	SPEAKERS
Smoke detector	AA	SMOKE
Light, heat, fan	BB	LHF

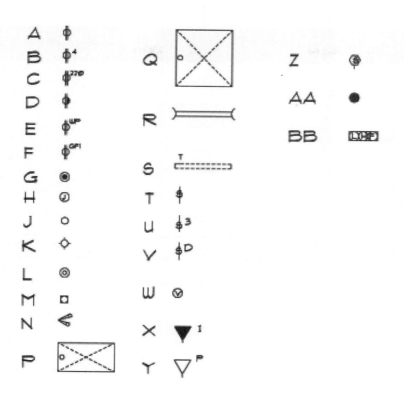

7. Use the following drawing as a guide to draw these symbols, and save them as a wblock in a folder titled **\PROTO\FLOOR\DOORS** using the designated block name. Create all blocks on the 0 layer using a 12" block unit.

DESCRIPTION	SYMBOL BLOCK	NAME
Exterior door	A	XDOOR
Exterior door with 1 sidelight	B	XDOOR1LITE
Exterior door with 2 sidelights	C	XDOOR2LITE
Pair of exterior doors with 2 sidelights	D	PR2DOOR2LITE
Interior door	E	INDOOR
Exterior sliding glass door	F	XSLDOOR
Interior by-pass door	G	BYPASS
Interior bi-fold door	H	BYFOLD
Interior folding door	J	FOLD
Window	K	WINDOW

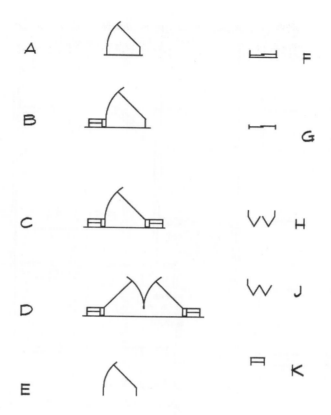

A

B

C

D

E

F

G

H

J

K

8. Open the appropriate template and use the following drawing as a guide to drawing the bathroom. Draw the exterior windows 6 feet wide, except for a 2-foot wide window at the toilet room, a 4-foot window over the counter, and 3-foot wide windows in the bedroom. Use a pair of 2-foot wide doors in the bedroom, a pair of 3-foot doors from the bedroom to the exterior, and a 28" pocket door into the toilet room. All other doors are to be 30" wide. Create the following layers: **WALLS, PLUMB, APPLIANCES, DIMEN, NOTES, DOORWIN** and **CABS**. Create and insert the required blocks to complete the drawing. Provide dimensions and label all fixtures. Save the drawing as a wblock titled **BATH**.

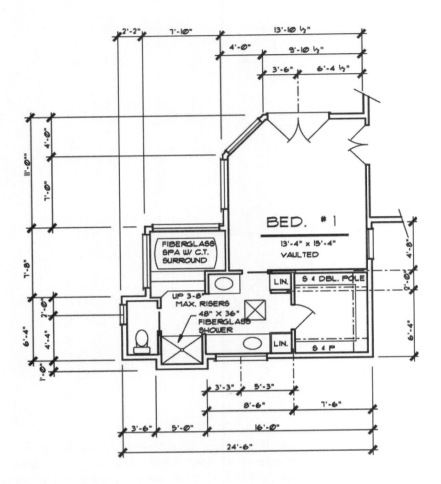

9. Open the appropriate template and use the following drawing to draw the kitchen. Design the door and window sizes to fit the space provided. Create layers that will logically separate information for future printing on floor, framing, and electrical plans. Insert the required blocks to complete the drawing. Align the bathroom with the utility room and provide a 42" fiberglass shower unit. Provide dimensions, determine the square footage of each room, and label the drawing. Turn off all layers containing notes and dimensions and design an electrical plan that assumes all appliances are electrical. Use 2 ceiling-mounted fixtures in the nook, a surface-mounted fixture in the utility room, 8 can lights in the kitchen and 2 in the hallway, a wall-mounted fixture over the bathroom sink, and a ceiling-mounted fixture in the center of the bath. Save the drawing as a wblock titled **KITCHEN**.

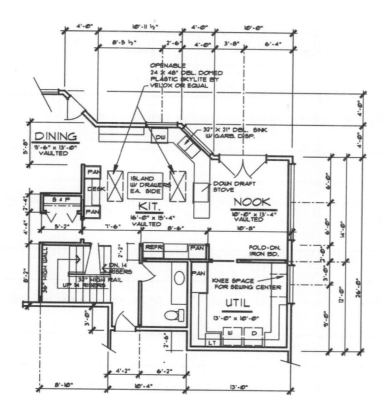

10. Open the **FOOTING** drawing and use it to create a wblock drawing titled **BASEFTG**. Save the file with an appropriate title.

11. Pick 5 different blocks from problem **E 18-5** and convert each block to a dynamic block. Assign a different parameter to each block.

CHAPTER 18 QUIZ

DIRECTIONS

Answer the following questions with short complete statements. Type your answers using a word processor.

1. Place your name, chapter number, and the date at the top of the sheet.

2. Type the question number and provide the answer.

Warning: Some of the questions have not been covered in the reading material and will require the use of the help menu. You may also have to do some exploring to answer the questions.

1. How can a wblock increase drawing speed?

_____

_____

2. Why is using blocks more efficient than copying an object?

_____

_____

3. If a block is created on the 0 layer, where will the block be located when it is inserted?

_____

_____

4. When the color of a block is set **BY BLOCK**, how will the block be affected?

_____

_____

5. What is the advantage of drawing blocks such as a window in a 12" box rather than full size?

_____

_____

6. List the process to create a block.

_____

_____

7. How does the way that the **Insertion Point** is specified when the block is defined affect the insertion point requested when the block is inserted?

_____

_____

8. What command will allow for multiple blocks to be inserted?

_____

_____

9. What command will allow for editing blocks?

_____

_____

10. What is the quickest method for editing the same mistake in several blocks?

_____

_____

11. Explain the difference between a block and a wblock.

_____

_____

12. List the process for inserting a wblock in a drawing.

_____

_____

13. How can a block be mirrored before it is inserted?

_____

_____

14. What method can be used to insert a block by selecting two points?

_____

_____

15. What is the effect of activating the **Explode** check box in the **Insert** dialog box?

_____

_____

EXERCISES

16. What is required to turn a block into a dynamic block?

_____

_____

17. What tool is used to create dynamic blocks?

_____

_____

18. Where are the controls for parameters located?

_____

_____

19. What does a parameter do to a dynamic block?

_____

_____

20. What does a value set do to a dynamic block?

_____

_____

21. A block will have linear parameters. What actions are available?

_____

_____

22. Are dynamic blocks required to have an action?

_____

_____

23. What are parameters sets?

_____

_____

24. What tools are used to manipulate dynamic blocks in the editor?

_____

_____

25. What options are available with the polar parameter?

_____

_____

Adding Attributes to Enhance Blocks

INTRODUCTION

Even with all the qualities that blocks and wblocks add to a drawing, they can still be enhanced. Many blocks need text to explain or provide a specification. Text can be added to a block using three methods. In the last chapter, text was added to the drawings using **Text**, and saved as part of the block. This works well if the text remains constant, such as the specification for a double kitchen sink. If text is added to the block, the block must be exploded if the text is to be edited. The **Explode** command allows the text to be edited, but characteristics of the block might be lost. Once a block has been inserted, text can be added to the drawing base to supplement the block by using the **Text** or **Mtext** command. This option works well only if a few blocks need written specifications.

This chapter will introduce a new method of adding text to a block by using attributes.

Commands to be introduced in this chapter include:

- **Attdef** (Attribute definition)
- **Attdisp** (Attribute display)
- **Attedit** (Attribute edit)
- **Attreq** (Attribute request)
- **Battman** (Block attribute manage)
- **Eattedit** (Enhanced attribute editor)
- **Eattext** (Attribute extraction)

THE BENEFITS OF ATTRIBUTES

An attribute is a grouping of objects that contains text. An example of a block with attributes can be seen in Figure 19.1. These text groupings are similar to a block, which is then attached to a drawing block. Attributes have several features that make them superior to text created with **Text** or **Mtext** and are well worth the effort to master. Once the benefits have been explored, thought must be given to defining the attribute text, controlling how the attributes will be displayed, changing the values, and extracting attributes from a drawing to compile a written listing of attributes.

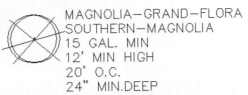

MAGNOLIA—GRAND—FLORA
SOUTHERN—MAGNOLIA
15 GAL. MIN
12' MIN HIGH
20' O.C.
24" MIN.DEEP

Figure 19.1 *Attributes can be added to a block to explain qualities that could vary with each insertion.*

The two major benefits of adding text through the use of attributes are control and retrieval of information.

CONTROLLING TEXT

Using attributes will allow for assigning prompts to a block to request information each time that the block is displayed. For instance, each time the door block is inserted in a drawing, these prompts could also be provided: size, thickness, type, manufacturer, finish, rough opening and hardware requirements.

Default values can be provided, but altered with each insertion. Using attributes also allows for correcting the text that is displayed without altering the block characteristics. These prompts might be hidden during the display or plotting process, even though the block they are attached to is displayed.

EXTRACTING ATTRIBUTE INFORMATION

In the example above, much of the information that will comprise the attribute is displayed in a plant schedule. Attributes can be extracted from the drawing, stored in a separate file, and printed using a third party database text editor, such as the Microsoft Notepad. A list of attributes also can be compiled and printed within the drawing.

PLANNING THE DEFINITION

It is easiest to add attributes to a drawing before the drawing has been saved as a block. Once the symbols have been drawn, a decision must be made about what specifications will be associated with the block. For best efficiency this should be done in a planning session prior to entering the **Attdef** (ATTribute DEFinition) command. For a block of a window, this might include the size, type, frame material, manufacturer, glazing, and rough opening. Other than the actual space limitations within the drawing, there is no limit to the amount of information that can be associated with a block. As the attributes are being defined, an option is even available to keep the attributes invisible.

After the material to be listed with the block has been determined, decide how the prompts for each attribute will be displayed. Attribute prompts are usually in the form of a question, but statements can also be used. For instance, the prompt might read "What is the window size?" or "Provide the window size." Determine the prompts for each attribute that you will be listing.

ENTERING THE ATTRIBUTE COMMANDS

Several commands will be used as you work with attributes. Once the block has been drawn and you've decided what attributes are to be assigned, the process of attaching text to the block is started by using the **Attdef** command.

CREATING ATTRIBUTES

Each of the qualities of an attribute can be controlled using the **Attribute Definition** dialog box or at the command line. Consult the **Help** menu if you would rather use the slower, harder Command prompt method. To access the dialog box, use one of the following methods:

- Type **ATTDEF** ENTER at the Command prompt.

- Select **Define Attributes** from **Block** on the **Draw** menu.

Each method will display the dialog box shown in Figure 19.2. Major elements of the dialog box include the **Mode, Attribute, Insertion Point**, and **Text Settings** areas.

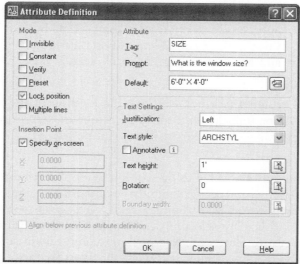

Figure 19.2 *Typing **ATTDEF** ENTER at the Command prompt or selecting **Define Attributes** from **Block** on the **Draw** menu will produce the **Attribute Definition** dialog box. The box can be used to design the tag, prompt, value, and insertion point of each attribute, as well as controlling how each will be displayed.*

ATTRIBUTE DISPLAY

The **Mode** area allows the display options for the attribute to be set. These include **Invisible, Constant, Verify, Preset**, Lock position, and Multiple lines.

Invisible

Selecting the **Invisible** check box will make the attributes invisible. Attributes will still be attached to the block when it is inserted, but they will not be displayed. If a separate window schedule is to be created, values can be invisible so they are not displayed twice. The **Attdisp** command, which will be introduced shortly, can be used to override the **Invisible** mode setting.

Constant

Selecting the **Constant** check box indicates that all uses of the block will have the same attribute value. The value is entered as the attribute is being defined and set throughout the use of the block. No prompt for new values will be given when the block is inserted with this box active.

Verify

By defining an attribute with this mode active, you will be given an opportunity to verify that the attribute value is correct during the insertion process.

Preset

This mode will allow you to create attributes that are variable but not requested each time the block is inserted. When a block containing a preset attribute is inserted, the attribute value is not requested and is automatically set to its default value.

Lock Position

Selecting the Lock Position check box will lock the location of the attribute attached to the block reference. When unlocked, the attribute can be moved relative to the rest of the block using grip editing, and multiline attributes can be resized.

Multiple Lines

This mode allows the attribute value to be multiple lines of text rather than a single line. With this mode active, you can specify a boundary width for the attribute. When you select this mode, the **Default** text box is disabled and you're given access to the **Multiline Attribute editor** dialog box by selecting the **Open Multilne Editor** button [...]. The **Multiline Attribute editior** allows you to type in text, adjust boundary width, insert fields, underline and overline. Right-clicking will open the shortcut menu providing additional editing options. The options include: Import text, Background color and Autocaps.

DEFINING ATTRIBUTES

Once the mode has been set, attributes can be defined. The three **Attribute** area edit boxes are **Tag**, **Prompt**, and **Default**, and they can be seen in Figure 19.3.

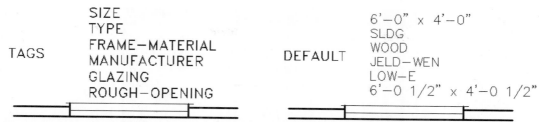

Figure 19.3 *The attribute tags are displayed as the definition is added. The values are displayed once the block has been inserted.*

Attribute Tag

The tag is used to identify an attribute definition in a similar way that a specific name is used to identify a block. The tag will identify each occurrence of the attribute throughout the drawing. During the initial planning stage for the windows, you might decide that information regarding the size, type, frame material, manufacturer, glazing, and rough opening should be associated with each window block. Each of these listings can be used as an attribute tag, but each tag must be entered individually with the prompt that follows used to control only this tag. As the tag is entered, it can be composed of any letters or characters except a blank space. A hyphen can be used between words to create a space. To name a tag, type a tag such as **SIZE** in the **Tag** edit box.

Attribute Prompt

The attribute prompt is the prompt you will see when the block is inserted. As you plan the prompts, questions or statements should be used. To add the prompt, type a prompt such as **WHAT IS THE WINDOW SIZE?** in the **Prompt** edit box. If the **Prompt** edit box is left blank, the tag (SIZE) will be used as the attribute prompt. If the **Constant** mode is active, the **Prompt** option will be inactive.

Attribute Default

Once the tag and the prompt have been specified, provide a value in the **Default** edit box. The value that is entered at this prompt will be displayed as the default value as the block is inserted. A specific value such as **6'-0"x4'-0"** can be entered, or the null response can be used. If the **Constant** mode for the attribute has been selected, a request for the prompt will not be made. Proceed and type a value such as **6'-0"× 4'-0"** in the **Default** edit box.

TEXT OPTIONS

The **Text Options** area contains settings to control the text appearance. Options include edit lists for justification and style, annotative check box and edit boxes for text height, rotation and boundary width values

Justification

Selecting the **Justification** edit arrow will display the list shown in Figure 19.4. The options are the same as those used with the **Text** command. To replace the Left option, select the desired option, and the list will be closed.

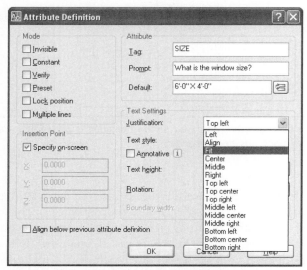

Figure 19.4 *Options for justification can be displayed by selecting the **Justification** edit arrow. The options are the same as those of the **Text** command.*

Style

Selecting the **Text** style edit arrow will produce a list of current drawing text styles. To alter the style, select the name of the desired style. The selected style name will be placed in the text box, and the new style will become current.

Annotative

This mode specifies that the attribute is annotative. If the block is annotative, the attribute will match the direction of the block. Refer to chapter 23 for complete functions associated with annotative objects.

Height

The **Height** edit box allows the text height to be set. You can edit the existing value or select the **Height** button. If the **Height** button is selected, the **Attribute Definition** dialog box is removed and a prompt is displayed to allow points to be selected to indicate the text height. You can enter a value at the Command prompt instead of selecting two points. When the points or value have been provided, the dialog box will be restored and the indicated height will be displayed in the **Height** edit box.

Rotation

The rotation angle for attributes is set in the same manner as the height. Enter the desired angle in the text box or select the Rotation button to select the angle with the cursor.

Boundary Width

The boundary width specifies the maximum length of the lines of text in a multiple-line attribute before text wraps to the next line. A value of 0.0 indicates that there is no restriction on the length of the line of text. This is not an option for single-line attributes.

SELECTING THE INSERTION POINT

The **Insertion Point** area allows the location of the attribute to be selected. Selecting the **Pick Point** button removes the dialog box and allows the insertion point to be selected using the cursor. If the exact location is known, coordinates can be entered using the **X**, **Y**, and **Z** text boxes.

ALIGNING ADDITIONAL ATTRIBUTES

The **Align below previous attribute definition** check box will allow additional attributes to be placed below previous attributes, using the same justification. The option is not active when the first attribute is written. With the option active, the **Text Options** and **Insertion Point** areas will be inactive.

Once all of the options have been selected, clicking the **OK** button will close the dialog box and display the tag at the selected insertion point, similar to the display shown in Figure 19.5. Don't be discouraged at seeing the tags displayed as the command is completed. The tags will not be displayed as the block is inserted. If you are not satisfied with the tags, they can be altered using the **Properties** or **Ddedit** command.

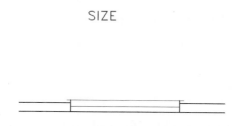

SIZE

Figure 19.5   *The window block with the **SIZE** attribute tag displayed.*

REPEATING THE PROCESS

To add additional attributes to the block, press ENTER to repeat the previous command, and the entire process can be repeated. By repeating the process five more times, the tags for type, frame material, manufacturer, glazing, and rough opening can be added to the block. Activating the **Align below previous attribute definition** check box will automatically place additional tags directly below existing tags with the proper spacing. Repeating the entire sequence will produce a display similar to the one in Figure 19.6.

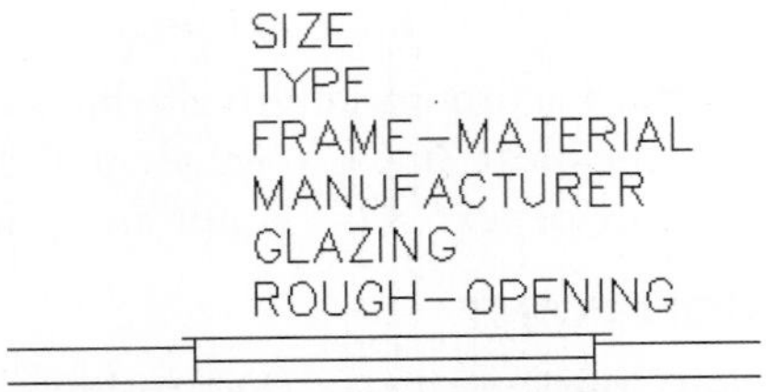

```
SIZE
TYPE
FRAME—MATERIAL
MANUFACTURER
GLAZING
ROUGH—OPENING
```

Figure 19.6 *Repeating the procedure used to create the **SIZE** tag allows an unlimited number of tags to be attached to a block.*

ADDING ATTRIBUTES TO EXISTING BLOCKS

Attributes can still be added if a block has already been created. The original block name must be changed, however, to avoid an error message. If the existing **DOOR** block is inserted in a drawing and attributes are added to it, it cannot be saved using the original name of **DOOR**. Enter a new name to use the existing block with the desired attributes. Many companies will use a name such as **DOOR-A**, using the -A to indicate that the block contains attributes. It's much easier to add the attribute prior to saving the block.

CONTROLLING ATTRIBUTE DISPLAY

Rather than wasting disk space by storing **DOOR** and **DOOR-A** to distinguish between blocks with and without attributes, the **Attdisp** (ATTribute DISPlay) command can be used to hide attributes if you decide that they should not be displayed. To access the command, use one of the following methods:

- Select **Attribute Display** from **Display** on the **View** menu.
- Type **ATTDISP** ENTER at the command line.

This command sequence is as follows:

```
Command: ATTDISP ENTER
Enter attribute visibility setting [Normal/ON/OFF] <Normal>:
```

Normal will display all attributes, **ON** will display the variables, and **OFF** will make all attributes invisible.

The **Attreq** (ATTribute REQuest) command will allow attributes to be displayed but will suppress the attribute prompts. Some blocks, such as kitchen sink, typically retain their default values with very little variation. When the prompt is not required for attributes, the **Attreq** value can be toggled from 1 to 0. The 0 setting will display all attributes in their default setting.

ALTERING ATTRIBUTE VALUES PRIOR TO INSERTION

An **Edit Attribute Definition** dialog box similar to Figure 19.7 can be displayed prior to saving the drawing as a block. To access the display:

- Type **ED** ENTER at the Command prompt.

This dialog box can be used to display and alter attribute values. The default values are listed to the right side of the current prompts. Alter values using methods similar to the methods used for other dialog boxes.

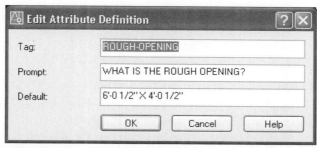

Figure 19.7　*The **Edit Attribute Definition** dialog box can be used to alter the tag, prompt, and default value for an attribute before it is saved within a block. Access the dialog box by typing **ED** ENTER at the Command prompt.*

The **Properties** palette can also be used to edit attributes. Display the palette by selecting the **Properties** button on the **Standard** toolbar. The **Properties** palette can be used to edit the color, layer, linetype, and each of the options defined as the attribute was created.

ATTACHING ATTRIBUTES TO A BLOCK

Attributes can be attached to a block when a block is defined or redefined (review chapter 18). Select the attributes and the object when prompted to select the objects that will comprise the block. The order that attributes are selected when the block is created will determine the order of display when the block is inserted.

Inserting a Block with Attributes

The **Insert** command introduced in chapter 18 can be used to insert blocks with attributes in a drawing. Blocks with attributes are inserted in a method similar to inserting blocks with no attributes. Once the block to be inserted is selected, the prompt will be given for the insertion point. Each of the options for a block can now be adjusted from the command line. The final aspect of inserting the block will be the display of the prompts defined as the attributes were assigned. Figure 19.8 shows the tags and the values for a window block.

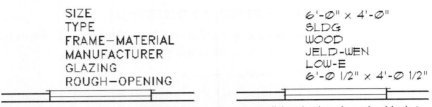

```
SIZE                    6'-0" × 4'-0"
TYPE                    SLDG
FRAME-MATERIAL          WOOD
MANUFACTURER            JELD-WEN
GLAZING                 LOW-E
ROUGH-OPENING           6'-0 1/2" × 4'-0 1/2"
```

Figure 19.8   *The window on the left shows how the tags will be displayed as the block is created. The window on the right shows how the values will be displayed once the block is inserted in a drawing.*

EDITING ATTRIBUTES

Once the values for an attribute have been assigned to a block, they can be altered using the **Edit Attributes** dialog box. To display the dialog box:

- Type **ATTEDIT** ENTER at the Command prompt.

As the command is entered, you'll be prompted to first select a block reference to be edited. Once the block is selected, a dialog box similar to Figure 19.9a is displayed to allow each attribute value to be altered. Alter the text by typing in the desired edit box. Clicking the **OK** button will alter the values displayed on the screen similar to Figure 19.9b. Clicking the **Cancel** button will remove the dialog box with no changes made to the drawing base.

There are two tools available that will allow modification of the attribute once it has been

Edit Attributes

Block name: 6' WINDOW

WHO MAKES THE WINDOW	JELD-WEN
WHAT IS THE ROUGH OPENING	6'-0 1/2" x 5'-1 1/2"
WHAT IS THE GLAZING MATERIAL	LOW-E
WHAT IS THE FRAME MATERIAL	VINYL
WHAT IS THE WINDOW TYPE	PICTURE/SLDG
WHAT IS THE SIZE	6'-0" x 5'-0"

OK   Cancel   Previous   Next   Help

Figure 19.9a   *Attributes can be edited using the **Edit Attributes** dialog box. Access the dialog box by typing **ATTEDIT** ENTER at the Command prompt.*

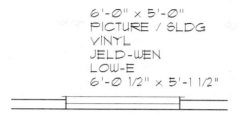

6'-0" × 5'-0"
PICTURE / SLDG
VINYL
JELD-WEN
LOW-E
6'-0 1/2" × 5'-1 1/2"

Figure 19.9b *The effects of altering the values in the **Edit Attributes** dialog box in Figure 19.9a.*

placed in a block. The **Block Attribute Manager** (**Battman**), and the **Enhanced Attribute Editor** (**Eattedit**). The **Block Attribute Manager** should be used if you want to modify the block definition of the attribute. The **Enhanced Attribute Editor** is the choice if you wish to edit the attributes in an individual block.

THE BLOCK ATTRIBUTE MANAGER

The **Block Attribute Manager** will allow you to easily modify the attribute definition within a block without having to explode or redefine the block. All changes are immediately reflected in the existing block insertions. This is useful when a minor change is needed and you do not wish to completely redefine the block containing the attribute. To access the **Block Attribute Manager**, use one of the following methods:

- Click **Block Attribute Manager** from the **Modify II** toolbar.

- Select **Block Attribute Manager** from the **Attribute** listing of **Object** on the **Modify** menu.

- Type **BATTMAN** ENTER at the Command prompt.

Figure 19.10 shows an example of the **Block Attribute Manager** dialog box.

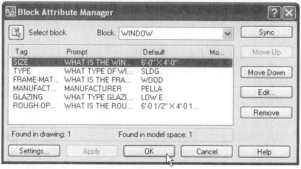

Figure 19.10 *The **Block Attribute Manager** dialog box allows you to easily modify the attribute definition within a block without having to explode or redefine the block.*

The **Block Attribute Manager** dialog box allows you to easily manipulate block attribute properties without redefining the block in the following ways:

- Remove an attribute completely from a block definition.

- Change the order of attributes. This affects the order in which the attributes appear when using the **Edit Attributes** dialog box (**DDATTE** or **ATTEDIT**) and the order in which the attribute prompts appear for input during the block insertion.

- Force the attributes in existing block references to be updated to match their block table definition. This updating process does not affect the attribute value.

- Change an attribute's tag name, prompt string, default value, or mode.

- Change an attribute's text settings, such as the size or font.

- Change an attribute's properties, such as color or layer.

- Customize the main attribute list to display selected columns of information and highlight any redundant tag names.

By default, all changes made to a block with attributes using the **Block Attribute Manager** dialog box are then assigned to all the existing references of that block in the drawing file.

Select Block

The **Select block** button enables you to select blocks with attributes in the current drawing for editing. This name is then highlighted in the **Block** list with the attribute information from the block. The information will then be displayed in the attribute list. If you selected an attribute in the block object, that attribute will be the one that is highlighted in the attribute list.

Block

The **Block** list will display the names of all the blocks with attributes in the block table. The list will not include blocks that do not contain attributes. When a block is selected, the attribute list is updated to display the attributes that are contained in the block that is selected. If the **Block Attribute Manager** dialog box has previously been accessed in the current drawing, the last block and attribute edited will be displayed by default. If the last block edited no longer exists, the selection will default to the first block in the list.

Attribute List

The attribute list displays information about all the attributes defined in the block selected in the **Block** list, including the tag, prompt, default value, modes, and annotative. This list can be customized to display specific columns of information and to highlight redundant tag names by using the **Settings** button.

Sync

The **Sync** button forces all of the existing references to the block displayed in the **Block** list to be updated to the new settings to match the block table definition. This update does not affect attribute values. This operation will automatically occur when the block is edited.

Edit

Selecting the **Edit** button will display the **Edit Attribute** dialog box for the highlighted attribute and allow the information to be modified.

THE EDIT ATTRIBUTE DIALOG BOX

The **Edit Attribute** dialog box consists of three tabs for editing different aspects of the highlighted attributed (see Figure 19.11). When **Auto preview changes** is checked, all edits that affect the appearance of the attribute will automatically be displayed in the editor. This permits you to see how edits being made will actually affect the attributed block. Tabs include the following:

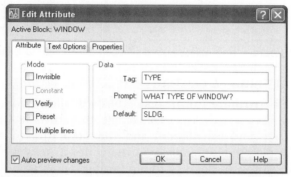

Figure 19.11 *The **Edit Attribute** dialog box shows the three tabs available for editing the high-lighted attribute.*

Attribute—On the **Attribute** tab, you can edit the attributes tag, prompt, default value, and modes.

Text Options—Allows changes to values that affect the text-based properties of the attribute, as in figure 19.12.

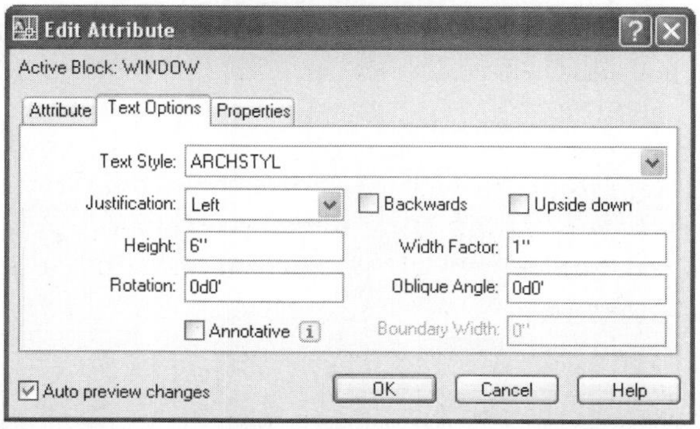

Figure 19.12 *The **Text Options** tab allows editing of the text properties of the attribute.*

Properties—Allows editing of values that affect the general appearance of the attribute, either in the editor or during plotting. See figure 19.13.

Figure 19.13 *The **Properties** tab will change the appearance of the attribute.*

THE ENHANCED ATTRIBUTE EDITOR

The **Enhanced Attribute Editor** dialog box will allow editing of the attributes in the block that is selected. You can change the attribute value, any text-related settings like size and orientation, and the properties of the attribute. All of the changes will be displayed automatically in the editor as they are made. To access the **Enhanced Attribute Editor,** use one of the following methods:

- Click **Edit Attribute** from the **Modify II** toolbar.
- Select **Attribute** from **Object** on the **Modify** menu and then select **Single**.
- Type **EATTEDIT** ENTER at the Command prompt.

The **Enhanced Attribute Editor** dialog box has three tabs available to allow editing the different features of the selected attribute in a block, as well as a button for selecting the block, as shown in Figure 19.14.

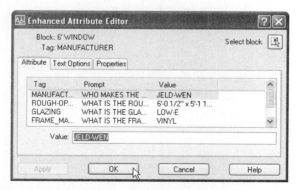

Figure 19.14 *The **Enhanced Attribute Editor** dialog box contains three tabs for changing features of the attribute.*

Select block—The **Select block** button allows you to select a block with attributes in the current drawing for editing. The name of the block appears at the top of the dialog box, and the attribute information assigned to that block will be added to the attribute list.

Attribute—This tab shows the information about all of the attributes that are a part of the selected block. When an attribute is selected from the list, the value can be modified.

Text Options—This tab allows editing of those values that will apply to the text properties of the attribute (see Figure 19.15).

Properties—This tab allows you to modify the appearance of the attribute either in the editor itself or at the time of plotting (see Figure 19.16).

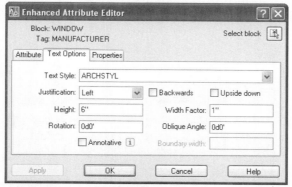

Figure 19.15 *The **Text Options** tab will modify the attribute text properties.*

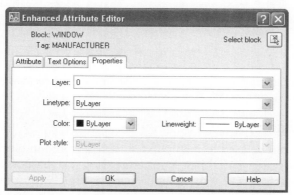

Figure 19.16 *The **Properties** tab will change the appearance of the attribute.*

EXTRACTING ATTRIBUTES

One of the most useful benefits of using attributes is the ability to extract information stored as an attribute and assemble it in a written list. Window, door, and finish schedules are typically associated with architectural drawings. Pier, steel, and beam schedules are but a few of the types of lists in engineering drawings. Figure 19.17 shows portions of a floor plan and the doors that will be extracted. The doors shown are a fraction of the total number of doors that would be represented on a complete multilevel structure. Each door was inserted in the floor plan as a block, with one visible and four invisible attributes. Attributes have been assigned to represent symbol, size, style, manufacturer, price, and quantity. Because these values are assigned as attributes, an accurate schedule can be maintained and quickly updated. With the **Eattext** command, a complete listing of all of the doors used throughout the project can be written to a file.

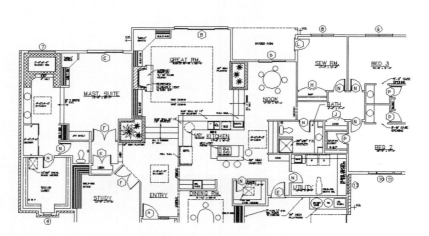

Figure 19.17 *Extracting attributes can be very useful on drawings containing multiple blocks with attributes such as this floor plan.*

USING A DATABASE

A database program can be used to manipulate the data stored in block attributes. A listing of doors or windows for a multilevel structure is a database that has essential data associated with it—such as size, type, rough opening, etc. Two terms generally associated with each database are *record* and *field*.

- A list of each of the construction components related to doors would be a record. The door, frame material, hinge type, and finishing method are each an example of a record. A door might be swinging, sliding, folding, or overhead. Each listing of a door is a record.

- The swing information about a door is a field. Swinging doors can be divided by interior, exterior, raised panel, slab, 1-lite, or 10-lite.

It is possible to take a listing of doors that are listed by size, and by using a database, reorder the listing by type. AutoCAD allows a list to be generated that contains all of the objects to be manipulated by the database.

EXTRACTING THE ATTRIBUTES OF A BLOCK

The **Attribute Extraction Wizard** allows the easy extraction of block attribute data to comma-separated text (.csv), Excel (.xls), and Access (.mdb) formats. You can attach alias names to blocks and attributes, publish data from multiple drawings and xref attachments, and save templates of selected blocks, attributes, and alias information for reuse with any drawing or set of drawings. To extract attributes in this manner you need to enter **ATTEXT** on the command line. For more Information about this feature refer to the **Help** menu. Attributes are typically put in tables and the tables are then added to the drawing. Tables were described in chapter 15. The **Attribute Extraction Wizard** will guide you through the process of attribute extraction. To access the wizard, use one of the following methods:

- Click the **Attribute Extract** button on the **Modify II** toolbar.

- Type **EATTEXT** ENTER at the command line.

The **Attribute Extraction begin** dialog box allows the selection of a template file. Template files store information related to which attributes and blocks are to be included in the output and the view from which data is extracted. The template will be created as the criteria are assigned with the wizard. You can then save the template for use in future attribute extractions. Figure 19.18 shows how to choose an existing template file.

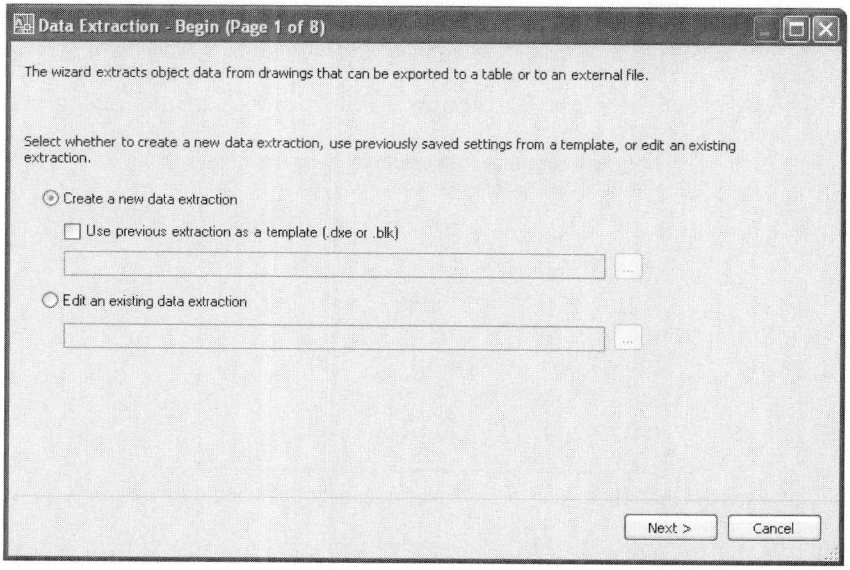

Figure 19.18   *A template can be selected from a previous extraction.*

The **Data Extraction – Define Data Source** dialog box determines the source of the attribute data. It will allow you to create a selection set of objects from the current drawing, select all of the blocks in the current drawing, or select multiple drawings to extract the information from. Figure 19.19 shows a sample of how the source data is to be selected.

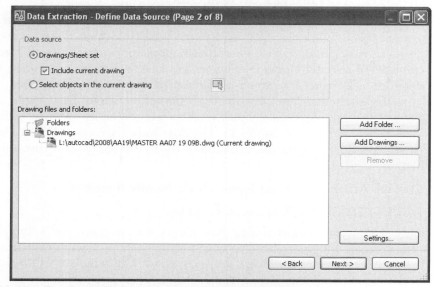

Figure 19.19   *The **Data Extraction - Define Data Source** dialog box determines the source of the attribute data.*

The **Settings** Button will display a dialog box allowing you to include external reference files (xrefs) and nested blocks in your search. (Nested blocks were introduced in chapter 18.) Figure 19.20 shows how the **Data Extraction – Additional Settings** dialog box appears.

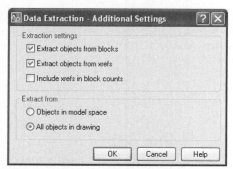

Figure 19.20   *External references and nested blocks can be included in the search.*

Selecting the **NEXT** button will display the **Data Extraction - Select Objects** dialog box. This dialog box consists of two sections:

- The **Objects** list will show the list of objects to extract data from in the selected drawing or drawings. Select the object by checking the box located next the object name.

- The **Display Options** provides different options regarding the display of the objects. Options include: **Display all object types, Display blocks only, Display non-blocks only, Display blocks with attributes only,** and **Display objects currently in-use only.** These options will prove to be valuable as you work on large files with many blocks and attributes.

Figure 19.21 Shows an example of the **Data Extraction - Select Objects** dialog box.

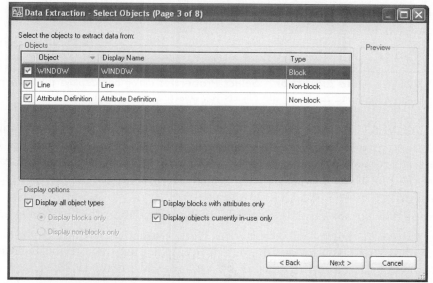

Figure 19.21 *The **Data Extraction - Select Objects** dialog box contains a block list and an attribute list.*

The **Data Extraction - Select Properties** dialog box shown in Figure 19.22 will now show a list of the properties that were found in the objects. Select the appropriate category filter to alter the properties.

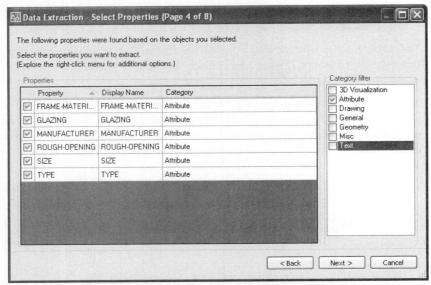

Figure 19.22 *The **Data Extraction - Select Properties** dialog box displays the properties that were found based on the objects you selected.*

Selecting the *NEXT* button will display the **Data Extraction - Refine Data** dialog box. This dialog box enables you to reorder, sort columns, filter results, add formula columns, and create external data links. Clicking the *NEXT* button will display the **Data Extraction – Choose Output** dialog box. This dialog box allows you to specify the output type for the extraction.

View Output dialog box displays a sample of the information that will be extracted.

Selecting the *NEXT* button will display the **Data Extraction - Table Style** dialog box. This dialog box allows you to save the output to a table that was previously created. This table can then be placed in the drawing file as shown in Figure 19.23.

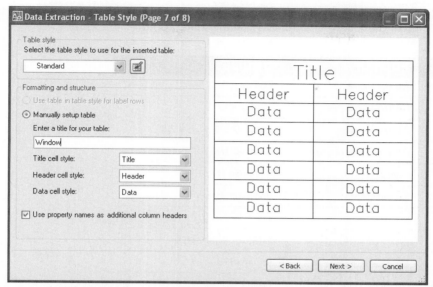

Figure 19.23 *The **Data Extraction - Table Style** dialog box displays a sample of the table that will be extracted.*

The final dialog box Data **Extraction - Finish** will display upon completion of the process.

CHAPTER 19 EXERCISES

1. Draw an 18" diameter pier using dashed lines and show a 4x4 post in the center of the pier. Use 5" high text to install the following attributes.

TAG	PROMPT	VALUE
Dia.	What is the diameter?	18" dia.
Depth	What is the pier depth?	12" min. into grade
Strength	What is the conc. strength?	2500 p.s.i.
Rebar	Will the pier be reinforced?	10 × 10-4 × 4 wwm 3" up
Post	Size of post to be supported?	4 × 4 DFL #1
Anchor	Will an anchor be used?	Simpson CB44 base

Save the drawing as a wblock on a diskette as **FNDPIER**.

2. Draw an elevation of a 3'–0" × 4'–6" double-hung window and use 5" high text to install the following attributes. Set the mode to invisible display.

TAG	PROMPT	VALUE
Manuf.	List the manuf.	POZZI
Frame	List the frame size	3'–0" × 4'–6"
Unit	List the unit size	UN=3'–2 5/8" × 4'–8 7/8"
R.O.	List the rough opening	RO=3'–0 1/2" × 4'–6 1/2"
Frame	Frame material	Kiln-dried fir
Exterior	Exterior finish	Prime ext. face
Sash	Sash material	1 11/16" thick sash
Glaze	Glazing material	9/16" air space
Weather	Weather stripping	foam-backed
Hardware	Hardware color	Bronze
Screen	Screen material	Alum.-bronze tint
Assembly	Where assembled	Site assembled
Shipping	Destination	Seattle, WA.

Prior to creating a wblock of this material, change the Frame material to Kiln-dried Western pine. Change the hardware to WHITE. Ship from Bend, OR. Change the Assembly point to Factory. Save the drawing as a wblock with a name of **EL3-46DH**.

3. Open a new drawing and draw a 24" × 24" skylight in plan view. Add the following attribute tags, prompts, and values.

TAG	PROMPT	VALUE
Size	Unit size	24" × 24"
Manuf.	Who is the manuf?	Velux
Model	Model number	SF
Frame	Frame material	Alum.-clad frame
Finish	Finish material	Lacquered finish
Glass	Glazing	Double-tempered glazing
Flashing	Flash material	32ga. lacquered alum.

Save the drawing on a diskette as a wblock titled **SKYLIGHT**. Insert the skylight in the drawing three different times. Change the size so that the skylight will be 24 × 24, 24 × 48, and 48 × 48. Change the size value of each skylight to reflect the proper size. Use global editing to change the model value to FS, change the frame to Alum-clad Frame, and change the finish to 22ga. Save the drawing as **E-19-3**.

4. Open drawing **E-18-5** and insert the **POTTY** block. EXPLODE the block and add the following attribute tags, prompts, and values.

TAG	PROMPT	VALUE
Manuf.	Who is the manuf?	Am. Std.
Cat-no	What is the Catalog number?	2006.014
Descrip.	What is the description?	Elon. Lexington
Color	What color will be used?	White
Supplier	Who will provide fixture?	General cont.
Installer	Who will install fixture?	Plumber
Hardware	Who will supply hardware?	Owner

Save the drawing as a wblock titled **POTTY**.

CHAPTER 19 QUIZ

DIRECTIONS

Answer the following questions with short complete statements. Type your answers using a word processor.

1. Place your name, chapter number, and the date at the top of the sheet.

2. Type the question number and provide the answer in the form of a statement that includes part of the question. You do not need to write out the entire question.

 Warning: Some of the questions have not been covered in the reading material and will require the use of the help menu. You may also have to do some exploring to answer the questions.

1. How are attributes saved?

2. What commands can be used to add a block or wblock to a drawing?

_____

_____

3. What must be done to add attributes to an existing block?

_____

_____

4. Describe what an attribute prompt is.

_____

_____

5. What is an attribute tag?

_____

_____

6. List the four modes for attribute display and give a brief explanation of each.

_____

_____

7. What is the initial step to assign attributes to blocks?

_____

_____

8. List the commands that produce dialog boxes for use with attributes.

_____

_____

9. List two methods to edit an attribute.

_____

_____

10. What **Attdia** setting is required to produce dialog box displays?

_____

_____

11. What two commands allow for attribute editing?

_____

_____

12. How can global editing of attributes be limited?

_____

_____

13. How do the prompts "**Enter string to change**" and "**Enter new string**" affect each other?

_____

_____

14. What options are provided when attributes are edited individually?

_____

_____

15. How will the **Attreq** command affect an attribute?

_____

_____

Oblique and Isometric Drawings

INTRODUCTION

The drawings that have been examined throughout this text have been orthographic drawings. These are drawings that show an object as if you were looking directly at it. Objects are assumed to be located in a glass box with the sides of the object parallel to the sides of the glass box. Orthographic drawings comprise the majority of construction drawings. Pictorial drawings, including oblique and isometric drawings, are used in some disciplines of engineering and architectural drawings to show specific details. This chapter will introduce:

- The theory of oblique and isometric drawings.

- The creation of oblique and isometric drawings.

Commands to be explored in this chapter include:

- **Snap**

- **Isoplane**

- **Ellipse**

- **Dimali** (Dimension aligned)

- **Dimedit** (Dimension edit)

ABOUT OBLIQUE AND ISOMETRIC DRAWINGS

Before examining what oblique and isometric drawings are, it is important to understand what they are not. These drawings appear to be three-dimensional, but are really two-dimensional drawings using lines that are created using X and Y coordinates.

3D DRAWINGS

One of the major features of AutoCAD is its ability to create true 3D drawings. Three-dimensional drawings require creation of a 3D model. For instance, in drawing a floor plan, heights are provided in addition to the normal information that would be drawn. Once the model is created, AutoCAD can rotate the model in space, allowing the viewing point to be altered. You've most likely seen commercials on television with structures growing out of a floor plan and progressing through the entire construction process. These drawings could be created starting with 3D models. Individual points, lines, and planes comprising the drawings are stored and become part of the drawing base. Figure 20.1 shows an example of a 3D drawing created using AutoCAD and a third party software program. Although these drawings are both beautiful and very useful, how they are created will not be discussed in this text. See the CAD listings at www.delmarlearning.com for a listing of books dealing with 3D drawings.

Figure 20.1 *3D drawings rendered in AutoCAD often resemble a photograph.*

OBLIQUE DRAWINGS

Oblique drawings show height, width, and depth of an object. Three common methods of oblique drawings include cavalier, cabinet, and general drawings. Examples of each can be seen in Figure 20.2. Each system places one or more surfaces parallel to the viewing plane and shows the receding surface at an angle. Each drawing method can be used at any scale.

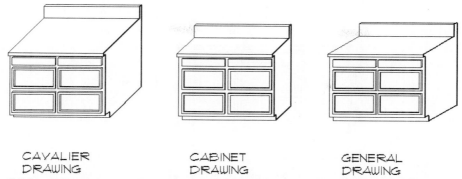

CAVALIER
DRAWING

CABINET
DRAWING

GENERAL
DRAWING

Figure 20.2   *Although lacking the details of 3D drawings, three types of oblique drawings are often used to show height, width, and depth of an object.*

Cavalier Drawings

Cavalier drawings use the same scale for the front and receding surfaces. The front side is drawn parallel to the viewing plane. The receding surface is drawn at a 45° angle to show depth. See Figure 20.3.

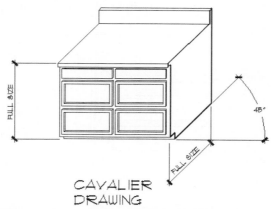

CAVALIER
DRAWING

Figure 20.3   *Cavalier drawings show all surfaces in full scale with the receding side drawn at a 45° angle.*

Cabinet Drawings

Cabinet drawings are similar to cavalier drawings but show the receding surfaces using a scale that is half of the scale used on the front surfaces. The receding surfaces are also shown at a 45° angle. See Figure 20.4.

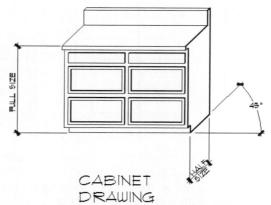

Figure 20.4 *Cabinet drawings show the front view at full scale with the receding side shown at half scale and a 45° angle.*

General Drawings

General drawings represent the receding surfaces using a scale that is three-quarters of the scale used to draw the front surfaces. The angle used to create the receding edge can be any angle except 45°. Figure 20.5 shows an example of a general oblique drawing.

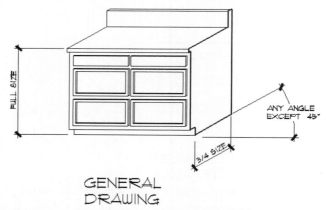

Figure 20.5 *General drawings show the front view at full scale with the receding side shown at three-quarter scale and any angle except 45°.*

ISOMETRIC DRAWINGS

Isometric drawings are a common form for showing three surfaces of an object. Rather than having one surface resting squarely on a base, the object is rotated at 35° 16' from a horizontal plane. This results in a view of the object from the front where each edge is drawn at 30° to a horizontal base. All vertical lines remain vertical, and an equal scale is used to represent all surfaces. Figure 20.6 shows the basic construction methods for an isometric drawing. Lines that are parallel to one of the three axes will remain parallel.

Inclined surfaces that are not parallel to an axis will need to be projected from a point that lies on a line that is parallel to an axis. Figure 20.7 shows the steps for blocking out an irregularly shaped detail.

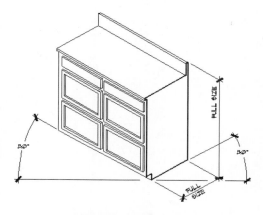

ISOMETRIC DRAWING

Figure 20.6 *Isometric drawings show all sides at full scale, and each side at a 30° angle.*

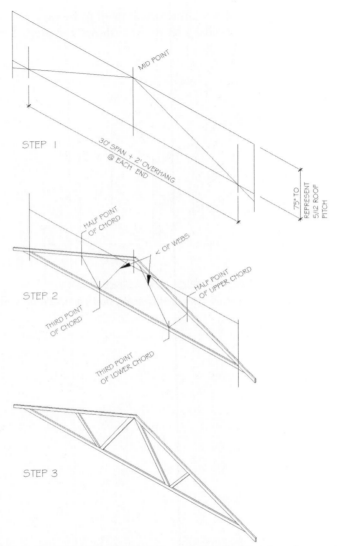

Figure 20.7 *Blocking irregularly shaped objects in rectangles will aid in drawing layout.*

DRAWING TOOLS FOR PICTORIAL DRAWINGS

The drawing tools used to create two-dimensional drawings also can be used to create pictorial drawings. Commands such as **Snap, Grid, Isoplane,** and **Ellipse** can be used to aid in drawing setup.

SNAP

Pictorial drawings can be created by altering the Snap setting. The setting can be adjusted using the following steps:

- Select the **Drafting Settings** from the **Tools** menu.
- Select the **Snap and Grid** tab of the **Drafting Settings** dialog box.
- Click the **Isometric Snap** button from the **Snap Type** area of the **Snap and Grid** tab.
- Clicking the **OK** button will close the dialog box, alter the snap setting to isometric, and restore the drawing area.

Returning to the drawing display will display an isometric grid and cursor, as seen in Figure 20.8. Activating SNAP and GRID will greatly aid in creating isometric drawings.

CROSSHAIRS ORIENTATION

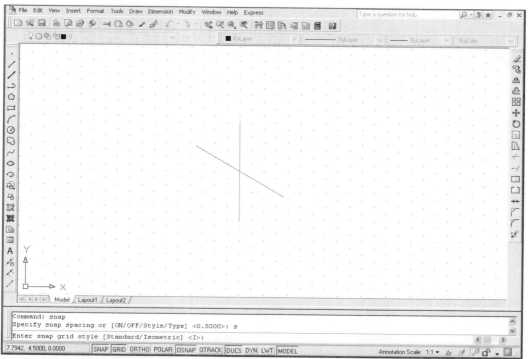

Figure 20.8   *An isometric grid can be established in the **Drafting Settings** dialog box (select **Drafting Settings** from the **Tools** menu). Select **Isometric Snap** from the **Snap Type** area of the **Snap and Grid** tab.*

Three surfaces are seen in pictorial drawings and are referred to by AutoCAD as Isoplane Left, Isoplane Top, and Isoplane Right. These three orientations are also available for the crosshairs to ease drawing construction. Figure 20.9 shows the three isometric planes and the options for crosshairs positioning. The two fastest methods of changing the orientation of the crosshairs are by using a function key or a CTRL combination.

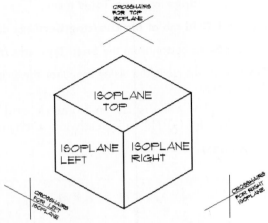

Figure 20.9 *AutoCAD refers to the three isometric surfaces as Isoplane Left, Isoplane Top, and Isoplane Right.*

Function Key

The easiest method to toggle between the isoplane options is to use the F5 function key. Each time F5 is pressed, the crosshairs will toggle to the next isoplane and the current setting will be displayed at the Command prompt.

Control Key Control

The crosshairs can also be controlled by pressing CTRL+E. Repeated pressing of CTRL+E will toggle the crosshairs between the Left, Top, and Right settings.

DRAWING ISOMETRIC CIRCLES

Circles in pictorial drawings are represented by ellipses. With the **Isometric Snap** set, start the **Ellipse** command and select the Isocircle option; an ellipse will be drawn automatically in the current Isoplane setting, as seen in Figure 20.10. The command sequence is as follows:

```
Click the Ellipse button (Or type EL ENTER.)
Specify axis endpoint of ellipse or [Arc/Center/Isocircle]:
    I ENTER
Specify center of isocircle: (Specify desired point.)
Specify radius of isocircle or [Diameter]: (Specify desired
    diameter or select a midpoint.)
```

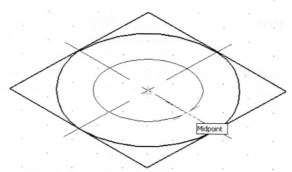

Figure 20.10 *An isometric ellipse will be drawn automatically in the current isoplane setting using the Isocircle option of the Ellipse command.*

A center point can be selected or a Radius or Diameter specified. One of the drawbacks to using pictorial drawings to show construction details is that circular objects often appear distorted. Round objects will appear as an ellipse in an isometric drawing. Figure 20.11 shows the proper orientation of circles for each plane of an isometric drawing. Notice that the centerline of the circle is parallel to the edge of one of the three surfaces.

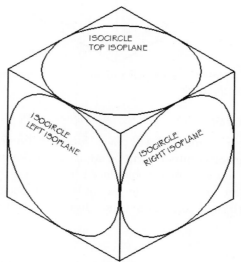

Figure 20.11 *Proper orientation of isometric circles.*

ISOMETRIC TEXT

No special commands are required to place text on pictorial drawings. On small details, text is best placed in its normal position and connected to the detail with leader lines. On larger drawings, text can be placed parallel to the isoplane on which it will appear by adjusting the **Rotation Angle** in the **Text** command. Figure 20.12 shows an example of text placed at 30°, –30°, 90°, and –90°. If large amounts of text need to be placed, create a text style for each angle.

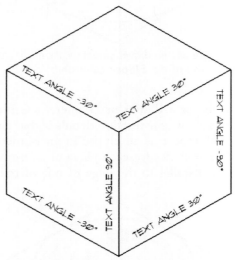

Figure 20.12 *Text can be placed in each plane of an isometric drawing using the **Text** command and rotating the angle.*

ISOMETRIC DIMENSIONS

Although no specific command will place isometric dimensions, dimensions created using the **Dimaligned** command can be adjusted quickly to conform to an isometric drawing. Aligned dimensions can be placed by clicking the **Aligned** button on the **Dimension** toolbar. Figure 20.13 shows an example of dimensions added to an isometric drawing using the **Dimaligned** command. Once dimension text is created, it can be edited using the **Dimedit** command. Select **Oblique** at the **Dimedit** prompt. You will be asked to select a dimension to be edited, and then prompted to specify an obliquing angle. Use either 30° or –30° for the angle. The command sequence is as follows:

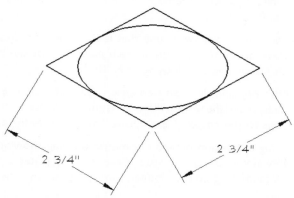

Figure 20.13 *Dimensions can be placed on an isometric drawing using the **Aligned** command. Click the **Aligned** dimension button.*

```
Specify first extension line origin or <select object>: (Select
    first point.)
Specify second extension line origin: (Select second point.)
Specify dimension line location or [Mtext/Text/Angle]: (Select
    point.)
Command: Click the Dimension Edit button.)
Enter type of dimension editing [Home/New/Rotate/Oblique] <Home>:
    O ENTER
Select objects: (Select dimension to edit.)
Select object: ENTER
Enter obliquing angle: 30 ENTER
```

The result of this sequence can be seen in Figure 20.14.

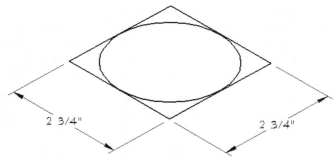

Figure 20.14 *Typing **O** ENTER for **Oblique** at the **Dimension edit** prompt will alter aligned dimensions to isometric dimensions.*

CHAPTER 20 EXERCISES

1. Start a new drawing and draw a cube that is 8' wide, 6' high, and 4'–6" deep using each of the oblique drawing methods. Place each drawing on a separate layer. Save the drawing as **E-20-1**.

2. Start a new drawing and draw a cube that is 8' in each direction using isometric techniques. Place an isometric circle in each plane so that the circle is tangent to the edges of each plane. Save the drawing as **E-20-2**.

3. Start a new drawing and draw an isometric cube that is 6' in each direction. Show the bottom, left, and right planes. Place an isometric circle in each plane so that the circle is tangent to the edges of each plane. Save the drawing as **E-20-3**.

4. Use the drawing below and draw an isometric drawing showing the framing of a 5'–0" × 3'–6" window. Use 2 × 6 studs. Label the top plates, a 4 × 8 header, 2 × 6 sill, and 2 × 6 studs, king studs, trimmers, and cripples. Save the drawing as **E-20-4**.

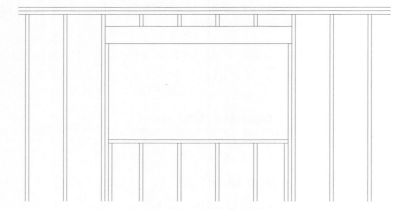

5. A window supplier would like to update its catalogue. Provide an isometric drawing of a 4'–0" wide × 3'–6" high vinyl window with a half-round window above. Show the window with a grid in each slider, and a radial pattern in the upper window. Draw the frame as 1" aluminum, and the grid as 1/2" wide material. Place the grid on a layer separate from the window layer. Save the drawing as **E-20-5**.

6. Use the floor plan on the next page and draw an isometric drawing showing the left wall with the refrigerator and double oven. Design and show drawers and cabinet doors. Use glass in some of the doors and show the shelving. Consult vendor catalogues for specific sizes of appliances and cabinets. Omit the refrigerator. Show 8'–0" high ceilings. Save the drawing as **E-20-6**.

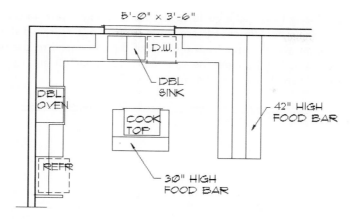

7. Use the floor plan for drawing **E-20-6** and the drawing started in Exercise 6 to draw an isometric of the complete kitchen. Draw the cabinets over the food bar as 30" high. Save the drawing as **E-20-7**.

8. Use the drawing below to draw an isometric drawing showing the metal beam seat. Show the bracket, but omit the beam. Make the side plates from 1/2" steel and the base plate from 7/8" × 7" wide steel plates. Place bolts at 2" in from each edge and at 3" o.c. Dimension the size of the steel plates. Save the drawing as **E-20-8**.

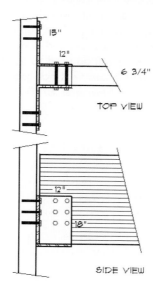

9. Use the drawing below as a guide to draw an isometric view of the truss. Assume that the top and bottom chords are made from two 2 × 3s, and the webs are 1" in diameter. Show the distance from the top of the top chord to the bottom of the metal bracket as 3 1/2". Show the total depth as 36 inches, and lay out the webs at 45° from the top chord. Save the drawing as **E-20-9**.

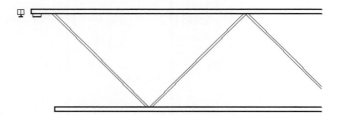

10. Use the drawing on the next page as a guide to draw an isometric view of the 6 3/4" × 13 1/2" glu-lam beam supported on two 6x6 posts at 10 feet o.c. on 18" diameter concrete piers. Show piers extending 8" minimum above the finish grade. Show the wall framed with 2 × 6 studs at 16" o.c. Frame the floor with 2 × 10 floor joists at 12" o.c. 12' above the grade covered with 3/4" plywood floor sheathing and an 18" cantilever. Use a 13" × 8" metal hanger with 2" × 10" legs to post. Show two bolts through the hanger into each beam, and two bolts through the hanger legs and post. Place bolts 1 1/2" from steel edges and 3" o.c. Use 1/2" steel cables with turn-buckles forming an "X' between each post. Attach the cable to each post with eye-bolts 6" up and from the bottom. Place the top eyebolt 3" below the bottom hanger's leg bolt. Label major materials and dimension the cantilever and post spacing. Because your drawing will be used to assemble this project and not to manufacture each item, great accuracy is not required on items such as the turn-buckle and eyebolts. Save the drawing as **E-20-10**.

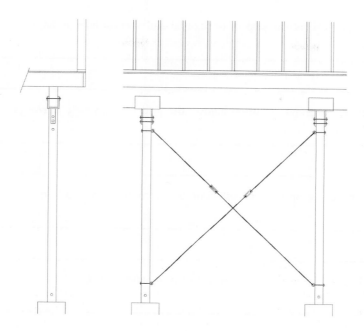

CHAPTER 20 QUIZ

DIRECTIONS

Answer the following questions with short complete statements. Type your answers using a word processor.

    1. Place your name, chapter number, and the date at the top of the sheet.

    2. Type the question number and provide the answer.

 Warning: Some of the questions have not been covered in the reading material and will require the use of the help menu and for you to do some exploring to answer the questions.

    1. What are the differences between cabinet and general oblique drawings?

    _____

    _____

    2. Compare the differences between cavalier and cabinet drawings.

    _____

    _____

3. Describe the relationship of horizontal and vertical lines in an isometric drawing.

_____

_____

4. What command and option adjusts the grid for isometric drawings?

_____

_____

5. List the three options for selecting the isometric snap options.

_____

_____

6. Explain the easiest way to construct isometric circles.

_____

_____

7. List common angles for placing text on isometric drawings.

_____

_____

8. Sketch and label the three positions of the crosshairs in isometric drawings.

_____

_____

9. What is the major difference between oblique and 3D drawings?

_____

_____

10. Explain the process for placing dimensions on an isometric drawing.

_____

_____

Working with Multiple Drawings

INTRODUCTION

This chapter will introduce:

- General considerations for multiple document environments
- Methods of moving information between two or more drawing files
- Features of DesignCenter
- Features of Tool Palettes

Commands to be introduced include:

- **Cutclip (Cut)**
- **Copyclip (Copy)**
- **Pasteclip (Paste)**
- **Copybase** (Copy with Base Point)
- **Pasteblock** (Paste as Block)
- **Pasteorig** (Paste to Original Coordinates)
- **Adcenter** (DesignCenter)

THE MULTIPLE DOCUMENT ENVIRONMENT

Template drawings have been used throughout this text to store frequently used styles, settings, and objects for use in similar drawings. This chapter will introduce methods for moving objects and information between two or more drawing files. Objects or drawing properties can be moved between drawings using cut/copy and paste, object dragging, the **Match Properties** command, and concurrent command execution.

GENERAL CONSIDERATIONS FOR MULTIPLE DOCUMENT ENVIRONMENTS

A powerful feature of AutoCAD is the capability to work with multiple documents in one AutoCAD session. Just as you can have several programs open on your desktop at once and rapidly switch from one program to another, AutoCAD allows you to have several drawings open at the same time. Chapter 1 introduced the techniques for displaying multiple drawings. Start AutoCAD and select **Open**. Select an existing drawing to be opened. While holding down CTRL, select three other drawings to be opened, and then click the **Open** button. You have now used one of the most powerful tools of AutoCAD and have four drawings opened in one drawing session.

Now select **Cascade** from the **Window** menu. Your drawing display will resemble Figure 21.1. The title bar of each drawing will be displayed. The drawing with the blue title bar on top of the stack is the active drawing. Apart from being on top of the pile, it is the only drawing that can currently be used. Return to the **Window** menu and select **Tile Vertical**. Now select **Tile Horizontal** instead from the **Window** menu. The drawing display will now resemble Figure 21.2. With four drawings open, the horizontal and vertical displays will appear alike. With an odd number of drawings open, they will be displayed in vertical rows with the **Tile Vertical** option. Important considerations to keep in mind when multiple drawings are open are as follows:

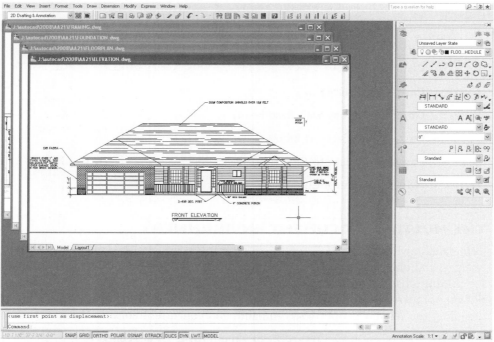

Figure 21.1 *Selecting **Cascade** from the **Window** menu will display the title bar of each drawing. The drawing with the blue title bar on top of the stack is the active drawing.*

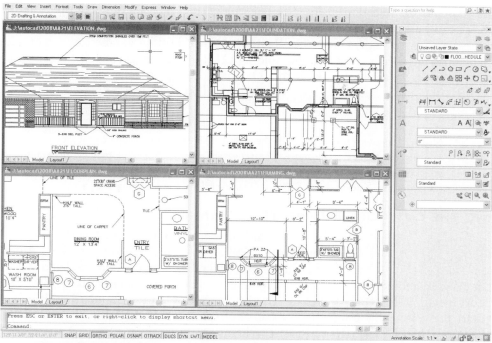

Figure 21.2 *Selecting **Tile Horizontal** from the **Window** menu will display each open drawing. In this orientation, any one of the four drawings can be made active. Notice that each small drawing area has all of the components of a normal AutoCAD drawing.*

- In any **Window** display selection, any one of the four drawings can be made active. Notice that each small drawing area has all of the components of a normal AutoCAD drawing.

- Any one of the four drawings can be maximized or minimized to ease viewing, or all can be left open and you can switch from drawing to drawing as needed.

- You can start a command in one drawing, switch to another drawing and perform a different command, and then return to the original drawing and complete the command in progress. As you return to the original drawing, click anywhere in the drawing area, and the original command will be continued.

- Each drawing is saved as a separate drawing file, independent of the other open files. You can close a file without affecting the contents of the remaining open files.

Opening Multiple Drawings from the Select File Dialog Box

In addition to being opened from within AutoCAD, multiple drawings can also be opened from the standard file selection dialog box. To start the process, open the standard file selection dialog box and double-click the drawing to be opened. Once the drawing is opened, minimize the drawing and reduce the size of the AutoCAD drawing session by clicking the **Restore Down** button at the top right side of the toolbar so that the drawings in the standard file selection dialog box can still be seen. With a drawing open, additional drawings can be dragged from the standard file selection dialog box and dropped into the current drawing session. Click the **Restore Down** button at the top right side of the toolbar. Drag files by pressing and holding CTRL and the left mouse button while the desired drawing is selected. Once the Drawing icon is in the drawing area, release the CTRL and select button to open the drawing. Drawings will be placed in a cascading display similar to Figure 21.1.

 Note: If a drawing is dragged from the standard file selection dialog box into a current drawing, the **Insert** command will be activated. Instead of opening multiple drawings, you'll be inserting a drawing in the current drawing. You can open multiple files at the same time by holding down CTRL and right-clicking and selecting **Open**.

USING THE CUT, COPY, AND PASTE COMMANDS

 In addition to standard procedures that are allowed using the **Cut**, **Copy**, and **Paste** commands of any Windows software, AutoCAD allows drawing objects and properties to be transferred directly from one drawing to another using these same commands. Each command can be performed using the **Edit** menu by shortcut menu or by selecting the button on the **Standard** toolbar. You can also start the commands using the keyboard by typing **CUTCLIP** ENTER, **COPYCLIP** ENTER, or **PASTECLIP** ENTER. **Cut** will remove objects from a drawing and move them to the Clipboard, from where they can be placed in a new drawing using **Paste**. **Copy** allows an object to be reproduced in another location. The object to be copied is reproduced in its new location using the **Paste** command. As the object is pasted to the new drawing, so are any new layers, linetypes, or other properties associated with the object. Figures 21.3a and 21.3b show examples of using **Copy** to place the footing detail into the second drawing. The footing was copied by using the following sequence:

From the Source Drawing:

```
Command: Copyclip ENTER
Select objects: (Select object to be copied using any selection
    method.)
Select objects: ENTER
```

In the Second Drawing:

```
Command: Pasteclip ENTER
Specify insertion point: (Select the location for the new
    object.)
```

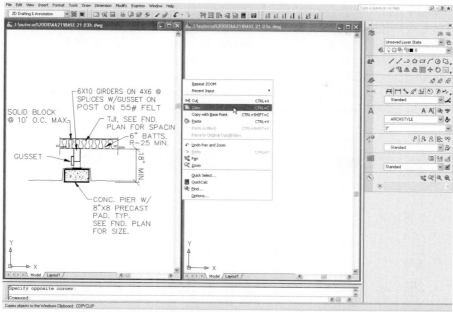

Figure 21.3a   *The footing in the left drawing can be copied into the right drawing by selecting* **Copy** *from the shortcut menu.*

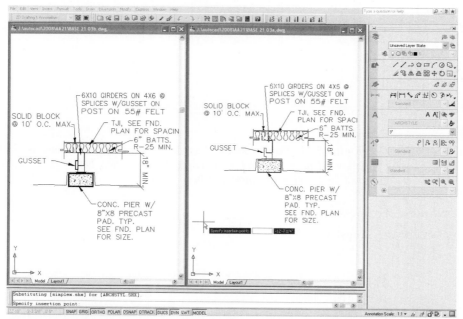

Figure 21.3b   *Once the object has been selected to be copied, move the cursor to the desired drawing, and single-click to make that drawing active. Right-click and choose* **Paste** *from the shortcut menu, and then select an insertion point.*

Copy with Base Point

The **Copy with Base Point** command on the shortcut menu is similar to the **Copyclip** command. You can also start the command by typing **COPYBASE** ENTER at the Command prompt. With this option, before you select objects to copy, you'll be prompted to provide a base point. This option works well when objects need to be inserted accurately. The roof detail in Figure 21.4 was copied into the new drawing using the **Copy with Base Point** command. The roof detail was copied using the following sequence:

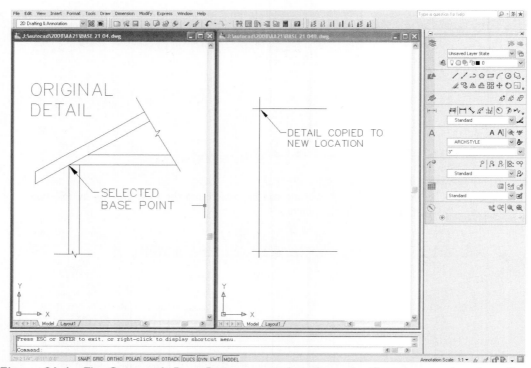

Figure 21.4 *The Copy with Base Point command, similar to the Copy command, works well when objects need to be inserted accurately. With this option, before you select objects to copy, you'll be prompted to provide a base point.*

From the Source Drawing:

```
Command: Copybase ENTER
Specify base point: (Select the base point for the objects to be
    copied.)
Select objects: (Select object to be copied using any selection
    method.)
Select objects: ENTER
```

In the Second Drawing:

```
Command: Pasteclip ENTER
Specify insertion point: (Select the location for the new
      object.)
```

Paste as Block

The **Paste as Block** command can be used to paste a block into a different drawing. It functions similarly to the **Paste** command. Select the objects to be copied in the original drawing and select the **Copy** command. Selected objects do not need to be a block because the **Paste as Block** command will turn them into a block. Next, activate the new drawing and right-click to select the **Paste as Block** command. The copied objects are now a block. To be edited, they must be exploded. You can also start the command by typing **PASTEBLOCK** ENTER at the Command prompt. The command sequence is the same as for the **Paste** command. You will notice that blocks that are copied will be renamed. A name such as **ROOFWALL** will be assigned a name such as **A$C2AA9F67C4** by AutoCAD. The new block can be renamed using the **Rename** command.

Paste to Original Coordinates

This command can be used to copy an object in one drawing to the exact same location in another drawing. The command could be useful in copying an object from one apartment unit to the same location in another unit. The command can be started by using the shortcut menu or by typing **PASTEORIG** ENTER at the Command prompt. The command is active only when the Clipboard contains AutoCAD data from a drawing other than the current drawing. Selecting an object to be copied starts the command sequence, and it is completed by making another drawing active and then selecting the **Paste to Original Coordinates** command from the shortcut menu.

DRAGGING OBJECTS

AutoCAD allows objects to be moved between drawings using a drag and drop sequence. Dragging allows the following operations:

- Move or copy objects to a new location in the same drawing
- Move or copy objects to another drawing in the same drawing session
- Copy objects to a drawing in another session
- Copy objects to another application

Dragging can be carried out using either the left or right mouse buttons. Slightly different results and procedures will be required using each method.

Selecting Objects to Drag with Left-Click

The procedure to drag using the left mouse button begins by selecting the objects to be moved or copied. Any object selection method can be used to select objects. Once the set is selected, move the cursor over one of the selected objects and press and hold down the left mouse button. Moving the cursor while the left button is held down will show the effect of moving the selected objects. Release the mouse button when the objects are in the desired location. If you're working with a single drawing, the selected objects will be moved. Objects can be copied in a single drawing by pressing CTRL prior to pressing the left mouse button to drag the object. (Objects will always be copied when dragged into a different drawing.) You are not allowed to drag objects into a drawing that has an active command. The unavailable cursor—a circle with a diagonal line through it—is displayed if you attempt to drag objects into a drawing with an active command. Figure 21.5 shows the steps to copy a detail into another drawing using the left mouse button.

 Note: After you select your object(s) to move or copy, be sure you don't hold the left mouse button over a hot grips box. This will activate the **Stretch** command instead of the desired **Move** or **Copy** command. (Refer to chapter 10 for more information on grips.)

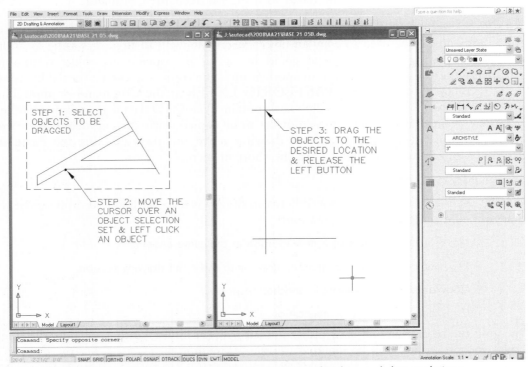

Figure 21.5 *Objects can be moved or copied easily using the drag and drop technique.*

Selecting Objects to Drag with Right-Click

The procedure to drag using the right mouse button begins by selecting the objects to be moved or copied. Any object selection method can be used to select objects. Once the set is selected, move the cursor over one of the selected objects and press the right mouse button. Moving the cursor while the button is held down will show the effect of moving the selected objects. Release the mouse button when the objects are in the desired location. As the right button is released, a menu will be displayed. The menu will depend on what type of drawing you're working in. The menu options are as follows:

Drag in the Same Drawing
- Move Here
- Copy Here
- Paste as Block
- Cancel

Drag in a Different Drawing
- Copy Here
- Paste as Block
- Paste to Orig Coords
- Cancel

Select a menu option to end the drag and drop sequence.

MATCHING PROPERTIES

The **Match Properties** command cannot be used to move or copy objects, but it can be used to copy properties from one object to another. The **Match Properties** command was introduced in chapter 5 for transferring properties within a drawing. The command can also be used to transfer properties from an active drawing to other drawings. The process to change the layer and linetype of objects in a drawing to match another drawing can be completed using the following process:

1. Select the source object using any selection method.

2. Select **Match Properties** on the **Standard** toolbar.

3. Move to the target drawing and make it the current drawing. The **Match Properties** cursor will be displayed by the selection box.

4. Select the target objects. The command can be continued within the current drawing, or another drawing can be made current, and the properties of the original object can be transferred to additional objects.

EXPLORING THE AUTOCAD DESIGNCENTER

The DesignCenter can be used to locate, organize, and customize drawing information. It can be used to directly transfer data such as blocks, layers, and external references from one drawing to another without using the clipboard. Drawing content can be moved from other open files in AutoCAD or from files stored on your machine, on a network drive, or on an Internet site. Access the DesignCenter by selecting the **DesignCenter** button on the **Standard** toolbar, or by typing **ADCENTER** ENTER at the Command prompt. Either method will display a DesignCenter similar to Figure 21.6. The DesignCenter can also be displayed or closed using CTRL+2.

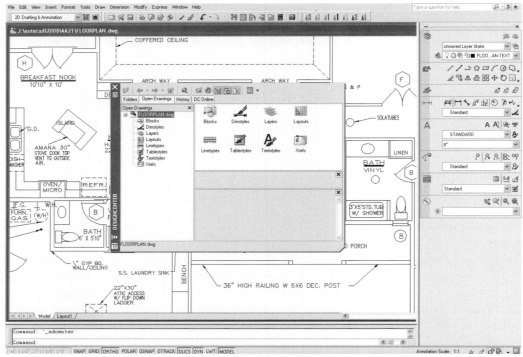

Figure 21.6 *Access the DesignCenter by selecting the **DesignCenter** button on the **Standard** toolbar or by typing **ADCENTER** ENTER at the Command prompt. The DesignCenter can be used to directly transfer data such as blocks, layers, and external references from one drawing to another.*

The default position of DesignCenter is a modeless dialog box (palette) floating in the center of the drawing screen. You can resize the DesignCenter by grabbing a corner and moving the cursor in a diagonal direction. If you are using the **DesignCenter**, the **Properties** palette, and a drawing, you might be tempted to shrink the size of the DesignCenter window. Resist shrinking the DesignCenter past the point where two rows of icons can't be seen in the palette display. The display can be reduced so no palette is displayed, but then you've limited the usefulness of the display. Clicking and dragging the title bar will float the DesignCenter, allowing it to be moved to any location in the drawing area. You can dock the DesignCenter by double-clicking the title bar of the DesignCenter display or by dragging it to the left or right side of the drawing area. Figure 21.7 shows an example of a docked DesignCenter. To maximize your drawing area, the DesignCenter palette can be hidden when not in use. To activate the Auto-hide mode, click the **Auto-hide** icon at the bottom of the palette sidebar. The DesignCenter can be redisplayed by moving the mouse over the **Auto-hide** icon on the palette sidebar.

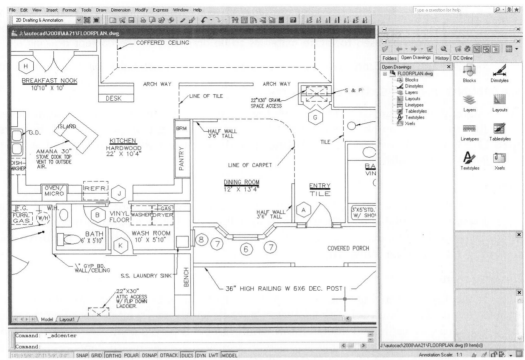

Figure 21.7 *Clicking and dragging the title bar will float the DesignCenter, allowing it to be moved to any location in the drawing area. You can dock the DesignCenter by double-clicking the title bar of the DesignCenter or by dragging it to the left or right side of the drawing area.*

FEATURES OF DESIGNCENTER

The DesignCenter provides direct access to any drawing file that you have on your machine, network, or the Internet. Files stored on diskette, the hard drive, network or Internet can be accessed and reused as a library for retrieving or coping blocks, referenced drawings, layouts, layers, linetypes, text styles, dimension styles or table styles. Other operations of the DesignCenter include the following:

- Create a Favorites folder to create a shortcut to frequently accessed drawings, folders, or Internet locations.

- View, attach, insert, or copy and paste layer and block definitions of any file you have access to into the current drawing. This feature could be used to update a drawing supplied by a subcontractor to match current standards.

- Open drawing files by dragging the file from the tree view into the current drawing area.

- Find drawing contents stored in any folder, drive, network, or from any Internet site that you have access to, based on search criteria for the source. Search tools such as a name or date can be used to locate materials to be dragged into the current drawing file.

- Use the DesignCenter palette to view aspects of the object, block, or drawing content.

EXAMINING THE DESIGNCENTER

The DesignCenter features a toolbar across the top of the palette and four tabs right below the toolbar to provide quick access to files. These four tabs are, from left to right, **Folders**, **Open Drawings**, **History**, and **DC Online**. The default tab active when DesignCenter is first opened is the **Folders** tab.

Folders

When the **Folders** tab is active, the display will resemble the standard file selection dialog box. All resources available to your workstation will be displayed, including internal drives, and local networks. The exact display will depend on the contents of the selected drive and the setting for the **Tree View Toggle** button.

Open Drawings

Selecting the **Open Drawings** tab will display a list of drawings that are currently opened, similar to Figure 21.8. Notice that the drawings have either a + or – symbol preceding the title. The + symbol indicates that additional information can be displayed about the drawing file or folder. Double-clicking the drawing name will display the drawing contents and double-clicking a drawing with a full display (–) will close the contents of that file.

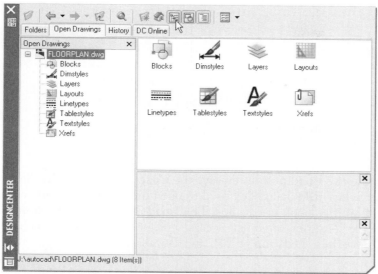

Figure 21.8 *Selecting the **Open Drawings** tab will display a list of drawings that are currently opened. Notice that the drawings have either a + or − symbol preceding the title. The + symbol indicates that additional information can be displayed about the drawing file or folder. Double-clicking the drawing name will display the drawing contents and double-clicking a drawing with a full display (−) will close the contents of that file.*

History

Selecting this tab will display a listing of the last twenty locations accessed through the DesignCenter. Selecting a drawing file within the **History** tab and right-clicking allows you to explore the contents of the file or delete the file from the history list.

Dc Online

This tab accesses the DesignCenter online web page. After you obtain a web connection, two panes are displayed on the welcome page. The left side contains folders that provide access to symbol libraries, manufacturers product information, and additional content libraries. These files can be selected and downloaded into your drawing.

The **DesignCenter** toolbar is located across the top of the DesignCenter. Before examining the contents, you might need to adjust the size of the DesignCenter window to be sure that all of the contents are displayed. The toolbar consists of eleven buttons for controlling the contents of the DesignCenter. These buttons, from left to right, are **Load, Back, Forward, Up, Search, Favorites, Home, Tree View Toggle, Preview, Description,** and **Views**. Each can be seen in Figure 21.9.

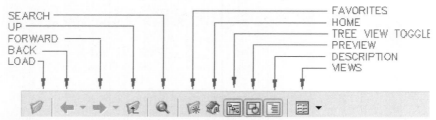

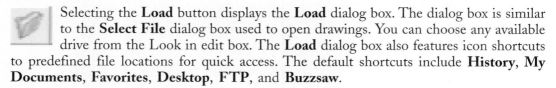

Figure 21.9 *The buttons on the DesignCenter toolbar.*

Load

Selecting the **Load** button displays the **Load** dialog box. The dialog box is similar to the **Select File** dialog box used to open drawings. You can choose any available drive from the Look in edit box. The **Load** dialog box also features icon shortcuts to predefined file locations for quick access. The default shortcuts include **History**, **My Documents**, **Favorites**, **Desktop**, **FTP**, and **Buzzsaw**.

Back

The **Back** button returns to the previous file location. Selecting the down arrow next to the **Back** button provides a listing of the last files selected.

Forward

The **Forward** button returns to the file you selected prior to using the Back button. Selecting the down arrow next to the **Forward** button displays the file that was selected prior to using the **Back** button.

Up

The **Up** button is used to move through the contents of the DesignCenter. The **Up** button performs the same function here as it does in the **Select File** dialog box. Each time the button is selected, the display is moved up one level in the current path tree.

Search

In addition to selecting information from the desktop, you can search for drawings or parts of drawings using **Search**. Selecting the **Search** button displays a **Search** dialog box similar to Figure 21.10. By default, the search will be conducted for drawings. Selecting the **Look for** edit box allows the search to be limited to Blocks, Dimstyles, Drawings, Drawing and Blocks, Hatch pattern files, Hatch patterns, Layers, Layouts, Linetypes, Table Styles, Text Styles, or Xrefs. Depending on the field, the tabs in the dialog box will be altered. The tabs for finding drawings include **Drawings**, **Date Modified**, and **Advanced**.

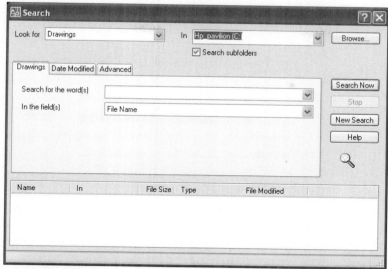

Figure 21.10   *Selecting the **Search** button displays a **Search** dialog box. The dialog box can be used to find Blocks, Dimstyles, Drawings, Drawing and Blocks, Hatch pattern files, Hatch patterns, Layers, Layouts, Linetypes, Text Styles, or Xrefs.*

Drawings   This tab can be used to specify the name or text that will be the object of the search. In Figure 21.11, the name ELECT is entered in the **Search for the word(s)** edit box. The **In the field(s)** edit box can be used to limit the search criteria. In the current setting, the search will be conducted in file names. Alternatives include looking in titles, subjects, author, and keywords.

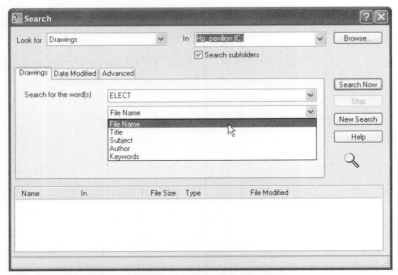

Figure 21.11 *Entering **ELECT** in the **Search for the word(s)** edit box and clicking the **Search Now** button will conduct a search for all file names containing the word **ELECT**.*

Date Modified The **Date Modified** tab allows a search to be conducted based on the date that the file was created or last modified. A date range specified in days or months can be used if you're not sure of the exact date the drawing was created.

Advanced The **Advanced** tab allows additional search parameters to be set based on text contained in a portion of a file or by the file size. A search can be defined using the **Containing** edit box to search block name, block and drawing descriptions, attribute tag, or attribute value for text containing, but not limited to, a specified text. The search parameters can also be limited to looking in files of a minimum or maximum size.

Favorites

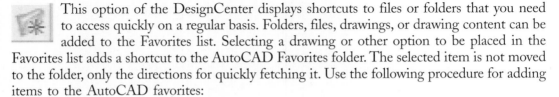

This option of the DesignCenter displays shortcuts to files or folders that you need to access quickly on a regular basis. Folders, files, drawings, or drawing content can be added to the Favorites list. Selecting a drawing or other option to be placed in the Favorites list adds a shortcut to the AutoCAD Favorites folder. The selected item is not moved to the folder, only the directions for quickly fetching it. Use the following procedure for adding items to the AutoCAD favorites:

1. Right-click the item to be listed as a favorite.

2. Select **Add to Favorites**. The selected item will be listed the next time the Favorites listing is displayed.

In addition to the toolbar icon, you can select the Favorites folder by right-clicking in the palette background and selecting Favorites. The shortcut saved in the Favorites folder can be copied, moved, or deleted using the Autodesk Explorer. The explorer is accessed by right-clicking in the palette and choosing Organize Favorites from the shortcut menu.

Home

The **Home** button moves you to the default folder named DesignCenter. This folder provides you with libraries of files with pre-defined blocks to choose from. You may choose to change the **Home** folder at any time. In the DesignCenter tree view, navigate to the folder that you want to set as your **Home** folder. Right-click on the folder and Select **Set as Home** from the shortcut menu. The next time you select the **Home** button on the **DesignCenter** toolbar, DesignCenter will automatically load this folder instead of the default.

Tree View Toggle

This button displays or hides the tree view in the left pane of the DesignCenter. Tree view displays open drawings, history, and files or folders for all of the drives available to your computer. Figure 21.8 shows a display using the tree view. Figure 21.12 shows a display with the tree view hidden. You can hide the tree view by clicking the **Tree View Toggle** button or by right-clicking in the palette and selecting **Tree** from the shortcut menu. With the tree view hidden, you can then restore the tree by clicking the button or by right-clicking in the palette and then clicking **Tree** from the shortcut menu.

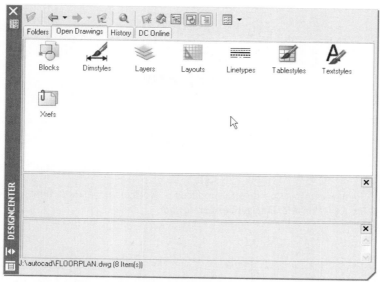

Figure 21.12 *The DesignCenter display with the tree view hidden.*

Preview

The **Preview** window displays an image of a selected file or block.

Description

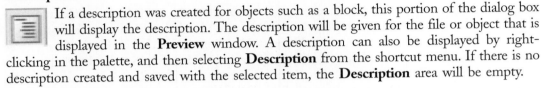
If a description was created for objects such as a block, this portion of the dialog box will display the description. The description will be given for the file or object that is displayed in the **Preview** window. A description can also be displayed by right-clicking in the palette, and then selecting **Description** from the shortcut menu. If there is no description created and saved with the selected item, the **Description** area will be empty.

Views

Selecting the **Views** button provides you with four options for displaying the format of the contents of the DesignCenter. Display options include **Large icons, Small icons, List,** and **Details.**

Shortcut Menus

Right-clicking in the palette produces a shortcut menu similar to Figure 21.13. The shortcut menu contains the same options as found on the **DesignCenter** toolbar.

Figure 21.13 *Right-clicking in the palette produces a shortcut menu that contains the same options as found on the **DesignCenter** toolbar.*

OPENING DRAWINGS USING THE DESIGNCENTER

The DesignCenter can be used to open drawings. Start AutoCAD using any of the options presented throughout this text. The DesignCenter can now be used to drag a drawing into an empty drawing area with the following steps:

1. Display the DesignCenter.

2. Select the drawing source from the Desktop.

3. Drag the icon of the drawing to be opened and drop it in the drawing area.

4. Provide the insertion point and responses to the scaling prompts.

5. Resize the drawing window as needed to display the drawing.

A drawing can also be opened by using the shortcut menu. From the DesignCenter, select the file to be opened, right-click, and select the **Open in Application Window** option from the shortcut menu.

ADDING CONTENT TO DRAWINGS USING THE DESIGNCENTER

One of the best uses for the DesignCenter is adding content from the palette to a new or existing drawing. Information can be selected from the palette or from **Search** and dragged directly into a drawing without opening the drawing containing the original. Chapter 18 introduced methods of moving blocks using DesignCenter. In addition to blocks, DesignCenter allows you to drag Dimstyles, Layers, Layouts, Linetypes, Textstyles, and Tablestyles from an existing drawing into your current drawing. The DesignCenter can also be used to attach external referenced drawings.

Inserting Blocks

The DesignCenter provides two methods for inserting blocks in a drawing: using the default scale and rotation of the block, or altering the block parameters as the block is dragged into the new drawing.

Providing Parameters for Block Insertion For most purposes this will be the preferred method of inserting a block using the DesignCenter. Use the following sequence to specify the insertion point, scale, and rotation angle as a block is inserted:

1. Display the DesignCenter.

2. Use the palette to select the block to be copied.

3. Use the right mouse button to drag the desired block to the new drawing.

4. Release the right button to drop the block into the drawing area. This will display a shortcut menu.

5. Choose **Insert Block** from the shortcut menu. This will display the **Insert** dialog box.

6. Enter the values for each parameter or choose the **Specify On-screen** option for each parameter.

7. Click the **OK** button to insert the block at the specified location.

Using the Default Scale When blocks are inserted using the default scaling and rotation option, the block is inserted using Autoscaling. Autoscaling compares the units of the drawing with those of the block. As the block is inserted, AutoCAD scales the block as needed, based on the ratio of the block to the new drawing. Blocks inserted using this method might not retain their true size. Use the following sequence to insert a block using the default parameters:

1. Display the DesignCenter.

2. Use the palette to select the block to be copied.

3. Use the left mouse button to drag the desired block to the new drawing.

4. Drop the block at the desired location.

Note: When the source file of a block is changed, block definitions inserted into the drawings are not automatically updated. With DesignCenter, you have the option to update the block definitions inserted into your drawing to reflect the new changes. The source file of a block definition can be a drawing file or a nested block in a drawing. Use the following procedure to update a block: Select the file you choose to redefine, right-click, and select **Insert and Redefine** or **Redefine only** from the shortcut menu.

Attaching External Referenced Drawings

The DesignCenter can be used to attach an external referenced drawing using steps similar to those used to attach a block and provide the parameters. Use the following procedure to bring an Xref into a new drawing.

1. Display the DesignCenter.

2. Use the palette to select the xref to be attached.

3. Use the right mouse button to drag the desired xref to the new drawing.

4. Release the right button to drop the xref into the drawing area. This will display a shortcut menu.

5. Choose **Attach as Xref** from the shortcut menu. This will display the **External Reference** dialog box.

6. Select **Attachment** or **Overlay** from the **Reference Type** box. See Chapter 22 for more information on the subject of Xrefs.

7. Enter the values for each parameter for the insertion point or choose the **Specify On-screen** option.

8. Click the **OK** button to complete the command.

Working with Layers

The DesignCenter can be used to copy layers from one drawing to another. Typically, a template drawing containing stock layers can be used when you create new drawing files. This option of the DesignCenter is useful when you're working with drawings created by a consulting firm and the drawing needs to conform to office standards. The DesignCenter can be used to drag layers from a template, or any other drawing, into the new drawing and ensure drawing consistency. Drag layers into a new drawing using the following command sequence:

1. Display the DesignCenter.

2. Use the palette or **Search** to select the drawing that contains the layers to be copied.

3. Double-click the name of the drawing to display the options of DesignCenter.

4. Double-click Layers.

5. Highlight the names of the layers to be added to the new drawing (holding down CTRL allows you to select several layers at once).

6. Press the right mouse button and drag the layer names to the new drawing.

7. When the cursor is in the new drawing area, release the right button to display the shortcut menu.

8. Select **Add Layers(s)** from the shortcut menu.

Note: In addition to layers, DesignCenter allows you to drag Dimstyles, Layouts, Linetypes, Textstyles, and Tablestyles from an existing drawing into your current drawing. Follow the same command sequence as described above to achieve this.

Tool Palettes

Tool palettes have proven to be an invaluable tool. Commands, gradient hatches, blocks, Xrefs, and flyouts can be added to a tool palette. Objects added to the tool palette will retain their existing properties. Objects can also be added from the DesignCenter to a Tool Palette. Utilizing tool palettes will minimize use of toolbars, maximize screen space, and save time by creating custom tool palettes tailored to your needs. To display your current tool palette, select the **Tool Palette** icon on the **Standard** toolbar. From the DesignCenter content area, you can drag file(s) to the current tool palette. When you add drawings to the tool palette, they are inserted as blocks when they are dragged into your drawing. Each tool palette is represented by a tab this allows you to better organize and access your blocks, commands, gradient hatches, and files. Figure 21.14 displays the tool palette with several tabs to organize all the blocks, commands, gradient hatches and drawing files.

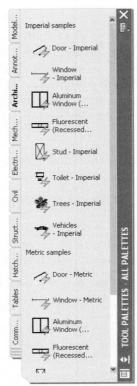

Figure 21.14 *The tool palette can be a timesaving tool when organized efficiently. The tabs created on the left side of the palette help to organize your blocks, commands, gradient hatches, and files.*

Creating a New Tool Palette

1. Place your cursor over an empty area of the tool palette and right-click. Select **New Palette** from the shortcut menu.

2. A new tool palette appears. Enter the name of your new tool palette and press ENTER. Each tab displayed on the left side represents a tool palette.

Creating a Hatch Tool

1. Right-click the title bar of your new tool palette. Make sure **Auto-hide** is not checked. The Tool Palettes window remains open when **Auto-hide** is off.

2. Click on an existing hatch pattern in your drawing to select it.

3. Drag the hatch pattern onto your new tool palette. Be sure that the cursor is not directly over the grip.

4. The new hatch tool is displayed on your tool palette. Right-click the hatch pattern and select **Properties** from the shortcut menu.

5. The **Tool Properties** dialog box appears. Modifications to the properties can be made here, as shown in Figure 21.15.

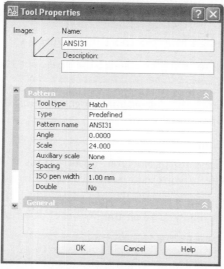

Figure 21.15   *The **Tool Properties** dialog box provides a quick way to modify the tool palette tools.*

Creating a Dimension Tool

1. Click on the existing dimension to select it.

2. Drag the dimension onto your new tool palette, remembering not to place the cursor over the grips. When you drag a dimension to the tool palette, the new tool contains a flyout (a set of nested commands). Click the small black arrow to view the set of nested commands (**Aligned Dimension, Radius Dimension, Angular Dimension**, etc.).

Creating a Command Tool

1. Right-click over an empty area on the new tool palette.

2. Select **Customize Commands** from the shortcut menu.

3. Hold down the left mouse button and drag the desired command from the toolbar onto the **Commands** tool palette.

4. Right-click the new tool to customize the properties. For example, if you dragged the **Line** command to your tool palette, the **Line** command would also have a tool flyout. You could specify the layer that the tool would draw objects on. When you work with flyouts, the same layer specified is for the entire set of nested commands. See the help menu for advanced customizing.

Customizing Tool Palette Icons

You can easily customize the icons associated with your tools on the tool palette. Use the following steps to customize the tool palette:

1. Right-click over the tool you would like to customize.

2. Select **Specify Image** from the shortcut menu.

3. In the **Select Image** dialog box, select the image for the icon associated with your tool.

4. If you wish to restore the default image, right-click on the customized tool icon and select **Remove Specifed Image**.

Creating Tool Palette Groups

Tool palettes can be organized into associated groups. This will maximize screen space by displaying only the tool palette you need. Use the following steps in creating tool palette groups:

1. Right-click over an empty area on the new tool palette.

2. Select **Customize** from the shortcut menu.

3. In the **Customize** dialog box, right-click under **Palette Groups** and select **New Group** on the shortcut menu.

4. A new folder appears; type in the tool palette group name. Follow steps 3 and 4 until you have created the desired groups (such as hatches, commands, dimensions, blocks).

5. Under **Palettes**, drag the tool palettes into the appropriate groups.

6. Select the tool palette you would like to set as current and right-click.

7. On the shortcut menu, click **Set Current** as shown in figure 21.16. The tool palette window now displays only the tool palettes in the group you selected.

8. You can quickly switch from one tool palette group to another. Right-click in the title bar of the tool palettes window. On the shortcut menu, click on the palette group you want to switch to.

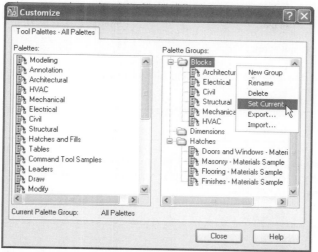

Figure 21.16 *Selecting **Set Current** displays only the tool palettes in that group.*

CHAPTER 21 EXERCISES

1. Open any three drawings and copy features from one drawing to the other two. Save each drawing as **E21-1a, 1b**, and **1c**.

2. Open the drawings created in Exercise 1. Copy one object from one of the drawings using the base point option so that the new object aligns with an existing intersection. Save the drawing as **E-21-2**.

3. Open one of the drawings from Exercise 2. Use the **Paste as Block** option to place two of the blocks created in chapter 18 in the drawing.

4. Open a template drawing and draw a 10' long x 5' wide bathroom. Show all walls as 4" wide. Use the DesignCenter and move blocks for a toilet, sink, and tub into the bathroom.

5. Open the floor plan that you've created and move five electrical symbols into the drawing. Save the drawing as **ELECT**.

CHAPTER 21 QUIZ

DIRECTIONS

Answer the following questions with short complete statements. Type your answers using a word processor.

1. Place your name, chapter number, and the date at the top of the sheet.

2. Type the question number and provide the answer.

Warning: Some of the questions may not have been covered in the reading material and will require the use of the help menu. You may also have to do some exploring to answer the questions.

1. Describe the options for inserting blocks using the DesignCenter.

_____

_____

2. Having multiple drawings opened in one drawing session is a powerful tool. Describe the different options used to display your drawings, and explain how to toggle from one drawing to another.

_____

_____

3. Explain the process to copy an object from one drawing to another in a multiple drawing session.

_____

_____

4. Explain the difference between **Paste with Original Coordinates** and **Paste as Block**.

_____

_____

5. How can the **Match Properties** command be used in a multiple drawing environment?

_____

_____

6. List two methods for opening the DesignCenter.

_____

_____

7. What determines whether an object will be moved or copied using drag and drop methods with the left mouse button?

_____

_____

8. Explain the effects of right-click on drag and drop methods.

_____

_____

9. How can the **History** tab of the DesignCenter be used?

_____

_____

10. Describe the four views that can be used in the DesignCenter.

_____

_____

11. You need to find a site drawing created in early September. Explain the process.

_____

_____

12. What do the + and − symbols mean when they are displayed by a folder or file name?

_____

_____

13. Explain the process to place a drawing in the Favorites folder.

_____

_____

14. A drawing template contains five of the best layer names ever created. How can these layers be transferred into a new drawing?

_____

_____

15. You've come to your senses and want to remove three of the layers that were just added to the drawing. How can this be done?

_____

_____

16. Explain why you would create multiple tabs in the tool palette and describe the process.

_____

_____

17. List five items that can be added to a tool palette.

_____

_____

18. What does the small black arrow represent on the tool palette?

_____

_____

19. How can customizing tool palettes enhance your drafting environment?

_____

_____

CHAPTER 22

Combining Drawings Using Xref

INTRODUCTION

This chapter will introduce:

- Merging a drawing into another drawing
- Externally referenced drawing options
- Methods for editing externally referenced drawings

Commands to be introduced include:

- **Xref** (External Reference)
- **Xbind** (External Reference Bind)
- **Xclip** (External Reference Clip)
- **Refedit** (External Reference Edit)
- **Refset** (External Reference Set)
- **Refclose** (External Reference Close)
- **Xopen** (External Reference Open)

Most of the drawings for a structure are worked on by teams of CAD drafters representing several firms. In addition to the architectural and engineering firms responsible for the design of a structure, CAD drafters often work for landscape and interior architects, as well as plumbing, mechanical, electrical, and civil firms. Each of these firms must have up-to-date drawings to ensure an error-free construction process. One way to provide these drawings to subcontractors is by the use of externally referenced drawings, or **Xrefs.** The use of the **Xref** command is instrumental in assuring quality drawings. The **Xref** command can also be used to compile a drawing sheet containing details or drawings completed at varied scales.

EXTERNAL REFERENCED DRAWINGS

In chapter 18 it was suggested that a wblock of a floor plan could be inserted in a drawing and used to form the base of that drawing. For example, the floor plan can form the base drawing for the electrical, mechanical, plumbing, and framing plans. This works well, but it does take up a lot of disk space. A more efficient method than inserting a wblock is to use an externally referenced drawing or Xref. An externally referenced drawing is similar to a wblock, in that it is displayed each time the master drawing is accessed. These drawings, however, are not stored as part of the master drawing file. Each time the base drawing is updated, all drawings that it is attached to will also be updated. Externally referenced drawings are created through the **Xref** command. Figure 22.1a shows a drawing that serves as the base drawing. Figures 22.1b and 22.1c show drawings that make use of the base drawing.

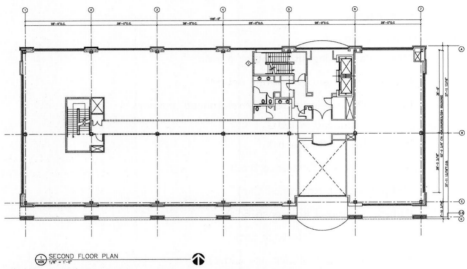

Figure 22.1a *The **Xref** command can be used to attach a drawing to another drawing. The information contained in the attached drawing is not stored in the base drawing. Each time the base drawing is updated, all drawings that it is attached to will also be updated. (Courtesy Peck, Smiley, Ettlin Architects.)*

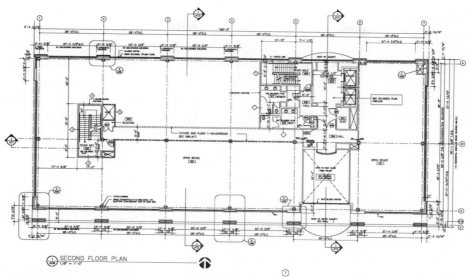

Figure 22.1b   *The drawing shown in Figure 22.1a is used as the base for the floor plan shown here and the drawings prepared by the mechanical, electrical, and plumbing subcontractors. (Courtesy Peck, Smiley, Ettlin Architects.)*

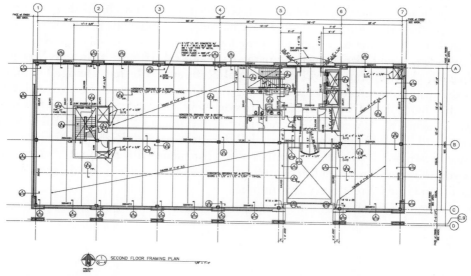

Figure 22.1c   *The framing plan is completed by an engineering firm that uses the base drawing created by the architectural team. The drawing is attached using **Xref**, and then the needed material is added to the drawing. (Courtesy Van Domelen/Looijeng/McGarrigle/Knauf Consulting Engineers.)*

SAVING SPACE

An externally-referenced drawing can be attached to the current drawing for viewing, but it does not become part of the current drawing. Depending on controls set by the originator, the Xref source file may or may not be able to be altered by subcontractors working in the drawing. An Xref is similar to a block, except that no drawing objects are recorded in the current drawing. Only the drawing name and a small amount of information needed to access the drawing are stored in the new drawing file.

AUTOMATIC UPDATING

Externally referenced drawings provide excellent benefits for drawing projects that are being developed by a team of engineers, architects, contractors, and drafters. Because of time constraints, subcontractors often need part of the drawings before they are complete. When Xref drawings are used, every time the host drawing is accessed, the most recently saved version of the external drawing is loaded. Changes made to the Xref source file will be updated automatically in every drawing where it is referenced. For example, if an engineering firm is working off a network and if one team member is working on the plan views and another is working on exterior elevations, every time the Xref floor plans are loaded for reference for the elevations, they will be updated. However, if you are working off a network, and the source file has changed but you have already loaded the file and are currently working on it, you may be working on an outdated source file. AutoCAD offers instant notification when an externally referenced drawing has changed. An icon on the status bar is displayed for any drawing that has an Xref attached. If one of the Xref files is changed, a balloon message appears in the lower right corner. If only one Xref file needs to be reloaded, it will display just the name of that **Xref**. If more than one Xref needs to be reloaded, the notification will be **External Reference file has changed**. Figure 22.2 shows the instant notification balloon listing the Xref that needs to be reloaded. Clicking on the file name listed in the balloon message will automatically update the modified Xref(s). Clicking the External Reference icon in the status bar, or typing **XR ENTER** at the command line will open the **External References** palette. Right-clicking the external reference drawing will provide you with a short-cut menu with the option to reload. This allows you the option to reload drawing(s) on a case-by-case basis. By default, AutoCAD checks for modified Xrefs every five minutes.

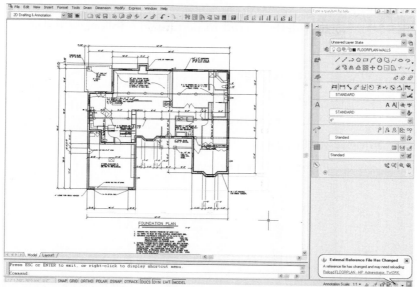

Figure 22.2 *If one of the Xref files attached to your drawing is changed, a balloon message appears in the lower-right corner. The message notifies you to reload Xrefs.*

ASSEMBLING MULTI-SCALED DETAILS

A common use for Xref drawings is to assemble details for a stock detail sheet. In most professional offices, stock libraries of details have been created for every possible construction alternative that is likely to be encountered. You can assemble a sheet of the appropriate details using externally referenced drawings with the knowledge that the most up-to-date version of the details is being provided. The next chapter will explore the process for assembling drawings to be plotted at different scales within the same drawing.

ATTACHING AN XREF TO A DRAWING

Although the **Xref** command can be completed from the command line, using the **External References** palette will greatly aid the use of the command. Access the palette by using one of the following methods:

- Click the **External Reference** button on the **Status Bar**.

- Type **XR** ENTER at the Command prompt.

Both methods will produce the **External References** palette shown in Figure 22.3. The palette allows a drawing to be referenced to the current drawing. It also displays, organizes, and controls referenced files. Like other AutoCAD palettes, the **External References** palette can be docked, and the size can be adjusted to fit your needs. When the palette is accessed, the reference name, status, size, type, date, and saved path will be displayed.

The info will be displayed as you scroll the bar over on the bottom on the **File Reference** section in the palette. The palette is used to perform the following tasks:

- Attach an Xref

- View and change the Xref path

- Detach an existing Xref

- Reload or unload an existing Xref

- Bind an entire Xref definition to the current drawing

- Open an Xref for editing in a new window

Figure 22.3 *The **External References** palette can be used to reference a drawing into a host drawing. Access the palette by clicking the **External Reference** button from the **Status Bar**, or by typing **XR** ENTER at the Command prompt.*

To attach a drawing as an External Reference, use one of the following methods:

- Click the **Attach Xref** button from the **Insert** toolbar.

- Click the **Attach DWG** button in the **External References** palette.

- Select **DWG Reference** from the **Insert** menu.

Each method will display the **Select Reference File** dialog box. Once a file to be referenced is selected, an **External Reference** dialog box similar to Figure 22.4 is displayed. The dialog box allows a drawing to be selected for external reference to the current file, much the way the Insert command works.

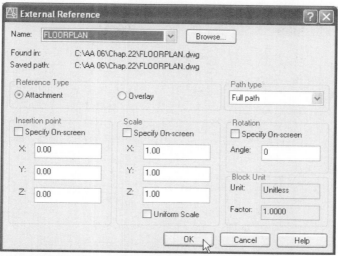

Figure 22.4 The **External Reference** *dialog box can be used to reference a drawing to another drawing using a procedure similar to the* **Insert** *command. Access the dialog box by selecting* **Attach Xref** *from the* **Insert** *toolbar, by selecting the* **Attach DWG** *button in the* **External References** *palette, or by selecting* **DWG Reference** *from the* **Insert** *drop-down menu.*

Figure 22.5a shows a floor plan that will be used as the base drawing in the following example. The base drawing is titled *C:\DRAWINGS\SCHOOL\ARCH\FLOOR.DWG*. Figure 22.5b shows the electrical information that will be added to the base drawing to make the electrical plan. This drawing is stored as *C:\DRAWINGS\SCHOOL\ARCH\ELECT.DWG*.

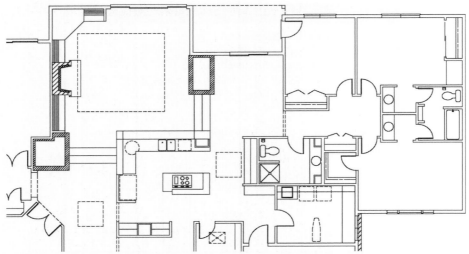

Figure 22.5a *This portion of a floor plan will serve as a base for the electrical plan shown in Figure 22.5b. The information on the base drawing will be displayed as the electrical plan is accessed. (Courtesy Aaron Michael Jefferis.)*

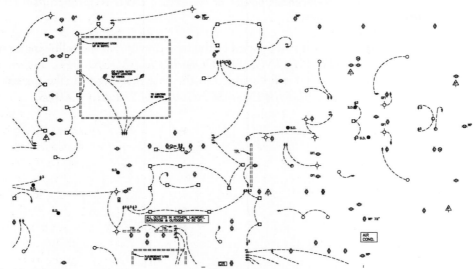

Figure 22.5b *By using referenced drawings, subcontractors from various firms can each complete their work without having to wait for others to finish. This electrical information was added to the base drawing shown in Figure 22.5a.*

To attach the floor plan information to the electrical drawing, display the **External References** palette and select the **Attach DWG** button. This will display the **Select Reference File** dialog box. The box is similar to each dialog box that is used to open a new drawing. Select the desired folder and file of the drawing to be attached. With the desired file highlighted, click the **Open** button to display the **External Reference** dialog box shown in Figure 22.4. This dialog box allows you to control how and where the attached drawing will be located in the base drawing by using the **Reference Type** and parameters areas.

EXTERNAL REFERENCE DIALOG BOX OPTIONS

So far, the **External Reference** dialog box shown in Figure 22.4 has been used only to attach one reference drawing to another drawing. Several parameters can be adjusted to alter its use.

NAME AND PATH OPTIONS

The **Name** edit box was empty as the dialog box was first displayed. Once a drawing has been attached, the name of the drawing will be displayed. When multiple drawings are attached, selecting the edit arrow will display a list of the referenced drawings. Selecting an attached drawing from the list will display the drawing name in the **Name** edit box, and its location will be displayed in the **Found in** and **Saved path** display. Selecting **Browse** will display the **Select Reference File** dialog box, allowing new Xrefs to be selected for the current drawing.

Path Type

The **Path type** edit box dictates whether the saved path to the Xref is the full path, a relative path, or no path. The current drawing must be saved before you can set the path type to **Relative path**. The **Relative path** option works well if you will be accessing the file from a network and a local drive as long as the file remains in the same folder directory. By default, the **Full path** option is active so that the path to the referenced drawing will be saved in the database of the host drawing. The **No path** option is often used when referenced drawings are shared with subcontractors who will work with the drawings. With this option selected, as AutoCAD loads the host drawing, it will only search for the referenced drawing in the path specified in the **Files** tab of the **Options** dialog box. If the **Full path** option is selected, it will search through only the file folders specified in that path. This will usually be the desired method. If you are working on several jobs at the same time, your drawing will not be overwritten by the wrong job. For example, if you are working on several jobs requiring an **ELECT.dwg** file to be attached, if you select the **Full path** option, it will only search *C:\PROGRAM FILES\PARKER\ELECT*, rather than *C:\PROGRAM FILES\ACAD2002\ELECT.dwg*. If you complete this search with **No path** specified, AutoCAD may find several different ELECT plans, but it would grab the first file found. It is best to be as specific as possible about the path. Make sure, as you create new drawings, that they are placed in the correct location.

REFERENCE TYPE

This portion of the **External Reference** dialog box provides the options of **Attachment** and **Overlay** for deciding how an external drawing will be attached to your drawing. Attached drawings allow you to build a drawing using other drawings.

Attaching a Drawing

With the default setting of **Attachment**, clicking the **OK** button will remove the dialog box, return you to the drawing area, and provide a prompt for an insertion point. As the point is selected with the mouse, the drawing will be attached and the command will be closed. Figure 22.6 shows the merged drawings from Figure 22.5a and 22.5b. The insertion point of the lower right corner was used with an Object Snap of Intersection to assure proper alignment of the two drawings.

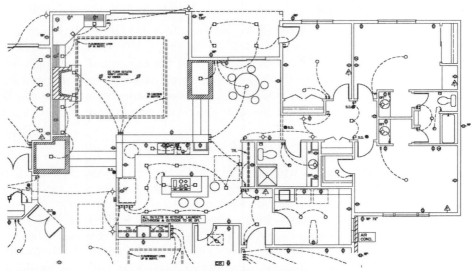

Figure 22.6   *The result of attaching the base drawing to the electrical drawing. (Courtesy Matthew & Megan Jefferis.)*

If a second drawing is to be attached to the host drawing, selecting the **Attach Xref** button will produce the **Select Reference File** dialog box. Once one drawing has been selected, when the **Open** button is selected, the **External Reference** dialog box is redisplayed. Once a drawing is attached, AutoCAD adds dependent symbols in the Xref to the base drawing. Dependent symbols are named items such as blocks, dimension styles, layers, linetypes, and text styles. Although these dependent symbols can be viewed, they can't be altered unless you preset values to allow the user to alter the drawing. This is one of the benefits of an Xref—you can give a referenced drawing to a subcontractor and not have to worry about the drawing being altered. When a drawing is referenced to a base drawing, you'll notice that the layer names of the Xref will be altered. Layer names for attached drawings will be displayed showing the name of the drawing, followed by a vertical bar and the layer name.

A name such as **FLOORPLAN|DOORS** would indicate a referenced drawing of **FLOORPLAN** and a layer title of **DOORS**.

Overlaying an External Drawing

Choosing the **Overlay** option will produce a drawing that looks like an attached drawing. The major difference between an attached and an overlaid drawing occurs when the referenced drawing contains nested reference drawings. A nested drawing is a drawing within another referenced drawing. When a referenced drawing containing nested drawings is overlaid, the nested drawings within the drawing will not be displayed. When a nested drawing is attached, all objects will be displayed. The **Overlay** option is typically used when you need to share information. An overlaid drawing will allow others to view how their portion of the project relates to the drawing. Rather than showing all of the drawings like the Mechanical and Electrical Xrefs, with the **Overlay** option, you can still benefit from the automatically updated files.

SETTING PARAMETERS

As the drawing in Figure 22.5a was attached to the drawing in Figure 22.5b, the insertion point was selected prior to attachment. Using the parameters portions of the **External Reference** dialog box allows the insertion point, scale factors, and rotation angle to be controlled for a referenced drawing.

Insertion Point

In the default setting, the **Specify On-screen** check box is active, allowing the insertion point to be selected with the cursor or by entering coordinates at the Command prompt. With the **Specify On-screen** check box deactivated, the text boxes for each option are activated. These coordinates allow specific X and Y coordinates to be specified for the insertion point of the referenced drawing. These options correspond to their counterparts in the Insertion Points for the **Insert** command.

 Note: For most firms, it is common to use an insertion point of 0,0. With an insertion point of 0,0, all drawings created by subcontractors will align perfectly. This is a much easier method than worrying about what corner of a structure might have been used as a base point.

Scale

This option specifies the scale factors for the selected referenced drawing as it is inserted in the host drawing. The default values for the X, Y, and Z scale factors are set as 1. Selecting the **Specify On-screen** check box for scale factors will deactivate the scale factor values and allow the values to be adjusted from the Command prompt as the drawing is attached.

Rotation

The **Rotation** portion of the dialog box is similar to that of the **Insert** dialog box. **Angle** allows the rotation angle of the referenced drawing to be controlled as it is inserted in the base drawing. In the default setting, a rotation angle can be entered in the edit box. With the **Specify On-screen** check box active, the rotation angle can be set at the Command prompt as the drawing is attached.

EXTERNAL REFERENCES PALETTE OPTIONS

So far, the **External References** palette shown in Figure 22.3 has been used only to attach one reference drawing to a host drawing. This palette can be used to list, attach, load, and modify referenced drawings contained in the current drawing. The display provided in the box will depend on the active button in the upper right corner of the palette. The buttons toggle between **List View** and **Tree View**. Change the display by selecting the inactive option or by pressing F3 or F4.

LIST VIEW OPTION

In the default setting, the dialog box is displayed in **List View**, with information related to the reference name, status, size, type, date, and saved path. Information can be provided in an alphabetical list of the referenced drawings contained in the current drawing. Notice in Figure 22.3 that not all of the information is visible. The reference name scrolls off the edge of the display field. Move the cursor to the line that divides the **Reference Name** area and the **Status** area. As the line is touched, the cursor turns to a double arrow, allowing the size of the box to be altered. This can be done to each box to allow a complete display. As the outside line of the palette is touched, the cursor turns into a double arrow, allowing the size of the palette to be altered. Expanding the palette displays all the Xref information. (See Figure 22.7a.). Click a list column heading to sort the referenced drawings by the column name. To sort referenced files by date, click the reference **Date** column.

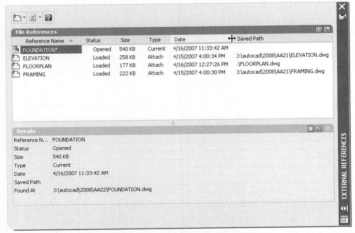

Figure 22.7a *The space between listings can be reduced to allow longer listings to be completely viewed. Place the cursor on the lines that divide titles to reduce or enlarge each display space.*

Reference Name

This column lists the names of the referenced drawings stored in the definition table of the drawing.

Status

This column shows the current status of referenced drawings. Options include the following:

> **Loaded**—The attached drawing is displayed.
>
> **Not Found**—The referenced drawing cannot be found.
>
> **Orphaned**—The drawing is attached to another referenced drawing that is unreferenced, unresolved, or not found.
>
> **Reload**—The referenced drawing is marked to be reloaded.
>
> **Unload**—The referenced drawing is marked to be unloaded from the drawing.
>
> **Unloaded**—The referenced drawing is not displayed and will not be displayed again until the reload option is selected.
>
> **Unreferenced**—An unreferenced drawing is attached to the drawing but erased.
>
> **Unresolved**—The referenced drawing cannot be read by AutoCAD.

Size

The **Size** column displays the size of the referenced drawing.

Type

The **Type** column indicates if the referenced drawing was placed using the **Attachment** or the **Overlay** option.

Date

The **Date** column lists the last date that the referenced drawing was modified.

Saved Path

The **Saved Path** column shows the saved path of the referenced drawing (this is not necessarily where the Xref is found).

TREE VIEW OPTION

With the **Tree View** active, a hierarchical representation of referenced drawings is shown similar to the manner that listings in a standard file selection dialog box are displayed. The **Tree View** method displays nested drawings in their relationship to the attached drawing. Figure 22.7b shows a listing using the **Tree View** option.

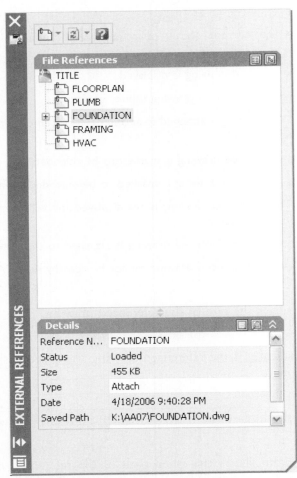

Figure 22.7b *Listings in the **External References** palette displayed using the **Tree View** option.*

EXTERNALLY REFERENCED DRAWING OPTIONS

Once the drawing is attached, each time the base drawing is opened, the attached drawing will be loaded and displayed as long as the path to the attached drawing is accessible. In addition to loading the referenced drawing, the other options of the **External References** palette will become available. Options are available by accessing the shortcut menu. The options include: **Open, Attach, Unload, Reload, Detach,** and **Bind.**

OPEN

When the **Open** option is selected, it activates the **Xopen** command, which opens the selected Xref in a new window allowing for editing. The easiest and most direct method for editing an Xref is to utilize the **Xopen** command. This command enables you to quickly open a referenced drawing into a new drawing window. Instead of browsing for the Xref using the **Select File** dialog box, you can access **Xopen** by clicking on the desired Xref, right-clicking to view the shortcut menu, and selecting **Open**. Holding down the CTRL key allows you to select multiple Xrefs at the same time, and each file will open in a separate window.

ATTACH

You've already explored the **Attach** option. Use the following summary of the steps to attach a drawing to your current drawing:

1. Choose the **External Reference** button on the **Reference** toolbar, or type **XR** ENTER at the Command prompt.

2. Select **Attach DWG** and choose a drawing to be attached to the current drawing from the **Select Reference File** dialog box. Select the **Open** button to continue. Selecting **Open** will display the **External Reference** dialog box.

3. Select the reference type to be used. Typically, this will be **Attachment**.

4. Specify the insertion point, scale, and rotation angle, or select the **Specify On-screen** option for each parameter.

5. Click the **OK** button to attach the referenced drawing to the current drawing.

UNLOAD

The **Unload** option allows a referenced drawing to be temporarily removed from the screen display without removing the Xref from the drawing base. The **Unload** option is similar to using the Freeze option of the **Layer** command. The referenced drawing is not displayed or considered in regeneration, increasing the drawing speed. The Xref can be reactivated by using the **Reload** option. The process to reload a drawing is similar to the Detach option. The **Unload** option is very handy if you are working on a very large drawing and you don't need certain information at the current time but will need the information at a later time.

RELOAD

This option allows an attached referenced drawing to be reloaded and updated in the middle of a drawing session. If a coworker revises the referenced drawing while you're working on a drawing that contains the Xref, your Xref is out of date. The **Reload** option allows the most current referenced drawing to be used without your having to exit the drawing session and reopening the master drawing. The process to reload a drawing is similar to the **Detach** option.

DETACH

The **Detach** option will remove unneeded referenced drawings from the host drawing. Similar to the way **Erase** removes a block, **Detach** removes copies of the referenced drawing and its definition. Detach a drawing by first selecting the name of the unneeded, and then selecting the **Detach** option from the shortcut menu. This will remove the listing from the palette. The selected item is removed from the screen and all references to the drawing will be removed from the host drawing. Detach an attached drawing file using the following procedure:

1. Choose the **External References** button on the **Reference** toolbar. To use the shortcut menu, select the referenced drawing to detach, right-click in the drawing area, and choose **External References** from the menu.

2. Select the drawing to be detached from the host drawing.

3. Select the **Detach** button from the shortcut menu. This will detach the drawing file from the current file.

4. Click the **OK** button to return to the drawing area.

BIND

The **Bind** option allows a referenced drawing to become a permanent part of the host drawing. When this option is used, referenced drawings function as a wblock rather than an **xref**. When a drawing is referenced to a host drawing, the drawing is displayed, but you can't modify the referenced drawing. Using the **Bind** option allows an external drawing to be edited, but the will no longer be updated as the original drawing is edited. The **Bind** option adds dependent symbols to the drawing base so that they can be used just like any other drawing object. Layer names such as FLOORPLAN|DOORS in an attached drawing will be altered to read FLOORPLAN0DOORS. This naming system allows you to quickly identify attached and bound layers. The **Bind** option should be used only when you know for sure that the review process for the referenced drawing is complete and the drawing is no longer subject to change. Because construction drawings are subject to change, **Bind** might not be an acceptable option. The **Xbind** command, introduced later in this chapter, allows specific symbols of a drawing to be permanently attached.

Binding a Drawing

Selecting the **Bind** option from the shortcut menu will produce the **Bind Xrefs** dialog box shown in Figure 22.8, providing the options of **Bind** and **Insert**.

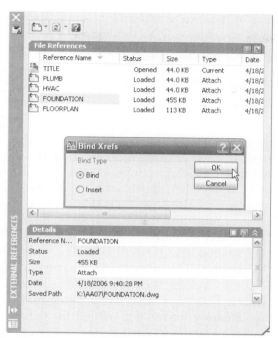

Figure 22.8 *The **Bind** option allows a referenced drawing to be permanently attached to a host drawing. Selecting the name of an attached drawing from the **External References** palette and right-clicking displays the shortcut menu with all the options. Selecting the **Bind** option displays the **Bind Xrefs** dialog box.*

Bind   The default option of **Bind** will permanently attach a drawing to the base. Clicking the **OK** button removes the dialog box, binds the attached drawing to the host drawing, and updates the names of dependent symbols. Symbols in the bound drawing can now be edited just as any object created in the host drawing.

Insert   If the **Insert** option of **Bind** is used, an attached drawing is bound to the base drawing as if a wblock had been inserted in a drawing. **Insert** will bind the referenced drawing to the current drawing in a method similar to detaching and inserting the reference drawing. Clicking the **OK** button removes the dialog box, inserts the attached drawing in the host drawing, and updates the names of dependent symbols. A layer of **FLOORPLAN|DOORS** would be renamed to **DOORS**.

THE XBIND COMMAND

The **Bind** option adds dependent symbols to the host drawing, so that they can be used. Automatic updating and other features of drawings are lost. This can be overcome with the **Xbind** command, which allows a portion of a referenced drawing—such as a block, layer, or linetype—to be permanently added to a current drawing, with the balance of the external drawing retaining its qualities. Access the **Xbind** command by using one of the following methods:

- Select **Xbind** on the **Reference** toolbar.
- Select **Object**, **External Reference**, **Bind** from the **Modify** menu.
- Type **XB** ENTER at the Command prompt.

Each option will produce the **Xbind** dialog box, shown in Figure 22.9, which lists referenced drawings that are attached to the host drawing. In Figure 22.9, six drawings are attached to the host drawing. Each drawing is preceded by the + symbol and the drawing file icon. The + indicates that the listing can be expanded. Selecting one of these options and double-clicking will display the contents of the drawing. In Figure 22.10a, the FLOOR listing was expanded, revealing that the drawing contains nested Blocks, Dimstyles, Layers, Linetypes, and Text styles. Notice in Figure 22.10b that the block **FLOOR10PBAC01** has been selected to be added, using the **Xbind** command. If the desired dependent symbol to bind, **FLOOR10PBAC01**, is selected and the **Add>** button is clicked, the name of the block will be moved from the listing to the **Definitions to Bind** display. (See Figure 22.10c.) Clicking the **OK** button will bind the block to the host drawing and close the **Xbind** dialog box. Selecting a listing from the **Definitions to Bind** display and selecting the **<Remove** button will unbind the definition and restore it to the **s** listing.

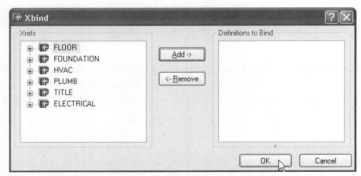

Figure 22.9 *The **Xbind** dialog box displays a listing of drawings that are attached to the current drawing. The + symbol preceding the drawing icon indicates that the listing can be expanded.*

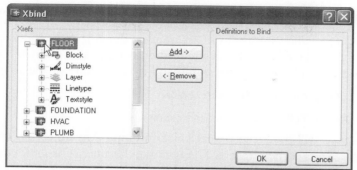

Figure 22.10a   *Selecting the FLOOR listing reveals nested blocks, dimstyles, layers, linetypes, and text styles.*

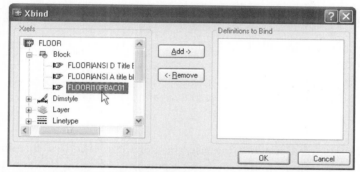

Figure 22.10b   *Selecting the **FLOOR10PBAC01** block for binding to the host drawing.*

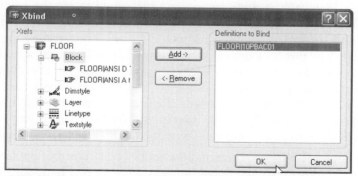

Figure 22.10c   *Selecting the name and selecting the **Add->** button will bind the selected definition to the host drawing.*

WORKING WITH XCLIP

The **Xclip** command allows portions of an attached drawing to be hidden. The command allows a boundary to be created in a referenced drawing so that all material that is outside the boundary is invisible. Access the command by using one of the following methods:

- Select **Xclip** on the **Reference** toolbar.
- Type **XC** ENTER at the Command prompt.

Each access method will produce a Select object prompt. Any object selection method can be used to select Xrefs to be clipped. The command sequence is as follows:

```
Select the Xclip button (Or type XC ENTER).
Select object: (Select Xrefs to be clipped.)
1 found
Select objects: ENTER
[ON/OFF/Clipdepth/Delete/generate Polyline/New boundary]: ENTER
```

Accepting the default value allows the clipping boundary to be selected and produces the following prompts:

```
Specify Clipping boundary or select invert option:
[Select polyline/Polygonal/Rectangular/Invert clip]: ENTER
Specify first corner: (Select first window corner.)
Specify opposite corner: (Select opposite window corner.)
```

Options for setting the boundary include the default method of forming a rectangular boundary with a selection window, selecting a polyline, selecting Polygonal, or by selecting Invert clip. Selecting Invert clip enables you to clip the inverse of a selected boundary. With this mode, using the rectangular option, you would select the area that you would want hidden. This option would work well if you were working on a office building and wanted the middle section to be excluded from the display but needed the rest of building to still be attached to the host drawing. Selecting Polygonal allows a boundary to be set by specifying points for the vertices of a polygon. Accepting the default by pressing ENTER produces prompts to select the First and Other corners. As the corners are selected, objects in the Xref that are outside the boundary will become invisible. This will be helpful if there are items on certain layers that you need visible and others that need to be invisible. Rather than altering the layers, you can exclude some objects with **Xclip**.

Figure 22.11a is an example of construction details that have been referenced to a host drawing. Figure 22.11b shows the detail on the right side of Figure 22.11a, hidden using the **Xclip** command.

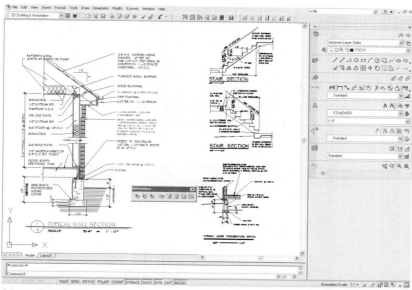

Figure 22.11a *The **Xclip** command allows information that has been attached to a host drawing to be hidden. The command requires that a boundary be selected. Anything outside the boundary will be made invisible.*

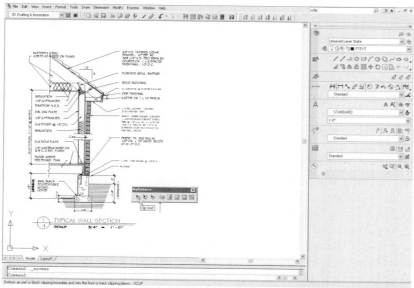

Figure 22.11b *The detail on the left side of Figure 22.11a was placed inside the boundary. With the **Xclip** command, the detail on the right side is still attached to the host drawing but is removed from the display set.*

EDITING REFERENCED DRAWINGS

AutoCAD allows externally referenced drawings that you've created to be edited, using in-place editing. In-place editing allows the referenced drawing to be revised. If the referenced drawing selected for editing contains attached Xrefs or block definitions, the reference and nested references are displayed in the **Reference Edit** dialog box. Display this dialog box by selecting the **Edit Reference In-Place** button on the **Refedit** toolbar. The dialog box can be seen in Figure 22.12. The **Refedit** toolbar presents five options for editing referenced drawings.

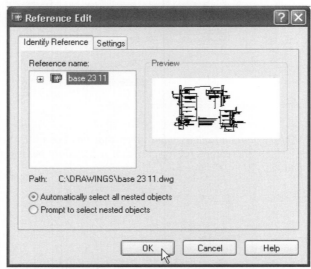

Figure 22.12 *AutoCAD allows referenced drawings to be edited using in-place editing with the* ***Reference Edit*** *dialog box. Display the box by selecting the* ***Edit Reference In-Place*** *button on the* ***Refedit*** *toolbar.*

EDIT BLOCK OR XREF

A referenced drawing will respond like a block if you attempt to edit it. Using the **Refedit** command allows individual objects in the referenced drawing to be altered. You can edit objects by selecting the **Edit Reference In-Place** button on the **Refedit** toolbar and then by choosing the referenced drawing to be edited. **Refedit** is a powerful command when you need to make minor changes; it allows you to make changes without having to go back and forth between drawings. If you need to make major changes to an Xref, open the Xref using the **Xopen** command introduced earlier in this chapter.

Using the Reference Edit Dialog Box

The **Reference Edit** dialog box allows a referenced drawing to be edited. Key elements of the dialog box are separated under two tabs: **Identify Reference** and **Settings**. Both tabs provide features and options for editing the selected item.

Identify Reference tab   This tab features the **Reference name** box, **Preview** window, **Path**, and options regarding nested objects.

> **Reference name**—This portion of the dialog box displays the referenced drawing that has been selected for in-place editing and any nested drawings or blocks within the selected drawing.
>
> **Preview**—The **Preview** window displays an image of the referenced drawing selected for editing.
>
> **Path**—Displays the file location of the selected referenced drawing. If the selected object is a block, no path will be displayed.
>
> **Automatically select all nested objects**—Selecting this option automatically includes nested objects in the reference editing session.
>
> **Prompt to select nested objects**—After you select this option, AutoCAD prompts you to select the specific objects in the referenced drawing that you want to edit.

Settings tab   This tab features options for editing references.

> **Create unique layer, style, and block names**—This option controls whether layer and symbol names of objects extracted from the referenced drawing will be unique or altered. In the active mode, symbol and layer names will be displayed with a prefix of #. The altered prefix allows for easy identification. With the option inactive, layer and symbols will appear as they do in the original drawing.
>
> **Display attribute definitions for editing**—This option controls whether variable attribute definitions in a block reference will be extracted and displayed during editing. Referenced drawings and blocks without definitions are not affected by this option. When active, attributes will be invisible but the object and the attribute definitions are available for editing. If changes are saved back to the block reference, the original reference remains unchanged. The new attribute definition will affect only future insertions of the block.
>
> **Lock objects not in working set**—This option provides the ability to lock objects not in the working set. This could save you many headaches, preventing you from accidentally editing objects in your current drawing while in a reference editing state. This option works much like the locked layer option in the **Layer Properties Manager**.

Once each option is set and you've selected a reference to edit, you'll be prompted to select specific objects in the referenced drawing to edit.

Editing Referenced Drawings Using Refedit

The **Refedit** command is used to select and edit a referenced drawing. Use the following procedure to edit individual objects of a referenced drawing:

> 1. Click the **Edit Reference In-Place** button on the **Refedit** toolbar, or type **REFEDIT** ENTER at the Command prompt.

2. Select the reference to be edited. This will highlight the referenced drawing and display the **Reference Edit** dialog box. If nested blocks or drawings are contained in the selected, all references available for selection are displayed in the dialog box.

3. If multiple drawings are listed in the dialog box, highlight the drawing to be edited.

4. Set the options of **Automatically select all nested objects**, **Prompt to select nested objects**, **Create unique layer, style and block names**, **Display attribute definitions for editing**, and **Lock objects not in working set**.

5. Click the **OK** button.

6. Select the objects to be edited.

This is your opportunity to select objects in the referenced drawing that will be edited. Once selected, objects in the referenced drawing can be edited just like any other drawing.

Adding or Removing Objects from the Working Set

To add objects to the working set or remove them from it as the referenced drawing is being edited, you can use the **Refset** command. The working set consists of what will be extracted, edited, and saved back to the drawing. You'll be editing and saving changes to the original drawing—even though you're working on a referenced drawing through a host drawing. You can supply a referenced floor plan to the HVAC subcontractor, and they can alter the plan as needed to allow for ducts and chases to be installed. As they save changes to the plan, you can now supply an up-to-date plan to other consultants. Start the command by selecting the appropriate button on the **Refedit** toolbar, by clicking **Add to Working set** or **Remove from Working set**, or by typing **REFSET** ENTER at the Command prompt.

Adding Objects to the Working Set To add objects to the working set, start the **Refset** command by clicking the **Add** button on the **Refedit** toolbar.

This will display the prompt:

    Select objects: *(Select objects to be added to the working set.)*

The command will vary slightly when entered by keyboard. You'll have to select Add or Remove at the prompt. (When using the toolbar, you've already made the decision by selecting the button.) Once objects are selected, a prompt will be displayed giving the number of objects added to the working set. The prompt will resemble the following:

    3 Added to working set

Removing Objects from the Working Set To remove objects from the working set, start the **Refset** command by clicking the **Remove** button on the **Refedit** toolbar.

This will display the prompt:

        Select objects: *(Select objects to be removed to the working
            set.)*

The command will vary slightly when entered by keyboard. Once objects are selected, a prompt will be displayed giving the number of objects removed from the working set. The prompt will resemble the following:

        3 Removed from working set

Objects removed from the working set will appear in a gray tone referred to as *dithered*. The intensity of the line can be controlled using the **Display** tab of the **Options** dialog box. (Select **Options** from the **Tools** menu.) Use the **Reference Edit Fading Intensity** setting to control the display of removed objects. A higher value provides greater contrast.

Saving and Discarding Changes with Refclose

Up to this point, you've referenced an external drawing to a drawing and edited a portion of the referenced drawing. If you try to edit other portions of the referenced drawing, AutoCAD will warn you:

Command not allowed, (drawing name) already checked out for editing

This is AutoCAD's way of not allowing two subcontractors to attempt to use and update a drawing at the same time. If you save the drawing, the new drawing will still contain your revisions, but the original drawing will remain unaltered. The **Refclose** command is used to either save back or discard changes made during the editing process of a referenced drawing. Once you've used **Refclose**, you can edit other referenced drawings or blocks. The command can be selected on the **Refedit** toolbar, shortcut menu, menu, or by keyboard. As the command is entered by keyboard, you'll be prompted to enter a letter for the Discard or Save option.

Close Reference   Selecting **Close Reference** will vaporize the working set so that the source drawing remains in its original state. Changes that you've made will remain in the current drawing, but the original drawing remains unchanged. You'll be given a warning, but proceeding will remove changes to the working set. **Undo** will restore the reference editing session if you change your mind after using Close Reference.

Save   Selecting the Save option will save the changes made to the working set back to the original drawing.

DENYING ACCESS TO REFERENCED DRAWINGS

Throughout the discussion of in-place editing, it has been assumed that you don't mind if someone alters your drawing. The idea of someone working for the HVAC contractor making changes to your floor plan might sound pretty good. The CAD operator for the

HVAC is saving you the work of incorporating their required changes into the base drawing. If for some reason you need to provide drawings to another firm, but you want to ensure accuracy, you can restrict the ability of others to edit the referenced drawing. Security can be maintained on the **Open/Save** tab of the **Options** dialog box (select **Options** from the **Tools** menu). By default, the **Allow other users to Refedit current drawing** check box in the **External References (s)** area is active. Making this option inactive will remove other users' ability to edit your source drawing.

Working with DGN Files

Along with DWG files, MicroStation V8 DGN files can also be attached to your AutoCAD drawings and managed using the External References palette. Typing **DGNATTACH** ENTER at the command line will open up the **Select DGN File** dialog box. The process is similar to the method of attaching an **xref** to your drawing. In addition to attaching, you can import and export DGN files as well. Launch the importing/exporting process by using the **DGNIMPORT** and **DGNEXPORT** command. For more in-depth information, refer to the help menu.

CHAPTER 22 EXERCISES

1. Open the **FOOTING** drawing and use it to create an drawing titled **BASEFTG**. Open drawing **FLOOR14** and attach **BASEFTG** to the drawing. Save the drawings as **XREFTEST**.

2. Open a drawing template containing the border and title block. Attach any four drawings in your library. Provide a title and the scale used by the original drawing.

3. Open any three drawings and attach them to a drawing base. Save the drawing as **XREF**.

CHAPTER 22 QUIZ

DIRECTIONS

Answer the following questions with short complete statements. Type your answers using a word processor.

1. Place your name, chapter number, and the date at the top of the sheet.

2. Type the question number and provide the answer.

 Warning: Some of the questions may not have been covered in the reading material and will require the use of the help menu. You may also have to do some exploring to answer the questions.

1. Describe how a referenced drawing can be used in an office to complete a project.

   _____

   _____

2. List and briefly define the six options.

   _____

   _____

3. List the command, the options required, and the prompts to attach a drawing to another drawing.

   _____

   _____

4. What is a dependent symbol and how does it affect a drawing?

   _____

   _____

5. Explain the effects of overlaying a drawing.

   _____

   _____

6. The Bind option has been used. How will the option affect the external drawing?

   _____

   _____

7. Explain the effects of **Xbind** on a referenced drawing.

   _____

   _____

8. How does the **Xclip** command affect an attached drawing?

   _____

   _____

9. How does the **Refedit** command affect a referenced drawing?

_____

_____

10. What is the difference between **Insert** and **Xref**?

_____

_____

Working with Layouts and Viewports

INTRODUCTION

This chapter will introduce:

- Comparisons between model space and paper space
- Creating and using layouts
- Working with tiled viewports
- Working with floating viewports

Commands to be introduced include:

- **Layout**
- **Vports** (Viewports)
- **Mspace** (Model space)
- **Pspace** (Paper space)
- **Vpclip** (Viewport clip)
- **Vpmax** (Viewport max)
- **Vpmin** (Viewport min)
- **Laywalk** (Layer walk)

Chapter 2 introduced the layout tabs of the AutoCAD template drawings. A layout is a paper space drawing environment that represents how your drawing created in model space will appear on a sheet of paper when plotted. Throughout this text you've used the model space tab of the Architectural template. The template has been edited to include your favorite linetypes, lineweights, text fonts, and dimensioning styles. This chapter will introduce you to using the layout tabs to provide a predictable plotting setup. Methods of using existing layouts and creating new layouts will be explored. By completing the steps in this chapter, you'll be able to take drawings that have been created in previous chapters, place them in a viewport within the layout, and prepare them for plotting. Methods for creating multiple viewports within a drawing will also be explored. Multiple viewports allow objects that must be displayed at different scales within the same drawing to be plotted easily.

COMPARING MODEL AND PAPER SPACE

As you begin to prepare a drawing for plotting, it's important that you understand the use of the terms model space, paper space, and layout. In chapter 2 you were introduced to drawing templates that were created in paper space. Throughout the text you've been drawing in model space. Model space is where you create the drawing; paper space is where you plot. A layout is a paper space tool that contains one or more viewports to aid in plotting. If you open the Architectural template and select the Architectural Title Block tab, your drawing is displayed in paper space. The paper space UCS icon is displayed in the lower left corner, and the architectural title block surrounds your drawing. If you click the **Paper** button on the status bar below the Command prompt, the UCS icon is changed to reflect model space and is placed inside the drawing template. Your drawing is now in a viewport or window that is in model space within a paper space border. Open a template or refer to chapter 2, and you'll notice that the architectural template contains a black line that overlays the border. This box is the viewport (see Figure 23.1). You can edit the drawing in the model space viewport without having to return to the model space tab. Later in this chapter you'll be introduced to methods for dividing model space into tiled viewports to represent different views of the drawing. Switch back to paper space by clicking the **Model** button, and the drawing is displayed in the paper space layout with viewports, ready for plotting.

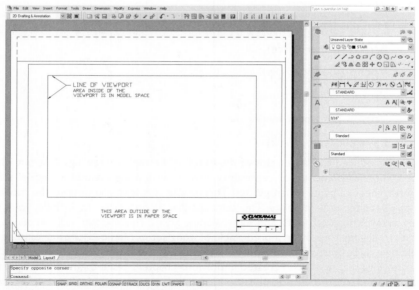

Figure 23.1 *The architectural template contains a black line that overlays the border. This box is the viewport.*

Note: As a general rule, NEVER alter the location of the drawing in paper space. When the drawing location is altered in paper space, it can affect layouts by other subcontractors who might be working on the project. Alter the drawing in model space and then return to paper space to display the drawing with the title block and borders.

AN OVERVIEW OF PAPER SPACE

To understand the process of viewing the floor plan drawing in Figure 23.2, visualize a sheet of 23' x 36' vellum with a title block and border printed on the sheet. Imagine a hole, the viewport, cut in the vellum that allows you to look through the paper and see the floor plan. This is a simplified version of what is required to display a drawing for plotting. Figure 23.3 shows the theory of displaying a drawing in model space to a layout in a template drawing created in paper space. The floor plan shown in Figure 23.2 is 90'–0" wide in model space. If you were to hold a sheet of D size vellum in front of the plan, the paper would seem minute. To make the floor plan fit inside the viewport on the paper, you're going to have to hold the paper a great distance away from the floor plan until the drawing is small enough to be seen through the hole. AutoCAD will automatically figure the distance. By entering the desired scale factor for plotting, you can reduce the floor plan to fit inside the viewport and maintain a scale typically used in the construction trade. Multiple viewports can be placed in a single layout, and multiple layouts can be created within a single drawing file.

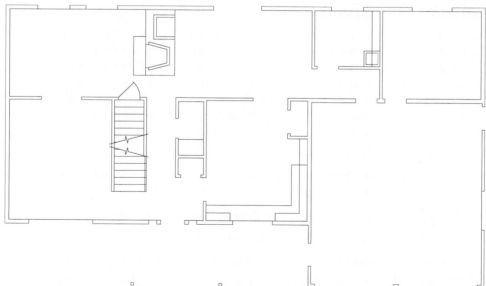

Figure 23.2 *A floor plan that is 90' long can be displayed in paper space using a paper space layout. (Courtesy Kyle Jones.)*

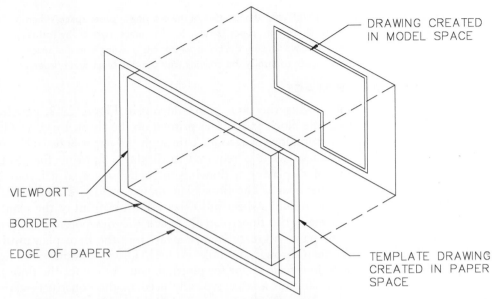

DRAWING CREATED
IN MODEL SPACE

VIEWPORT

BORDER

EDGE OF PAPER

TEMPLATE DRAWING
CREATED IN PAPER
SPACE

Figure 23.3 *A viewport is provided in each drawing layout created in paper space. Additional layouts can be created to allow for different plotting displays.*

WORKING WITH LAYOUTS

Throughout this text you've been using the Architectural template or a template that you've created to meet your needs. When your drawing is complete, you can select the layout tab to begin preparing the drawing for reproduction. If you're using the Architectural template, the layout tab is titled Architectural Title Block. Layout settings for plotting can be set once and stored with the drawing. For now, the goal is to help you create drawing displays that are ready for plotting. When your drawing is complete, use the following steps to adjust a drawing in the viewport in paper space:

1. Select the Architectural tab.

2. Click the **Paper** button on the status bar at the bottom of the screen to enter model space in the viewport.

3. Use the All or Extents option of **Zoom** to view the drawing.

4. Use **Pan** to center the drawing in the layout.

5. Click the **Model** button on the bottom of the screen to return to paper space.

The procedure to view the drawing in paper space is simple and should be familiar by now. In addition to using layouts for plotting, you can create several layouts for the same drawing so that different parameters can be emphasized in the plotting process. Different scales and paper sizes can be assigned to different layouts, or different plotting devices can be assigned to the layouts.

CREATING NEW LAYOUTS

The existing template and layout work well for displaying a single drawing for plotting. Additional paper space layouts can be created to plot various layers, or in some other way customize a drawing for a specific plotting need. Layouts can be assigned a unique name, and multiple paper space viewports can be assigned to one or more of the layouts. Independent settings can be assigned and stored in each layout for plotter and page setup. Once created, layouts can be copied, deleted, moved, or renamed. Selecting **Create Layout** from **Wizards** on the **Tools** menu allows a new layout to be created. The **Create Layout Wizard** contains a series of pages that walk you through the process of creating a new layout. Options are similar to the plotting options introduced in chapter 3. Each setting will be further discussed in the next chapter. Figure 23.4 shows an example of the beginning page of the **Create Layout Wizard**. This page is used to assign a name to the layout being created. Once the desired name is entered, selecting the **Next** button will produce a display for choosing a printer from a list of configured printing devices. Using multiple layouts with multiple plotter configurations will allow you to create check prints with a laser printer, or full size plots with a plotter, without having to adjust plotter settings each time you make a plot. Once the plotter to be assigned to the layout is selected, choosing **Next** will display a **Create Layout - Paper Size** dialog box similar to Figure 23.5. Depending on the configured plotter, this dialog box allows the paper size and drawing units to be selected.

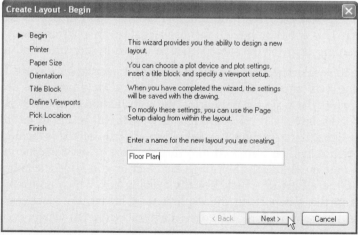

Figure 23.4   *The **Create Layout Wizard** contains a series of pages that walk you through the process of creating a new layout. Options are similar to the plotting options introduced in chapter 5. The beginning page of the **Create Layout Wizard** is used to assign a name to the layout being created.*

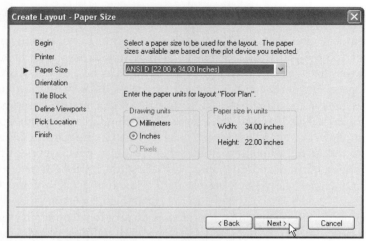

Figure 23.5 *Once the plotter to be assigned to the layout is selected, choosing **Next** will display a **Create Layout - Paper Size** dialog box that allows the paper size and drawing units to be selected.*

Selecting the **Next** button allows the orientation of the plot to the paper to be configured. Options include **Portrait** or **Landscape**. When finished, selecting the **Next** button allows a predefined template to be assigned to the layout. The name of the template selected will be displayed in the **Preview** area. By using multiple layouts, you can display various title blocks with the drawing. This could be helpful for displaying a floor plan drawn by the architectural team and the framing plan for the same structure drawn by the structural team.

Once a title block is selected, clicking the **Next** button displays a **Create Layout - Define Viewports** dialog box similar to Figure 23.6. This dialog box can be used to define a viewport and select the scale for presenting the information in the viewport. By default, a single viewport is displayed. Creating multiple viewports will be discussed in the next section of this chapter. The default setting for the viewport scale factor is **Scaled to Fit**. In this setting, a 90' structure will be crammed into the viewport. Most construction drawings need to be plotted at a measurable scale. Selecting the **Next** button after selecting the desired scale will display the **Create Layout - Pick Location** dialog box, which prompts you to select the corner locations of the viewport. In the Architectural layout, the viewport edges align with the top, right, and bottom borders. This option allows a different viewport size to be selected for the new layout. The size of the viewport will depend on the size of the drawing to be displayed, the desired scale, and the number of viewports to be created in the layout. As the size is selected, the final page of the wizard will be displayed. This page will display the name of the layout you've created and methods for altering the layout. Selecting the **Finish** button will place the layout in the drawing base as the current layout.

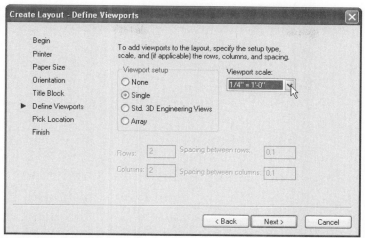

Figure 23.6　*The **Define Viewports** page allows the scale of the drawing in the viewport to be set and maintained in paper space.*

Summary of Steps in Creating a New Layout

In summary, the steps to creating a new layout with a wizard are as follows:

1. Select **Create Layout** from **Wizards** on the **Tools** menu.

2. Provide a name for the layout being created.

3. Select the plotting device.

4. Select the paper size and the drawing units.

5. Select the orientation.

6. Select the title block to be used.

7. Select the number of viewports and the scale to be used in the viewport.

8. Select the corners of the viewport.

9. Click the **Finish** button to return to the drawing area.

The new layout will now be displayed as the current layout, similar to Figure 23.7a. Notice that the viewport that was created is not large enough to display the full floor plan. Selecting the viewport border will display the grips for the viewport. Make one of the grips hot, and then stretch the viewport to the desired size, as shown in Figure 23.7b. Refer to chapter 10 to review grips. Pressing ESC will end the **Stretch** command and restore the viewport. Figure 23.7c shows the results of stretching the viewport. Other modifications can be made to the layout by using the **Page Setup** dialog box. Access this dialog box by selecting the **Page Setup** button on the **Layouts** toolbar or by right-clicking the Model tab or a layout tab and choosing **Page Setup**.

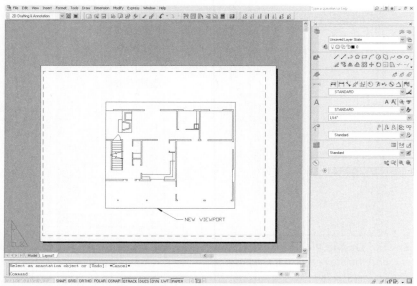

Figure 23.7a   *The viewport that was created might not be large enough to display the entire drawing.*

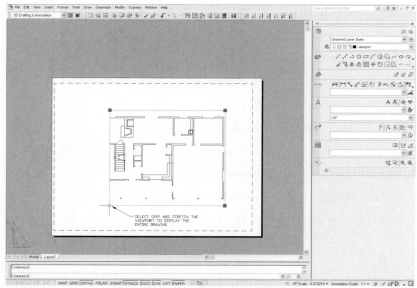

Figure 23.7b   *Selecting the viewport border will display the grips for the viewport. Make one of the grips hot, and then stretch the viewport to the desired size.*

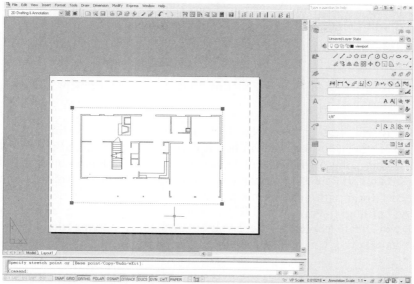

Figure 23.7c   *A viewport can be stretched to display the full drawing. Pressing ESC will end the **Stretch** command and restore the viewport.*

Note: As several layouts are created, all of the layout names might not appear. To display more of the layout tabs, select and hold the edge of the slide bar, and then move the cursor to the right. Figures 23.8a and 23.8b show the process. You can also display the other layouts by using the arrow keys directly to the left of the Model tab. This allows you to scroll through the layout tabs and select the layout you need. You can also use the CTRL+PAGE UP and CTRL+PAGE down keys to access the layouts quickly from the keyboard.

Figure 23.8a   *Depending on the size of the slide bar, not all the layout tabs might be displayed.*

Figure 23.8b   *Decrease the size of the slide bar to display more layout titles.*

ALTERING LAYOUTS

Once your layouts are created, they can easily be moved and put in a different order. You can drag and drop the layout tabs to put them in the desired order. You can create a new layout or alter an existing one by typing **LAYOUT** ENTER at the Command prompt. This method will present the following prompt at the command line:

```
Enter layout option [Copy/Delete/New/Template/Rename/ SAveas/
    Set/?]:
```

These options can be used to create, define, and save a new drawing layout.

Copy

Selecting Copy will create a new layout by making a copy of an existing layout. A prompt for the name of the layout to copy will be displayed at the Command prompt. The default name will be the name of the current layout—in this case, the Architectural Title Block is the default name. If a new layout name is not provided, the new layout assumes the name of the copied layout, with an incremental number added to the name. You can also copy a layout by holding down the CTRL key as you use the drag and drop sequence. AutoCAD will create copies of the layout(s) instead of moving them. The command sequence is as follows:

```
Enter layout option [Copy/Delete/New/Template/Rename/ SAveas/
    Set/?] C ENTER
Enter layout to copy <Architectural Title Block>: ENTER
Enter layout name for copy <Architectural Title Block (2)>: ENTER
```

Delete

This option will delete a layout. By default, the current layout will be deleted. (The Model tab cannot be deleted.) One or more layouts can be deleted. If all layouts are removed, an empty layout titled **Layout1** will be created and displayed.

New

This option creates a new layout tab without the use of the **Create Layout Wizard**. Selecting the New option will display the following prompts:

```
Enter layout option [Copy/Delete/New/Template/Rename/ SAveas/
    Set/?] <Set>: N ENTER
Enter new Layout name <Architectural Title Block>: COOL STUFF
    ENTER
```

You can also create a new layout tab by right-clicking on an existing layout tab and choosing **New layout** from the shortcut menu, or by clicking the **New layout** button on the **Layouts** toolbar.

Template

Selecting the **Template** option creates a new layout based on an existing template (.DWT) or drawing file (.DWG). The layouts from an existing template or drawing file can be inserted in a new drawing. In addition to the layout, objects contained in the layout can also be made part of the layout being created. The command sequence is as follows:

```
Enter layout option [Copy/Delete/New/Template/Rename/ SAveas/
    Set/?] T ENTER
```

Pressing ENTER will display the **Select Template from File** dialog box, allowing a template to be selected. Selecting the **File of type** edit box allows a drawing file to be selected. Once the template or file is selected, it will be displayed in the preview area. Selecting the **Open** button will close the **Select Template from File** dialog box and provide the **Insert Layout** dialog box that verifies the template to be added to the layout. Selecting **OK** will return the layout display with the template. You can also select a template by choosing the **Layout From Template** button on the **Layouts** toolbar, or by right-clicking and selecting **From template** on the shortcut menu.

Rename

This option can be used to rename a layout. The layout that was last used is shown as the default. Prompts will be displayed to select the layout to be renamed, and to select the new name. Prompts to rename the **JUNK** layout as **COOL STUFF2** are as follows:

```
Enter layout option [Copy/Delete/New/Template/Rename/ SAveas/
    Set/?] R ENTER
Enter Layout to rename <Architectural Title Block>: JUNK ENTER
Enter new layout name: COOL STUFF2 ENTER
```

You can also rename a layout by double-clicking on the layout tab. This will highlight the name and allow you to type in the new name. Layout names are not case sensitive and can be up to 255 characters long. Only 32 characters are displayed on the layout tab.

Saveas

As you complete the **Create Layout Wizard**, the new layout is stored as part of the current drawing. Any existing drawing can be saved and used as a template drawing. You can save all of the objects, properties and settings as a .DWT file (template) by selecting the Saveas option from the **Layout** command. New drawings can be created from drawing templates or new layouts can be created in the current drawing based on a layout in the template file.

Set

The Set option makes an existing layout the current layout. Clicking the tab of a layout does the same thing.

The ? Option

Selecting this option will list all of the layouts defined in a drawing. This can be helpful when all layout names cannot be displayed below the drawing. Pressing F2 will also display a list of the current layouts.

WORKING WITH VIEWPORTS

AutoCAD allows two types of viewports to be created. Floating viewports are used to create plotting layouts in paper space. Tiled viewports can be created in model space to aid in the display of the drawing. Up to four tiled viewports can be created, but only one viewport can be active at a time. Tiled viewports allow you to divide the screen into multiple areas so that you have multiple views of different parts of the same object. Figure 23.9 shows an example of a sheet of details. Creating multiple viewports would allow you to zoom in on a typical pier elevation in one viewport, section plan at pier in another viewport, and a display of the whole sheet of details in a third viewport, as shown in Figure 23.10. This multiple use of viewports allows for drawing and editing within the enlarged portions of the drawing without having to zoom in and out to move from one detail to another. It is important to remember that even though you are seeing multiple images on the screen, you're working with only one database. An object that is drawn or edited in one viewport will also affect the other viewports.

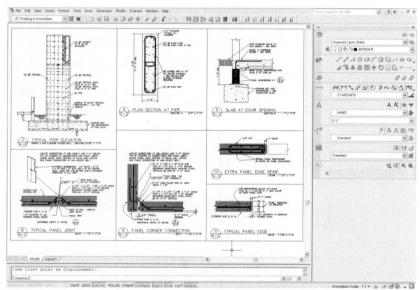

Figure 23.9 *When you work on a sheet of details, it is often hard to coordinate information from one detail to another.*

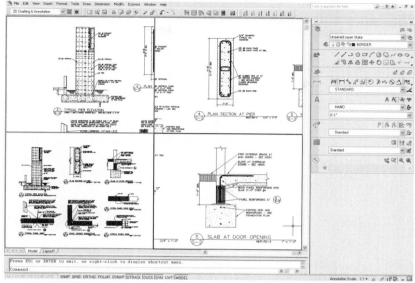

Figure 23.10 *Tiled viewports allow you to divide the screen into multiple images so that you have multiple zoom images of different parts of one drawing file. The details shown in Figure 23.9 can be easily viewed using multiple viewports.*

DISPLAYING MULTIPLE TILED VIEWPORTS

 Multiple tiled viewports can be displayed in several configurations that are controlled by the **Viewports** dialog box. Display the **Viewports** dialog box by using one of the following methods:

- Click the **Display Viewports Dialog** button on the **Viewports** toolbar.
- Select **New Viewports** from **Viewports** on the **View** menu.
- Type **VPORTS** ENTER at the Command prompt.

Each method will produce the display shown in Figure 23.11. In the current setting, one viewport is displayed. Selecting a viewport configuration from the standard viewports listing will alter the display in the preview to reflect the change. Figure 23.12 shows the display for each option. Use the following steps to display multiple tiled viewports in a new configuration:

1. Display the **Viewports** dialog box.
2. Select the name of the desired viewport configuration from the **Standard viewports** list.
3. Select Display as the current setting for the **Apply to** list.
4. Select 2D as the current setting for the **Setup** list.
5. Click the **OK** button to apply the configuration and close the dialog box.

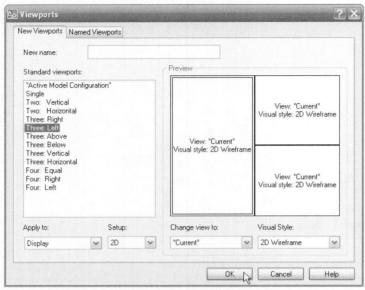

Figure 23.11 *To display the **Viewports** dialog box select the **Display Viewports Dialog** button on the **Viewports** toolbar, by selecting **New Viewports** from **Viewports** on the **View** menu, or by typing **VPORTS** ENTER at the Command prompt.*

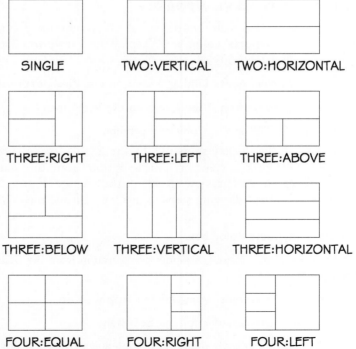

Figure 23.12 *Possible display options for tiled viewports.*

Notice, in Figure 23.10, that a bold black line surrounds the viewport on the right side. This viewport also contains the crosshairs. As you move the cursor around the screen, you'll notice that the crosshairs only work in the drawing on the right side. This view is called the active viewport. You can draw or edit in this viewport just as in any other drawing. As the cursor crosses the center of the drawing into the viewport on the left, the crosshairs change to an arrow.

To make a different viewport active, move the cursor to the desired viewport and click. To convert a drawing like Figure 23.9 to one similar to Figure 23.10, change between active viewports and use the **Zoom** command. Remember, once the viewport is active, all drawing and editing commands function normally. First make the upper left viewport active. Select the desired detail, and utilize the **Zoom Window** command to enlarge the details to fill the viewport. Repeat the steps for the remaining viewports to display the desired enlarged details. With these details enlarged, you can make changes to each without having to zoom from one detail to another.

Working in Multiple Viewports

In addition to easing viewing, viewports can ease working. A command such as **Line** can be started in one viewport and completed in another viewport. This could be especially helpful on large drawings, such as exterior elevations. Exterior elevations are typically created side by side so that heights can be projected from one view to another. If all views are to be displayed, each view will be tiny. Creating a viewport for each elevation allows for easy viewing. Being able to draw lines from one viewport to another will speed production. Use the following steps to draw a line between viewports:

1. Select the viewport that contains the start point, making it the active viewport.

2. Start the **Line** command sequence.

3. Select the viewport that will contain the ending point, making it the active viewport.

4. Select the ending point for the line. A line will now be displayed between the two selected points, similar to Figure 23.13.

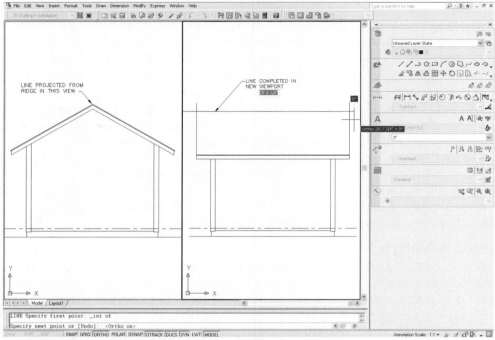

Figure 23.13 *A command can be started in one viewport and completed in another viewport. Select the viewport that contains the start point, making it the active viewport, and start the command sequence. Next select the viewport that will contain the endpoint, making it the active viewport, and select the endpoint for the line.*

WORKING WITH FLOATING VIEWPORTS

Each time you display a layout, you're working with a floating viewport. The viewport is referred to as floating because it can be moved and the size can be altered, just like a floating dialog box. The viewport allows you to look through the paper and see the drawing created in model space. The floating viewport is used to display the drawing in the layout for plotting. An unlimited number of viewports can be created in paper space, but only 64 viewports can be visible at one time. The balance of this chapter will explore floating viewports and how they can be used to prepare a drawing layout for plotting. A floating viewport can be considered as a drawing object that provides a view into model space. Several floating viewports can be created within a layout, and the viewports can be overlapped or separated from each other. A floating viewport can also be an irregular shape. Because floating viewports are considered objects, drawing objects displayed in the viewport cannot be edited. To edit objects within the viewport, model space must be restored. You can restore model space by choosing the Model tab or you can toggle between model space and paper space within the paper space layout. Methods for toggling between model and paper space within a paper space layout include the following:

- Double-clicking in a floating viewport will switch the display to model space.

Note: Remember, while editing in model space in a layout, you risk changing the drawing setup. This should only be used by experienced AutoCAD users or when in the initial setup of the view inside the viewport.

- Click the **Paper** or **Model** button on the status bar. Remember, when you're in a layout, the page is in paper space, but the viewport can be toggled to model space or paper space. When the **Paper** button is displayed, the viewport is a paper space display of the drawing in a paper space layout. Clicking **Paper** will change the button to Model and display the drawing in model space within a paper space layout.

- Type **MS** ENTER for **Mspace** or **PS** ENTER for **Pspace** at the Command prompt.

In additional to the methods listed above, the **VPMAX** and **VPMIN** commands greatly enhance your ability to quickly and accurately edit within a viewport. These toggle buttons provide a quick method of navigating through your viewports.

Maximizing

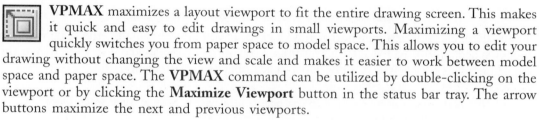 **VPMAX** maximizes a layout viewport to fit the entire drawing screen. This makes it quick and easy to edit drawings in small viewports. Maximizing a viewport quickly switches you from paper space to model space. This allows you to edit your drawing without changing the view and scale and makes it easier to work between model space and paper space. The **VPMAX** command can be utilized by double-clicking on the viewport or by clicking the **Maximize Viewport** button in the status bar tray. The arrow buttons maximize the next and previous viewports.

Minimizing

 After making changes to the model space drawing, you can type **VPMIN** at the command prompt to quickly return to your previous paper space view. Other options to minimize include clicking on the **Minimize Viewport** button in the status bar tray or selecting the **VPMIN** command from the shortcut menu.

Most professionals work in model space and arrange the final drawing for output in paper space. While you work in a layout, with the viewport in paper space, any material added to the drawing will be added to the layout, but not shown in the model display. This will prove useful as the final plotting layout is constructed.

ALTERING VIEWPORT PROPERTIES

 Because viewports are objects created by AutoCAD, they have properties such as color, layer, linetype, lineweight, and plot style. You can alter each of these properties using the **Properties** palette. As you create a viewport, a new layer can be created and displayed for the viewport. This layer should be frozen before the drawing is saved so that it is not plotted. You can also use the **Match Properties** command; this will allow you to "paint" properties from one viewport to another. For example, you can apply the layer and locking values from one viewport to another.

VIEWPORT SCALING

Most users of AutoCAD plot their drawings displayed in a layout at a scale of 1:1. To do this, the drawing must be inserted or attached to the viewport at a predetermined scale factor. AutoCAD has created an aid to do this. Earlier it was pointed out that the scale can be set as a drawing is placed in a viewport. The scale can be locked using the **Properties** palette. Locking the viewport scaling will prevent the scale from being altered if the drawing is zoomed once it is displayed in the viewport. The scale can be locked by the following procedure:

1. Display the layout containing the viewport to be locked.

2. Select the edge of the viewport so that it is highlighted.

3. Display the **Properties** palette by selecting the **Properties** button on the **Standard** toolbar.

4. Select **Display locked** from the **Properties** palette. (You may need to scroll down to view options.)

5. Select Yes from the **Display locked** edit list.

6. Close the **Properties** palette and return to the drawing area.

The scale of the selected viewport is now locked. If you attempt to alter the scale while in model space, AutoCAD displays "warning: Viewport is view-locked." If you change the zoom factor in the viewport, only paper space objects are affected.

 Note: It's been stressed not to edit objects in viewports while in paper space. Once you've placed a lock on the viewport, it is safe to edit.

CREATING MULTIPLE FLOATING VIEWPORTS

The **Viewports** toolbar shown in Figure 23.14 can be used to create floating viewports. Buttons include **Display Viewports Dialog**, **Single Viewport**, **Polygonal Viewport**, **Convert Objects to Viewport**, **Clip Existing Viewport**, and **Viewport Scale**. Up to this point, you've used layouts with viewports that take up the entire drawing area. Floating viewports can be created in a layout that fills the entire drawing area or only takes up a small portion of the drawing area. Floating viewports can be created using a process that is similar to the process for creating tiled viewports. Floating viewports are created using the **Viewports** dialog box shown in Figure 23.15. Compare the **Viewports** dialog box in Figure 23.15 with the **Viewports** dialog box in Figure 23.11. Although each dialog box is very similar, they create different types of viewports, depending on how they are accessed. The dialog box shown in Figure 23.11 was accessed while in model space. A process similar to placing tiled viewports in a layout can be used to place multiple floating viewports in an existing layout. Use the following steps to create a floating viewport:

Figure 23.14 *The **Viewports** toolbar can be used to control floating viewports. Buttons include **DisplayViewports** Dialog box, **Single Viewport**, **Polygonal Viewport**, **Convert Objects to Viewport**, **Clip Existing Viewport**, and **Viewport Scale**.*

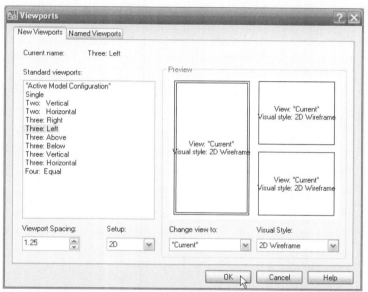

Figure 23.15 *The **Viewports** dialog box. Although this dialog box is very similar to the dialog box shown in Figure 23.11, they create different types of viewports, depending on how they are accessed.*

1. Display a layout by selecting an existing layout tab or by creating a new layout.

2. Display the **Viewports** dialog box by selecting the **Display Viewports Dialog** button on the **Viewports** toolbar, by selecting **New Viewports** from **Viewports** on the **View** menu, or by typing **VPORTS** ENTER at the Command prompt.

3. Select the desired viewport configuration from the **Standard viewports** list.

4. Enter the desired spacing to be maintained between viewports in the **Viewports Spacing** edit box.

5. Select 2D in the **Setup** edit box.

6. Click the OK button to apply the configuration and close the dialog box.

The command prompt will ask you to specify first corner and opposite corner of the new viewport configuration.

CREATING A SINGLE FLOATING VIEWPORT

Floating viewports provide an excellent means of assembling multi-scaled drawings for plotting. New floating viewports can be created without using the standard viewport configurations. Figure 23.16 shows a sheet of sections and details assembled for plotting using floating viewports. Construction projects typically comprise drawings that are drawn at a variety of scales, ranging from 3/8"=1'–0" to 1 1/2"=1' –0". This sheet of sections and details cannot be easily drawn or plotted without the use of multiple viewports and inserting or referencing each detail to a base drawing. As with drawings where a single object is attached, the use of scale factors to control text and dimension variables will be critical to the success of the final outcome of the drawing. The process for creating multiple viewports will be similar to the creation of one viewport.

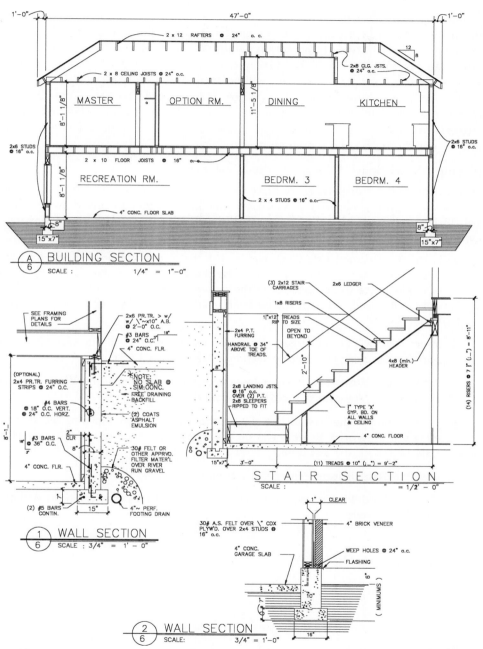

Figure 23.16 *Drawings to be attached to the base sheet. (Courtesy Piercy & Barclay Designers, Inc., A.I.B.D.)*

The following example will demonstrate the steps used to prepare the section and three section details shown in Figure 23.16 for plotting. The building section will be displayed at a scale of 1/4"=1"–0", two of the section details will be displayed at a scale of 3/4"=1"–0", and the stair section detail will be displayed at a scale of 1/2"=1"–0". For most users, drawings such as a floor plan or section will be created in a drawing template using model space. The following guidelines will walk you through the process of creating multiple viewports in a template and inserting or attaching multiple drawings. For this discussion, the **Insert** command will be used. The drawings could also be attached using **Xref** command, depending on the drawings needs.

Displaying Model Space Objects in the Viewport

The easiest method for displaying multiple drawings at multiple scales is by inserting each of the drawings into model space. With the details arranged in one drawing similar to Figure 23.16, switch to the Architectural Title Block layout. Use the following steps to display the drawings in multiple viewports:

1. In the Architectural Title Block layout, click the **Paper** button in the status bar to toggle the viewport to model space. Use the **Zoom All** option to enlarge the area of model space to be displayed in the existing viewport. Each of the four drawings will be displayed, but each will be at an unknown scale. Your drawing will resemble Figure 23.17.

2. Select what will be the largest drawing to work with first. In our example, the viewport for displaying the building section will be adjusted first.

3. Activate the **Viewports** toolbar by selecting **Toolbars** from the **View** menu and checking the **Viewports** toolbar check box.

4. Set the scale of the viewport to the appropriate value for displaying the section using the **Viewport Scale** menu on the **Viewports** toolbar. For this example a scale of 1/4"=1"–0" was used.

5. Use **Pan Realtime** to center the section in the viewport.

6. Click the **Model** button on the status bar to return to paper space in the Architectural Title Block layout.

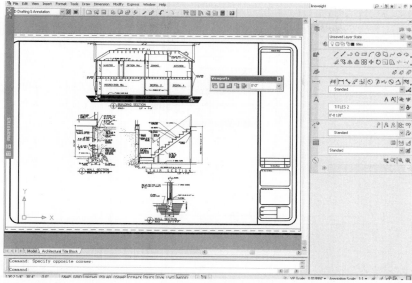

Figure 23.17 *Toggle to model space and use the **Zoom All** option to view all four drawings in your viewport, but each will be at an unknown scale.*

Adjusting the Existing Viewport

Once the viewport is created, the size of the viewport can be altered if the entire drawing can't be seen. To alter the viewport size, move the cursor to touch the viewport and activate the viewport grips. Select one of the grips to make it hot and then use the hot grip to drag the window to the desired size. The section has now been inserted in the template and is ready to be plotted at a scale of 1/4"=1"–0". The display would resemble Figure 23.18.

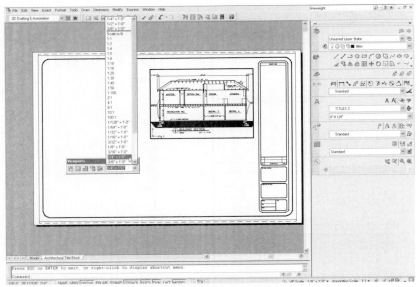

Figure 23.18 *Once the viewport is created, set the scale of the viewport to the appropriate value for displaying the drawing using the **Viewport Scale** menu on the **Viewports** toolbar. Next, alter the size of the viewport if the entire drawing can't be seen. The section is now ready to be plotted at a scale of 1/4"=1'-0".*

Note: All of the drawing commands can be used inside the viewport, but try to limit editing to changes that will affect only this plot. Changes to the drawing that will be needed on future drawings should be made to the original drawing. Remember, if the original is attached to this drawing by **Xref**, the changes made to the original drawing will be reflected in all uses of the referenced drawing.

Creating Additional Viewports

Use the following steps to prepare for additional viewports:

1. Set the current layer to Viewport. This layer is part of each template.

2. Select **Single Viewport** on the **Viewports** toolbar.

3. Select the corners for the new viewport. The coordinates for the viewport can be entered or specified by indicating the corners of the viewport—just as if it were a window used for determining a selection set.

The results of this command can be seen in Figure 23.19. Although you should try to size the viewport accurately, you can stretch it to enlarge or reduce its size once the scale has been set.

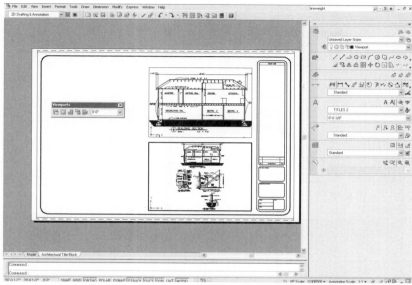

Figure 23.19 *If a new viewport is created, the display from the existing viewport will be shown in the new viewport as well.*

Altering the Second Viewport

In its current state, if a new viewport is created, the display from the existing viewport will be shown in the new viewport as well. Use the following steps to alter the display.

1. Click the **Paper** button on the status bar to toggle the viewport to model space.

2. Make the new viewport the current viewport by placing the cursor in the viewport and clicking.

3. Use the **Zoom All** option to display all of the model space contents in the second viewport.

4. Set the viewport scale to the appropriate value for displaying the section. For this example, a scale of 3/4"=1"–0" will be used to display the wall section.

5. Use the **Pan** command to center the wall section in the viewport.

6. Click the **Model** button in the status bar to toggle the drawings to paper space.

7. Select the edge of the viewport to display the viewport grips.

8. Make a grip hot and shrink the viewport so that only the wall section is shown.

9. Use the **Move** command to move the viewport to the desired position in the template.

10. Save the drawing for plotting.

The drawing should now resemble Figure 23.20

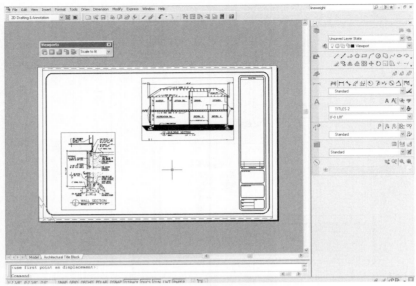

Figure 23.20 *Adjust the viewport scale to the desired value, and use the **Pan** command to move the drawings so that the desired information is centered in the viewport. Once it is centered, adjust the size of the viewport so only the desired material is displayed.*

Repeating the Process for Additional Viewports

Once the viewport for the wall section been adjusted, additional viewports can be created. The process to create the viewports to display the **BRICK1ST** and **STAIR 8FTWALL** drawings is the same process as used to create the viewport for the **WALL SECTION** drawing. If you have planned the sheet contents well, the viewports for **BRICK1ST** and STAIR can also be created. Placing all of the viewports at once will eliminate having to keep switching from paper space to model space. Once all of the desired details have been added to the layout, the viewport outlines can be frozen. Start the **Layer** command, select **VIEWPORT** as the layer to be edited, and then select the Freeze option. Freezing the viewports will eliminate the line of the viewport being produced as the drawing is reproduced. Any viewports that are still visible were created on a layer other than **VIEWPORT**. Change any existing viewports to the correct layer. The finished drawing, with the viewports frozen, will resemble Figure 23.21.

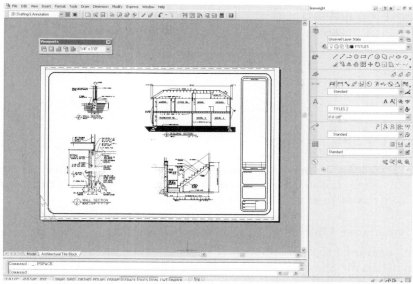

Figure 23.21 *The finished drawing, with the Viewports layer frozen, is ready for plotting.*

Displaying Annotative Objects in Viewports

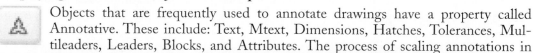

Objects that are frequently used to annotate drawings have a property called Annotative. These include: Text, Mtext, Dimensions, Hatches, Tolerances, Multileaders, Leaders, Blocks, and Attributes. The process of scaling annotations in different layout viewports and in model space is automated. Changing the workspace from the default **AutoCAD Classic** setting to **2D Drafting & Annotation** changes the workspaces so that you can work in a custom, task-oriented drawing environment. The workspace menu is shown in figure 23.22. The menus, toolbars, palettes, and dashboard control panels that are grouped and organized appropriately depending on which workspace you choose. The **2D Drafting & Annotation** workspace aids in the process of scaling annotations so they plot at the correct size on paper. This gives you the capability of creating two different viewports of the same drawing and viewing them at different scale factors in one drawing session. This would be a valuable tool if you were displaying one view of the section of a house and in a different viewport you zoomed into a specific area of the same section and displayed it at a larger scale to show important detail. Annotative scaling would allow for the text and leaders in both viewports to be plotted at the same height regardless of the different scale factors.

Figure 23.22 *Selecting 2D Drafting & Annotation will organize the workspace with the appropriate menus, toolbars, palettes, and dashboard control panels to aid in the process of scaling annotations so they plot at the correct size.*

Creating Annotative Objects

The dialog boxes used to create annotative objects contain an Annotative button; clicking this button makes the object annotative as shown in figure 23.23. When annotative objects are created, they are scaled based on the current annotation scale setting and automatically displayed at the correct size. This setting can be viewed and modified in the status bar as shown in figure 23.24. It's very important to set the annotation scale to the same scale as the viewports in which you want those objects to display prior to adding annotative objects to your drawing.

Figure 23.23 *Clicking the Annotative button in the dialog box makes the object annoatative.*

Figure 23.24 *It's very important to set the annotation scale to the same scale as the viewports in which you want those objects to display prior to adding annotative objects to your drawing.*

Convert existing objects into annotative objects

Existing objects can be turned into annotative objects by changing the annotative property on the **Properties** palette as shown in figure 23.25. When the Annotative property is turned on (set to Yes), these objects are now changed into annotative objects. You will also notice that there are three new descriptions added to the properties palette while working with annotative text. **Annotative scale**, **Paper text height**, and **model text height** have been added. These list the scale and size of text. 1/8" is the most common size used for paper size and depending on the scale factor the model size will vary.

Figure 23.25 *Existing objects can be turned into annotative objects by changing the annotative property on the Properties palette.*

Add/Delete Current Scale

   A triangle icon displays by the object when you move your mouse over the annotative object in the drawing session. Select the object by left-clicking, this will highlight the object. Right-clicking will display the short-cut menu with an option for **Annotative Object Scale**. Select this option and you will be provided three options: **Add Current Scale, Delete Current Scale, or Add/Delete Scales**. These options are also available by selecting **Annotative Object Scale** on the **Modify** menu, or by clicking the command buttons on the dashboard. Selecting **Add Current Scale** or **Delete Current Scale** will do just as it says. It will add or delete the current scale that is set. Selecting the **Add/Delete Scales** option will open the

Annotation Object Scale dialog box as shown in figure 23.26. This dialog box allows you to view and add or delete additional scales with a quick selection.

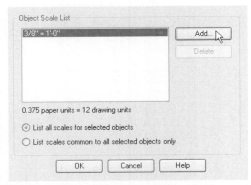

Figure 23.26   *Selecting the Add/Delete Scales option will open the Annotation Object Scale dialog box.*

Viewing and Modifying Annotative Objects in Viewports

The **Annotation Visibility** button on the status bar sets the display for annotative objects. This will be visible when you are in a layout tab rather than model space. This button will toggle between **Annotation Visibility: Show annotative objects for current scale only**, or **Show annotative objects for all scales**. This reduces the need to use multiple layers to manage the visibility of annotations for model space and layout viewports.

You can use grips to modify individual scale representations of annotative objects.

This gives you the capability to modify the placement of individual scale representations of the same object. For example, you may want the same text to display in two different places in two different viewports. Set the annotation scale for the scale representation you want to modify. Then use the grips to move the scale representation. This will only move the text in that specific scale representation.

After selecting and modifying annotative objects it is time to put this powerful command to full use. Select the **Layout tab** to view your drawing in paperspace. Use the following steps to set the annotative scale in your viewport:

- Click on **PAPER** on the status bar (this will put you in model space inside your viewport).
- Click the arrow next to the displayed annotation scale on the right side of the status bar.
- Select a scale from the list.

Saving to previous versions while preserving Annotative Objects

Annotative scaling is not available in previous releases of AutoCAD. There in an option in the **Open and Save** tab of the **Options** dialog box which allows you to keep the visual consistency of your annotative objects when saving to previous AutoCAD versions. With this option selected, each scale representation of the annotative objects in your drawing is saved as a normal (non-annotative) object on it's own layer. The appropriate layers are automatically frozen in each viewport to match the visibility of the original annotative objects in the AutoCAD 2008 drawing. If you choose not to keep the visual consistency of your annotative objects, all annotative objects will be saved based on the annotation scale of the Model tab or its last supported scale. When you save an AutoCAD 2008 drawing, Linetypes are automatically scaled by the viewports annotation scale. This option can be disabled by using the MSLTSCALE system variable.

CONTROLLING THE DISPLAY IN FLOATING VIEWPORTS

As you place drawings in viewports, the **Pan** and **Zoom** commands are used to alter what is displayed in each viewport. AutoCAD allows displayed objects to be altered by screening, turning viewports ON or OFF, and hiding lines. Hiding lines is typically associated with 3D drawings and will not be covered in this chapter. Consult the **Help** menu for additional information.

Screening Objects in Viewports

Screened objects use less ink when plotted, so they appear dimmer on screen and when plotted. Screening could be used to distinguish between objects, without plotting in color. On remodeling projects, existing objects can be screened to easily distinguish between new and existing work. To screen an object, you need to assign it a plot style and then assign the screening value to that plot style. (Plot styles will be discussed in the next chapter.) Values ranging from 0 to 100 can be assigned, with 100 being the default value. A value of 100 assigns no screening and will provide normal plotting values. A value of 0 will assign no ink to the object and have the same effect as freezing the viewport contents.

Displaying a Floating Viewport

By default, a viewport is ON as it is created. The performance of your machine can be affected if a large number of complex viewports are created. Turning viewports OFF can save regeneration time for large drawings. Toggle viewports ON or OFF using the **Properties** palette and the following steps:

1. Select the viewport to alter.
2. Open the **Properties** palette.
3. Select **On** from the **Misc** listing and then select No from the edit list.
4. Close the **Properties** palette.

As the viewport is toggled to No, the contents of the viewport will be removed from the drawing display. The line representing the viewport can be removed from the display by using the **Layer Properties Manager** dialog box to freeze the layer.

Controlling Layer Display with the Layer Properties Manager

The contents of specific layers in each viewport can be controlled using the **Layer Properties Manager**. The **Layer Properties Manager** can also be used to turn plotting ON or OFF for visible layers. A layer created for interoffice notes can be displayed on the screen, but frozen for plotting. Layers are frozen for plotting by selecting the **Plot** icon of the desired layer. Clicking the **Layer Properties Manager** button, or typing **layer** at the command line will open the **Layer Properties Manager** dialog box. Controlling layers in viewports proves to be valuable when an object will be displayed in two or more views. Specific aspects of the drawings, such as layers containing text, can be frozen in one viewport and displayed in another. Options for freezing layers in floating viewports include:

> **New Layer Frozen In All Viewports-** this option is accessed by clicking the New Layer Frozen In All Viewports button or by selecting the command from the short-cut menu. This option will create a new layer and freezes it in all existing layout viewports.

> **VP Freeze-** this option freezes the selected layers. You can freeze or thaw layers in the current viewport without affecting layer visibility in other viewports.

> **New VP Freeze-** this option freezes the selected layers in a new layout viewport without affecting existing layout viewports.

Objects in viewports can be displayed differently by also using the **Layer Properties Manager**. These options can only be accessed when the **Layer Properties Manager** is opened while in a layout tab rather than model space. Activating the viewport prior to accessing the **Layer Properties Manager** will allow you to select the different option overrides. Refer to figure 23.27 to view the four override variables.

> **VP Color-** this option will place an override for the color specified on the selected layer, such as the text layer is set to black however in the selected viewport you would like to override the specified color with a new color. This would only permit changes to the display outcome in the viewport without affecting the layers of the drawing.

> **VP Linetype-** this option will place an override for the linetype specified on the selected layer, such as the Footing layer is set to Continuous however in the selected viewport you would like to override the specified linetype with a linetype. This would only permit changes to the display outcome in the viewport without affecting the layers of the drawing.

> **VP Lineweight-** this option will place an override for the lineweight specified on the selected layer, such as the wall layer is set to a specific lineweight however in the selected viewport you would like to override the specified lineweight with a thicker lineweight. This would only permit changes to the display outcome in the viewport without affecting the layers of the drawing.

> **VP Plot Style-** this option will place an override for the plot style specified on the selected layer. This would only permit changes to the display outcome in the viewport without affecting the layers of the drawing.

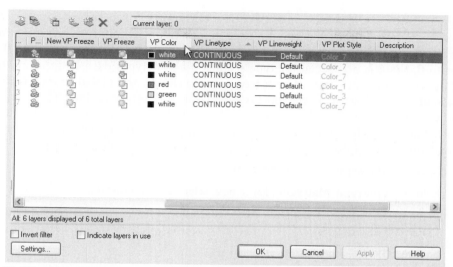

Figure 23.27 *Activating the viewport prior to accessing the Layer Properties Manager will allow you to select the different option overrides. The four override variables include VPColor, VP Lineype, VP Lineweight, and VP Plot Style.*

The viewport overrides will only change the display of object properties that are set to **BYLAYER**. When a viewport contains overrides, a **Viewport Overrides** property filter is created and the override properties are displayed with a different background color. You can select the **viewport overrides** filter to all the layers that contain overrides. The short-cut menu in the layer list allows you to remove viewport overrides and the **VPLAYEROVER-RIDESMODE** system variable allows you to temporarily ignore the overrides for viewing or plotting.

The command **Laywalk** (Layerwalk) can aid in the process of controlling the layers in both model space and paper space. The **LayerWalk** dialog box can be accessed by selecting the **Layer Walk** button on the **Layers II** toolbar or by typing **Laywalk** and pressing ENTER at the command line. The **LayerWalk** dialog box allows you the capability to isolate layer(s) and view the contents of the selected layers. It will display the objects on the layer(s) you selected, which will provide clarity on which layer(s) you would like displayed in your viewports and/or drawing. **Layerwalk** doesn't freeze any layers; it is simply a tool to use to view and separate the layers so that you know which ones you need to freeze and/or thaw while using the **Layer Properties Manager**. Explore the **Help** menu for advanced features in the **LayerWalk** dialog box.

Controlling Scaling in Viewports

Altering the size of a floating viewport does not change the scale of the drawing within the viewport. When the scale was adjusted for the viewport, you set a consistent scale for accurate display and plotting. Earlier in the text, selecting a scale for linetypes was discussed. If a hidden line is to be 1/8' long when plotted at 1/4"=1'–0", a scale factor must be applied to the line to create the desired result. The scale for linetypes can be based on the drawing units of the space where the object was created, or based on paper space units. The **Psltscale** system variable can be used to maintain a uniform linetype scaling for objects displayed at different zoom factors. The **Psltscale** variable can be set to **0** or **1**. If it is set to **0**, there is no special linetype scaling. The linetypes dash lengths are displayed according to the drawing units. The linetypes are scaled by the global Ltscale factor. With the **Psltscale** variable set to **1**, viewport scaling controls linetype scaling. The linetype dash lengths will be consistently the same regardless of the magnification of each individual viewport. The **Psltscale** variable can be set by typing **Psltscale** and pressing ENTER at the command line. The global scale factor can be set using the following steps:

1. Select **Linetype** from the **Format** menu.

2. In the **Linetype Manager** dialog box, select **Show Details**.

3. Enter the desired global scale factor to apply to the linetypes in the **Global Scale Factor** edit box.

4. Click **OK** to return to the drawing area.

CREATING VIEWPORTS WITH IRREGULAR SHAPES

AutoCAD allows viewports to be created in an irregular shape that best suits the object to be displayed. Create a nonrectangular viewport by selecting the **Polygonal Viewport** button, by clipping an existing viewport, or by converting an existing object to a viewport.

CREATING POLYGONAL VIEWPORTS

A polygonal viewport is created by using one of the following methods:

- Click the **Polygonal Viewport** button on the **Viewports** toolbar.

- Select **Polygonal Viewport** from **Viewports** on the **View** menu.

- Type **−VPORTS** ENTER and then **P** ENTER at the Command prompt.

The command is completed by selecting a series of points using a command sequence similar to the sequence for creating a polyline. The command sequence is as follows:

```
Command: (Select the Polygonal Viewport button on the Viewports
    toolbar.)
Specify start point: (Specify viewport starting point.) Specify
    next point or [Arc/Length/Undo]: (Select points to define
    polygon.)
Specify next point or [Arc/Close/Length/Undo]: (Select points
    until the polygon is defined.)
```

Figure 23.28 shows an example of an irregularly shaped viewport.

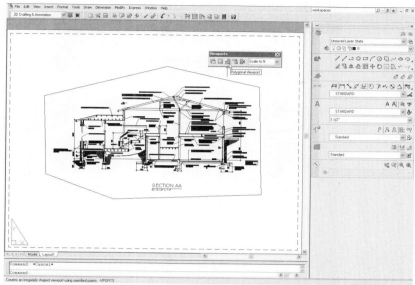

Figure 23.28 *A polygonal viewport can be created by clicking the Polygonal Viewport button on the Viewports toolbar.*

USING VPCLIP TO ADJUST THE VIEWPORT

The **Vpclip** command can be used to clip or redefine the shape of an existing viewport. Access the command by using one of the following methods:

- Click the **Clip Existing Viewport** button on the **Viewports** toolbar.

- Type **VPCLIP** ENTER at the Command prompt.

- Select the viewport to be accessed, right-click to display the shortcut menu, and then select **Viewport Clip**.

The command sequence is as follows:

```
Select the Clip Existing Viewport button (Or type VPCLIP ENTER)
Select viewport to clip: (Select desired viewport.)
Select Clipping object or [Polygonal/Delete]
Specify start point: (Specify viewport starting point.)
Specify next point or [Arc/Length/Undo]: (Select a minimum of
     three points to define polygon.)
Specify next point or [Arc/Close/Length/Undo]: (Select points
     until the polygon is defined.)
Command:
```

Objects that lie within the polygon will be displayed. Objects outside the polygon will be removed from the viewport and will not be displayed. The viewport in Figure 23.29 has been edited using **Vpclip**. Irregular viewports can be edited by selecting the edge of the viewport. With the viewport grips displayed, the viewport can be edited in the same way as any other object is altered with grips.

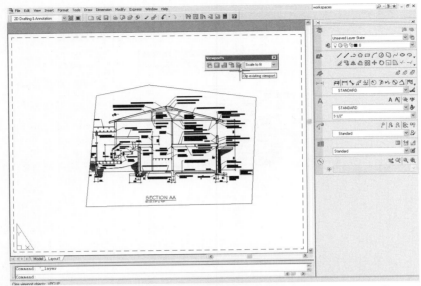

Figure 23.29 *The* **Vpclip** *command can be used to clip or redefine the shape of an existing viewport. You can also access the command by selecting the* **Clip Existing Viewport** *button on the* **Viewports** *toolbar.*

CONVERTING EXISTING OBJECTS TO A VIEWPORT

AutoCAD allows existing shapes to be converted to viewports. The object can be a circle, ellipse, region, spline, or closed polyline comprised of lines and arc segments. Access the command by clicking the **Convert Object to a Viewport** button on the **Viewports** toolbar. The command is completed using the following command sequence:

```
Click the Convert Object to a Viewport button on the Viewports toolbar
Select object to clip viewport: (Select desired object.)
```

Options include the following:

ON and OFF—These serve as toggle switches to display or hide the contents of the individual viewports. Selecting either option will provide a prompt to select objects so the desired viewport can be controlled. This option can be very useful when you plot stock sheets of details, if one or two attached details do not apply to the current use. Setting these viewports to OFF will eliminate the viewport from the plot. This option is also very useful for controlling **Regen** time. Inactive viewports can be set to OFF while the current viewport is edited. If all viewports are set to OFF, you will not be allowed to work in model space of a layout until a new viewport is created.

Fit—This option will create a viewport to fit the current screen display. The size of viewport will be determined by the dimensions of the paper space view. In the example used throughout this chapter, the objects were created first in base drawings and then drawings were inserted in a template. The Fit option will allow a viewport to be created around an existing drawing or a portion of a drawing.

Hideplot—This option will allow hidden lines to be removed from selected viewports when plotting in paper space. This option generally is not required in architectural and engineering drawings.

Lock—Locks the current viewport in a similar manner to locking a layer.

Object—Selects a circle, closed polyline, ellipse, region, or spline to be converted to a viewport.

Polygonal—Creates an irregularly shaped viewport using the same process that is used when the **Polygonal Viewport** button is selected.

Restore—This option can be used to restore saved viewport layouts. The prompt is:

> **Enter viewport configuration name or [?] <*Active>** The name of a saved view can be entered. Typing the question symbol (?) will provide a list of view names.

2, 3, 4—Selecting one of these options will create two, three, or four viewports without your having to reissue the command. The required number of viewports is entered, followed by specifications for how to arrange the viewport. Each option is similar to the arrangements used for tiled viewports.

CHAPTER 23 EXERCISES

1. Open the Architectural template and create a new layout titled COOL STUFF. Use the **Viewports** dialog box and create viewports using the Three: Above layout. Insert a drawing in model space. Adjust each viewport so that a different aspect of the drawing is centered in each viewport. Save the drawing as **E-23-1.**

2. Open the Architectural template and use the **Create Layout Wizard** to create a new layout titled VIEWPORT TEST. Set the printer as the default system printer, and if necessary select an appropriate paper size. Set the orientation to portrait. Assign a single viewport suitable for displaying a floor plan at 1/4"=1'-0". Select the corners of the viewport so that it will only fill half of the page. Save the drawing as **E-23-2.**

3. Open any three drawings and insert them in a template drawing. Create a floating viewport to display the drawings at a suitable scale. Save the drawing as **E-23-3**

4. Open the Architectural drawing template. Attach any four drawings from your library. Each drawing is to be shown at a different scale when plotted. Provide a title and the scale for each detail. Save the drawing as **E-23-4.**

5. Open drawing **E-23-4** and clip two of the drawings so that they are not displayed. Save the drawing as **E-23-5.**

CHAPTER 23 QUIZ

DIRECTIONS

Answer the following questions with short complete statements. Type your answers using a word processor.

1. Place your name, chapter number, and the date at the top of the sheet.

2. Type the question number and provide the answer.

 Caution: Some of the questions may not have been covered in the reading material and will require the use of the help menu. You may also have to do some exploring to answer the questions.

1. Describe how a referenced drawing can be used in an office to complete a project.

_____

_____

2. Use the **Help** menu and determine how and what the **Tilemode** setting affects.

_____

_____

3. What layer is added to the drawing base when you work with viewports? How will it affect the display?

_____

_____

4. What command or process is used to create a viewport?

_____

_____

5. What are the benefits of using layouts?

_____

_____

6. How can the display in the second viewport be altered so that it does not display the same drawing as another viewport?

_____

_____

7. How does paper space differ from model space?

_____

_____

8. A drawing has been inserted in a template in model space but it can't be seen. Describe how to show it.

_____

_____

9. List the steps to create a layout using the **Create Layout Wizard**.

_____

_____

10. What is an Annotative Object?

_____

_____

11. Explain the difference between floating and tiled viewports.

_____

_____

12. How is a scale applied to the display in a viewport?

_____

_____

13. What are the benefits of creating Annoative objects?

_____

_____

14. How do you change existing text so that it is annotative object?

_____

_____

15. List three options for controlling the display of the viewport.

_____

_____

16. A drawing is displayed at the proper scale in a viewport, but the entire drawing can't be seen. Describe how the problem can be fixed.

_____

_____

17. Three drawings are displayed in a viewport, but you would like to hide one from a meeting with a client. Describe how to hide the drawing.

_____

_____

18. Describe the process to create an irregularly shaped viewport.

_____

_____

19. List two ways to switch from paper space to model space.

_____

_____

20. Describe the process to view annotative objects in a viewport.

_____

_____

Working with Sheet Sets

INTRODUCTION

This chapter will introduce:

- Benefits of sheet sets
- Features of the Sheet Set Manager
- Methods for creating, modifying, and organizing sheet sets
- Process of creating a sheet index
- Methods for publishing, plotting, and creating transmittal sets

Commands to be introduced include:

- **Sheetset**
- **Newsheetset**
- **Opensheetset**
- **Sheetsethide**
- **Archive**

BENEFITS OF SHEET SETS

A sheet set is a series of related drawings. For example, you work for an engineering firm that provides systems for HVAC and plumbing. You would have a sheet set for the project and within the sheet set there would be subsets that pertain to the Architectural, Mechanical, Plumbing, and Electrical details of that project. Prior to using sheet sets, you would organize and manage all your files in folders and assign appropriate file names. The Sheet Set Manager enables you to organize and navigate through your drawing sheets within the AutoCAD program. This allows you to create file subsets that represent different disciplines of drawings in the same project. In the example above, the Architectural, Mechanical, Plumbing and Electrical drawing sets would be subsets of the project. The Sheet Set Manager provides the user the capability to efficiently create, manage, and share sheet sets. The Sheet Set Manager displays all your drawing sheets and subsets in a tree structure; this displays which drawings are available and which drawings are currently accessed. Utilizing the Sheet Set Manager allows you, within a single interface, to

- Create, view, organize, delete, and manage your sheets.
- Create layout views automatically.
- Automate sheet number and detail labeling.
- Link sheet set information into title blocks and plot stamps.
- Create a sheet index.
- Share sheet sets with a project team using plots, eTransmit, and publish to DWF files.

SHEET SET MANAGER

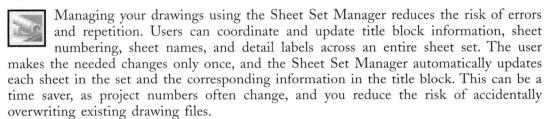

Managing your drawings using the Sheet Set Manager reduces the risk of errors and repetition. Users can coordinate and update title block information, sheet numbering, sheet names, and detail labels across an entire sheet set. The user makes the needed changes only once, and the Sheet Set Manager automatically updates each sheet in the set and the corresponding information in the title block. This can be a time saver, as project numbers often change, and you reduce the risk of accidentally overwriting existing drawing files.

Creating a Sheet Set Using a Wizard

AutoCAD features a wizard that will take you through the process of creating a new sheet set. You can access the **Create Sheet Set** wizard by using one of the following methods:

- Type **NEWSHEETSET** at the Command prompt.
- Select **New Sheet Set** from the pull-down menu in the **Sheet Set Manager**.
- Select **New Sheet Set** from the **File** menu on the **Standard** toolbar.

Figure 24.1 shows an example of the **Create Sheet Set** wizard. The wizard offers you the flexibility to create a new sheet set by using an example sheet set as a template or by importing layouts from existing drawings.

Figure 24.1 *Selecting New Sheet Set from the File menu or typing **newsheetset** ENTER opens the **Create Sheet Set** wizard.*

Creating a Sheet Set Using an Example Sheet Set (Template)   Creating a sheet set by using an example sheet set provides the new sheet set with the organizational structure and default settings from the template chosen. This is a valuable tool because CAD uniformity is essential in producing quality drawings. As shown in Figure 24.2, this option provides you with several example sheet sets to choose from, or you can browse and choose a template that you created. Choosing the Architectural Imperial Sheet Set and selecting the **Next>** button continues the process. Entering the sheet set details is the next step of the wizard. There are edit boxes that prompt for the name of the sheet set being created and a description (optional), and you can specify where the sheet set data file is stored. Keep in mind that the data file should be stored in a place where all members of the project team can access it. Figure 24.3 shows the **Sheet Set Details** page in the **Create Sheet Set** wizard. The final step in the wizard launches a preview of the new sheet set created. Selecting the **Finish** button closes the wizard, and the new sheet set is displayed in the Sheet Set Manager.

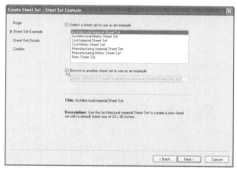

Figure 24.2   *The **Sheet Set Example** page provides default settings and organizational structure that will save time and create uniform drawings.*

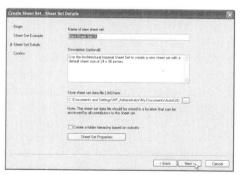

Figure 24.3   *The **Sheet Set Details** page prompts for the name of the sheet set being created and a description (optional), and you can specify where the sheet set data file is stored.*

Creating a Sheet Set Using Existing Drawings   Creating a sheet set by using existing drawings allows you to automatically import the layouts from existing files into the new sheet set. Selecting the **Next>** button brings you to the **Sheet Set Details** page, which prompts for a sheet set name and description (optional), and allows you to specify the location for the sheet set data file. The data file should be stored in a place where all members of the project team can access them. Selecting the **Next>** button accesses the **Choose Layouts** page, which allows you to browse and select the desired folder as shown in Figure 24.4. You can select one or more folders when importing files. Selecting the **Next>** button displays a preview of the new sheet set created. Selecting the **Finish** button closes the wizard, and the new sheet set is displayed in the Sheet Set Manager.

Figure 24.4 *The **Choose Layouts** page in the **Create Sheet Set** wizard allows you to browse and select the desired folder(s).*

Viewing and Modifying a Sheet Set

Now that you have created a sheet set, you can view and modify your sheet set using the Sheet Set Manager. Open the Sheet Set Manager by typing **SSM** at the command line, or by choosing **Open Sheet Set** from the **File** menu. The Sheet Set Manager has three tabs on the right side: **Sheet List**, **Sheet Views**, and **Model Views**. In this section we will be using the **Sheet List** tab to perform the different tasks. Multiple users can retrieve a sheet set at the same time. If someone attempts to change the sheet set, such as when adding a new sheet, the sheet set automatically locks to prevent anyone else from editing it. When a sheet set is locked, an icon in the upper left corner of the Sheet Set Manager specifies the status and who locked it. Even if the sheet set is locked, another user can view the contents of the sheet set, publish sheets, create transmittal sets, and even edit drawings that are referenced by the sheet set. After the changes are completed to the sheet set, AutoCAD automatically unlocks the sheet set, making it available for others to edit.

Opening a Sheet Set   In the Sheet Set Manager, a drop-down list enables you to open sheet sets. Clicking the arrow and selecting **Open** will access the **Open Sheet Set** dialog box. Browse through the appropriate folders, select your file, and click the **Open** button. This closes the dialog box and displays your selected sheet set in the Sheet Set Manager as shown in Figure 24.5.

Figure 24.5 *The Sheet Set Manager allows you to create, view, organize, delete, and manage your sheets.*

Viewing a Sheet In the Sheet Set Manager, under the Sheets heading, double-click the Architectural subset as shown in Figure 24.6. This will display the sheets included in the sheet set. Highlighting a sheet will display information in the Details box in the bottom section of the Sheet Set Manager. The details will include information on the status (accessible or locked), file name, location, file size, last saved and last edited by. In the Details section, click the **Preview** button to view a thumbnail preview of the selected sheet. Double-click the A1 - Office Plan sheet. The Office Plan sheet opens, with the Office Plan layout tab currently active. The sheet was created using a template. The template contains information shared by all sheets, such as sheet size and title block. In addition to viewing the details of a sheet, you can also view the properties of a sheet set. Right-clicking the sheet set and choosing **Properties** can achieve this.

Figure 24.6 *Your sheet set can contain many subsets to organize your project; your drawing files will be organized in the appropriate subsets.*

Adding/Deleting a Sheet You can easily add/delete sheets in a sheet set. Use the following steps to add a sheet:

1. Highlight the name of the sheet set or the subset that you would like the new sheet to reside in.

2. Right-click and select **New Sheet** from the shortcut menu.

3. Enter in sheet number, title, and file name and select **OK** as shown in Figure 24.7.

The Sheet Set Manager also features a way to create a new sheet while using an existing drawing. This gives you the benefit of incorporating existing drawings into a new project. You can add existing drawings and layouts to the active sheet set by using one of the following methods:

- Follow the steps above and select **Import Layout as Sheet** from the shortcut menu in place of **New Sheet** in step 2.

- Right-click on the Layout tab and select **Import Layout as Sheet** from the short-cut menu.

- Drag & drop the layout tab onto the active sheet set.

Figure 24.7   *The New Sheet dialog box provides a quick and easy way to add a new sheet to your project.*

Use the following steps to delete a sheet:

1. Highlight the name of the sheet that you would like to delete.

2. Right-click and select **Remove Sheet** from the shortcut menu.

3. Click **OK** in the **Remove Sheets** dialog box.

Creating/Deleting a Subset   In the Sheet Set Manager, you can organize your sheets and place them into subsets and nested subsets. The steps for creating and deleting subsets resemble the steps used in creating and deleting sheets. Keep in mind that when adding subsets, you can add nested subsets within subsets. This can be helpful if you are creating different phases in a project or revisions. If you are choosing to create a nested subset, highlight the subset you would like the new subset to be nested in prior to accessing the shortcut menu, as shown in Figure 24.8. If you are choosing to create a new subset within the sheet set, simply highlight the sheet set name at the top of the tree structure and right-click.

Figure 24.8 *Highlight the subset you would like the new subset to be nested in prior to accessing the New Subset command.*

Sheet Views

After you create your sheets, the Sheet Set Manager can be used to place views of a drawing on a sheet. The Sheet Set Manager automates this procedure, because you no longer need to manually attach an xref, create a viewport, zoom, and scale the viewport. The Sheet Set Manager has three tabs on the right side: **Sheet List**, **Sheet Views**, and **Model Views**. The **Sheet List** tab displays all the sheets and subsets in your project, and the **Sheet Views** tab shows a list of all the views you have added to your sheets, along with the category it was assigned. The **Model Views** tab is a compilation of the drawings you will use to import the views from. Prior to adding views, you will need to add resource drawings to your project. For example, if you have a sheet of details and would like to include only the foundation footing detail, you would add the drawing file in your **Model Views** tab. You would then create a named view to isolate the one detail from the others. We will utilize all three tabs as we work with views. Use the following steps to add views to your sheet:

Adding Resource Drawings

1. Prior to adding views to your sheets, you need to first open the sheet set project you want to work on.

2. After your sheet set is open, select the **Model Views** tab.

3. Double-click **Add New Location**, select the appropriate folder, and click **Open**. You can access any number of resource drawings by adding their folder locations to the **Model Views** tab. Remember, adding files to the **Model Views** tab does not automatically turn them into views; it's a database of drawings you will use to create views.

Placing a View on a Sheet

1. On the **Model Views** tab, under Locations, highlight the desired file. Under **Details**, click the **Preview** button to view a thumbnail image of the file as shown in Figure 24.9.

2. Right-click to display the shortcut menu and select **Place on Sheet**.

3. To modify scale, right-click anywhere on screen, and select the desired scale.

4. Click on the screen to place the view in the desired location.

Figure 24.9 *Highlight the file and click the **Preview** button to view a thumbnail image of the file.*

If you want to edit the file before placing the view on your sheet, you can use the shortcut menu to open the resource drawing. When you add a view using a resource drawing, the new view automatically attaches the resource drawing as an xref. You can modify the size, scale, and location of the new view using standard AutoCAD commands. Notice, as you placed a view on your sheet, that a label block was automatically inserted with the view. The next section will walk through the process on renaming the view title.

Renaming/Renumbering a Sheet View After a view is placed on a sheet, the view is automatically listed as a paper space view on the **Sheet Views** tab. Use the following steps to rename a sheet view:

1. Click the **Sheet Views** tab in the Sheet Set Manager.

2. Right-click the view you want to rename.

3. Select **Rename & Renumber** from the shortcut menu.

4. Enter new name/number in the **Rename & Renumber View** dialog box, and select **OK**.

5. Select **Regen** from the **View** menu; the label block is updated reflecting the new changes.

Creating/Placing a Named View on a Sheet **Create a Named View** provides flexibility when you attach views to your sheet. This option works well when you want to place a specific view on your sheet instead of a view zoomed to extents. Named views are stored in the **Model Views** tab until the view is actually placed on a sheet, and then you can view it in the **Sheet Views** tab. Use the following steps to create and place a named view on a sheet:

1. Using the Sheet Set Manager, select the **Model Views** tab.

2. Right-click the desired file and select **Open** from the shortcut menu.

3. Zoom in on the area you want to include in your view.

4. Select **Named Views** from the **View** menu.

5. Click **New** in the **View** dialog box.

6. Type in the view name in the **New View** dialog box.

7. Type in the view category (this is the folder in which the named views will be listed in under the **Model Views** tab in the Sheet Set Manager) and click **OK** to close the **New View** dialog box.

8. Click **OK** to close the **View** dialog box.

9. Save and close the drawing you used to capture the named view from. Named views are displayed on the **Model Views** tab; they can be viewed by clicking the + sign next to the drawing that the named view originates from.

10. Right-click the desired view and select **Place on Sheet** from the shortcut menu.

11. To modify the scale, right-click anywhere on screen, and select the desired scale as shown in Figure 24.10.

12. Click on the screen to place the view in the desired location.

13. Follow the steps in "Renaming/Renumbering a Sheet View" to rename the label block.

You can place named views on a layout based on a model space view created with the **View** command. Views will retain Layer status. If a layer is frozen when the view is created, the layer is set to be frozen when the view is used. Along with a stored Layer status, a view also can have a thumbnail associated with it. The **View** dialog box features an **Edit Boundary** button that allows you to easily define what part of the drawing is displayed with the view.

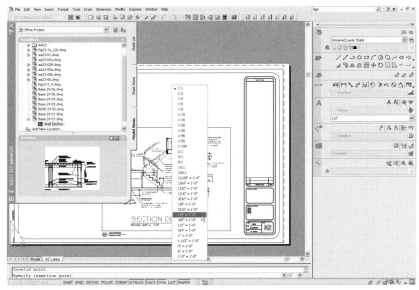

Figure 24.10 *Once you select **Place on Sheet** from the shortcut menu, the scale can be modified by right-clicking on the screen and selecting the desired scale.*

Organizing Sheet Views into View Categories Sheet views are named paper space views that are recorded and tracked in the Sheet Set Manager. Using the **Sheet Views** tab, you can organize, view, and add a view label or callout block. The **Sheet Views** tab displays all the views used in the sheet set. The views can be organized into view categories and assigned different callout blocks to each view category. For example, you may want different callout blocks for different types of information. For section views you want section bubbles, and for elevations and details you would want a different callout block. The view categories are created using the same steps used to create subsets. However, the view categories are organized and created in the **Sheet Views** tab rather than the **Sheet List** tab of the Sheet Set Manager.

Managing Sheet Views The Sheet Set Manager automatically manages sheet views. The valuable tool references between sheets and views, and updates each detailed drawing, reference tag, label, and callout blocks. We have already learned how to rename and renumber sheets and labels; we will now focus on callout blocks. A callout block is a symbol that indicates a reference to another view such as an elevation, detail, or section from another sheet. Use the following steps to place a callout block on a sheet:

1. Click the **Sheet Views** tab in the **Sheet Set Manager**.

2. Right-click the view where you would like to place a callout block.

3. Select **Place Callout Block** from the shortcut menu.

4. Choose desired symbol.

5. Click in the drawing area to place the callout block.

6. Zoom in to view the text. The text is cross-referenced to the other sheets in the project. If the other numbers change, the **Sheet Set Manager** updates the callout blocks to match the new changes.

Creating a Sheet List Table (Index)

A sheet list table (index) can be inserted on the title sheet, which automatically lists all the sheets in the set. The sheet list table is managed through the **Sheet Set Manager**, which ensures that it contains the most current data. Given that the sheet list table is generated from the sheet list, if you make any manual edits they will be lost whenever you update the sheet list table. Use the following steps to create a sheet list table:

1. Open your sheet set project and double-click the Title sheet.

2. Using the **Sheet List** tab, right-click the name of your sheet set (this will be at the top of the tree structure).

3. Select **Insert Sheet List Table** from the shortcut menu.

4. Make any adjustments to the settings and click **OK**.

A Sheet List Table alert box appears to remind you that the table is generated from the sheet list. Any manual changes will only be temporary because the table will be updated when you select **Update Sheet List Table** from the shortcut menu. When you make changes in the Sheet Set Manager and would like your Table to reflect new changes, click the outside edge of the sheet list table and right-click. On the shortcut menu, select **Update Sheet List Table**. The Sheet List Table also provides the capability to open a sheet from the table. You can quickly open any sheet listed in the sheet list table by simply holding down CTRL and clicking the sheet name. The sheet selected opens in a new window.

Subheaders for Subsets   If your sheet set includes subsets to organize your project, you may want to include these as subheaders in the index. In the **Insert Sheet List Table** dialog box, there is a box to check if you would like subheaders added to the index. This is located in the bottom left side of the dialog box. You can view the changes made in the preview box.

Modifying/Adding Columns   Columns can easily be added and modified in your table. Click the **Add** button in the **Insert Sheet List Table** dialog box. A new column will be added with the default data as Sheet Number. Double-click the default data, and from the drop-down list select the appropriate choice. In the text box next to the drop-down list, type in the Heading text that you want to appear in the new column. Existing columns can also be modified. The procedure is also done by double-clicking the data and changing the heading text or data type. Utilize the **Add**, **Remove**, **Move Up**, and **Move Down** buttons to change the order of columns.

Table Style Settings   The table settings can be adjusted, and this controls how information in the table is displayed. Access the **Table Style** dialog box by clicking the [...] button to the right of the **Table Style name** in the **Insert Sheet List Table** dialog box as shown in Figure 24.11. In the **Table Style** dialog box, click the **Modify** button, and this will open the **Modify Table Style** dialog box shown in Figure 24.12. Some of the settings that can be adjusted include, but are not limited to, text style, height, color, alignment, border properties, and margins.

Figure 24.11 *There are many settings that can be tailored to customize your table; clicking the [...] button accesses the **Table Style** dialog box.*

Figure 24.12 *The **Modify Table Style** dialog box contains many settings that can be adjusted to customize your table.*

Archiving Sheet Sets

You can archive an entire sheet set. Access the **Archive a Sheet Set** dialog box from the shortcut menu in the Sheet Set Manager. The archive function allows you to retrieve the saved project data at vital milestones throughout the process of the project in addition to the completed project. Archiving your project data instead of creating a copy in another folder reduces the risk of costly mistakes caused by storing multiple drawings with the same file name, which may overwrite one another. Archiving also allows you to continue productivity as you await approval. Let's say you submit a set of drawings and details to your client. Upon waiting for approval, you can create an archive of the submitted sheet set prior to continuing your design. This allows you to continue productivity with the ability to refer back to the original submitted set. Use the following steps to archive a sheet set:

1. Right-click the name of your sheet set in the **Sheet Set Manager**.

2. Select **Archive** from the shortcut menu.

3. The **Archive a Sheet Set** dialog box appears. By default, all files, tables, templates, and plot configurations are included as part of the archive. Click the boxes with the check mark to deselect any files you would want to exclude in the archive. Notice

that there are three tabs to scroll through: **Sheets, Files Tree** and **Files Table**, as shown in Figure 24.13.

4. Click **Add a File** to add any additional files not listed in the sheet set.

5. Click **Modify Archive Setup** to adjust additional settings. These include package type, file format, archive file folder, archive file name, and additional archive options.

6. Click **OK** to continue and close the dialog boxes.

Figure 24.13 *Access the **Archive a Sheet Set** dialog box by right-clicking on the sheet set and selecting **Archive** from the shortcut menu.*

Publishing an Electronic Sheet Set (Dwf)

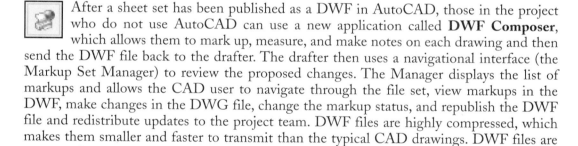 After a sheet set has been published as a DWF in AutoCAD, those in the project who do not use AutoCAD can use a new application called **DWF Composer**, which allows them to mark up, measure, and make notes on each drawing and then send the DWF file back to the drafter. The drafter then uses a navigational interface (the Markup Set Manager) to review the proposed changes. The Manager displays the list of markups and allows the CAD user to navigate through the file set, view markups in the DWF, make changes in the DWG file, change the markup status, and republish the DWF file and redistribute updates to the project team. DWF files are highly compressed, which makes them smaller and faster to transmit than the typical CAD drawings. DWF files are not a replacement for DWG files. Data cannot be edited within the DWF file. Use the following steps to publish to DWF:

1. Open the appropriate sheet set in the **Sheet Set Manager** and double-click on a sheet name that will be included in the published set.

2. Highlight the sheet set name at the top of the tree structure to select entire sheet set or select the desired files you wish to publish to DWF by holding down CTRL and highlighting only the specific files.

3. Right-click and choose **Publish to DWF** from the shortcut menu as shown in Figure 24.14 or click the **Publish to DWF** icon.

4. Type in file name in the **Select DWF File** dialog box and click **Select.**

5. The DWF publish job will take several seconds; a balloon message in the status bar tray will appear. Click the link to view plot and publish details and Select **Close** to close the **Plot and Publish Details** dialog box.

6. Right-click the **Plot and Publish** icon in the status bar tray and select **View DWF File** from the shortcut menu.

 Note: The file will not open if you do not have an external viewer to view Design Web Format (DWF) files. You can download an Autodesk DWF viewer from the web at http://www.autodesk.com/viewers.

Figure 24.14 *Right-click and choose **Publish to DWF** from the shortcut menu.*

Including Layer Information By default, layer information is not included. When the layer information is included, it allows the DWF viewer control of the layer visibility. Use the following steps to change the layer information status:

1. Follow Steps 1 and 2 from above, and then right-click and choose **Sheet Set Publish Options** from the shortcut menu.

2. In the **Sheet Set Publish Options** dialog box under DWF Data click **Layer information**.

3. Select **Include** from the drop-down menu.

Adding a Password A password-protected DWF file can provide confidentiality when needed. You can control who can open and view your drawing set. DWF passwords are case sensitive and can be made up of letters, numbers, and punctuation. If you lose or forget the password, it cannot be recovered. The password-protected option is also accessed in the **Sheet Set Publish Options** dialog box:

1. Click **Password Protect Published DWF** in the **Sheet Set Publish Options** dialog box.

2. Click **Specify Password**, enter desired password, and click **OK**.

3. Reenter your password in the **Confirm DWF Password** dialog box, and click **OK**.

4. Now that changes have been made in the **Sheet Set Publish Options** dialog box follow the steps above to publish to DWF with the modified settings.

Printing a Sheet Set

The Sheet Set Manager automates the process of plotting as an entire sheet set, or selected sheets can be published to the plotter in a single batch. You have the option to publish your sheet set in reverse order. This option can be chosen by selecting **Publish in Reverse Order** on the short-cut menu. Select this option before selecting **Publish to Plotter** to accomplish reverse order publishing. **PUBLISHCOLLATE** is a system variable which allows you to stop identical prints from plotting. This allows plots from other drawings or sheet sets to enter the plot spool. The core difference is whether the sheet set can be interrupted by other plot jobs. Setting the system variable to **1** allows the sheet set to be published as a single job without any interruptions. Setting the system variable to **0** allows other drawings or sheets sets the capability to enter the plot spool.

Use the following steps to publish a sheet set to a plotter:

1. Open the sheet set in the Sheet Set Manager and double-click on a sheet name that will be included in the published set.

2. Highlight the sheet set name at the top of the tree structure to select entire sheet set or select the desired files you wish to publish to plotter by holding down CTRL and highlighting only the specific files.

3. Right-click and choose **Publish to Plotter** from the shortcut menu or click the **Publish** icon as shown in Figure 24.15 and select **Publish to Plotter**.

4. The plot job will take several seconds; a balloon message in the status bar tray will appear. Click the link to view plot and publish details and Select **Close** to close the **Plot and Publish Details** dialog box.

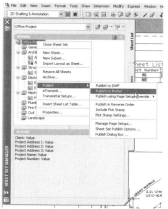

Figure 24.15 *Click the **Publish** icon and select **Publish to Plotter**.*

Creating Transmittal Sets

The eTransmit feature in AutoCAD enables you to assemble and send a transmittal package to clients or consultants. Using the Sheet Set Manager, with just a few clicks an entire project can be sent. When you create a transmittal package, referenced files are automatically included. These include but are not limited to sheet set data files, xrefs, plot configuration files, and font files. A transmittal package can also include Word documents, Excel spreadsheets, PDFs, etc. Use the following steps to create a transmittal package:

1. Open the sheet set in the Sheet Set Manager.

2. Highlight the Sheet(s) or subset(s) you would like to include.

3. Right-click and select **eTransmit**.

4. This opens the **Create Transmittal** dialog box as shown in Figure 24.16. All the files included in the transmittal are organized into three tabs. The **Sheets** tab shows the sheets included, the **Files Tree** tab lists all the xrefs, font files, plot configurations, template files, etc., and the **Files Table** tab lists all the files included, in a table format instead of a tree structure.

5. To add additional files (such as a Word document, Excel spreadsheet, or PDF, for example), in the **Files Tree** tab click the **Add File** button. Choose file(s) and click **Open**.

6. Click the **OK** button to create the transmittal package and select **Save** to complete the process. The package is saved as a ZIP file and is ready to be sent via email.

Figure 24.16 *The* **Create Transmittal** *dialog box allows you to create a transmittal package in just a few clicks.*

There are several modifications that can be made when creating a transmittal package. In the **Create Transmittal** dialog box, selecting **Transmittal Setups** allows you to define how the transmittal is packaged. Selecting **Modify** opens the **Modify Transmittal Setup** dialog box. In the **Modify Transmittal Setup** dialog box as shown in Figure 24.17, you can define the package type (ZIP, EXE file, or folder), transmittal file name and location, and transmittal (email) options. The **Create Transmittal** dialog box also features a **View Report** selection. This feature creates a report including information regarding all the files included in the transmittal package.

Figure 24.17 *The **Modify Transmittal Setup** dialog box allows you to define how the transmittal is packaged.*

CHAPTER 24 QUIZ

DIRECTIONS

Answer the following questions with short complete statements.

Type your answers using a word processor.

    1. Place your name, chapter number, and the date at the top of the sheet.

    2. Place the question number and provide the answer.

 Caution: Some of the questions may not have been covered in the reading material and will require the use of the help menu. You may have also to do some exploring to answer the questions.

1. List three different tasks that you can perform using the Sheet Set Manager.

_____

_____

2. Describe the relationship between a sheet set and subset.

_____

_____

3. What needs to be done prior to adding a view on a sheet?

_____

_____

4. Explain the process to create a sheet list table.

_____

_____

5. How do you open a sheet from the sheet list table (index)?

_____

_____

6. List five things that can be changed in your table format.

_____

_____

7. What would be a benefit of archiving a sheet set?

_____

_____

8. Why would you publish an electronic sheet set (DWF)?

_____

_____

9. How do you plot a sheet set in a single batch?

_____

_____

10. When you create a transmittal package, name three different types of files that can be included.

_____

_____

CHAPTER 25

Controlling Output

INTRODUCTION

Chapter 24 walked you through the process of taking a drawing created in model space and placing it in a layout for plotting. As computers become more common in contractor and subcontractor offices, more construction projects will be transmitted between offices electronically. AutoCAD allows an electronic plot to be created that is accessible over the Internet. This allows the designers for the subcontractors to have access to areas of the project that require contributions from them. In addition to providing access to the drawings, electronic communication can save many hours by eliminating the need to produce paper copies. This chapter will introduce methods for:

- Controlling plotting
- Controlling the plotting device
- Producing a plot
- Using a Plot Stamp
- Assigning plot styles
- Configuring the drawing to be plotted

The commands to be introduced in this chapter include:

- **Plot**
- **Stylesmanager**
- **Pagesetup**

AN OVERVIEW OF PLOTTING

The commands that control plot devices and settings can be accessed from the **Plot** dialog box. The dialog box contains settings for controlling the plotter, paper size, and other basic plotting parameters. The table can be expanded to reveal additional plotting features. To access the **Plot** dialog box, use one of the following methods:

- Click the **Plot** button on the **Standard** toolbar.

- Type **PLOT** ENTER at the Command prompt.
- Select **Plot** from the **File** menu.

Each method will produce a **Plot** dialog box similar to Figure 25.1. The **Plot** dialog box can be used to control either a printer or a plotter.

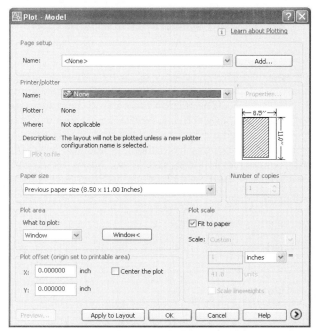

Figure 25.1   *The commands to control plotting can be accessed from the **Plot** dialog box. The dialog box is used to select plotters, set plotting parameters, and plot style tables. Access the **Plot** dialog box by clicking the **Plot** button on the **Standard** toolbar, by typing **PLOT** ENTER at the Command prompt, or by selecting **Plot** from the **File** menu. The dialog box can be expanded or collapsed using the expand/collapse button in the lower right corner.*

 Note: It's important to keep sight of the goal as you begin this chapter. Your goal is to produce a plot at a useable scale. Since chapter 3, you've been making scaled plots using the **Fit to Paper** option. In the industry, drawings are plotted at a scale such as 1/4"=1'–0".

REVIEWING THE PLOTTING PROCESS

A scaled plot can be created using the **Plot** dialog box with the same procedure that was introduced in chapter 3. As the dialog box is displayed, you'll be prompted to specify the parameters that will define the plot. Once the **Plot** dialog box is displayed, the plotting process will include the following steps:

1. If a named page setup has been previously defined and saved, it can be selected from the **Page setup name** list.

2. Verify that the device listed in the **Printer/plotter** edit box is current.

3. Verify the settings for **Paper size**, **Plot area**, **Drawing orientation**, and **Plot scale**, at the desired settings.

4. Select the **preview** option.

5. Press ENTER. If the preview does not provide the expected results, redefine the plotting parameters as needed and repeat the preview process.

6. When the plot preview meets your expectation, select the **OK** button to produce the plot.

EXPLORING THE PLOT DIALOG BOX

The **Plot** dialog box consists of the base display and an expandable portion used for advanced controls. Major features of the box include areas that describe **Page setup name**, **Printer /Plotter** controls, **paper size**, **plot area**, **plot scale** and **plot offset**. The basic **Plot** dialog box is shown in Figure 25.1. The expanded **Plot** dialog box can be seen in Figure 25.2

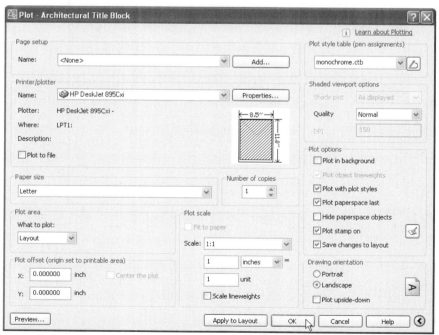

Figure 25.2 *Selecting the **Expand** button in the lower right corner expands the **Plot** dialog box. Once expanded, select the Collapse button to restore the box to its original display.*

Page Setup

The **Page Setup Manager** dialog box was first introduced when layout tabs were first examined. It should be used to define plotting parameters. The dialog box can be seen in Figure 25.3. The **Page Setup Manager** can be used to apply a named page setup to a layout. The **Page Setup Manager** is accessed by selecting **Page Setup Manager** from the **File** menu or by typing **Pagesetup** at the command prompt. The setup can specify one of many plotters that might be connected to a network, and other setting that will speed the plotting process. A page setup can be imported as a named page setup from another drawing and applied to layouts in the current drawing. When the drawing is complete, it will be plotted based on the parameters specified during the page setup. As you gain experience, plotting parameters can be specified and saved within a layout. Plotting parameters assigned in the **Plot** dialog box will override those set in the **Page Setup** dialog box. Options in the **Page setup** manager include **Set Current**, **New**, **Modify**, and **Import**.

Figure 25.3 *The **Page Setup Manager** can be used to save existing, or create new plot settings to control plotting of a specific layout.*

Page Setup Name This portion of the **Plot** dialog box displays a listing of named and saved page setups with a default listing of **None**. Attempting to plot a drawing displayed in the Architectural title block would include the following options: Previous Plot, import and Architectural English-no output device.

Add Selecting the **Add** button will display the **Add Page Setup** dialog box, which is used to name the plot setup that is to be saved to the layout. In addition to naming the current setup, this option can be used to rename, delete, or import user-defined setups and apply them to the current drawing.

Selecting a Printer/Plotter

This portion of the dialog box displays the configured plotter, the current paper size, the printable area, and toggles from inches to metric. The plotting devices that are configured for your workstation or network can be used by AutoCAD to plot drawings. As check prints have been made from model space, AutoCAD has used the plotter specified in the **Options** dialog box to determine the default plotter. As you plot using a layout, the plotting device specified by **Printer/plotter** is used. By default, **None** is listed as the selected plotter, so a plotter will need to be selected before proceeding. Options for controlling the plotter in this are of the dialog box include **Name, Properties, Plotter, Where,** and **Description**.

Name   Selecting the **Name** edit box will display a list similar to Figure 25.4. The list shows each available plotter that is connected to either the workstation or to a network that connects multiple computers and plotters. Notice that an icon is displayed by each listing to distinguish between system devices and those stored with a PC3 file name. A PC3 file is a file created by AutoCAD that is drawing independent. PC3 files are just like any other file in that they are portable and can be shared between different workstations. A PC3 file contains all of the information on what to plot and how to plot, including the specified plotting device, paper size, and pen information. A system device refers to a printer or plotter that is configured in the **Printer and other Hardware** folder of **Control Panel**, or added by selecting **Printers** from the **Settings** menu. Selecting the **Add Printer** option from the Windows **Printers and Other Hardware** menu will produce a prompt for the **Add Printer** wizard.

In addition to selecting a printer, the **Name** edit box allows the drawing to be saved as a PDF file. Because PDF viewers such as Adobe Reader can be downloaded from the Adobe website without cost, drawings can be shared with virtually anyone who may not have AutoCAD installed on their computers.

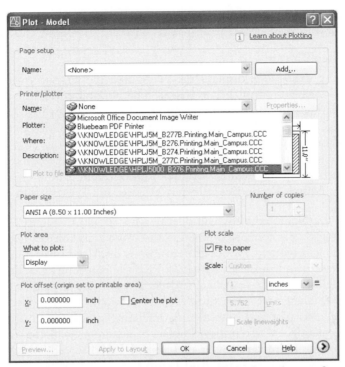

Figure 25.4 *Selecting the arrow in the **Name** edit bar will display a listing of available printers and plotters. This display shows a portion of the available plotters connected to a computer on a network.*

Plotter Selecting the name of a plotter from the menu in the **Name** edit display will list the plotter to be used for plotting the current drawing. If a plotter is selected that can't print on a paper size that is suitable for the current drawing, the warning shown in Figure 25.5 will be displayed. To remedy the problem, select the **OK** button to remove the warning, and select a plotter capable of using a suitable paper size, or adjust the scale to be used for plotting. Altering the plotting scale will be discussed later in this chapter.

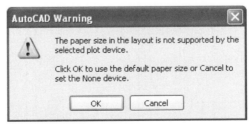

Figure 25.5 *AutoCAD will display a warning if the selected plotter can't support the current paper size required for the current layout.*

Properties Selecting the **Properties** button displays the **Plotter Configuration Editor** similar to Figure 25.6. This dialog box can be used to modify the current plotter configuration, ports, device and media settings. No changes are usually required to these settings.

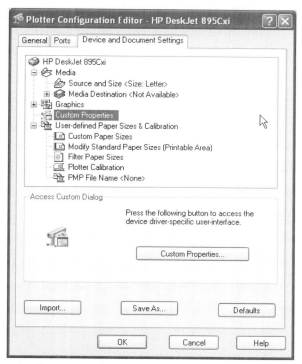

Figure 25.6 *Selecting the **Properties** button will display a listing of the properties of the current plotter.*

Where Once a plotter has been selected, for computers connected to a network, the physical location of the device can be displayed in this area of the dialog box. For an individual computer connected to a plotter, the display will list a connection port such as LPT1.

Description Descriptive text related to a specific plotter can be place here to provide information relevant to a specific plotter. Information regarding paper capacity, color capacity, or department usage can be posted here using the **Plotter Configuration Editor**. Adjustments to this area are not required by the cad technician to produce a plot. The system administrator usually supplies the description as the plotter is initially configured.

Partial Preview The partial preview provides a visual display of the plot area relative to the paper size and the printable area.

Plot to File When you complete the required options in the dialog box, the information will be sent to the plotting device and the information will be processed. The time required to produce the plot will depend on the size of the drawing file and the type of plotter. If the **Plot to file** check box is selected, the plot will be stored as a PLT file instead of being sent to the plotting device. Saving a plot file allows the drawing to be plotted at a more convenient time. If the **Plot to file** check box is selected, the **Number of Copies** option will become inactive. Once the plot settings are adjusted, select the **OK** button to save the drawings as a PLT file. This will display the **Browse for Plot file** dialog box, and allow a storage location to be selected. In the **File name** edit box, a name such as **Floor-Layout1** will be displayed, indicating the drawing and layout names, and the location of the current drawing. The PLT file will be stored in the same location as the drawing to be plotted. Selecting the **Save in** edit box will produce a display similar to the **Explorer** display. The dialog box allows the storage location of the plot file to be specified using methods similar to that used by other dialog boxes for saving files. The file will be stored on disk with the indicated name followed by the *.plt* extension. Any storage device in your machine, accessible to the network or available on the World Wide Web, can be used to store the plot file.

Paper Size

This portion of the dialog box displays a listing of available paper sizes for the configured plotter. Selecting the edit arrow will display a list similar to Figure 25.7. The list will vary, based on the configured plotter. The paper size to be used for the plot can be selected from the list. The sizes displayed are the actual paper sizes. Sizes are listed as width (X)×height (Y). Once the desired value is selected, the **Printable area** will be updated. This is the maximum area available based on the current paper size.

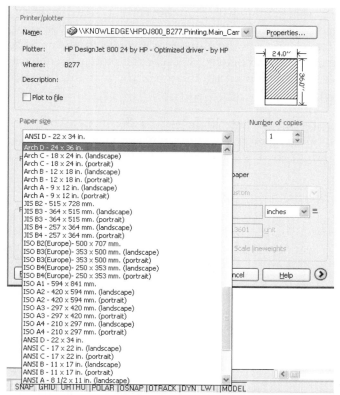

Figure 25.7 *Selecting the **Paper size** edit box will display a list of available paper sizes that are available to the select plotted.*

Number of Copies

The **Number of Copies** edit box allows the number of copies to be plotted to be set. This option is inactive when the **Plot to file** option is active.

Plot Area

This area of the dialog box provides options for describing what portion of the current drawing will be plotted. The plot area consists of the **What to plot** selection menu. If **Window** is the default setting as the dialog box is opened, the **Windows** selection button will be active allowing a window to be used to select what is to be printed. Because the entire drawing often does not need to be plotted, five options are given to decide what portion of the drawing will be plotted: **Display, Extents, Layout, View,** and **Window**.

Display   This setting plots only the material that is currently displayed on the screen. If you have zoomed into a drawing, only that portion of the drawing will be plotted. A scale appropriate to the use of the plot and the selected paper size will need to be selected.

Extents Choosing this option will plot the current drawing with the extent of the drawing objects as the maximum limits that will be displayed in the plot. The option is similar to the **Extents** option of the **Zoom** command. Using this option might eliminate objects at the perimeter of the drawing because the plotter cannot reach the edge of the plotting surface. A scale appropriate to the use of the plot and the selected paper size will need to be selected.

Layout This option will plot everything within the margins of the selected paper size. This will be the option of choice when the entire contents of the layout need to be plotted. When plotting a layout use a scale of 1/1.

View If specific views have been saved using the **View** command, the **View** option for plotting will be activated. This option will allow specific views to be plotted as they were saved. In the initial setting, the **View** radio button is inactive. When a drawing contains named views, the **View** button will be active and a listing showing each saved view name can be displayed. Selecting the desired view to be plotted will highlight that view, place the name in the edit box, close the list, and allow the selections to continue.

Window This option will allow a specific area of a drawing to be defined for plotting. In the same way that a window is used to zoom or select objects for editing, a window can be specified here to define the selection plot. If the **Window** option is selected, the **Plot** dialog box will be removed from the display area, and prompts for defining the window will be displayed. The prompts are as follows:

```
Specify first corner: (Select first window corner.)
Specify other corner: (Select opposite window corner.)
```

Once the window is defined, the **Plot** dialog box will be redisplayed.

 Note: Use the window option if you're making a plot of information displayed in model space. This option is a poor choice for plotting material that has been placed in a layout.

Plot Offset

The **Plot offset** portion of the dialog box contains the **Center the plot** check box and a display for X and Y coordinates. Activating the **Center the plot** option will automatically center the plot in the center of the plotting paper. The **Plot offset** value specifies an offset distance of the drawing to be plotted from the lower left corner of the paper. If you're plotting a layout, the lower left corner of the plotting area is positioned at the lower left edge of the plotting material. You can offset the origin of the plot from the paper edge by entering a positive or negative value. The plotter unit values are in inches or millimeters on the paper. Entering a value in the **X** edit box specifies the plot origin in the X direction. Entering a value in the **Y** edit box specifies the plot origin in the Y direction. The default setting for plotting origin is X=0–Y=0. For most plotters, this will be the lower left corner of the paper.

By selecting the X or Y origin box, new values can be entered. Entering an X value of 12 and a Y value of 10 will place the new origin 12 units to the right and 10 units above the original plotting origin.

Plot Scale

The **Plot scale** area of the dialog box controls settings for the scale and lineweight scale. As drawings have been created throughout this text, they've been drawn at real size rather than at a reduced scale. In chapter 23 you were introduced to methods of creating layouts for plotting. If you're plotting a layout, the scale was set as the layout was created. This will allow the scale factor to be at 1:1 here. If you want to make a plot of a drawing created in model space, you'll need to provide a scale factor for plotting. Select the **Scale** edit arrow and the desired scale from the list to provide a plotting scale. Figure 25.8 shows a portion of the **Scale** list showing common scale factors that can be used to produce a plot at a given size. Dimensions and text were both multiplied by a scale factor in anticipation of the scale at which the drawing would be plotted.

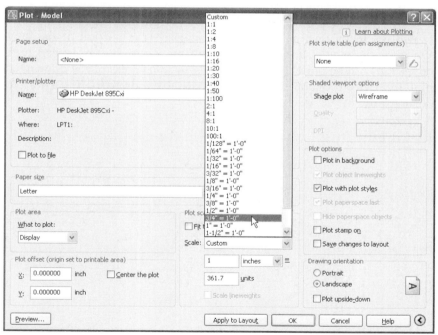

Figure 25.8 *Select the **Scale** edit arrow to display a list of common scale factors that can be used to produce a plot at a given size.*

Custom Scales An alternative to selecting a precise scale is to activate the **Fit to Paper** box in the **Plot scale** portion of the dialog box. Letting AutoCAD determine the scale with the **Fit to paper** option will allow the desired drawing to be sized to fit within the parameters of the paper. The resulting scale will be displayed in the **Custom** display box.

Drawings that are scaled to fit are useful as check prints to determine completeness, but they are not suitable for most projects within the construction industry. The **Custom** box is better suited for entering a scale that is not listed. The scale is created by entering the number of inches or millimeters equal to the number of drawing units.

Scale Lineweights   Use the **Scale lineweights** check box to scale lineweights in proportion to the plotted scale. Lineweights selected during drawing construction normally specify the line width of printed objects and are plotted with the specified width regardless of the plot scale.

Apply to Layout

Selecting this option will save the current settings in the **Plot** dialog box to the current layout.

Preview

Choosing this option will produce a graphic display of the drawing as it will appear when plotted. Figure 25.9 shows an example of a full preview. A few seconds will elapse between the selection of the **Preview** button and the display of the proposed plot. An outline of the paper size will be drawn on the screen, which displays the portion of the drawing to be plotted. Once the drawing is displayed, the **Realtime Zoom** button is displayed. The **Pan** and **Zoom** options can be quite useful in viewing detailed areas of the drawing a final time prior to plotting. Once the indicated **Pan** or **Zoom** is performed, press ENTER to restore the **Plot** dialog box.

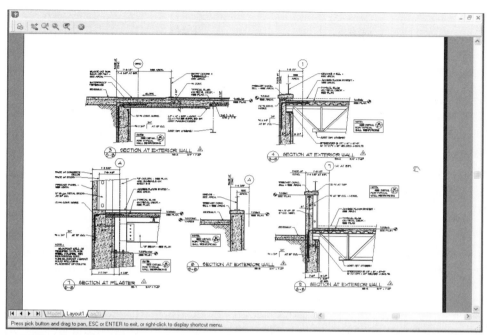

Figure 25.9 *A preview shows each drawing object as it will appear in the finished plot.*

Plot Options

In addition to the options that you've just explored, the expanded **Plot** dialog box offers four additional areas for controlling a plot. This area includes: the **Plot style table, Shaded viewport** options, **Plot options,** and **Drawing Orientation.**

Plot Style Table (Pen Assignments)   Just as a style can be assigned to text or dimension, AutoCAD allows a style to be assigned to each layout and model space view. This portion of the **Plot** dialog box is used to create, edit, or set the plot style table assigned to the current model and layout tabs. Selecting the **Name** edit box displays a list of available styles. Later in this chapter you'll be introduced to the procedure to create and plot with plot styles. Selecting the **New** button will start the **Add Color-Dependent Plot Style Table Wizard**. The wizard can be used to create a new plot style table. Once a plot style table has been created, it will be listed in the **Name** box. Notice that in Figure 25.10, the listing shows several plot styles, even though none have been defined. These are plot styles defined by AutoCAD. For most uses, **Monochrome** is the preferred selection. This setting will make a black and white print of a color drawing. Other options will be explored as **Plot Styles** are discussed.

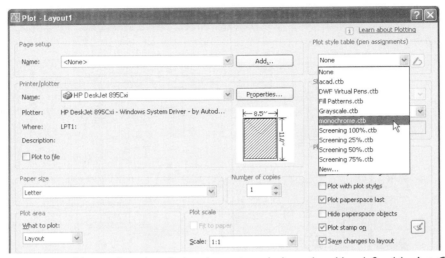

Figure 25.10 *The **Plot style name** listing shows several plot style tables defined by AutoCAD. As you create new styles they will be added to the list.*

Shaded Viewport Options This portion of the dialog box is used to specify how shaded and rendered viewports are plotted. No adjustments need to be made in this area for plotting 2D drawings. Use the **Help** menu for assistance if you need to add shaded areas to 3D drawings.

Plot Options This portion of the dialog box contains seven check boxes for controlling the outcome of the plot. The options control lineweight, plot styles, and the current plot style table. Plot styles will be introduced in the next portion of this chapter.

Plot in background—Making this option active allows your drawing to plot or publish in the background while you continue working on the drawing. This is beneficial for extremely large plot jobs. You can control background plotting from the **Plot and Publish** tab of the **Options** dialog box. If you disable background plotting, you are able to monitor the progress of your current plot job on screen with the Progress dialog box.

A **Plot** icon is displayed in the status tray, providing quick access to your plot job. The shortcut menu features various plot-related options, including the ability to cancel the entire job. When the plot job is complete, a bubble notification is placed in the lower right corner of the drawing area. You can select the link to view the details in the **Plot and Publish Details** dialog box as shown in Figure 25.11.

Plot object lineweights—With this box active, lineweights specified in the drawing will be plotted.

Plot paperspace last—With this option active, model space geometry will be plotted first.

Plot with plot styles—With this option active, plot styles will be applied to objects that are defined in the plot style table. All style definitions with different property characteristics are stored in the plot style tables.

Hide paperspace objects—This option is only available from a layout tab. With this option active, hidden lines will be removed from layouts. Viewports can be used to remove hidden lines for model space drawings.

Plot stamp on—Activating this option will display the **Plot Stamp Settings** icon and attach a plot stamp on each drawing. The plot stamp includes the drawing name, path, production date and time of the plot. Plot stamps are created using the **Plot Stamp** dialog box. Plot stamps will be introduced in the next section of this chapter.

Save changes to layout—Activating this option saves changes that have been made in the **Plot** dialog box to the layout.

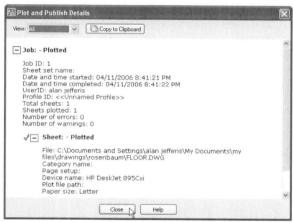

Figure 25.11   *Selecting the **Plot** icon from the status tray during a plot displays the **Plot and Publish Details** display. The display allows you to monitor the progress of a very large plotting job on screen.*

Drawing Orientation

This area of the dialog box allows the drawing orientation to be specified. The box on the right side of the display is the orientation icon and not a selection box. The current orientation of the box is based on the configured plotter and indicates the drawing will be plotted in the normal orientation for construction drawing. This setting is referred to as *landscape*. When **Landscape** is used, the A will be placed on its side, indicating the print will be read from the long edge. When the plot is rotated so that it is read from the title block end, the plot is referred to as a *portrait* plot. The A will be in the vertical position indicating the print will be read looking from the title block end. A portrait plot is more likely to be used with laser printers. The orientation of the drawing to the paper can be adjusted by selecting the appropriate radio button. A check box is also available for plotting the drawing upside-down. Selecting this option will rotate the portrait or landscape layout 180°.

USING PLOT STAMPS

With the **Plot stamp on** box activated in the **Plot options** area, the **Plot stamp settings** icon is displayed. Selecting the icon will display the **Plot Stamp** dialog box shown in Figure 25.12. The dialog box consists of four major areas that are used to describe what information will be plotted in the drawing stamp. The areas include the **Plot stamp fields, User defined fields, Plot stamp parameter fields**, and the **Preview** area.

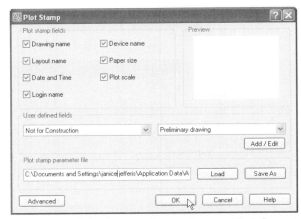

Figure 25.12 *The **Plot Stamp** dialog box can be used to provide information about the plot production in one of four corners of the plotted drawings.*

PLOT STAMP FIELDS

This area contains seven toggle settings for information that can be included in the drawing plot stamp. As each option is selected, a check will be placed in the box indicating that the option is active. Activating an option will include that information in the plot stamp. Selecting all options would produce a display similar to the following display:

```
C:\ Documents and Settings\janice jefferis\My Documents\my
     files\school\my projects, steel details, layout 1, 08/16/06,
     10:59:40 AM, HP5000, 22×34, 1:1
```

USER DEFINED FIELDS

This option allows for optional text to be placed in the plot stamp. Other options allow information in this field to be logged for future reference. This would be a useful area for company to denote information such as Preliminary drawing, Not for Construction, Permit Approved or Corrections Noted.

Information is added by selecting the **Add/Edit** button. This will display the **User Defined Fields** dialog box. Selecting the **Add** button allows the desired information to be entered. Selecting the **OK** button closes the dialog box and restores the **Plot Stamp** dialog box The Edit box will continue to display **None** until one of the new options has been selected. With the **None** option active, no information will be plotted.

PLOT STAMP PARAMETER FIELDS

This portion of the dialog box stores the plot stamp information in a file with a *.pss* extension. This will allow multiple users within a company to access the file and stamp the plot based on company standards.

PREVIEW AREA

The preview area provides a visual display of the plot stamp location, not of the plot stamp contents. No preview is displayed until setting have been entered using the Advanced setting are adjusted. Selecting the **Advanced** button displays the **Advanced Options** dialog box shown in Figure 25.13. The dialog box is used to control the location, text properties and units of the plot stamp. The default location for the stamp is in the lower left corner. The selected location will be indicated with a very faint display of **10-10-10** in the selected corner of the preview window. A different location for the stamp can be made using the **Location** menu. Other options of the **Advanced** options can be explored using the **Help** menu.

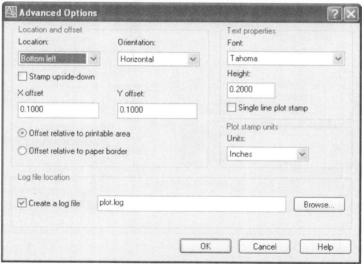

Figure 25.13 *The **Advanced Options** dialog box can be used to provide information to control the location of the plot stamp.*

PRODUCING A PLOT

Each option of the **Plot** dialog box has now been examined. Even though the **Preview** option was presented before the dialog box was expanded, you can preview the plot settings at any point of process. Before pressing the **OK** button, select the **Preview** button one last time. If you change your mind about the need for a plot, select the **Cancel** button. The dialog box will be terminated, and the drawing will be redisplayed. If you are satisfied with the plotting parameters that have been established, select the **OK** button. The dialog box will be removed from the screen, the drawing will be returned, and the plot will be created. The completed plot can be seen in Figure 25.14.

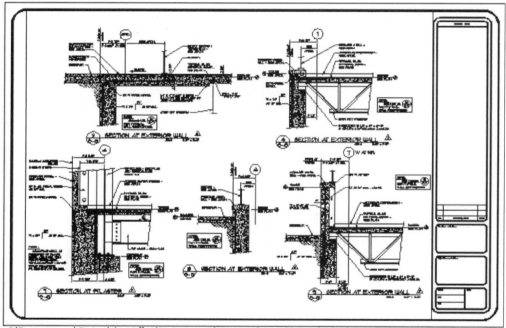

Q:\Documents and Settings\alan jefferis\My Documents\Drawing1.dwg, Architectural Title Block, 05/07/2005 12:16:57 PM, alan jefferis, HP DeskJet 895Cxi, Letter, 1:3.3356, Not for Construction

Figure 25.14 *The plot of the drawing previewed in Figure 25.9.*

WORKING WITH PLOT STYLES

In addition to the plot styles provided with AutoCAD, plot styles can be created and saved as a file. Plot styles are object properties assigned to all drawing objects to control the appearance of the finished plot. Plot styles can be used as the drawing is reproduced to override the object properties assigned as the drawing was created. They can be used as overrides for color, dithering, gray scale, linetype, lineweight, pen assignments, screening, end styles, join styles, and fill styles. They can be assigned layer-by-layer or object-by-object. Each aspect of the style is defined in a plot style table that can be attached to the model or layout tab. By assigning different plot styles to the same layout, you can reproduce the drawing in different ways to meet the needs of different clients or subcontractors. The three steps required to use a plot style are as follows:

1. Define the plot style using a plot style table.

2. Attach the plot style table to the layout.

3. Assign the plot style to an object or layer in the drawing.

The effects of the attached plot style can be viewed using the **Plot Style Table Setup** dialog box shown in Figure 25.15. The dialog box is viewed by selecting the **Plot Style Table Settings** button on the **Plot and Publish** tab of the **Options** dialog box.

Figure 25.15   *The **Plot Style Table Setup** dialog box lists the two types of modes used for all drawings in AutoCAD. These modes are color-dependent and named plot styles.*

EXPLORING PLOT STYLE MODES

The **Plot Style Table Setup** dialog box shown in Figure 25.15 lists the two types of modes used for all drawings in AutoCAD. These modes are *color-dependent* and *named* plot styles.

Color-Dependent Plot Styles

A color-dependent plot style controls plotting based on the color assigned to objects in the drawing. There are 256 color-dependent plot styles based on each of the colors in the AutoCAD color index. Key characteristics of a color-dependent plot style include the following:

- Plot characteristics are assigned to each color to determine how the drawing will be plotted.

- Color-dependent plot styles cannot be added, deleted or renamed in a style table.

Using a color-dependent plot style, you can alter the appearance of all red drawing objects by changing the plot style that corresponds to RED. For most of the drawings that you've done in this text, a color-dependent plot style will work well. All red lines could be assigned to a specific line width, all blue lines could be assigned to a different width. However, as your drawings become more complex, using the object color could limit the use of color in a drawing. By associating color with a specific quality, you lose the ability to work with a color independent of lineweight or linetype. Creating color-dependent plot styles will be explored as the use of the **Add Plot Style Table Wizard** is explored.

Named Plot Styles

A named plot style depends on a user-assigned name rather than the color of an object. Named styles are independent of object colors in the style. Key characteristics of a named plot style include the following:

- The **Normal** style assigned by AutoCAD cannot be modified.

- Each defined style in the plot style table has a name and specific plotting characteristics.

- A named plot style can be assigned to layers or objects.

Using named plot styles allows drawing objects to have the same color but be plotted with different properties based on the name of the assigned plot style.

CREATING A PLOT STYLE TABLE

One of the most common uses of a plot style table is to plot a layout containing a wide variety of colors on a plotter using black lines. The drawing can be plotted without using a plot style table, but each line might be displayed in a gray tone rather than black. By creating a plot style table, you can assign each color the color black for plotting, with the same color intensity. A plot style table can be created using the **Stylesmanager** command. To access the command, use one of the following methods:

- Select **Plot Style Manager** from **File** on the **Standard** toolbar.

- Type **STYLESMANAGER** ENTER at the Command prompt.

Each method will display the **Plot Styles** dialog box shown in Figure 25.16. Each of the available plot style tables is displayed in the **Plot Styles** dialog box by a name and icon. Selecting the **Add-A-Plot Style Table** Wizard option will start the wizard. After reading the introductory text, click the **Next** button.

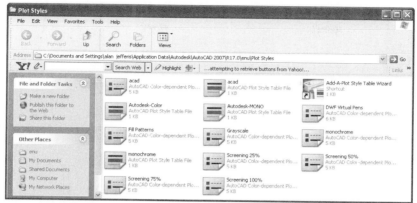

Figure 25.16   *A plot style table can be created using the **Stylesmanager** command. Access the command by selecting **Plot Style Manager** from **File** on the **Standard** toolbar or by typing **STYLESMANAGER** ENTER at the Command prompt.*

Figure 25.17 shows the second dialog box prompt of the **Add Plot Style Table** wizard. By default, you'll be starting the plot table from scratch. The dialog box also allows options for the following:

Use an Existing Plot style table—This option will copy an existing plot style table to be used for the style to be created. If this option is selected, the **Table Type** mode will be skipped as the wizard advances through the process of creating a new table.

Use My R14 Plotter Configuration (CFG)—This option allows a new plot style table to be created using pen assignments stored in an R14 drawing, but you don't have access to the PCP or PCP2 files associated to the drawing.

Use a PCP or PC2 file—This option allows a new plot style table to be created using pen assignments stored in *acadr14.cfg* files.

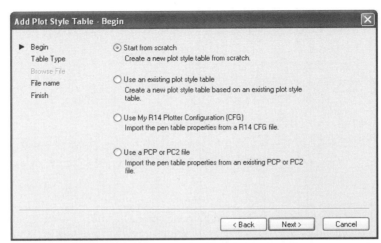

Figure 25.17 *The **Add Plot Style Table - Begin** dialog box shows the options for selecting a starting point for creating a plot style table. By default, you'll be starting the plot table from scratch.*

As you proceed through the initial viewing of the wizard, accept the default option and select the **Next** button. The next dialog box of the **Add Plot Style Table** wizard will display a prompt for the type of table to be used. Options include:

- **Color-Dependent Plot Style Table**
- **Named Plot Style Table**

Select the **Named Plot Style Table** button. Selecting the **Next** button will display the **File Name** dialog box. Enter a name to describe the plot style in the **Name** edit box. For this example a name of MONO will be used. Once a name has been entered in the edit box, select the **Next** button to display the final page of the wizard.

Up to this point, you've started a plot style table named MONO. The final page of the setup wizard allows the naming process to be completed or the plot style to be edited. Selecting the **Plot Style Editor** button will display a **Plot Style Table Editor** dialog box similar to Figure 25.18. This table can be used to alter the properties of the drawing to be plotted. Later sections of this chapter will deal with altering drawing properties. Select the **Finish** button to create a named plot style table and exit the wizard. Selecting **Finish** will create a *.stb* file that is listed in the **Plot Styles** dialog box. Selecting the **Use this plot style table for layouts in new drawings and pre-AutoCAD 2007 drawings** option will make the newly created plot style the default plot style for future drawings.

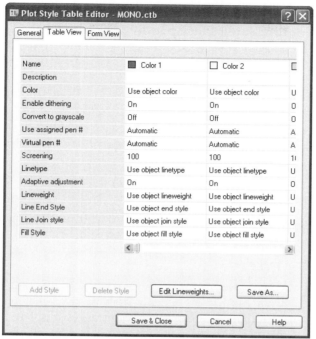

Figure 25.18   *Selecting the **Plot Style Editor** button will display a **Plot Style Table Editor** dialog box. This dialog box can be used to alter the properties of the drawing to be plotted.*

Creating a Color-Dependent Plot Style Table

The process to create a color-dependent plot style is similar to the process of creating a named style. The steps to create a color-dependent include the following:

1. Select the **Add A Plot Style Table Wizard** button in the **Plot Styles** dialog box.

2. Select the **Next** button after reading the introductory material.

3. Select the option to be used to start the plotting style table and select the **Next** button.

4. Choose the type of plot style table to be created (Color dependent) and select the **Next** button.

5. Enter the name of the plot style table (My Colors) and choose the **Next** button.

6. Set the options on the Finish page and click the **Finish** button.

The color-dependent plot style will be stored as a *.ctb* file and listed in the **Plot Styles** dialog box.

ATTACHING A PLOT STYLE TABLE TO A LAYOUT

A plot style created with the **Add Plot Style Table Wizard** can be attached to the model or layout tab using the **Page Setup** dialog box. Use the following steps to attach a plot style to a layout tab:

1. Select the desired layout tab to apply the plot style.

2. Select **Page Setup Manager** from the **File** menu.

3. Using the **Page Setup Manager** dialog box, select the page setup to be modified. Select **Modify**, and the **page setup Manager** dialog box will be replaced with the appropriate **Plot Style** page.

4. Select the desired plot style to be applied in the **Plot Styles Table** portion of the **Page Setup** page.

5. Select **Yes** when the Assign this Plot Style to all layouts prompt is displayed.

The settings assigned to the plot style can now be applied to the selected layout.

VIEWING A PLOT STYLE

Once a plot style has been created, it can be previewed before it is plotted. Changes made to object properties in a layout can be seen when a drawing is regenerated, or you can use the **Preview** option in the **Plot** dialog box. Plot styles can be displayed in a drawing using the following steps:

1. Select **Page Setup** from the **File** menu.

2. Select **Display Plot Styles** from the **Plot style table (pen assignments)** in the Page **Setup** dialog box.

3. Select the **OK** button.

4. Select the **Close** button on the **Page Setup Manager** to return to the drawing.

The properties of the plot style will now be displayed. If the changes are not displayed, type **REGENALL** ENTER at the Command prompt to display plot styles in a layout.

ADDING NAMED PLOT STYLES

Once you've created your own plot style, it must be added to a plot style table before it can be applied to the layout to be plotted. **Named** plot styles are added to the plot style table using the **Plot Style Table Editor**. Plot styles are added using the following sequence:

1. Select **Plot Style Manager** from **File** on the **Standard** toolbar.

2. Use one of the following methods to display the **Plot Style Table Editor**:

   • Double-click an STB file.

   • Right-click a STB file and select **Open** from the shortcut menu.

3. Select the **Table View** tab or **Form View** tab in the **Plot Style Table Editor** dialog box to view the style settings.

4. Select the **Add Style** button.

A style titled **Style1** will be added to the table. While the new style is still highlighted, change the name to better describe the style contents. Figure 25.19 shows an example of a style that was added. Adjust the options as needed, and select the **Save & Close** button to return to the **Plot Styles** dialog box.

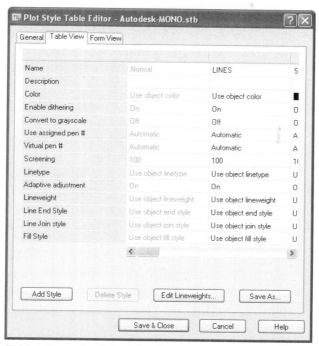

Figure 25.19 *The **Table View** tab of the **Plot Style Table Editor** dialog box can be used to add a plot style.*

DELETING A NAMED PLOT STYLE

As plotting requirements change throughout the life of a project, a plot style can be deleted. The procedure to delete a style is similar to the procedure used to add the style. Use the following steps to delete a file:

1. Select **Plot Style Manager** from **File** on the **Standard** toolbar.

2. Double-click the STB file to be edited.

3. Use the **Plot Style Table Editor** dialog box to select the plot style to be deleted.

4. With the style highlighted, click the **Delete Style** button.

5. Select the **Save & Close** button or the **Cancel** button to return to the **Plot Styles** dialog box.

DEFINING A PLOT STYLE

Figure 25.19 shows an example of the **LINES** plot style. Properties of a style can be defined or edited, using similar methods used to alter a layer color in the **Layer Properties Manager**. These settings will provide overrides to be used while plotting for values defined in the drawing. Settings can be defined using the **Table View** or **Form View** tab of the **Plot Style Table Editor** dialog box. The **Table View** tab shown in Figure 25.19 works well if only a few styles are assigned to the table. Figure 25.20 shows an example of the **Form View** tab. The **Form View** tab works well if several styles have been assigned to the table. Notice that the style names are listed on the left side, and the values for the current style are listed on the right side.

To define the color for the plot style requires the following steps:

1. Using the **Table View** tab of the **Plot Style Table Editor** dialog box, select the current color setting for the plot style to be altered. As **Use object color** is selected, the standard color list will be displayed.

2. Select the desired color. As the color is selected, the list will be closed, allowing other options to be defined.

Other options can be defined using similar methods.

Dither

Dithering is used to approximate colors using a dot pattern. If the plotter does not support dithering, this option will be inactive. If the plotter supports dithering, dithering can be turned on to provide dim colors more visibility. With dithering **OFF**, AutoCAD maps colors to the nearest color.

Figure 25.20   *The **Form View** tab works well if several styles have been assigned to the table. Notice that the style names are listed on the left side, and the values for the current style are listed on the right side.*

Grayscale

With this option active and a plotter that supports grayscale, AutoCAD will convert colors to grayscale.

Pen Numbers

This option specifies which pen will be used with each plot style. The default setting is automatic. In the **Plotter Configuration Editor**, you can specify that a certain object will be plotted using pen #1 as black and .02", and pen #2 as blue and .031". With this option, pen #1 can be assigned to one plot style and pen #2 can be assigned a different plot style. This can be a very useful tool if a subcontractor has used pen assignments different from the desired standard. Using the **Automatic** or **0** setting, AutoCAD will select the pen closest in color to the color of the object you are plotting. The value can't be altered if a color-dependent plot table is being used or if the plot style color setting of **Use object color** is active. Values can be entered by keyboard or by using the range arrow. As the box is selected, arrows are displayed to adjust the range. With a setting of 0, only the up arrow is active. A pen value from 1 to 32 can be specified.

Virtual Pen Number

AutoCAD will specify a color from the **AutoCAD Color Index** closest to the color of the object to be plotted. A virtual pen number between 1 and 255 can be provided and used by plotters without pens. Using virtual pens, all other style settings are ignored and only the virtual pen settings are used. Values can be entered by keyboard or by using the range arrow. As the box is selected, arrows are displayed to adjust the range.

Screening

This setting allows a color intensity to be specified by controlling the amount of ink to be placed during the plot. A value between 0 and 100 can be entered. Entering 0 will have the same effect as freezing the assigned objects, and a setting of 100 will display the color at its full intensity. Values can be entered by keyboard or by using the range arrow. As the box is selected, arrows are displayed to adjust the range. With a setting of 100, only the down arrow is active.

Linetype

Selecting the **Linetype** option will display a linetype menu, showing a list of available linetypes. The default setting for plot style linetype is **Use object linetype**. Selecting a linetype from the list will override the linetypes specified in the drawing.

Adaptive Adjustments

This setting adjusts the linetype scale to be used to display the linetype pattern. This setting should be **OFF** when the display of the linetype is important.

Lineweight

The **Lineweight** box displays a sample of the selected lineweight and the numeric value of the lines to be used with the style. An override value can be selected from the drop-down list, displayed as the box is selected.

Line End Style

In addition to using the line end style specified for drawing objects, AutoCAD allows four options to be assigned to override the line end style of drawing objects. Selecting this field will produce a list with four options in addition to the default option of **Use object end style.** Other options and their styles can be seen in Figure 25.21.

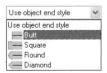

Figure 25.21 *Selecting the **Line End Style** option allows the plot style to override the object's line end style as the drawing is plotted.*

Line Join Style

AutoCAD allows four options to be assigned to override the line join style specified for drawing objects. Selecting this field will produce a menu with the four options in addition to the default **Use object join style.** Other options and their styles can be seen in Figure 25.22.

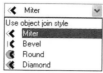

Figure 25.22   *Selecting the **Line Join Style** option allows the plot style to override the object's line join style as the drawing is plotted.*

Fill Style

AutoCAD allows fill patterns to be assigned to override the fill style specified for drawing objects. The patterns will be applied to donuts, plines, and solids. Selecting this field will produce a list with the options. The default setting is **Use object fill style.** Other options and their styles can be seen in Figure 25.23.

Figure 25.23   *Selecting the **Fill Style** option allows the plot style to override the object's fill style as the drawing is plotted.*

EDITING PLOT STYLES

A plot style can be edited using the same methods that were used to define it. Either the **Table View** or the **Form View** tab can be used to edit the plot style. Qualities in a plot style can be edited using the same methods that were used to define the style. In addition to altering the qualities, the **Plot Style Table Editor** dialog box can be used to add, delete, copy, paste, and modify plot styles.

Copying a Plot Style

A plot style can be copied using the **Table View** tab of the **Plot Style Table Editor** dialog box. A plot style can be moved from one table to another table by using the drag and drop technique. In addition, one or more settings from a plot style can be copied to another style using the following steps:

1. Open the **Plot Style Manager**.

2. Double-click the desired CTB or STB file to be opened.

3. Make the **Table View** tab current.

4. Select the plot style to be copied by selecting the gray area above the plot style name. (You can copy individual settings by selecting individual fields.)

5. Right-click and choose **Copy** from the shortcut menu.

6. Switch to the plot style table to be altered.

7. Select the plot style to be replaced by selecting the gray area above the plot style name.

8. Right-click and choose **Paste** from the shortcut menu.

Renaming Plot Styles

A named plot style can be renamed using the **Rename** command. Before a style can be renamed, it must exist in the plot style table associated with the layout. The new plot style name must be a name that is not currently used by any object or layer. Use the following steps to rename a named plot style:

1. Open the **Plot Style Manager**.

2. Double-click the desired CTB or STB file to be opened.

3. Make the **Form View** tab current.

4. Select the plot style to be renamed.

5. Right-click and choose **Rename Style** from the short-cut menu.

6. Type in the new name.

CHAPTER 25 EXERCISES

1. Use a laser printer or plotter and set the required parameters to plot **E-18-8** at a scale of 1/4"=1'–0".

2. Open drawing **E-13-4**. Plot the drawing at a scale of 3/8"=1'–0".

3. Open drawing **E-14-10**. Use the **Copy** command and create a second copy of the elevation. Design a second elevation using a different architectural style and

materials. Save the drawing file as **E-25-3**. Plot the drawing containing your border and title block to a C size sheet of vellum.

4. Plot your drawing template containing the border and title block on a D size sheet of paper. Create a new base drawing that will fit within the limits of the borders. Attach any four drawings in your library. Each drawing is to be shown at a different scale when plotted. Provide a title and the scale used by the original drawing prior to plotting.

CHAPTER 25 QUIZ

DIRECTIONS

Answer the following questions with short complete statements.

Type your answers using a word processor.

1. Place your name, chapter number, and the date at the top of the sheet.

2. Type the question number and provide the answer.

Warning: Some of the questions have not been covered in the reading material and will require the use of the help menu. You may also have to do some exploring to answer the questions.

1. How many colors can be configured for plotting?

_____

_____

2. Is it possible to have more than one plotting device connected to a single computer? Explain your answer.

_____

_____

3. What options are available to preview the information to be plotted prior to plotting, and how are they accessed?

_____

_____

4. Describe the meaning of the display used in a partial plot preview of the **Plot** dialog box.

_____

_____

5. You're about to plot a floor plan that was setup for plotting at a scale of 1/4"=1'-0". The drawing has been inserted into an architectural template. What scale should be selected in the Plot dialog box?

_____

_____

6. What can be used to control how much of the current drawing will be displayed?

_____

_____

7. List five options for controlling what portion of the drawing will be plotted.

_____

_____

8. Describe the typical origin location for a plotter and a printer.

_____

_____

9. What is the advantage of saving the plot information with **Plot to file**?

_____

_____

10. Which takes precedence, the page setup or the plot style?

_____

_____

11. How does AutoCAD deal with pen assignments?

_____

_____

12. Explain the difference between portrait and landscape.

_____

_____

13. What plot style should be used to plot a drawing with 46 different colors using only black?

_____

_____

14. What command is used to create a plot style?

_____

_____

15. List two types of plot style tables and explain which will most typically meet your needs.

_____

_____

16. Explain the steps to attach a plot style to a layout.

_____

_____

17. What is the purpose of a plot style?

_____

_____

18. You've inserted five details in a drawing template. Each of the details was created for plotting at a different scale. Each was inserted in the template using the proper methods. What scale should be used to plot this drawing?

_____

_____

19. How are plotters configured using AutoCAD?

_____

_____

20. What does the **What to plot** area control?

_____

_____

AutoCAD and the Internet

INTRODUCTION

The Internet is an important way to exchange files and information. AutoCAD allows you to access and store drawings and related files through the Internet. AutoCAD also allow hyperlinks to be added to drawings, so that a variety of related documents can easily be accessed. AutoCAD will create drawings to be displayed in Web format (DWF) files that can be viewed with an Internet browser. DWF files can be viewed by anyone with **Volo View Express**, provided by Autodesk to allow clients that do not have access to AutoCAD to view AutoCAD drawings. This chapter will introduce methods for:

- Accessing and working with drawings on the Internet
- Launching a Web browser
- Using a hyperlink, or Uniform Resource Locator (URL)
- Using the Drawing Web Format (DWF) to display drawings on the Internet
- Publish drawings to the Web

Commands to be introduced include:

- **Browser**
- **Inetlocation**
- **Hyperlink**
- **Pasteashyperlink**
- **Publishtoweb**
- **Etransmit**

INTERNET ACCESS

You're probably familiar with the best-known uses for the Internet: email (electronic mail) and the Web (short for the World Wide Web). Email lets users exchange messages and data at very low cost. The Web brings together text, graphics, audio, and movies in an easy-to-use format. To use the features of AutoCAD requires Internet access and an Internet browser to be installed on your workstation. To view and plot

AutoCAD DWF files from an Internet browser requires **Volo Viewer Express,** which can be downloaded from autodesk.com. With the required software loaded, AutoCAD can open, insert, and save drawings to and from the Internet. To save files to an Internet location requires access rights to the directory where the files are stored.

AutoCAD is able to launch a Web browser from within AutoCAD using the **Browser** command. The browser is used for viewing drawings in 2D format on Web pages. The **Hyperlink** command can be used to link a drawing with other documents on your computer or the Internet. For example, a window or door schedule created on a spreadsheet can be linked to a floor plan created in AutoCAD. The electronic plot option of the **Plot** command can create DWF files for viewing 2D drawing on Web pages. With the **Open**, **Insert** and **Saveas** commands, AutoCAD can import or export drawings to and from the Internet.

WORKING WITH UNIFORM RESOURCE LOCATORS

A uniform resource locator (URL) is the file naming system of the Internet. This system allows files such as a text file, a Web page, a program file, an audio file, or movie clip to be located on the Internet. A typical URL would resemble http://www.autodesk.com. It is usually not required to enter the prefix http://. Most Web browsers automatically add in the routing prefix, which saves you a few keystrokes. Common URLs you should be familiar with include the following:

URL: Meaning http://www.autodesk.com	Autodesk Primary Web site
http://www.autodeskpress.com	Autodesk Press Web site

URLs can access several different kinds of resources such as Web sites, email, or news groups. No matter the resource, the URL always take on the same general format: "scheme://networkaddress."

The scheme accesses the specific resource on the Internet, such as

file://	File located on your computer's hard drive or local network
ftp://	File Transfer Protocol (used for downloading files)
http://	HyperText Transfer Protocol (the basis of Web sites)
mailto://	Electronic mail (email)
news://	Usenet news (news groups)
telnet://	Telnet protocol
gopher://	Gopher protocol

The characters :// indicate a network address. Autodesk recommends these formats for specifying URL-style file names:

Web Site	http://servername/pathname/filename
FTP Site	ftp://servername/pathname/filename

Local File	file:///drive:/pathname/filename
or	file:///drive/pathname\filename
or	file://\\localPC\pathname\filename
or	file://\\localPC\pathname\filename
or	file:////localPC/pathname/filename
Network File	file://localhost/drive:\\pathname/filename
or	\\localhost\drive:\pathname\filename
or	file://localhost/drive

The following definitions will help to clarify this confusing terminology.

Term	Meaning
Servername	A name or location of a computer on the Internet such as www.autodesk.com
Pathname	The same as a subdirectory or folder name
Drive	Your local drive
localPC	A file located on your computer
localhost	The name of the network host computer

If you are not sure of the name of the network host computer, use Windows Explorer to check the Network Neighborhood for the network names of computers.

OPENING AND SAVING INTERNET FILES

The **Browser** command lets you start a Web browser while inside AutoCAD. The two most common browsers are Microsoft Internet Explorer and Netscape Navigator. By default, the **Browser** command uses the Web browser program that is registered in your computer's Windows operating system. AutoCAD lists the name of the browser before prompting you for the uniform resource locator (URL). The URL is the Web site address, such as http://www.autodesk.com. The **Browser** command is used to launch a Web browser from within AutoCAD and requires that Netscape Navigator or Microsoft Internet Explorer be installed on your computer. The command can also be used in scripts, toolbar or menu macros, and AutoLISP routines to automatically access the Internet. To access the command, use one of the following methods:

- Click the **Browse the Web** button from the **Web** toolbar.

- Type **Browser** ENTER at the Command prompt.

Each method will produce the following command sequence:

```
Select the Browse the Web button (Or BROWSER ENTER)
Enter Web location (URL) < http://www.autodesk.com>: (Enter URL.)
```

Pressing ENTER will accept the default URL, and an HTML file will be added to your computer during the installation of AutoCAD. This URL is Autodesk's own Web site. After you type the URL and press ENTER, AutoCAD launches the Web browser and contacts the Web site, such as the Autodesk Web site shown in Figure 26.1.

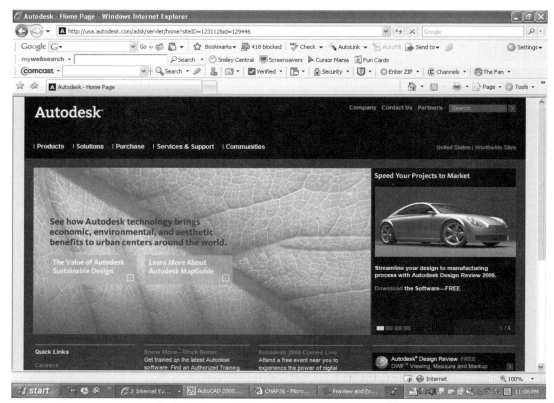

Figure 26.1 *Display the Autodesk Web site by selecting the **Browse the Web** button from the **Web** toolbar or by typing **BROWSER** ENTER at the command prompt and then entering the name of the URL.*

CHANGING THE DEFAULT WEB SITE

Altering the **Inetlocation** system variable setting can change the default Web page that your browser starts with from within AutoCAD. The variable stores the URL used by the **Browser** command and the **Browse the Web** dialog box. Make the change as follows:

```
Command: INETLOCATION ENTER
Enter new value for INETLOCATION "http://www.autodesk.com"
    >:(Type URL.)
```

DRAWINGS ON THE INTERNET

When a drawing is stored on the Internet, you access it from within AutoCAD using the standard **Open, Insert,** and **Save** commands. Instead of specifying the file's location with the usual drive-subdirectory-file name format, such as *C:\my drawings\filename.dwg,* use the URL format. The URL is the universal file naming system used by the Internet to access any file located on any computer hooked up to the Internet.

OPENING DRAWINGS FROM THE INTERNET

Use the **Open** command from the **File** menu to open a drawing from the Internet. The options available are to search the **History,** open from My **Documents, Favorites, Desktop, FTP,** or the **Buzzsaw** site shown in figure 26.2.

> **History**—This will open the Internet History to access previously utilized drawings.
>
> **My Documents**—With this option you can access the **My Documents** folder on the computer.
>
> **Favorites**—Open the **Favorites folder**, which is the equivalent to bookmarks that store Web addresses.
>
> **Desktop**—This area will show the files that are on the Windows Desktop.
>
> **FTP**—Choosing this option will list any FTP sites that have been accessed.
>
> **Buzzsaw**—Browse for files at the Buzzsaw Web site.

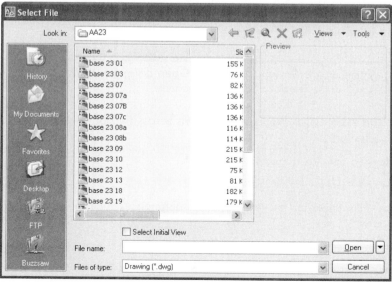

Figure 26.2 *The **Select File** dialog box contains the options to aid in working with the Internet.*

When you click the **Search the Web** button, AutoCAD opens the **Browse the Web** dialog box. This dialog box is a simplified version of a Web browser. The purpose of this dialog box is to allow you to browse files at a Web site. By default, the **Browse the Web** dialog box displays the contents of the URL stored in the **Inetlocation** system variable (see Figure 26.3). The URL can easily be changed to another folder or Web site as noted earlier.

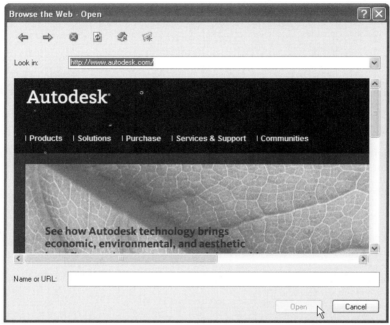

Figure 26.3 *By default, the **Browse the Web - Open** dialog box displays the contents of the URL stored in the **Inetlocation** system variable.*

The **Browse the Web** dialog box has six buttons across the top to aid in working with files. These buttons include:

Back—Go back to the previous URL.

Forward—Go forward to the next URL.

Stop—Halt displaying the Web page. This option is useful if the connection is slow or the page is very large.

Refresh—Redisplay the current Web page.

Home—Return to the location specified by the Inetlocation system variable.

Favorites—List stored URLs (hyperlinks) or bookmarks. If you have previously used Internet Explorer, you will find all your favorites listed here. Favorites are stored in the*Windows\Favorites* folder on your computer.

The **Look in** edit box allows the name of a URL to be entered by keyboard. Alternatively, click the down arrow to select a previous destination. If a Web site address has been stored in the **Favorites** folder, then a URL from that list can be selected by double-clicking a file name in the window or by typing a URL name in the edit box. The following table gives the templates and examples for typing the URL to open a drawing file:

Drawing Location	Template URL
Web or HTTP Site	http://servername/pathname/filename.dwg
	http://practicewrench.autodeskpress.com/wrench.dwg
FTP Site	ftp://servername/pathname/filename.dwg
	ftp://ftp.autodesk.com
Local File	drive:\pathname\filename.dwg
	c:\acad 2000\sample\tablet2000.dwg
Network File	\\localhost\drive:\pathname\filename.dwg
	\\upstairs\e:\install\sample.dwg

When you open a drawing over the Internet, it will take much longer than when opening a file on your hard drive. During the file transfer, AutoCAD displays a dialog box to report the progress. If your computer uses a modem, you should allow about five to ten minutes per megabyte of drawing file size. If your computer has access to a faster T1 connection to the Internet, you should expect a transfer speed of about one minute per megabyte.

The **Open** command will place the drawing in the current drawing file.

 Note: The **Locate** and **Find File** buttons in the **Select File** dialog box do not work for locating files on the Internet.

INSERTING A BLOCK FROM THE INTERNET

When a block (symbol) is stored on the Internet, you can access it from within AutoCAD using the **Insert** command. When the **Insert** dialog box appears, click the **Browse** button to display the **Select Drawing File** dialog box. This is identical to the dialog box discussed earlier in this chapter. After the file is selected, AutoCAD downloads the file and continues with the **Insert** command's familiar prompts.

The process is identical for accessing external reference (xref) and raster image files. Other files that AutoCAD can access over the Internet include 3D Studio, SAT (ACIS solid modeling), DXB (drawing exchange binary), and WMF (Windows metafile). All of these options are found on the Insert menu.

ACCESSING OTHER FILES ON THE INTERNET

Most other file-related dialog boxes allow you to access files from the Internet. This allows your firm or agency to have a central location that stores drawing standards. When you need to use a linetype or hatch pattern, for example, you can access the LIN or PAT file over the Internet. More than likely, you would have the location of those files stored in the **Favorites** list.

Some examples include the following:

> **Linetypes**—Select **Linetype** from the **Format** menu. In the **Linetype Manager** dialog box, click the **Load**, **File**, and **Look in Favorites** buttons.

> **Hatch Patterns**—Use the **Web browser** to copy PAT files from a remote location to your computer.

> **Layer Name**—Select **Layer Properties Manager** from **Layers**. In the **Layer Manager** dialog box, click the **Import** and **Look in Favorites** buttons.

> **LISP and ARX Applications**—Select **Load Applications** from the **Tools** menu.

> **Scripts**—Select **Run Scripts** from the **Tools** menu.

> **Menus**—Select **Customize Menus** from the **Tools** menu. In the **Menu Customization** dialog box, click the **Browse** and **Look in Favorites** buttons.

> **Images**—Select **View** from **Image Displays** on the **Tools** menu.

SAVING THE DRAWING TO THE INTERNET

When you are finished editing a drawing in AutoCAD, you can save it to a file server on the Internet with the **Save** command. If you inserted the drawing from the Internet in the default drawing using **Insert**, AutoCAD insists you first save the drawing to your computer's hard drive.

When a drawing of the same name already exists at that location, AutoCAD warns you just as it does when you use the **Saveas** command. Recall from the **Open** command that AutoCAD uses your computer system's temporary subdirectory, hence the reference to it in the dialog box.

USING HYPERLINKS WITH AUTOCAD

AutoCAD allows you to employ URLs in two ways: directly within an AutoCAD drawing, and indirectly in DWF files displayed by a Web browser. (URLs are also known as hyperlinks, the term we will use from now on.) Hyperlinks are created, edited, and removed using the **Hyperlink** command. To access the command, use one of the following methods:

- Select the **Hyperlink** option from the **Insert** menu.

- Type **Hyperlink** at the Command prompt.

- Press CTRL+K.

Each method will produce a prompt to "Select objects." Once the desired objects have been selected, the **Insert Hyperlink** dialog box shown in Figure 26.4 will be displayed.

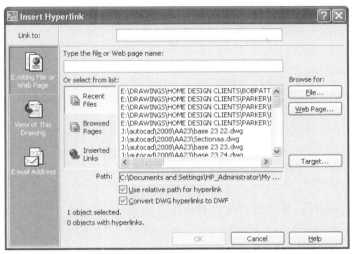

Figure 26.4   *The **Insert Hyperlink** dialog box.*

As an alternative, you can use the **-Hyperlink** command. This command displays its prompts at the command line, and is useful for scripts and AutoLISP routines. The **-Hyperlink** command has the following syntax:

```
Command: -HYPERLINK ENTER
Enter an option [Remove/Insert] <Insert>: ENTER
Enter hyperlink insert option [Area/Object] <Object>: ENTER
Select objects: (Pick an object.)
1 found Select objects: ENTER
Enter hyperlink <current drawing>: (Enter the name of the
     document or Web site.)
Enter named location <none>: (Enter the name of a bookmark or
     AutoCAD view.)
Enter description <none>: (Enter a description of the hyperlink.)
```

Notice that the command also allows you to remove a hyperlink. It does not, however, allow you to edit a hyperlink. To do this, use the **Insert** option and re-specify the hyperlink data.

In addition, this command allows you to create hyperlink areas, which are rectangular areas that can be thought of as a 2D hyperlink. The dialog box–based **Hyperlink** command does not create hyperlink areas. When you select the Area option, the rectangle is placed automatically on layer URLLAYER and colored red.

In the following sections, you'll learn how to apply and use hyperlinks in an AutoCAD drawing and in a Web browser through the dialog box **Hyperlink** command.

HYPERLINKS INSIDE AUTOCAD

AutoCAD allows a hyperlink to be added to any object in the drawing. An object is only permitted to have a single hyperlink, but a single hyperlink can be applied to a several sets of objects. Objects that have a hyperlink can be identified as the cursor passes over the object. The cursor displays the Hyperlink (the linked Earth) icon, as well as a tooltip describing the link, similar to Figure 26.5.

Figure 26.5 *An example of a hyperlink.*

The hyperlink cursor can be set to **OFF** if you find the display distracting. The display is controlled by activating the **Hyperlinks** button on the **User Preferences** tab of the **Options** dialog box found on the **Tools** menu.

When you work with hyperlinks in AutoCAD, three limitations are often encountered:

- AutoCAD does not check that the URL you type is valid.

- If you attach a hyperlink to a block, the hyperlink data is lost when you scale the block unevenly, stretch the block, or explode it.

- Wide polylines and rectangular hyperlink areas are only "sensitive" on their outline.

Pasting as Hyperlink

AutoCAD has a shortcut method for creating hyperlinks in the drawing. The **Pasteashyperlink** command pastes any text in the Windows Clipboard as a hyperlink to any object in the drawing. Here is how it works:

1. In a word processor, select some text and copy it to the **Clipboard** using either CTRL+C or the **Copy** command from the **Edit** menu. The text can be a URL such as http://www.autodeskpress.com or any other text.

2. Switch to AutoCAD and select **Paste as Hyperlink** from the **Edit** menu. Note that this command does not work (grayed out) if anything else is in the **Clipboard,** such as a picture.

3. Select one or more objects, as prompted:

   ```
   Command: _PASTEASHYPERLINK ENTER
   Select objects: (Select an object.)
   1 found Select objects: ENTER
   ```

4. Pass the cursor over the object and note the hyperlink cursor and tooltip. The tooltip displays the same text that you copied from the document.

If the text you copy to the Clipboard is very long, AutoCAD displays only portions of it in the tooltip, using ellipses (...) to shorten the text. You cannot select text in the AutoCAD drawing to paste as a hyperlink. You can, however, copy the hyperlink from one object to another. Select the object, right-click, and select **Copy Hyperlink** from the **Hyperlink** shortcut menu. The hyperlink is copied to the Clipboard. You can now paste the hyperlink into another document, or use AutoCAD's **Paste as Hyperlink** command to attach the hyperlink to another object in the drawing.

Highlighting Objects with URLs

Although you can see the rectangle of area URLs, the hyperlinks themselves are invisible. To see them, you can use the **Qselect** command, which highlights all objects that match specifications. Select **Quick Select** from the **Tools** menu. AutoCAD displays a **Quick Select** dialog box similar to Figure 26.6.

Figure 26.6 *Selecting **Quick Select** from the **Tools** menu displays the **Quick Select** dialog box.*

Enter these specifications: Choose **Entire Drawing** in the **Apply to** edit box. Choose **Multiple** for **Object type** and **Hyperlink** for **Properties**. Choose **Select All** as the **Operator**.

Click **OK**, and AutoCAD highlights all objects that have a hyperlink. Depending on your computer's display system, the highlighting shows up as dashed lines or as another color.

Editing Hyperlinks

Now that you know where the objects with hyperlinks are located, you can use the **Hyperlink** command to edit the hyperlinks and related data. Select the hyperlinked object and start the **Hyperlink** command. When the **Edit Hyperlink** dialog box appears, make the changes and click **OK**.

 Note: The **Edit Hyperlink** dialog box appears identical to the **Insert Hyperlink** dialog box.

Removing Hyperlinks from Objects

To remove a hyperlink from an object, use the **Hyperlink** command with the object selected. When the **Edit Hyperlink** dialog box appears, click the **Remove Link** button. To remove a rectangular area hyperlink, you can simply use the **Erase** command. Start the command and select the rectangle, and AutoCAD erases the rectangle.

THE DRAWING WEB FORMAT (DWF)

To display AutoCAD drawings on the Internet, Autodesk developed a file format called drawing Web format (DWF). The DWF file has several benefits over DWG files and some drawbacks. The DWF file is compressed to make it smaller than the original DWG drawing file, so that it takes less time to transmit over the Internet, particularly with the relatively slow connections. The DWF format is more secure, because the original drawing is not being displayed, so another user cannot tamper with the original DWG file.

Volo View Express is a stand-alone viewer that views and prints DWG, DWF, and DXF files. Volo View Express can be downloaded free from the Autodesk Web site.

You can also attach a DWF file to a DWG file as an external reference file using the **Dwfattach** command. The command is accessed by entering **DWFATTACH** ENTER at the command prompt. After you select the DWF file in the **Select DWF File** dialog box, the **Attach DWF Underlay** dialog box is displayed. In this dialog box, you can define the parameters and details of the attached DWF underlay. Once a DWF file is attached as an underlay, you're allowed to move, rotate, and scale the DWF file, but the geometry in the DWF file cannot be modified. Placing drawings on the Internet typically lies beyond the needs of a beginning CAD operator, so refer to the **Help** menu for further assistance.

PUBLISHING DRAWINGS TO THE WEB

Publishing drawings to the Web allows the files to be accessed across the Internet. The drawings can be posted on a Web site in a number of different file formats. A wizard is available in AutoCAD to make this a simple process for any user. To begin to use the wizard, choose **Publish to Web** from the **File** menu. Other AutoCAD features that will allow files to be shared across the Web include the I-drop technology and **eTransmit**.

PUBLISH TO THE WEB

The **Publish to Web** command provides a method for creating a Web page that contains images of AutoCAD drawings. This tool eliminates the need to write programming code, which means that no previous Web development knowledge will be necessary with this method. To begin publishing your drawings to the Web, choose **Publish to Web** from the **File** menu. The Publish to Web instructions will walk you through the process.

ETRANSMIT

The **eTransmit** feature in AutoCAD enables you to pack currently open files with all of the associated files and x-refs into a transmittal set that can be sent to the desired location. This is very similar to sending email messages across the Web. This will allow files to be shared both internally in a company and externally to any location on the Web. To begin a transmittal, select **eTransmit** from the **File** menu and follow the instructions in the dialog box. See the AutoCAD help area for more information.

CHAPTER 26 QUIZ

DIRECTIONS

Answer the following questions with short complete statements.

Type your answers using a word processor.

1. Place your name, chapter number, and the date at the top of the sheet.

2. Type the question number and provide the answer in the form of a statement that includes part of the question. You do not need to write out the entire question.

Warning: Some of the questions have not been covered in the reading material and will require the use of the help menu. You may also have to do some exploring to answer the questions.

1. How can you launch a Web browser from within AutoCAD?

_____

_____

2. What does **DWF** mean?

_____

_____

3. What is the purpose of **DWF** files?

_____

_____

4. What is **URL** short for?

_____

_____

5. Which of the following URLs are valid?

    a. www.autodesk.com

    b. http://www.autodesk.com

    c. Both of the above

    d. Neither of the above

6. What is the purpose of a **URL**?

_____

_____

7. What is **FTP** short for?

_____

_____

8. What is a "local host"?

_____

_____

9. Can URLs be used in an AutoCAD drawing?

_____

_____

10. The purpose of URLs is to let you create _____ between files.

11. When you attach a **URL** to a block, the **URL** data is _____ when you scale the block unevenly, stretch the block, or explode it.

12. Can you attach a **URL** to rays and Xlines?

_____

_____

13. The **Attachurl** command allows you to attach a URL to _____ and
_____ .

14. To see the location of URLs in a drawing, use the _____ command.

15. Rectangular URLs are stored on layer _____ .

16. The Listurl command tells you _____ .

17. The Detachurl command removes a _____ from an object.

18. Compression in the **DWF** file causes it to take (less, more, the same) time to transmit over the Internet.

_____

_____

19. A **DWF** is created from a _____ file using the _____
or _____ command.

20. Can a Web browser view DWG drawing files over the Internet?

_____

_____

21. _____ is an **HTML** tag for embedding graphics in a Web page.

22. A file being transmitted over the Internet via a 28.8 Kbps modem takes about
_____ minutes per megabyte.

23. To open a drawing located on the Internet, use the _____ command.

INDEX

SYMBOLS

2D Annotation workspace, 7
3D drawings, 767–768

A

absolute coordinates, 88–90
AIA (American Institute of Architects), file management and, 136
Align command, 354–356
aligned dimensions, 612–613
aligning objects, 354–356
angular dimensions, 621–621
annotations, linetypes, 156
Annotative controls, 13–14
aperture box, 251–252
  aperture color, 253–254
  size, 252–253
Apparent Intersection mode, object snaps, 265–266
arcs, 223–224
  center, start
    angle, 231–232
    end, 230–231
    length of chord, 232
  Continue option, 233
  start, center
    angle, 226
    end, 225–226
    length, 226–227
  start, end
    angle, 227–228
    direction, 228–229
    radius, 229–230
  three-point, 224–225
Area command, 490–495
Array command, 324
arrays
  polar arrays, 332
    center point, 333
    method, 333–335
    object base point, 336
    rotating, 335
  rectangular, 326
    angles, 331
    column offset, 329
    negative offset values, 330
    object selection, 327
    positive offset values, 330
    row offset, 329
    row specification, 327
ATTEDIT command, 750
Attribute Definition dialog box, 743
attributes, 741–742
  additional, 747
  alignment, 747
  Block Attribute Manager, 751–753
  commands, 743
  creating, 743
  defining, 744–745
  definition, 742–750
  display, 743–744
    controlling, 748
  Edit Attributes dialog box, 753–754
  editing, 750–755
  Enhanced Attribute Editor, 754–755
  existing blocks, 748
  extracting
    blocks, 757–761
    database and, 756–757
  insertion point, 747
  text
    boundaries, 747
    height, 746
    justification, 745
    rotation, 746
    style, 746
  values
    altering prior to insertion, 749
    attaching to blocks, 749–750
AutoCAD
  leaving, 134–135
  starting, 3–6
AutoSnap
  magnet, 255
  marker, 254–255
  tooltip, 255–256
AutoTrack, Polar Tracking Vector, 256–257

B

backup files, 133
backward lettering, 537
base point, blocks, 701–703
baseline dimensions, 624–625
Block Attribute Manager, 751–753
Block command, 698
Block Definition dialog box, 698–699
blocks
  attributes, extracting, 757–761
  base point, specifying, 701–703
  benefits of, 694–695
  creating, 698–701
  DesignCenter and, 711–712
  dynamic
    creating, 727
    tools, 720–727
  editing, definitions, 713–714
  entire drawing as, 709
  exploding, 715–716
  hyperlinks, 712
  insertion, 707–708
    from Internet, 937
    preparing for, 704–707
  multiple copies, 708
  multiple drawings, 710
  nesting, 709–710
  objects for, 696–698
  redefining, 714–715
  uses, 695–696
  Wblocks, 716–717
    creating, 717–720
    editing, 720
    inserting, 720
Break at point command, 377
Break command, 374–377